Advances in Human Palaeopathology

Advances in Human Palaeopathology

Edited by

Ron Pinhasi

Department of Archaeology, University College Cork, Cork, Ireland

Simon Mays

Centre for Archaeology, English Heritage, UK

John Wiley & Sons, Ltd

Telephone (+44) 1243 779777

Email (for orders and customer service enquiries): cs-books@wiley.co.uk
Visit our Home Page on www.wiley.com

Other Wiley Editorial Offices

John Wiley & Sons Inc., 111 River Street, Hoboken, NJ 07030, USA

Jossey-Bass, 989 Market Street, San Francisco, CA 94103-1741, USA

Wiley-VCH Verlag GmbH, Boschstr. 12, D-69469 Weinheim, Germany

John Wiley & Sons Australia Ltd, 42 McDougall Street, Milton, Queensland 4064, Australia

John Wiley & Sons (Asia) Pte Ltd, 2 Clementi Loop #02-01, Jin Xing Distripark, Singapore 129809

John Wiley & Sons Canada Ltd, 6045 Freemont Blvd, Mississauga, ONT, L5R 4J3, Canada

Wiley also publishes its books in a variety of electronic formats. Some content that appears in print may not be available in electronic books.

Library of Congress Cataloging in Publication Data

Advances in human palaeopathology / edited by Ron Pinhasi, Simon Mays.
p. ; cm.
Includes bibliographical references and index.
ISBN 978-0-470-03602-0 (alk. paper)
1. Paleopathology. I. Pinhasi, Ron. II. Mays, Simon. III. Title: Advances in human paleopathology.
[DNLM: 1. Paleopathology. QZ 11.5 A244 2008]
R134.8.A38 2008
616.07—dc22

2007026251

British Library Cataloguing in Publication Data

A catalogue record for this book is available from the British Library

ISBN 978-0-470-03602-0

Typeset in 10/12pt Times by Integra Software Services Pvt. Ltd, Pondicherry, India
Printed and bound in Great Britain by Antony Rowe Ltd, Chippenham, Wiltshire
This book is printed on acid-free paper responsibly manufactured from sustainable forestry in which at least two trees are planted for each one used for paper production.

To Hattie and Maralyn
Simon Mays

To Rita
Ron Pinhasi

Contents

Preface ix
Simon Mays and Ron Pinhasi

Contributors xiii

Part 1: Analytical Approaches in Palaeopathology

1. The Chemical and Microbial Degradation of Bones and Teeth **3**
Gordon Turner-Walker

2. How Representative Are Human Skeletal Assemblages for Population Analysis? **31**
Ron Pinhasi and Chryssi Bourbou

3. Epidemiological Approaches in Palaeopathology **45**
Ron Pinhasi and Katy Turner

4. Macroscopic Analysis and Data Collection in Palaeopathology **57**
Anne L. Grauer

5. Radiography and Allied Techniques in the Palaeopathology of Skeletal Remains **77**
Simon Mays

6. Computed Tomography Scanning and Three-Dimensional Visualization of Mummies and Bog Bodies **101**
Niels Lynnerup

7. Histological Studies on Ancient Bone **121**
Gordon Turner-Walker and Simon Mays

8. Molecular Palaeopathology of Human Infectious Disease **147**
Helen D. Donoghue

9. Databases **177**
William White

Part 2: Diagnosis and Interpretation of Disease in Human Remains

10. Differential Diagnosis of Skeletal Lesions in Infectious Disease **191**
Donald J. Ortner

11. Metabolic Bone Disease **215**
Simon Mays

12. Tumours and Tumour-like Processes **253**
Don Brothwell

13. Advances in the Palaeopathology of Teeth and Jaws **283**
Alan Ogden

14. Trauma **309**
Pia Bennike

15. Congenital Anomalies **329**
Ethne Barnes

16. Growth in Archaeological Populations **363**
Ron Pinhasi

Index **381**

Preface

Traditionally, in palaeopathology the principal emphasis was on descriptions of individual cases, principally in order to demonstrate the diagnosis of specific conditions and to help establish the antiquity of various diseases. In recent decades, however, although the case study still has a place, there has been a greater emphasis on population studies. In part, this reflects a move away from a medico-historical orientation to one where addressing archaeological questions takes precedence. A dominant theme is now evaluating disease frequencies at a population level and integrating this with cultural data pertaining to the populations under study from archaeological (or historical) sources in order to address questions of broader archaeological interest.

Several factors underpin this approach in palaeopathology, among which are the following.

1. An understanding of biases and limitations of the skeletal record caused by differential skeletal survival and other factors.
2. Rigorous quantification of disease or lesion frequency in fragmentary and incomplete remains.
3. The accurate ascription of causation to bony pathologies (be it diagnosis of specific disorders or ascription of more general causes to non-specific lesions, such as dental enamel hypoplasias).

As regards the first of these, our understanding of factors affecting skeletal decomposition in the burial environment and the mechanisms of diagenesis has been greatly increased in the last 15 years by the application of physical, chemical and microscopic analyses to ancient bone. As regards the second point, we are increasingly coming to grips with the problems of quantifying lesions and diseases in incomplete and fragmentary skeletons, and the potential of applying epidemiological methodologies to ancient remains has begun to be appreciated. There is also an increasing realization that we may need to go beyond recording of simple prevalence rates in order to unlock more fully the information on earlier human populations contained in skeletal pathology. More workers are now attempting to record differences in severity of lesions objectively, and it is becoming increasingly common for workers to record whether lesions were active or healed at time of death. As regards the third point, the increasing use of medical imaging techniques, microscopic examination of lesions and biomolecular analyses has been a major aid to the description and/or diagnosis of disease in human remains. In addition, more workers are integrating the study of specific or non-specific disease with aspects such as growth and are examining associations between the occurrence of different types of lesion or disease. These studies aid the interpretation of skeletal disease. It was in the light of these developments that we conceived the current volume.

We made the decision to concentrate mainly (but not exclusively) on skeletal remains rather than preserved soft tissue. This is simply because in most instances the skeleton is all that survives. We have organized this volume into two parts. Part 1 deals with analytical issues in palaeopathology. The first contribution, by Gordon Turner-Walker, deals with the diagenesis of buried skeletal tissues. He describes the changes wrought by chemical and microbial agents in the organic and inorganic components of skeletal tissues. He considers some of the determinants of the rate of diagenesis; chief among these is the availability of water in the burial environment, together with its pH and the presence or otherwise of dissolved ionic species. As Turner-Walker points out, a sound understanding of post-depositional changes to hard tissues is essential when attempting to interpret pathological conditions in skeletons. Developing this theme in Chapter 2, Pinhasi and Bourbou discuss how skeletal survival, as well as other factors such as excavation methods and ancient burial practices, may bias a skeletal sample and, hence, complicate the interpretation of disease at a population level. They also emphasize the importance of controlling for age at death in population studies in palaeopathology and provide a case study to illustrate one approach to this.

The third contribution, by Pinhasi and Turner, discusses some analytical approaches to the quantitative study of disease frequency in palaeopopulations: palaeoepidemiology. They discuss key paleoepidemiological concepts and provide hypothetical examples to illustrate the application of some of these concepts to skeletal data.

Chapters 4–8 focus on techniques for examining pathological changes in ancient human remains. Anne Grauer discusses macroscopic data collection in skeletal palaeopathology in Chapter 4. She notes that, despite the advent of technologically advanced techniques, gross visual examination of specimens remains the foundation of palaeopathological investigation. She discusses the historical background of study and evaluates attempts toward standardizing terminology and data collection. She identifies a number of problems and issues germane to this area, and offers suggestions as to how these might be resolved.

The next four chapters concentrate on the application of medical imaging and histological and biomolecular techniques in palaeopathology. Radiography is the oldest and still the most frequently used augment to visual examination of specimens in palaeopathology. In Chapter 5, Simon Mays discusses plain-film radiography, quantitation of cortical bone thickness from radiographs (radiogrammetry) and various radiological methods of measuring bone density. The principles of these techniques are explained and their contribution to palaeopathological description and diagnosis discussed. In Chapter 6, Lynnerup discusses the imaging by CT of hard and soft tissues in mummies and bog-bodies. An important focus of both Mays' and Lynnerup's contributions is on the issues raised and problems encountered when applying imaging techniques developed for medical application to ancient human remains.

Turner-Walker and Mays discuss the microscopic study of disease in skeletal remains in Chapter 7. Focusing principally on light and electron microscopy, they discuss sample preparation techniques, the effects of diagenesis on the histological appearance of buried bone and the role of microscopic studies of skeletal lesions in palaeopathology. Histological structures may be studied in a qualitative manner and any abnormalities noted may assist in diagnosis of disease. They may also be studied quantitatively (histomorphometry) to investigate the extent of progressive metabolic conditions such as osteoporosis or to estimate age at death. The contribution concludes with a discussion of the potential role of newer microscopic techniques.

Donoghue covers the fast-moving field of biomolecular study of ancient infectious disease in Chapter 8. Focusing principally on the study of DNA, the degradation and authentication of ancient DNA are discussed and the contribution of biomolecular study to the palaeopathology of various specific infections is evaluated. The potentially important contribution to be made by ancient DNA studies to our understanding of the evolution of disease-causing microorganisms is also considered.

The systematic gathering of large amounts of osteological data raises questions of how best to organize these data and make them available to the wider scholastic community. In the last chapter in this section, Bill White reviews issues concerned with the establishment and maintenance of computerized databases of human remains. He identifies several different types of database, ranging from simple inventories to help researchers locate archived collections, to those which include considerable osteological detail with the intent that scholars use the data directly in their research. He presents an evaluation of the strengths and weaknesses of some of the major extant databases of human remains, and discusses possible future directions.

In Part 2 we concentrate on the diagnosis and interpretation of various classes of disease. We have not attempted to be comprehensive in our coverage of the different categories of disease, but rather have endeavoured to select those where recent advances have made themselves most felt.

Don Ortner discusses diagnostic issues in the evaluation of skeletal infectious disease in Chapter 10, with an emphasis on macroscopic study. He reviews the major infectious diseases that can be identified on the skeleton and provides an extensive photographic illustration of lesions, and emphasizes the need for careful description of lesions and rigorous differential diagnosis.

In his chapter on metabolic bone disease, Simon Mays reviews the pathophysiology, palaeopathological diagnosis and interpretation of vitamin D deficiency, vitamin C deficiency, osteoporosis and Paget's disease of bone. Most studies of the former two conditions have been conducted in order to investigate biocultural questions concerning living conditions and diet. Palaeopopulation studies of the latter two have been mainly orientated toward increasing our understanding of the risk factors for these poorly understood conditions which continue to be important contributors to morbidity and mortality today.

A review of tumours and tumour like processes is provided by Don Brothwell in Chapter 12. He gives a wide-ranging review of the archaeological evidence for both benign and malignant tumours in hard and soft tissues, and considers the potential for relating changes in frequency through time to environmental or cultural factors. The need for collation of widely scattered data and for rigorous statistical analysis is emphasized.

In his chapter on dental disease, Alan Ogden concentrates on some key recent developments in our understanding. He describes a new type of dental enamel hypoplasia, discusses diagnostic criteria for periodontal disease and presents a simple scoring system. He then proceeds to discuss the categorization and significance of periapical voids in alveolar bone.

Pia Bennike discusses trauma in skeletal remains in Chapter 13. She outlines various fracture types and their recognition and quantification in skeletal populations. She also considers the significance of decapitation and mass graves. To illustrate her points, she draws particularly on examples from Denmark and other parts of Scandinavia.

In her chapter on congenital anomalies, Ethne Barnes gives an account of the morphogenesis of different areas of the skeleton and the anomalies which arise from disturbances to that process. Because genetic factors are important causes of most of the anomalies discussed, their biocultural significance lies chiefly with what they can reveal of relationships between populations and between individuals. She illustrates this last point with examples from the palaeopathological literature.

In the final contribution, Pinhasi discusses the value of growth studies of past populations. He considers some of the methodological issues pivotal to such studies. He emphasizes the value of the study of multiple skeletal elements in order to provide a fuller picture of bone growth, and of the potential of studies that attempt to ascertain the effect of disease on growth in past populations. He illustrates these points with reference to key palaeopathological publications.

Although each contribution reflects the author or authors' own unique perspective, a number of common themes do seem to emerge. The quantification of lesions and disease frequency in archaeological skeletal remains continues to present challenges. The benefits of collating data generated by different authors are clear; but, in reality, it is often difficult to compare data between publications, not least because of the sometimes rapid developments and advances in recording methodologies and diagnostic criteria. Gross study of skeletal lesions is likely to remain the foundation of palaeopathology, and there is a continued emphasis on the development and refinement of macroscopic diagnostic criteria. However, although diagenesis complicates the interpretation of medical imaging and histological and biomolecular analyses of ancient human remains, these techniques are likely to play an increasing role in future. The increasing use of technologically advanced laboratory techniques, together with the increased appreciation of the value of analytical models from other disciplines such as epidemiology, and the need to integrate palaeopathological data with historical and or archaeological data, means that collaboration with other disciplines is more vital than ever for the continued development of palaeopathology as a field of study.

Simon Mays and Ron Pinhasi

Contributors

Ethne Barnes
Tucson, Arizona, USA

Ethne Barnes is a physical anthropologist and palaeopathologist consultant and independent researcher, based in Tucson, Arizona. She is recognized for establishing the morphogenetic approach to analysing developmental defects of the skeleton in palaeopathology. She holds a PhD in physical anthropology from Arizona State University (1991), an MA in anthropology (1977) and BSN (1974) from Wichita State University. She has former clinical and teaching experiences prior to becoming consultant to the Corinth Excavations of the American School of Classical Studies at Athens in 1994, and with INAH excavations in Sonora, NW Mexico, in 1998. Research and consultations also include archaeological projects in other parts of Greece, Turkey, China, South America and North America. Major publications include *Developmental Defects of the Axial Skeleton in Paleopathology* (University of Colorado Press, 1994) and *Diseases and Human Evolution* (University of New Mexico Press, 2005).

Pia Bennike
Laboratory of Biological Anthropology, Institute of Forensic Medicine, University of Copenhagen, Blegdamsvej 3, DK-2200 Copenhagen, Denmark

Pia Bennike received her PhD from the University of Copenhagen, Denmark, in 1984. She is currently associated Professor at the Laboratory of Biological Anthropology, Department of Forensic Medicine, University of Copenhagen. She is teaching skeletal biology for archaeologists and palaeopathology (PhD courses, and EAA Summer School), and is supervisor for a number of medical and archaeological students. Her research encompasses most areas of human osteoarchaeology and especially palaeopathology. Key publications include: *Palaeopathology of Danish Skeletons* (Akademisk Forlag, 1985); 'Ancient trepanations and differential diagnosis' in *Trepanation: History, Discovery, Theory*, Arnott R, Finger S, Smith CUM (eds) (Routledge, 2003); 'Rebellion, Combat, and massacre: a medieval mass grave at Sandbjerg near Næstved' in *Warfare and Society: Archaeological and Social Anthropological Perspectives*, Otto T, Thrane H, Vandkilde H (eds) (Aarhus Universitetsforlag 2006); and various publications in Danish.

She is currently vice-president of the European Anthropological Association (former president 2000–2004) and president elect of the Paleopathology Association.

Chryssi Bourbou
28th Ephorate of Byzantine Antiquities, Chania, Crete, Greece

Dr Chryssi Bourbou is a Research Associate at the 28th Ephorate of Byzantine Antiquities (Hellenic Ministry of Culture). Her main research interests focus on the bioarchaeology of medieval Greek populations, with special emphasis on subadult mortality and reconstruction of dietary, breastfeeding and weaning patterns. She has participated in many excavations and research projects in Greece and elsewhere, and offered lectures and organized workshops on bioarchaeology and palaeopathology at institutions and universities worldwide. She has participated in various national and international conferences and has a number of publications, including: 'Infant mortality: the complexity of it all!', *Eulimene*, 2001; 'Health patterns of proto-Byzantine populations (6th–7th centuries AD) in South Greece: the cases of Eleutherna (Crete) and Messene (Peloponnese)', *International Journal of Osteoarchaeology*, 2003; *The People of Early Byzantine Eleutherna and Messene (6th–7th Centuries A.D.): A Bioarchaeological Approach* (Crete University Press, 2004) and is a co-editor of the volume *New Directions in the Skeletal Biology of Greece* (OWLS, forthcoming). She is currently a post-doctoral fellow of the State Fellowship Foundation in Greece (2005–2007) and has recently received a Dumbarton Oaks Project Grant for Byzantine Studies (2006–2007).

Don Brothwell
Department of Archaeology, University of York, King's Manor, York YO1 7EP, UK

Don Brothwell is now emeritus professor of human palaeoecology in the University of York. He has a doctorate from the University of Stockholm, has taught in the universities of Cambridge, London, and York, and for a period was Head of the Sub-Department of Anthropology (now extinct) in the British Museum of Natural History. He still teaches, and currently researches on fossil hominins and mammoths, recent *Microtus*, and the palaeopathology of humans and other mammals. Recent publications include a chapter in *The Myth of Syphilis: The Natural History of Treponematosis in North America*, Powell M, Cook D (eds) (University Press of Florida, 2005) and 'Skeletal atrophy and the problem of the differential diagnosis of conditions causing paralysis', *Anthropologia Portuguesa*, 2000. He is currently trying to find time to return to his first love, art (being originally an art school dropout).

Helen D. Donoghue
Centre for Infectious Diseases and International Health, Department of Infection, University College London, London, UK

Helen Donoghue received her PhD from the University of Bristol. She spent 6 years at the MRC Dental Unit in Bristol, investigating oral microflora. For 4 years she was Lecturer in Medical Microbiology at the University of Bradford and is now Senior Lecturer at University College London. Her recent research has focused on DNA from pathogenic microorganisms in archaeological material, using PCR. Most work has been done on ancient tuberculosis, leprosy and, more recently, parasites such as schistosoma and leishmania. Key publications include: 'Widespread occurrence of *Mycobacterium tuberculosis* DNA from 18th–19th century Hungarians' (with Fletcher, Holton, Pap and Spigelman), *American Journal of Physical Anthropology*, 2003; 'Molecular analysis of *Mycobacterium tuberculosis* from a family of 18th century Hungarians' (with Fletcher, Taylor, Van Der Zanden, and Spigelman), *Microbiology*, 2003; and 'Co-infection of *Mycobacterium tuberculosis* and *Mycobacterium leprae* in human archaeological samples – a possible explanation for the historical decline of leprosy' (with Marcsik, Matheson, Vernon, Nuorala, Molto, Greenblatt, and Spigelman), *Proceedings of the Royal Society of London, Series B*, 2005. She is a Fellow of the Royal Society for Tropical Medicine and Hygiene, a member of numerous microbiological societies and of the Paleopathology Association.

Anne L. Grauer
Department of Anthropology, Loyola University of Chicago, 6525 N. Sheridan Road, Chicago, IL 60626, USA

Anne Grauer received her PhD from the University of Massachusetts–Amherst in 1989. She is currently a Professor in the Department of Anthropology at Loyola University of Chicago. Her research focuses on exploring issues of health and disease in medieval England and within non-Native American groups in the USA. Of particular interest is the use of documentary evidence, combined with skeletal analyses, to understand the lives of women. In 1993, she was awarded the National Science Foundation Presidential Faculty Fellowship to support her incorporation of undergraduate students into scientific research. Key publications include: *Bodies of Evidence: Reconstructing History Through Skeletal Analysis* (editor, Wiley-Liss, 1995); *Sex and Gender in Paleopathological Perspective* (co-editor Stuart-Macadam, Cambridge University Press, 1999); 'Where were the women?' in *Human Biologists in the Archives*, Herring DA, Swedlund AC (eds) (Cambridge University Press, 2003). She has recently served on the editorial board of the *American Journal of Physical Anthropology*, the Executive Board of the American Association of Physical Anthropologists, and as Treasurer and Webmaster of the Paleopathology Association.

Niels Lynnerup
Laboratory of Biological Anthropology, The Panum Institute, Blegdamsvej 3, DK-2200 Copenhagen, Denmark

Niels Lynnerup received his PhD from the University of Copenhagen, Denmark, in 1995. He is currently head of the Laboratory of Biological Anthropology at the Institute of Forensic Medicine, University of Copenhagen. His biological anthropological research comprises both the living (photogrammetry, gait analyses) and the dead (mass-grave analyses, palaeodemography, stable isotopes, aDNA, and the utilization of CT-scanning and 3D visualization techniques to study mummies and bog bodies). Key publications include: 'Life and death in Norse Greenland' in *Vikings: The North Atlantic Saga* (Smithsonian Institution Press, 2000); '3-D CAT-scan: anthropology, archaeology and virtual reality' in *Archaeological Informatics, Pushing the Envelope*, (British Archaeological Reports, 2002); 'Age and fractal dimensions of human sagittal and coronal sutures' (with Brings Jakobsen), *American Journal of Physical Anthropology*, 2003; and 'Person identification by gait analysis and photogrammetry' (with Vedel), *Journal of Forensic Sciences*, 2005. He is a member of the board of the Forensic Anthropology Society Europe (FASE).

Simon Mays
English Heritage Centre for Archaeology, Fort Cumberland, Eastney, Portsmouth PO4 9LD, UK

Simon Mays received his PhD from the University of Southampton, England, in 1987. He is currently Human Skeletal Biologist for English Heritage and is a Visiting Lecturer at the University of Southampton. His research encompasses most areas of human osteoarchaeology. Key publications include: *The Archaeology of Human Bones* (Routledge, 1998); *Human Osteology in Archaeology and Forensic Science* (Greenwich Medical Media, 2000, co-edited with M. Cox); 'Palaeopathological and biomolecular study of tuberculosis in a mediaeval skeletal collection from England' (with Taylor, Legge, Shaw & Turner-Walker), *American*

Journal of Physical Anthropology, 2001; 'Skeletal manifestations of rickets in infants and young children in an historic population from England' (with Brickley and Ives), *American Journal of Physical Anthropology*, 2006. He is a member of the managing committee of the British Association for Biological Anthropology and Osteoarchaeology (BABAO), of the Human Remains Advisory Panel of the UK Governmental Department of Culture, Media and Sport, and is Secretary of the Advisory Panel on the Archaeology of Christian Burials in England.

Alan Ogden
Biological Anthropology Research Centre, Department of Archaeological Sciences, University of Bradford, Bradford BD7 1DP, UK

Alan Ogden was trained as a dental surgeon, and was a Lecturer/Associate Specialist in Restorative Dentistry at Leeds Dental Institute for 20 years, with an especial interest in implants, and in 1988 was awarded his doctorate. He then trained in osteology and palaeopathology at Bradford and has been a Research Fellow/Contract Osteologist in the Bioanthropology Research Centre at the University of Bradford since 2001. He has run the postgraduate courses on musculo-skeletal anatomy in 2003 and archaeology of human remains in 2004 and regularly lectures on the anatomy and palaeopathology of teeth and jaws. He has produced skeletal reports on more than 2000 medieval individuals from Norton Priory, Chichester Leper Hospital and Hereford. He has reported on 60 middle Bronze Age burials from a British Museum dig in the Lebanon, and he is currently re-examining 50 Neolithic individuals excavated in the 1860s from barrows in the Yorkshire Wolds. Recent joint publications include 'Tallow Hill Cemetery, Worcester: the importance of detailed study of post-mediaeval graveyards' and 'A study of Paget's disease at Norton Priory, Cheshire, England' *British Archaeological Reports* International Series 2005; 'Morbidity, rickets, and long-bone growth in post-medieval Britain', *Annals of Human Biology*, 2006; 'Gross enamel hypoplasia in subadults from a 16th–18th Century London graveyard', *American Journal of Physical Anthropology* (in press). A former Curator and Honorary Member of the British Society for the Study of Prosthetic Dentistry, he is a member of the American Association of Physical Anthropologists, the Paleopathology Association and the British Association for Biological Anthropology and Osteoarchaeology.

Donald J. Ortner
Department of Anthropology, National Museum of Natural History, Smithsonian Institution, Washington, DC 20560, USA

Donald J Ortner holds a PhD in physical anthropology from the University of Kansas, USA, and an honorary DSc degree from the University of Bradford, England. He is a biological anthropologist in the Department of Anthropology, National Museum of Natural History, Smithsonian Institution, USA, where he has worked during most of his professional career. In 1988 he was appointed a Visiting Professor in the Department of Archaeological Sciences at the University of Bradford, England. His major research interest is in human adaptation, but he has a specific interest in calcified tissue biology and the effect of disease on human evolution during the Holocene. The latter interest includes a focus on the impact of major developments in human society, such as urbanism, on human health. Recent publications include: the second edition of *Identification of Pathological Conditions in Human Skeletal Remains* (Academic Press, 2003); 'Skeletal manifestations of hypothyroidism from

Switzerland' (with Hotz), *American Journal of Physical Anthropology,* 2005; 'Infectious disease and human evolution', *2006 McGraw-Hill Yearbook of Science & Technology* (The McGraw-Hill Companies, 2005). Administrative posts include 4 years as chairman of the Department of Anthropology and 2 years as acting director of the National Museum of Natural History, USA. He has done fieldwork in Jordan and has conducted research projects in the USA, Europe, and Australia. He has served on several boards and review panels. From 1999 to 2001, he was president of the Paleopathology Association. He is a member of the American Association of Physical Anthropologists, the Paleopathology Association and the International Skeletal Society.

Ron Pinhasi
Lecturer in Prehistoric Archaeology, Department of Archaeology, University College Cork, Cork, Ireland

Ron Pinhasi received his PhD from the University of Cambridge, England in 2003. He spent two years in a Lise Meitner postdoctoral position at the Natural History museum, Vienna, examining the health status of early medieval Austrian populations. He is currently a lecturer in Archaeology, University College Cork, Ireland. His research focuses on growth and development in past populations, the origin and spread of leprosy in Eurasia, and the origins and spread of farming in the Near East. He carries out fieldwork in Israel and directs prehistoric excavations in Armenia. Key publications include 'Morbidity, rickets, and long bone growth in post-medieval Britain – a cross-population analysis' (with Shaw, White and Ogden), *Annals of Human Biology*, 2006; 'A cross-population analysis of the growth of long bones and the os coxae of three early medieval Austrian populations' (with Teschler-Nicola, Knaus and Shaw), *American Journal of Human Biology*, 2005; 'Tracing the origin and spread of agriculture in Europe' (with Fort and Ammerman), *PLoS Biology*, 2005; and 'A regional biological approach to the spread of farming in Europe: Anatolia, the Levant, south-eastern Europe, and the Mediterranean' (with Pluciennik), *Current Anthropology*, 2004. He is a member of the European Archaeological Association, and the Paleopathology Association.

Katy Turner
Research Associate, Division of Epidemiology, Public Health and Primary Care, Imperial College, Praed Street, St Mary's Campus, London W2 1PG, UK

Katy Turner received her PhD from Imperial College London in 2004. She spent 18 months at the Health Protection Agency, developing mathematical models to investigate the potential impact of the National Chlamydia Screening Programme in England. She is currently a research associate at Imperial College, investigating the evolutionary biology of human pathogens. Her research focuses on the complex interactions between bacterial populations (intra- and inter-species) and between host and pathogen, using a variety of mathematical modelling and analytical approaches, together with the best available empirical data, to address important public health questions. Key publications include: 'The impact of the phase of an epidemic of sexually transmitted infection on the evolution of the organism' (with Garnett), *Sexually Transmitted Infections,* 2002; 'Modelling the effectiveness of chlamydia screening in England' (with Adams, Lamontagne, Emmett, Baster and Edmunds), *Sexually Transmitted Infections*, 2006; and 'Modelling bacterial speciation' (with Hanage, Spratt and Fraser), *Philosophical Transactions of the Royal Society of London, Series B: Biological Sciences*, 2006.

Gordon Turner-Walker
School of Cultural Heritage Conservation, National Yunlin University of Science and Technology, 123 University Road Sec. 3, Touliou, 640 Yunlin, Taiwan (ROC)

Gordon Turner-Walker was awarded a PhD by the University of Durham, England, in 1993. He worked as the archaeological conservator for Norfolk Museums Service before taking up a 3-year postdoctoral fellowship at the Norwegian University of Science and Technology investigating osteoporosis in the medieval population of Trondheim. He is currently Associate Professor of cultural heritage conservation at National Yunlin University of Science and Technology, Taiwan. His main areas of research are post-mortem alterations to bone chemistry and microstructure, the archaeology of osteoporosis and the degradation of cultural materials in marine and terrestrial environments. Key publications include: 'The West Runton fossil elephant: a pre-conservation evaluation of its condition and burial environment' *The Conservator*, 1998; 'Quantifying histological changes in archaeological bones using BSE-SEM image analysis' (with Syversen), *Archaeometry*, 2002; 'Sub-micron spongiform porosity is the major ultra-structural alteration occurring in archaeological bone' (with Nielsen-Marsh, Syversen, Kars and Collins), *International Journal of Osteoarchaeology*, 2002; 'Osteoporosis in a population from medieval Norway' (with Mays and Syversen), *American Journal of Physical Anthropology*, 2006. He is a Fellow of the International Institute for Conservation of Historic and Artistic Works.

William White
Centre for Human Bioarchaeology, The Museum of London, 150 London Wall, London EC2Y 5HN, UK

Bill White CChem, FRSC, FSA, is the Senior Curator of Human Remains at the Museum of London's Centre for Human Bioarchaeology. He began his career as an organic chemist working in the pharmaceutical industry. Early on he developed an interest in archaeology and obtained a Diploma in Archaeology at the University of London, England. After undertaking the university post-diploma course in 'Human Remains in Archaeology' with Theya Molleson of the Natural History Museum, South Kensington, he began to prepare the first of what was to become a long series of bone reports, chiefly from archaeological sites excavated in London. During the past 20 years or more he has analysed thousands of human skeletons, albeit using continuously changing recording media. In 2003 Bill was appointed the inaugural Curator of Human remains at the Museum of London, responsible *inter alia* for booking in and invigilating postgraduate and postdoctoral researchers working on the huge collection of archaeological skeletons at the museum. He also headed the team of osteoarchaeologists who recorded 5000 of these skeletons onto an electronic relational database, the Wellcome Osteological Research Database, under a grant from the Wellcome Trust and which went online in 2007.

PART 1

Analytical Approaches in Palaeopathology

1

The Chemical and Microbial Degradation of Bones and Teeth

Gordon Turner-Walker

School of Cultural Heritage Conservation, National Yunlin University of Science and Technology, 123 University Road Sec. 3, Touliou, 640 Yunlin, Taiwan (ROC)

INTRODUCTION

The physical survival of bone is integral to any kind of palaeopathological study. Not only must the skeleton survive in the burial environment or tomb, it must retain sufficient strength to be excavated, lifted, archived and studied. When assessing skeletal remains for pathological conditions, it is also important to distinguish successfully between bone lesions that arose ante- or peri-mortem as a result of disease or trauma, and damage caused by post-mortem processes taking place in the burial environment. A sound understanding of post-mortem changes to mineralized tissues is, therefore, essential when attempting to interpret pathological conditions in skeletons, particularly those (the majority) that have been buried in soils for centuries or millennia. Unlike some gross post-mortem patterns of destruction caused by root action, insects or rodents, which are frequently visible on the outer surfaces of the specimens, microbial and chemical degradation is microscopic in nature and can influence the interiors of the bones as well as their surfaces. This unseen deterioration not only contributes to the fragility of archaeological bones, but by altering the chemistry and microstructure of the tissues it can also have a serious impact on chemical or radiological analyses and on the radiocarbon dating of skeletons (Lee-Thorpe and van der Merwe, 1987; van Klinken, 1999; Mays, 2000; Petchley and Higham, 2000; Dupras and Schwarcz, 2001). The potential for leaching and the movement of soluble salts into and from the bone structure also has a bearing on the interpretation of radiodensitometry (Mays, Chapter 5 this volume) and measurements of bone density using clinical techniques such as dual energy X-ray absorptiometry (Agarwal and Grynpas, 1996; Mays, 1999; Mays *et al.*, 2006). Thus, changes to skeletal tissues arising from their interaction with the burial

Advances in Human Palaeopathology Edited by Ron Pinhasi and Simon Mays

environment and from the actions of soil microorganisms have an impact on almost all aspects of palaeopathological study and the value of human skeletons as a source of information about the past.

In recent decades, rapid developments in the field of biomolecular archaeology have demonstrated that physical and microscopic integrity is no longer enough when considering the research potential of an individual skeleton or assemblage. The integrity of any isotopic and molecular evidence contained within bone and tooth tissues is equally important (Muyzer *et al.*, 1992; Cattaneo *et al.*, 1995; Evershed *et al.*, 1995; Baron *et al.*, 1996; Taylor *et al.*, 1996; Weser *et al.*, 1996; Braun *et al.*, 1998; Stott *et al.*, 1999; Götherström *et al.*, 2002; Geigl, 2002). Recognition of this has driven much of the research into how and why skeletal tissues degrade in the soil, and the progress made in the understanding of these diagenetic processes during the last decade of the 20th century and early years of the 21st century has been almost as dramatic as the huge strides made in the analyses of DNA, lipid and protein residues over the same period.

Compared with other scientific studies of archaeological and fossil bones, the study of bone deterioration is relatively young. The term *taphonomy*, to describe post-mortem processes influencing bone survival, was introduced nearly 70 years ago by Efremov (1940), and these 'laws of burial' were invoked to help interpret fossil and archaeological bone assemblages. In its broadest sense, taphonomy concerns all aspects of the passage of organisms from the *biosphere* (the living world) to the *lithosphere* or Earth's crust (Olson, 1980). The primary goal of taphonomic studies is to work backwards from the surviving bone assemblages to the composition, structure and dynamics of the parent populations (human or animal) using evidence recovered from the bones themselves, the nature of their contexts and an understanding of post-mortem processes (Olsen, 1980). The geological term *diagenesis* is defined as the processes by which sediment is transformed into sedimentary rock under conditions of low temperature and pressure. In recent years, this term has been adopted to describe the changes undergone by skeletal tissues in the burial environment. These changes may involve dissolution of bone tissue or its cementation by exogenic minerals, recrystallization of bone mineral or its replacement by other mineral species. These alterations to bone tissue are often crudely referred to as fossilization (Behrensmeyer and Hill, 1980) and a combination of taphonomic and diagenetic processes determine whether a bone decays and ultimately disappears or persists throughout the course of archaeological or geological time.

As early as the middle years of the 19th century, microscopic examination of ancient bones had identified the potential importance of microorganisms in the destruction and degradation of bone tissues. In 1864, Wedl examined thin sections of ancient bones under the light microscope and described small channels or tunnels penetrating the bone tissues (Wedl, 1864). Roux, working in the late 19th century, also identified these features in fossil bones and termed them bored channels or *Bohrkanäle* (Roux, 1887). The presence of fine, brown filaments visible in these tunnels suggested to him the action of fungi in their formation. Thus, from the outset, the action of fungi was implicated as the principal causal factor in the destruction of dead bone tissues – an assumption that persisted for more than 100 years and remains contentious today.

By the middle of the 20th century, chemical analysis of ancient skeletal tissues was being used as a means of absolute dating, initially with the introduction of fluorine-content dating and later followed by the radiocarbon revolution in archaeology. One of the earlier successes for carbon-14 dating was the confirmation of the Piltdown find of an 'English ape-man' as a modern hoax (de Vries and Oakley, 1959). Suspicions had already been voiced after the failure to find the significant levels of fluoride in the bones that would be expected for a find of geological age. As a result of these developments, together with the introduction of uranium-series dating, calcium-41 dating and amino acid racemization dating, scientists became increasingly aware of the importance of understanding changes in the structure and composition of bones and teeth. These problems were later underlined during attempts to isolate faint dietary signatures, in trace element concentrations or in stable isotope variations, from larger diagenetic chemical alterations.

Before discussing post-mortem changes to skeletal tissues it is necessary to take a closer look at the nature of bones and teeth.

THE CHEMISTRY, ULTRASTRUCTURE AND MICROSTRUCTURE OF SKELETAL TISSUES

Skeletal tissues have a very ancient ancestry in the evolutionary record. Work on a group of fossil elements called *conodonts* has confirmed that these tooth-like structures represent the grasping mouthparts of primitive marine animals resembling eels (Briggs, 1992). These tiny fossils, measuring between 0.2 and 2 mm in length, are composed of the calcium phosphate mineral carbonate fluorapatite, and investigations of their microstructure have shown that they bear many features in common with the hard tissues (such as calcified cartilage, bones and teeth) of more advanced vertebrates (Sansom *et al.*, 1992; Schultze, 1996). These discoveries push back the origin of bony tissues, and consequently our ultimate ancestors, to the late Cambrian period, over 500 million years ago.

The basic chemistry of the calcified tissues bone, antler and tooth dentine (including ivory) is fundamentally the same, although they differ in their mode of growth and microstructure. Tooth enamel is rather specialized and differs from the other calcified tissues in that it is more crystalline and has a negligible organic content. Since bone is by far the most common calcified tissue, it is perhaps appropriate to consider it first before outlining the ways in which other tissues differ from it.

Bone

Living bone consists of three major components: organic matter, principally proteins; mineral in the form of calcium phosphates; and water. Here, the inclusion of water as a major constituent may seem pedantic, but the water contents of buried bones and the sediments that surround them play as important a role in their future integrity over archaeological time-scales as the chemistry and availability of biological fluids do during life. The organic matter in dry bone accounts for approximately 22–23 % by weight (Turner-Walker, 1993) and 40 % by volume (Nielsen-Marsh and Hedges, 2000a). About 90 % of this component is made up of long fibrils of Type I collagen that give living bones their tensile strength and a small degree of flexibility. Type I collagen molecules are highly organized, comprising three stretched helical amino acid chains which are themselves twisted into a triple helix. Collagen is characterized by a high glycine content, which makes up every third amino acid (33 %), with high levels of proline and hydroxyproline, which together account for a further 20 %. Each triplet is approximately 300 nm in length and 1.5 nm in diameter (Yamamoto *et al.*, 2000, De Cupere *et al.*, 2003).

The individual collagen molecules self-assemble or aggregate extracellularly and assume a hierarchical architecture with triplets organizing into bundles, called microfibrils, which ultimately form into fibrils and fibres. These fibre bundles align themselves with a quasi-hexagonal packing (Figure 1.1). Type I collagen is insoluble under normal physical and physiological conditions because of this well-ordered three-dimensional arrangement of the fibres, the ionic and hydrophobic interactions between adjacent amino acid chains, and a degree of cross-linking between the molecules. Strong aldehyde cross-links form between the lysine and hydroxylysine of adjacent collagen molecules and the microfibril is further stabilized by numerous intramolecular hydrogen bonds. Newly formed microfibrils are about 20 nm in diameter but grow in size with maturity up to approximately 90 nm, with an average microfibril diameter in young adults of 75 nm (Sarathchandra *et al.*, 1999). The unmineralized collagen network or organic matrix also contains non-collagenous proteins (including osteocalcin) and mucopolysaccharides which make up the remaining 10% by weight (Tuross, 2003). Some of these non-collagenous proteins can be extremely stable over geological time-scales, strongly suggesting an intimate association with the mineral phase (Muyzer *et al.*, 1992; Smith *et al.*, 2005).

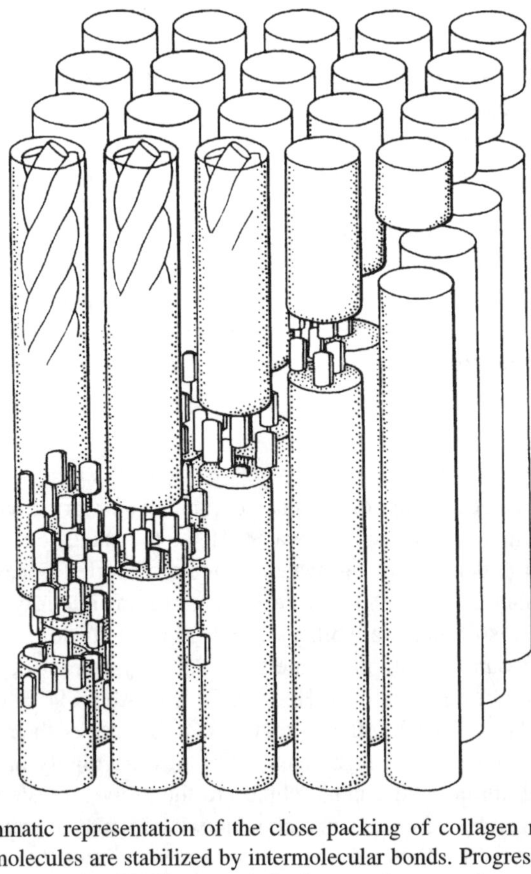

Figure 1.1 Diagrammatic representation of the close packing of collagen molecules (triplets) into fibrils. In reality the molecules are stabilized by intermolecular bonds. Progressive mineralization with small platelets of hydroxyapatite (HAP) proceeds in the gaps between the ends of the molecules and between adjacent triplets

The compressive strength of bone tissues is provided by the mineral component, which is generally accepted to be a stoichiometrically imperfect, carbonate-containing HAP analogue with a composition approximating to $Ca_{10}(PO_4)_6(OH)_2$, also called bioapatite. This mineral phase also includes traces of other anionic and cationic species that variously adsorb on crystal surfaces or substitute for Ca^{2+}, PO_4^{2-} and hydroxyl ions in the lattice. The exact nature of these mineral – ion interactions is not relevant to this discussion, but it is important to understand that they are closely related to the small sizes of the bioapatite crystals and their total available surface area. HAP crystals are plate-like in morphology and have currently accepted dimensions of approximately 35 nm by 5 nm and with a thickness of about 2–3 nm (Lowenstam and Weiner, 1989; Nielsen-Marsh *et al.*, 2000). It is widely recognized that the average sizes of the HAP crystals in bone increase with the maturity of the tissue. The extreme small sizes of the individual bone crystals, or more properly crystallites, present an enormous active surface area for bone mineral, estimated at between 100 and 200 $m^2\ g^{-1}$ (Posner, 1985; Newesely, 1989). It is unlikely, however, that this large active area is ever realized, because of the intimate association between the collagen matrix and the HAP. It has long been known that bone sections exhibit birefringence in polarized light, and this optical property arises from the orientation of both the collagen fibres and the HAP crystallites (Figure 1.2). These crystallites are embedded in the collagen matrix with their *c*-axes aligned parallel to the long axes of the fibres. These fibres are aligned in lamellae in which the fibre orientation in successive layers is rotated to give a plywood-like structure (Giraud-Guille, 1988; Weiner and Traub, 1992). Evidence points to initial deposition of HAP crystallites (primary mineralization) within gaps in the closely grouped collagen fibrils (Figure 1.1), with the bulk of the mineral load progressively filling the interfibrillar spaces (secondary mineralization), a process that may take several weeks or months. This results in greater variability in mineral density between mature and more recent bone tissues in older individuals, especially in osteonal or Haversian bone (Ortner and Turner-Walker, 2003). There is an intimate association between the collagen molecules and HAP, and this chemical affinity is strengthened by the non-collagenous protein osteocalcin, which makes up 2 %

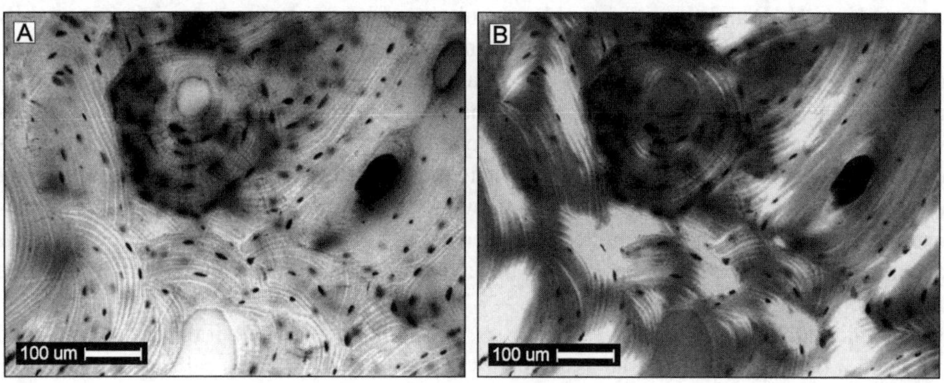

Figure 1.2 (a) Transmitted light image of medieval human bone from Trondheim, Norway. Histological preservation is excellent, but staining around the central osteon illustrates the fine canalicular network that connects the tissues with the soil environment. (b) The section viewed in polarized light with a quarter-lambda plate. The spectacular birefringence arises from the alignment of collagen fibrils and HAP in the bone lamellae

by weight of dry bone (Smith *et al.*, 2005). Osteocalcin is known to bind both to HAP and to collagen, and this relatively small protein plays an important role in the primary mineralization of skeletal tissues.

Dry, fresh bone contains about 8 % water that is loosely bound and can be driven off by heating in air at 105°C (Eastoe and Eastoe, 1954). However, for materials like bone with a high microporosity, the total amount of bound water held by a sample depends strongly on both the temperature and local relative humidity. For very small pores, quite high temperatures are required to drive off all the liquid water held in small capillaries, and even higher temperatures are necessary for chemically bound water. Determination of total bound water in fresh bone is further complicated because, in thermogravimetric measurements, weight losses at elevated temperatures are compounded by thermal decomposition of organic matter and loss of bound carbonates from the bone mineral.

Measurements undertaken by Nielsen-Marsh and Hedges (2000a) of pore volumes for fresh bone using calibrated relative humidities indicated that the macroporosity (those pores with radii between 4 and 20 nm) and microporosity (pores less than 4 nm in radius) were 0.075 $cm^3\ g^{-1}$ and 0.059 $cm^3\ g^{-1}$ respectively, giving a total pore volume below 20 nm of 0.134 $cm^3\ g^{-1}$. This figure compares well with measurements of total pore volume for fresh, compact bovine bone, which lie in the range 21–26 % by volume or 0.110–0.158 $cm^3\ g^{-1}$ (data from Turner-Walker and Parry (1995)). These latter measurements (made from liquid water absorption) included larger pores attributable to vascular channels and voids left by degraded bone cells (osteocyte lacunae). More recently, mercury intrusion porosimetry has refined the interpretation of bone porosity in the range 2 nm to 100 μm, and this technique has had a significant bearing on current understanding of bone diagenesis (Nielsen-Marsh and Hedges, 1999; Turner-Walker *et al.*, 2002; Jans *et al.*, 2004).

Bone is a physiologically active tissue, repairing itself when damaged – either at a macroscopic scale, as during the healing of a fracture, or microscopically, as in the constant remodelling and replacement of bone to remove the microfractures that accumulate through normal activity. Bone is also involved in calcium homeostasis, releasing or absorbing Ca^{2+} ions to maintain serum calcium levels within physiological limits. This requirement for skeletal bone mineral to be immediately accessible hinges on both the large available surface area of bone HAP and the considerable vascularity of bones. Living bone is penetrated by numerous channels (Haversian canals and canals of Volkmann) averaging about 50 μm in diameter, through which pass blood vessels and nerves (Figure 1.3). The branching architecture of these vessels provides a pathway between the countless bone cells or osteocytes within the bone tissues and the circulating blood. A large number of cytoplasmic processes extend from each osteocyte, connecting to neighbouring cells via canaliculi with a diameter of approximately 200 nm. This extended network of fine channels penetrating bone allows chemical messages to be transmitted throughout the tissue, as well as permitting nutrients and mineral ions to be supplied to the bone matrix and metabolic waste products to be removed (Figure 1.2a).

The microarchitecture of bone tissue varies, depending upon where it forms and the speed at which it develops. Bone tissue associated with very rapid growth is called woven or fibre bone. Fibre bone is not as dense or as well organized as other types of bone associated with slower growth rates. The collagen microfibrils are irregular in thickness and lack the linear orientation typical of later stages of bone development. Fibre bone forms early in the growing skeleton but may be found in later life in abnormal bone tissue, such as fracture callus and neoplasms (cancers) or beneath the periosteum as a response to infection. Mature

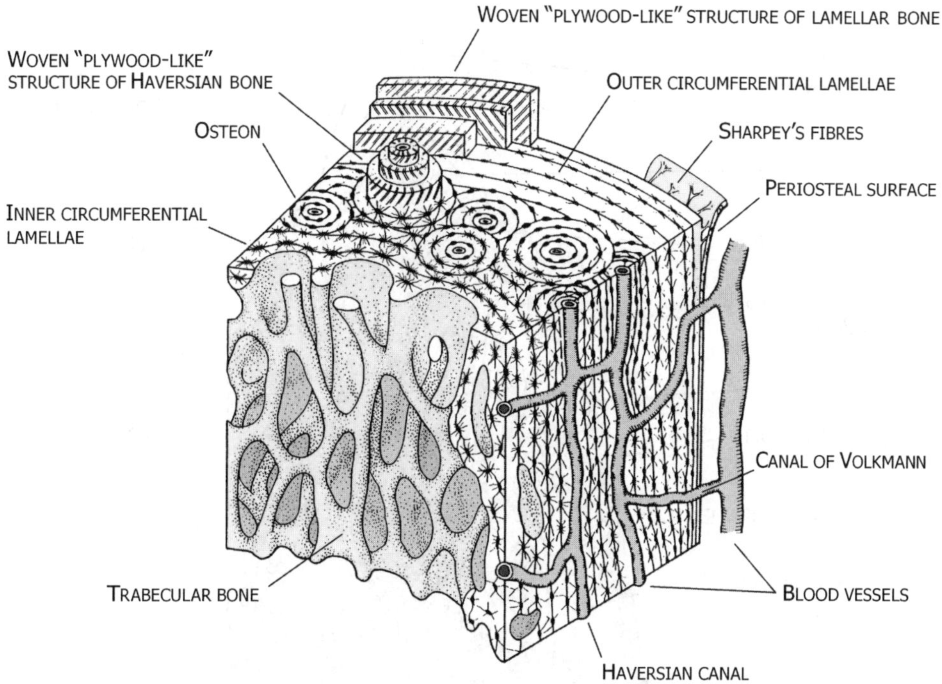

Figure 1.3 Three-dimensional representation of the micro-architecture of compact bone

bone has a more lamellar structure, forming either by apposition on the periosteal surface (circumferential lamellar bone) or by remodelling of the interiors (Haversian or osteonal bone). The microarchitecture of bone tissues clearly influences its mechanical properties, porosity and, ultimately, its resistance to post-mortem degradation. However, a detailed description of bone development and physiology lies outside the purposes of this chapter. For a fuller account of the biology of skeletal tissues the reader is referred to Ortner and Turner-Walker (2003) and Tuross (2003).

Tooth Dentine and Enamel

Teeth are complex structures that have properties that represent a trade-off between the need for a hard, resistant material that can efficiently withstand many years of biting or grinding tough foods and one that has good resistance to fracture. Good teeth are fundamental to the survival of any animal, and nature has perfected many different designs to suit different diets and feeding strategies. Unlike bones, which grow *in situ* and remain surrounded by soft tissues, teeth form within the jaw and are later erupted through the gum into the mouth, where they are in frequent and intimate contact with the outside world. Once in place, any remodelling or repair of damage is strictly limited because the tooth is effectively removed from the cellular apparatus that formed it. By way of compensation, humans develop two sets of teeth, the milk or deciduous teeth of infancy and the permanent teeth which gradually replace the deciduous teeth. The permanent dentition is complete by about 18 years of age.

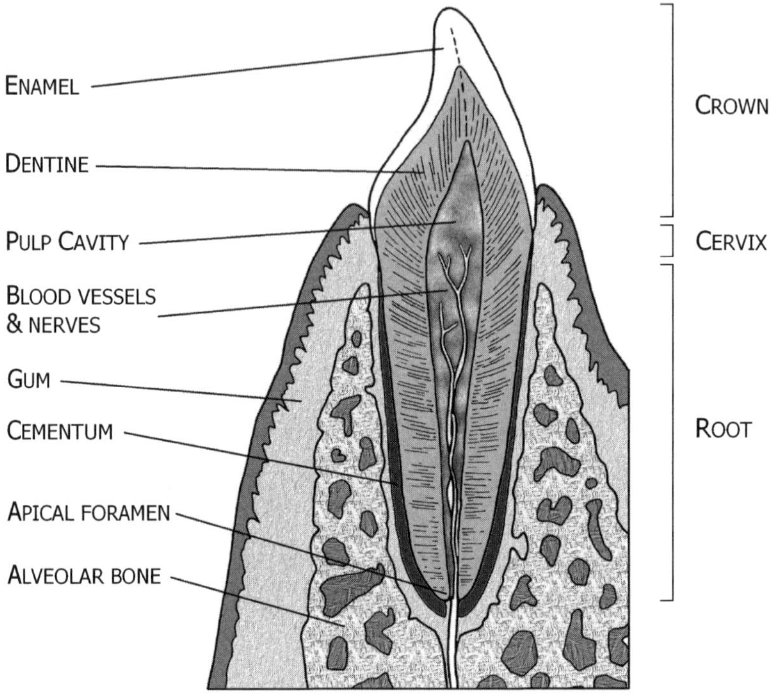

Figure 1.4 Simplified cross-section of a tooth (incisor) and jaw

The mature human tooth can be divided into three parts: the crown, which is the part visible above the gum; one or more roots, which anchor the tooth into the jaw; and the neck or cervix, where the crown meets the root and which lies between the gum-line and the socket (Figure 1.4). The bulk of the tooth is composed of dentine, which forms the underlying load-bearing structure. Unlike bone, dentine is an avascular tissue with no blood supply. It is also largely acellular and the living part of the tooth is restricted to the pulp cavity, which extends from a small hole or foramen in the base of the root into the body of the tooth. The pulp cavity contains blood vessels and nerves and is lined with cells called odontoblasts. Numerous, tightly packed dental tubules extend radially out from the pulp cavity towards the outer surfaces of the tooth. These tubules reflect the developmental growth of the tooth (in the growing tooth, dentine is laid down on the interior surface of the enamel and proceeds inwards) and provide a sensory mechanism for detecting loads on the teeth. The crown of the tooth is encased in hard enamel, which is made up of parallel prisms composed of almost pure HAP. Enamel has negligible organic content and is more crystalline than bone HAP as a result of a larger crystallite size and their parallel alignment within prisms. Once enamel is damaged by tooth wear or dental disease (caries) there is no natural mechanism for effective repair. The outer surface of the tooth root is covered in a type of woven bone called cementum which, together with the periodontal ligament, anchors the tooth in the socket (Mays, 1998; Ortner and Turner-Walker, 2003). Healthy enamel has zero porosity, apart from occasional growth defects. Although there has been little or no investigation of the porosity of tooth dentine, it is clear that its porosity is low compared with that of bone.

Because of the absence of a vascular network in tooth dentine, its relatively low porosity and the hard shell of impervious enamel that covers the exposed crown, it is generally accepted that teeth are less susceptible to diagenesis than bones and, therefore, that they represent a more reliable source of ancient DNA and other biomolecular information. Recent evidence (Götherström *et al.*, 2002; Wandeler *et al.*, 2003; Gilbert *et al.*, 2005, 2006) supports the view that the potential for post-mortem and post-excavation contamination of dentine is much lower than for bones (which are frequently handled by archaeologists and researchers). Nevertheless, teeth are by no means immune to the diagenetic forces that affect bone tissues.

CHEMICAL DIAGENESIS OF BONES AND TEETH

It is common knowledge that bone tissue degrades in the soil. Bone-meal (ground bone) has been used by gardeners as a fertilizer for centuries. This makes good sense, since bone is rich in both nitrogen and phosphorus. Anecdotal evidence has suggested that water and temperature play important roles in the deterioration of human corpses. For example, in Act V: Scene 1 of Shakespeare's *Hamlet* the sexton refers to the destructive power of water on interred corpses: '... your water is a sore decayer of your whoreson dead body'. Also, in some countries of northern Europe it was common practice in the past to pack wood shavings around corpses before sealing the coffins if it was anticipated that the grave may have to be reopened within the year to add the body of a close relative. The additional insulation presumably raised the temperature of the body and accelerated decomposition of the soft tissues, thus reducing the smell of decay. More rigorous research has confirmed the importance of both soil hydrology (Pike, 1993; Hedges and Millard, 1995; Nielsen-Marsh, 1997; Pike *et al.*, 2001; Nielsen-Marsh and Hedges, 2000a) and soil temperature (Collins *et al.*, 1995; Gernaey *et al.*, 2001) on the deterioration of archaeological bones.

The role of water

The availability and movement of water within the soil, and hence through and around archaeological bones, has an immense influence on their potential for survival. Water is the medium of almost all chemical reactions that take place in the soil, and the presence of water also supports microbial metabolism. Whilst in the body, bone mineral lies within a relatively closed system and is surrounded by fluids that have a strictly controlled pH and are approximately saturated with respect to HAP. *In vivo* dissolution and recrystallization of bone mineral is mediated by bone cells which are themselves stimulated by a complex web of systemic and local chemical signals, including physical stimuli, growth factors, parathyroid hormone, and calcitriol – the active form of vitamin D (Ortner and Turner-Walker, 2003). In sharp contrast to this, the soil represents an open system that is far from saturated in calcium and phosphate ions (except perhaps in the case of deeply cut charnel pits containing many hundreds of tightly jumbled bones). Bone mineral, therefore, is vulnerable to dissolution in soil water, which can also bring in exogenous ions that may bind to the surface of the HAP or substitute for Ca^{2+}, PO_4^{2-} or CO_3^{2-} ions within the crystal lattice (Hedges and Millard, 1995).

Analyses of archaeological bones from different environments have demonstrated that those bones that come from soil horizons where there is considerable fluctuation in the groundwater content, i.e. water repeatedly flows around and through the bones, exhibit very poor preservation compared with those that lie in permanently saturated conditions, i.e. below the water table (Nielsen-Marsh *et al.*, 2000: Figure 2). Clearly, susceptibility to dissolution depends on many variables, but one obvious factor is the porosity of the bone. Water does not act solely on the outer surfaces of bones; it penetrates the interiors via the network of interconnecting vascular channels. However, once these small pores are filled and the pore water is saturated in Ca^{2+} and PO_4^{2-} ions there may be considerable resistance to the flow of water out of the bone, and dissolution of HAP is limited by a local diffusion gradient. Diffusion rates through fine pore networks are generally very slow, leading to very limited dissolution of bone mineral. These diffusive environments can be found in waterlogged deposits or in sediments that resist movement of soil water, such as clays; and bones from these environments frequently exhibit exceptionally good preservation.

If bones lie in an environment where there are repeated cycles of wetting and drying, then, as the surrounding soil dries out, a hydraulic potential is generated that will draw water (saturated in Ca^{2+} and PO_4^{2-} ions) out from the interiors of the bones. After heavy rain, the bones experience a recharge regime in which the pores are refilled with water that is no longer saturated in Ca^{2+} and PO_4^{2-} ions (Hedges and Millard, 1995). After many wetting and drying cycles, the successive losses of calcium and phosphorus from the bone matrix cause further increases in the porosity of the bones, which then become locked into a positive feedback loop. Bones that are excavated from shallow, free-draining soils are generally less well preserved than those from deep, waterlogged sites.

Bones buried in well-drained soils overlying sands or gravels that never become saturated in water are particularly susceptible to leaching of the mineral phase. The rate and volume of water flow through a bone in such a soil depends upon the relative hydraulic conductivities of the soil and the bone (i.e. their relative porosities), and the total volume of water available to flow (i.e. the amount of rainfall). This flow regime is potentially the worst situation for the survival of archaeological bones, and in extreme cases (such as inhumations cut through sands and gravels) can lead to total leaching of the skeleton, leaving only a soil silhouette or 'sand body' (Keeley *et al.*, 1977; Bethel and Carver, 1987; Carver, 2005).

The solubility of apatites in groundwaters is heavily dependent on the water's pH and the presence of other dissolved ionic species. As a general rule, all the calcium phosphates become increasingly soluble as pH falls. Solid HAP will reach an equilibrium with stationary water in contact with its surface according to

$$Ca_{10}(PO_4)_6(OH)_2 + 14H^+ \rightleftharpoons 10Ca^{2+} + 6H_2PO_4^- + 2H_2O$$

Thus, an increase in the hydrogen ion concentration (decrease in pH or increase in acidity) will drive the equilibrium to the right and HAP dissolves. In alkaline soils, therefore, bone mineral will tend to be protected from dissolution, unless there is a high dissolved carbon dioxide concentration, in which case Ca^{2+} can precipitate out as bicarbonate or carbonate. Removal of calcium ions, therefore, will also drive the equation to the right and HAP once again can dissolve. HAP, therefore, acts as a buffer helping to stabilize local pH variations. From the equation above it is also clear that soils low in phosphate may also lead to demineralization of bones. HAP is most stable at pH 7.8 (Nielsen-Marsh *et al.*, 2000).

HAP is thermodynamically one of the most stable forms of solid calcium phosphate. Both bone mineral and synthetic HAP can be dissolved in mineral acids and reprecipitated once more to a poorly crystalline HAP on the addition of alkali. In fact, carnivores that consume large quantities of bone routinely excrete finely divided HAP in their faeces. X-ray diffraction (XRD) studies on modern and ancient hyena coprolites have demonstrated that their spectra are essentially the same as those of bone apatite (Horowitz and Goldberg, 1989). Although calcium and phosphorus are excreted via the gut of all vertebrates, the bulk of the apatite in these coprolites will derive directly from bone dissolved in the stomach. Therefore, it is thermodynamically favourable that bone mineral solubilized in the burial environment, either by the active intervention of microorganisms or by the movement of groundwaters over bone, will be reprecipitated as a poorly crystalline HAP (or carbonate apatite depending upon the local availability of dissolved carbon dioxide) when the pH rises or when the solubility product of either calcium or phosphate is exceeded (Nielsen-Mash and Hedges, 2000b). Several authors have reported brushite ($CaHPO_4{\cdot}2H_2O$) as a product of diagenesis in archaeological bones (Hassan and Ortner, 1977; Newesely, 1989; Piepenbrink, 1989). Brushite has also been identified as a product formed in fresh bone after prolonged immersion in an acidic solution (Lee-Thorpe, 1991; Nielsen-Marsh, 1997). However, since brushite is more readily soluble than HAP, it seems unlikely that it forms a major component of degraded bones in normal archaeological soils except in special circumstances, such as skeletons interred in stone vaults, where acid decomposition products may be expected to accumulate and soluble species will not be leached away from the bones.

The pH of the groundwater also determines which other ions are present in solution and, therefore, available for ion exchange with the bone mineral. Thus, bones from acidic burial environments tend to be brown in colour because transition-metal ions (such as iron and manganese) and humic acids are soluble in the local groundwater. Conversely, bones from alkaline soils tend to be white or cream coloured because many of the metal ions are locked up as insoluble oxyhydroxides or carbonates. Several studies have been made on the trace elements in human bone, usually with a view to addressing whether trace elements can be used as dietary discriminants (Lambert *et al.*, 1979, 1982, 1983) and to what extent diagenesis has altered the original chemical composition (Badone and Farquhar, 1982; Lambert *et al.*, 1985a,b; Price *et al.*, 1985; Buikstra *et al.*, 1989; El-Kammar *et al.*, 1989; Grupe and Piepenbrink, 1989; Pate *et al.*, 1989). A comprehensive review of trace element and isotopic analyses is given in Price (1989).

One of the first changes to be recorded in archaeological and fossil bones was an increase in their 'crystallinity' compared with that of fresh bone. This increase can be detected using XRD or through studies of the infrared spectra of bones and is often expressed as the 'crystallinity index' (Bartsiokas and Middleton, 1992; Hedges *et al.*, 1995) or infrared splitting factor (Weiner and Bar Yosef, 1990; Nielsen-Marsh and Hedges, 2000a,b). Increases in crystallinity have been found to be strongly correlated with other diagenetic parameters, including a reduction in the carbonate content of archaeological bone (Hedges *et al.*, 1995; Nielsen-Marsh and Hedges, 2000a). There are several probable explanations for these observed increases in crystallinity: dissolution and loss of the smallest crystals; dissolution and subsequent recrystallization to larger and thermodynamically more stable crystals; a reordering of the internal crystal structure; and slow growth of existing crystals by apposition. While all these mechanisms are possible and may play some role in diagenesis of ancient bones and teeth, it is the second, i.e. dissolution and recrystallization, that on present evidence appears to dominate. Clearly, this process carries with it the likelihood that exogenous ions (e.g. Sr^{2+}, Zn^{2+}, CO_3^{2-}, F^-, etc.) or various isotopes (e.g. C, O) may be incorporated in the reprecipitating crystals, with all that implies for the interpretation of subsequent chemical analyses. Of course, the implications of increases in crystallinity and the incorporation of exogenous mineral species are not always negative. These processes fall under the general heading of fossilization, and it is undoubtedly true that the vast majority of bones simply would not survive over geological time-scales without a certain level of permineralization and the infilling of internal voids with calcite or silicates.

The mineral phase is not the only component of bone that is susceptible to chemical degradation over time. The protein contents of archaeological bones are generally very low compared with fresh bone, and in truly fossil bones the levels of organics are reduced to chemical traces of amino acids and osteocalcin. The survival of proteinaceous material in the archaeological record is generally restricted to very special circumstances and environments. Mummified remains of humans are, of course, found, but most often in environments with very restricted liquid water, such as arid deserts, frozen soils and ice, mountain caves, etc. In deep cultural layers that lie beneath the water table, tanned leather can survive for many centuries, and the natural tanning effects found in sphagnum bogs are in part responsible for the spectacular preservation of 'bog bodies' such as Tollund Man, Grauballe Man and Lindow Man. In most environments, however, liquid water not only permits the growth of microbes, but it can also accelerate loss of protein via hydrolysis. In normal soils, unmineralized collagen degrades rapidly via biological degradation in which microorganisms use extracellular proteolytic enzymes to break the long collagen molecule into smaller peptides that can be assimilated by bacteria and fungi. In mineralized collagen, the intimate association between the protein and mineral has a powerful stabilizing effect that influences both microbial and chemical degradation of bones. The resistance of bone to microbial attack arises from the absence of microscopic pores larger than 8 nm. Microbial collagenases are large molecules with sizes ranging between 60 and 130 kDa (Bond and van Wart, 1984) and they are unable to penetrate the smaller pores between the HAP and the collagen (Nielsen-Marsh *et al.*, 2000; Gernaey *et al.*, 2001). Enzymatic hydrolysis of mineralized collagen would require that the mineral be removed first, and this is indeed what happens in the microbial degradation of bones and teeth (see below).

In the absence of enzymatic degradation, collagen can persist in bones for many hundreds or even thousands of years. Bones recovered from deep gravel quarries at the Pleistocene site of Shropham in Norfolk, UK, retain up to 85 % of their original collagen after more than 120 ky (Turner-Walker, unpublished data). In fact, mineralized collagen resists chemical hydrolysis far longer than kinetic studies of unmineralized collagen suggest (Collins *et al.*, 1995). The chemical affinity between collagen and HAP is such that the presence of the mineral not only excludes any molecule larger than water, it also physically constrains (straitjackets) the collagen helix to a far greater degree than is achieved by tanning for example. Unmineralized collagen will shrink or melt at a temperature of about 68°C. For tanned leathers this temperature is typically in the range 75–85°C, whereas for mineralized collagen this transformation does not occur until over 150°C (Nielsen-Marsh *et al.*, 2000). The straightjacketing effect of the HAP restricts the ability of the helix to expand. However, since there is always some water held in the microporosity of bones (even in dry soils), collagen will undergo slow chemical hydrolysis that can be accelerated by either an increase or reduction in local pH and by increasing the temperature. Collagen is most stable against hydrolysis when the pH lies in the range pH 3–7.5. At pH 1 the rate of hydrolysis is 10 times faster than at neutral pH, and at pH 12 the rate is 100 times faster (Collins *et al.*, 1995: Figure 1). Even though chemical hydrolysis may cause chain scissioning in the collagen molecules, cutting the long fibrils into shorter peptide units, the strong affinity between

HAP and collagen combined with the small sizes of the micropores will severely restrict diffusion of the fragments out of the bone structure. Therefore, unless there is an infiltration of humics, which can affect cross-links between the damaged fibrils, a situation arises in which archaeological bones can retain relatively high collagen content and low macroporosity but have considerably reduced mechanical strength (Collins *et al.*, 1995; Turner-Walker and Parry, 1995). Recent transmission electron microscope studies of degraded collagen from bones excavated from experimental burials have shown that the structure of the fibrils begins to break down after only a few years, exhibiting localized swellings and an apparent unravelling of the tightly packed collagen molecules. This damage is limited to short sections in the middle and ends of the fibrils and is also seen in cooked bones (Koon, 2006).

In the absence of microbial attack then, collagen and bone mineral are locked into a state of mutual protection. The HAP protects the organic fraction from microbial enzymolysis and retards the rate of chemical hydrolysis. The collagen in turn surrounds the tiny crystallites of HAP and inhibits their dissolution by percolating groundwaters. Once one of the components begins to break down, however, bone begins to degrade in the burial environment. Loss of protein increases the microporosity and allows water to penetrate further into the mineral phase. If the protein fraction is stripped from fresh bone using hydrazine hydrate ($NH_2NH_2 \cdot H_2O$) then the porosity changes dramatically: macroporosity shows an almost fourfold increase from 0.075 $cm^3\ g^{-1}$ to 0.300 $cm^3\ g^{-1}$ and there is a corresponding decrease in the microporosity from 0.059 $cm^3\ g^{-1}$ to 0.031 $cm^3\ g^{-1}$ – nearly half of its initial value (Nielsen-Marsh and Hedges, 2000a). Because mineral makes up the bulk of the volume of fresh bone, any loss through dissolution has a large influence on the degraded bone's porosity. In the case of demineralization, however, the bone loses rigidity and tensile strength and is susceptible to shrinking, warping and cracking. More crucially, unless there is some component in the groundwater either to inhibit microbial enzymolysis or to induce cross-linking (tanning) in the exposed collagen matrix, the organic matter is quickly degraded by soil microorganisms. In the case of many bog bodies, one of the components of sphagnum peat, i.e. *sphagnan*, is responsible for both the loss of the HAP and, simultaneously, tanning of the collagen and deactivation of microbial enzymes (Painter, 1983, 1991a,b; 1995, 1998; Turner-Walker and Peacock, in press).

MICROBIAL DIAGENESIS OF BONES AND TEETH

During the decomposition of a corpse, the role of microorganisms is dominant and loss of soft tissues, i.e. skeletonization, is largely mediated by bacteria and fungi, although autolysis also plays an important role in the early stages of decay of the body. When a cell dies, a cocktail of enzymes (proteases and DNases) are released which quickly break down the surrounding cell components and tissues. The onset of this autolysis is very rapid, but it is short-lived. Thereafter, bacterially mediated tissue destruction takes over with large numbers of microorganisms being released from the gut into the abdominal cavity. The sequence of autolytic decomposition follows that of tissues with the highest rates of synthesis of adenosine triphosphate, the fuel that drives the body's metabolism. Thus, the intestines, stomach, liver and organs related to digestion are the first to deteriorate, together with the heart, blood and circulatory systems. These are followed by the lungs, kidneys and bladder, brain and nervous tissues, and later the skeletal muscles. Connective tissues, which are predominantly collagen, are highly resistant to autolysis (Gill-King, 1997).

As the autolysis phase draws to an end, an almost entirely anaerobic environment is created that is favourable to the proliferation of bacteria liberated by the decomposition of the gut and, to a lesser extent, the local soil bacteria. In a healthy adult colon, 96–99 % of the microbial florae are anaerobes and these work quickly on the body tissues, the fermentation releasing the decomposition gases characteristic of putrefaction (Gill-King, 1997). There has been some suggestion that the early release of microorganisms from the gut causes more rapid degradation of the bones located around the abdomen (Child, 1995), but this is not always borne out by examining skeletal element survival rate in large assemblages of skeletons (Waldron, 1987). It has also been noted that diagenesis in bone from domesticated animals that were slaughtered and butchered is often less pronounced than in equivalent human bone from inhumations, raising the possibility that the early stages of putrefaction have some bearing on later degradation by bacteria (Jans *et al.*, 2004). However, it is equally likely that these differences arise from the relative proportions of Haversian bone in humans and animals. Domestic animals are typically slaughtered soon after reaching sexual maturity and, therefore, have proportionally higher primary lamellar bone than humans, who typically have more porous secondary or Haversian bone (Turner-Walker *et al.*, 2002: Figure 5).

Once reduced to a skeleton, the diagenesis of bones is mediated almost entirely by microorganisms, the presence of which has a profound influence on their preservation potential. Of course, local groundwater, oxygen availability, pH and temperature will not only influence what kinds of microorganisms are present, but also how quickly they multiply. From the earliest histological investigations of ancient bones, fungi were implicated in the post-mortem destruction of bone tissues. Certainly, fungi can readily be found on excavated bones, which are frequently washed in contaminated water and often are relegated to low-priority storage facilities where damp and poor air circulation encourage mould growth. Marchiafava *et al.* (1974), Hackett (1981) and Piepenbrink (1986) all conducted experiments in fungal attack on buried bone in an attempt to replicate tunnelling and other features associated with diagenesis. Marchiafava *et al.* (1974) compared experimentally buried fresh human vertebrae with Neanderthal specimens using transmission electron microscopy and optical microscopy. Mould specimens that developed around and within the vertebrae were cultivated on agar for identification and subsequent inoculation into both sterilized soil and bone autoclaved at 200°C for 20 min. Only one fungus, *Mucor*, was successfully cultivated in isolation on inoculated sterile bone buried in sterilized earth. In retrospect, this study would appear to have been fundamentally flawed. Autoclaving at 200°C for 20 min is approximately equivalent to boiling the bone for over 300 h. This would reduce the collagen to a hydrolysed gelatine mass that would make an ideal food for a wide range of microorganisms, but which formed a poor model for uncooked archaeological bones.

Hackett (1981) experimented in the reproduction of what he termed *microscopical focal destruction* in bone using samples of sterilized compact bone which he had buried in garden soil at room temperature for 1 year. On excavation and microscopic examination, at least two of these showed evidence of tunnelling and dissolution and reprecipitation of bone mineral. The results of this experiment were ultimately inconclusive, however, since the most promising specimens failed to show tunnelling in subsequent experiments. Towards the end of his paper, Hackett suggested that the narrow Wedl tunnels found in exhumed and fossil bone may result from the activity of certain bacteria, deriving their nourishment from the debris left by fungi.

Piepenbrink (1986) also investigated the fungal degradation of buried bones using a wide variety of analytical techniques, including histology, microradiography and microbiological incubation. He identified and isolated several species of fungi from stained areas in exhumed bones. These fungi were subsequently found to colonize sterilized bone rapidly, but none produced tunnelling or any of the other features associated with diagenetic alteration of bone, such as loss of birefringence. As part of an investigation of the effects of microbial degradation on trace element concentrations, Grupe and co-workers (Grupe and Piepenbrink, 1989; Grupe *et al.*, 1993) inoculated fresh, irradiation-sterilized pig bone with several species of fungi and bacteria. However, they also reported that no tunnelling could subsequently be detected in any of the samples examined, although some superficial staining and loss of birefringence was seen in the periosteal layers.

Research undertaken subsequent to the 1990s has shifted the focus away from fungi; now, bacteria are recognized as playing a fundamental role in the destruction of bone tissues in archaeological contexts. At the time of writing, there was no smoking gun that convincingly identified a particular species of soil organism as being responsible for destruction of histology in bones or teeth. However, despite the numerous classifications for different types of destruction seen in histological sections – Wedl or centrifugal tunnelling, linear longitudinal tunnelling, budded tunnelling and lamellate tunnelling (Hackett, 1981; Trueman and Martill, 2002; Jans *et al.*, 2004) – only two broad classes of bacteria seem to be involved: aerobic bacteria in normal archaeological soils (Nielsen-Marsh and Hedges, 1999; Turner-Walker *et al.*, 2002; Turner-Walker and Syversen, 2002) and cyanobacteria in marine environments (Bell *et al.*, 1991). Much of this shift from fungi to bacteria has arisen as a result of applying more powerful techniques to the problem of diagenesis and using microscopy techniques with higher resolutions than available earlier, particularly backscatter scanning electron microscopy (BSEM), which has replaced microradiography and, to a large extent, optical microscopy as the technique of choice (Bell, 1990; Bell *et al.*, 1991, 1996; Turner-Walker *et al.*, 2002; Turner-Walker and Syversen, 2002).

The higher resolution of current SEM techniques over previous light microscopy of thin sections has revealed a fine structure to the microscopical focal destruction described by other researchers. The tunnels identified by Hackett and other researchers, and which appeared to have diameters around 5–10 μm, are actually comprised of numerous smaller pores with diameters that range from 0.1–1.0 μm (Figure 1.5). In BSEM images these pores can be seen to be confined to localized zones, each 10–40 μm across. These zones are frequently surrounded by an electron-dense region that either delineates the extent of the tissue destruction or completely fills the intervening area between the small pores (Figure 1.5d). In other places the pores lie within a general area of lower electron density than the surrounding unaffected bone. This patchwork of demineralized and hypermineralized zones is responsible for many of the features seen in light microscopy of archaeological bone sections. The development of the sub-micrometre pore network is also responsible for changes in the optical properties of ancient bones viewed in this section. The threadlike tunnels created by the bacteria also disrupt the optical properties of bone tissues, reducing its transparency in affected regions and causing them to appear opaque when viewed in polarized light (Figure 1.5b). Staining from soil water may leave the affected bone black or dark brown in thin sections (Turner-Walker and Syversen, 2002).

The soil bacteria responsible for the destruction of bone tissues infiltrate the interior of the bones via the network of vascular channels but appear to be inhibited from attacking the periosteal surfaces in many environments by the presence of humic substances in the soil

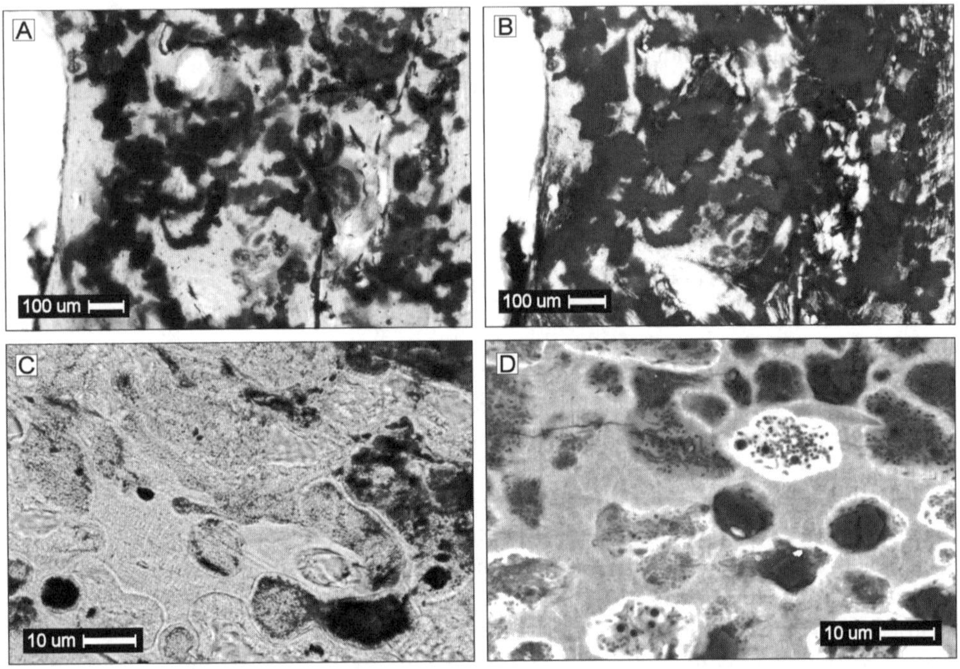

Figure 1.5 (a) Transmitted light image of medieval human bone from Wharram Percy, UK. Most of the tissue is affected by diagenetic degradation, which obscures much of the histological features (compare with Figure 1.2a and b) Section viewed in polarized light. Almost all the birefringence is obscured or lost, demonstrating a disruption of the collagen – HAP bond. (c) High-magnification image showing details of the affected bone. (d) An equivalent BSEM image of similar bone from Wharram Percy. The affected bone is revealed as penetrated by numerous pores or tunnels. Demineralization and reprecipitation of HAP is also evident

water (Figure 1.6a). These humics may act either on the collagen molecules, creating cross-links that reduce the effectiveness of bacterial enzymes (Hedges, 2002), or by deactivating the collagenases themselves (Jans *et al.*, 2004). The penetration of bacteria through the compact bone tissue is influenced by the microarchitecture of the tissues. For example, bacteria seem unable to cross the cement lines that mark the boundaries between secondary osteons (Haversian systems) and the surrounding primary lamellar bone or that mark the reversal of resorption in remodelled bone (Figure 1.6b). In cross-sections of affected bone, some bacterial colonies can be seen to tunnel normal to the plane of the section (i.e. along the long axis of the bone), whereas in other places they create meandering tunnels that stream parallel to the plane of the section (Figure 1.6c and d respectively). This suggests that the bacteria follow the orientation of the collagen fibres in different parts of the bone and are able to exploit planes of weakness in the tissues.

This preferred orientation in the spread of bacteria through calcified tissues is much more marked in longitudinal sections of diagenetically altered teeth. Even in cases where bacterial destruction obscures much of the histological details of the tissues it is possible to distinguish the boundary between the dentine, where elongated destructive foci are aligned along the dentinal tubules, and the cementum, where the destructive foci are larger and more globular,

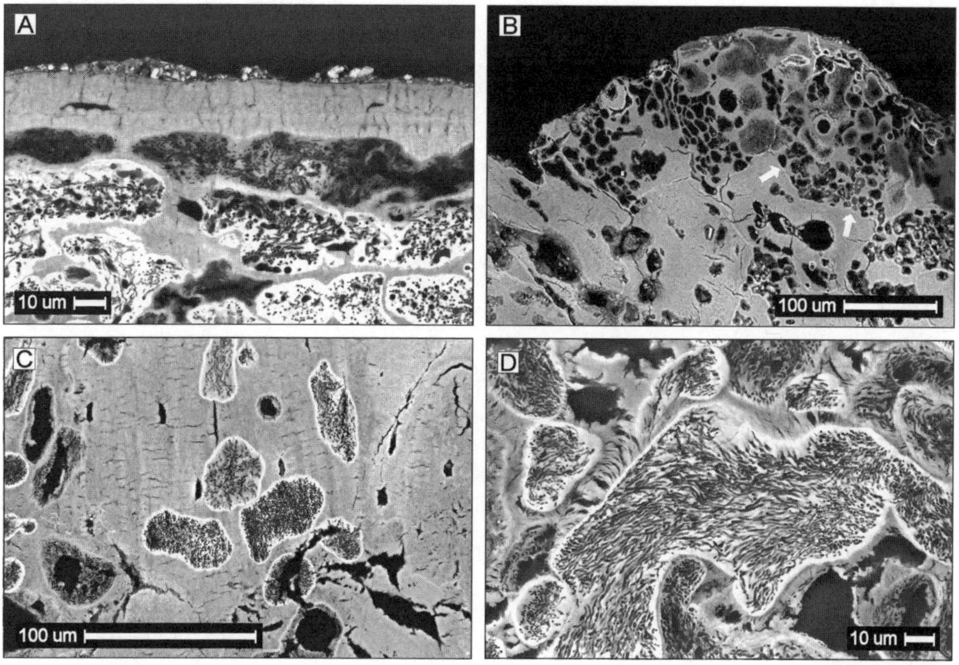

Figure 1.6 BSEM images of bones from various sites. (a) Inhibition of bacterial attack at the periosteal surface. (b) Bacterial tunnelling stopped by cement line surrounding osteon (arrowed). (c) Bacterial tunnelling passing normal to the plane of the cross-section. Note the bright borders of hypermineralized HAP. (d) Meandering pathways of bacterial tunnelling parallel to plane of cross-section

showing no preferred orientation (Figure 1.7a and b). Where recent and archaeological teeth show evidence of dental caries there is no similarity with the kinds of diagenetic tissue destruction seen in bone, cementum or dentine. Rather, there is a general demineralization of the dentine that reveals the growth patterns of dentinal tubules (Figure 1.7c), but none of the tunnelling or reprecipitated mineral seen in typical bacterial alteration of skeletal tissues. Similarly, demineralization resulting from caries highlights the internal structure of enamel (Figure 1.7d).

The observation that bone mineral is dissolved and reprecipitated in the zones affected by bacterial attack supports theoretical considerations that microbial enzymes are unable to degrade mineralized collagen. Further support for the necessity to demineralize bone tissues prior to enzymatic degradation is provided by studies of remodelling in living bone. Bone resorption is undertaken by mature osteoclasts using a combination of acid dissolution of bone mineral and destruction of organic matrix by proteolytic enzymes. This essential initial step, i.e. removal of bone mineral, determines the rate and extent of bone removed from the resorption pit (Ortner and Turner-Walker, 2003). In bone that has been recovered from acid or free-draining leaching soils, the reprecipitated HAP that is no longer protected by its intimate association with collagen is susceptible to dissolution and loss. The ragged holes left behind erase any trace of the smaller porosity and leave the bone extremely fragile. In extreme cases the bone may disappear entirely from the archaeological record.

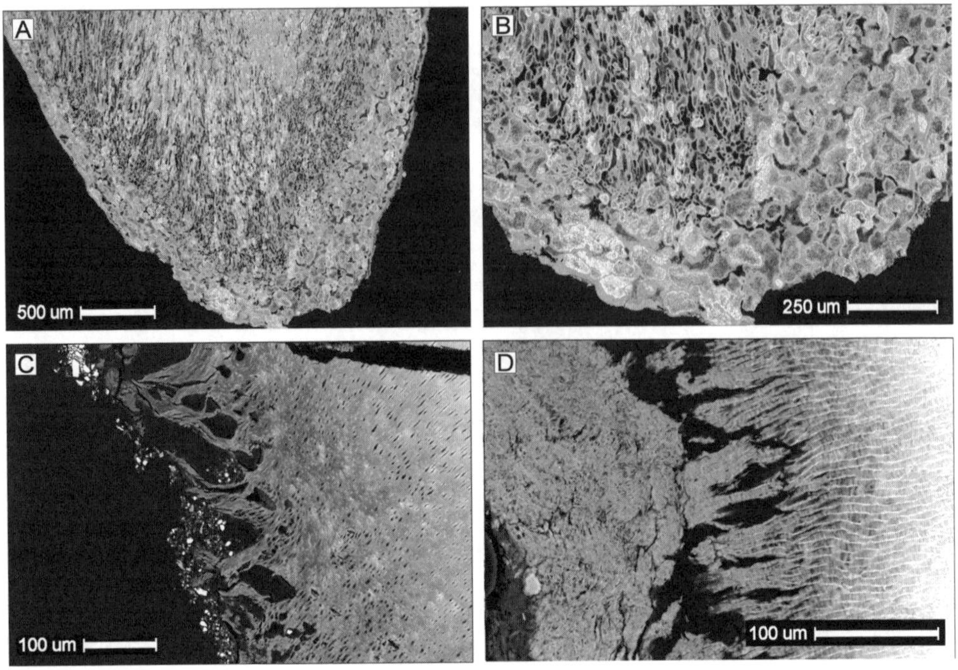

Figure 1.7 Diagenetic degradation of teeth. (a) Longitudinal section of root of archaeological human incisor from Wharram Percy, UK. The pattern of bacterial tunnelling shows clearly the border between the dentine and the overlying cementum. (b) Detail of image in (a). (c) Demineralized dentine in a modern tooth with dental caries. This is easily distinguished from degradation by soil bacterial in archaeological specimens. (d) Demineralized enamel in modern dental caries specimen

Of the bone that does survive and is recovered from excavation sites, the bacterial attack described in the previous paragraphs is almost ubiquitous and can be found in bones of almost all ages, from decades to millions of years. Only in bones recovered from contexts representing rapid burial in anoxic sediments or those from very cold climatic regions are these features absent. In medieval skeletons from Trondheim in mid-Norway, for example, bacterial destruction of bone tissues is not in evidence. A combination of low average soil temperatures and graves cut into waterlogged, organic-rich or clay soils has led to spectacular preservation of the bones, which consequently have a high residual collagen content and excellent preservation of lipids and other biomolecules (Figure 1.8a). These observations suggest two conclusions. First, the bacterial attack seen in so many bones derives from the action of aerobic soil bacteria. Second, since the early stages of putrefaction of the corpses in Trondheim presumably followed a similar path as those in more temperate regions, the influence of gut bacteria on subsequent destruction of bone tissue may not be as important as has been suggested by some researchers. Bones from waterlogged anoxic environments often contain pyrite framboids in their internal porosity (Figure 1.8b). These clusters of finely divided iron sulphides are a characteristic by-product of the metabolism of certain anaerobic sulphate-reducing bacteria (SRB). These SRB are primitive organisms that are incapable of metabolizing large organic molecules such as peptide fragments and, thus, cannot destroy mineral tissues directly. Instead, they use the sulphate ion as an oxidizing agent for simple

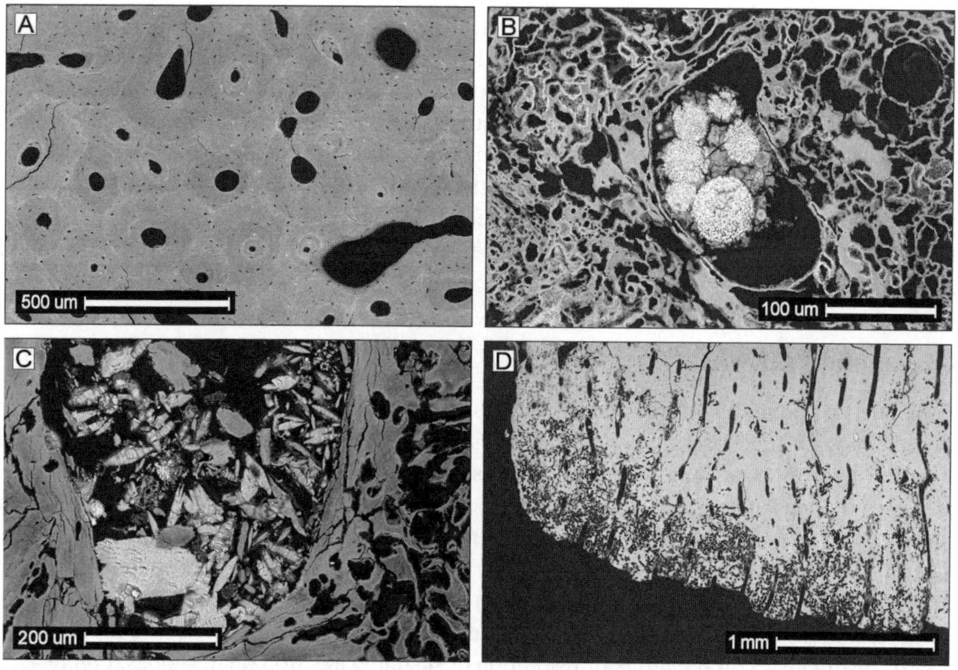

Figure 1.8 (a) Medieval human femur from Trondheim, Norway. The histology is perfectly preserved with no evidence of bacterial degradation. Note the actively resorbing osteons and numerous cement lines showing several remodelling episodes. (b) Bone from a Neolithic cemetery in Ypenburg, the Netherlands. A rise in the local water table in antiquity has caused loss of the bacterially degraded tissues, and anoxic conditions have favoured the colonization of the pore spaces by SRB. Note the well-developed pyrite framboids. (c) Animal bone from a Mesolithic site in the Vale of Pickering, UK. This is similar to the bone from Ypenburg, but oxidation of the pyrite to sulphuric acid has given rise to crystallization of lenticular gypsum crystals in the pore spaces. (d) Animal bone from the late Neolithic site of Aartswoud, the Netherlands. The settlement was on salt marshes and tidal flats, and this marine environment is reflected in the characteristic tunnelling by cyanobacteria. Note tunnelling is limited to the outer millimetres

organic compounds, such as acetate, lactate and propionate, found in decaying organic matter. There is a corresponding reduction of the sulphate ion to sulphide, which combines with metal ions, such as iron to give iron sulphides. If there is a change in the burial environment to more oxidizing conditions, then this finely divided pyrite can undergo oxidation with the consequent release of sulphate and hydrogen ions. The resulting fall in pH can cause local dissolution of HAP and give rise to deposition of gypsum ($CaSO_4 \cdot 2H_2O$, Figure 1.8c; Turner-Walker, 1998a,b) or vivianite ($Fe_3(PO_4)_2 \cdot 8H_2O$) within and on the bones (Mann *et al.*, 1998; Maritan and Mazzoli, 2004).

The other group of organisms associated with tunnelling destruction of bone tissues are the cyanobacteria. These are phototrophic organisms and, consequently, are restricted in their habitats. Nevertheless, they are implicated in destruction of bones from tidal or estuarine deposits, where they may thrive down to depths of several metres of water. Bone destruction in specimens from these sites exhibit a different pattern of attack (Figure 1.8d). Destruction proceeds from the periosteal surface inwards, or sometimes from the larger physiological pores. The size (5–10 μm), close spacing and tortuous branching habit of these tunnels are very similar to those made by the endolithic filamentous cyanobacterium *Mastigocoleus testarum*, responsible for bioerosion in marine shells and vertebrate skeletons (Davis, 1997; Kaehler, 1999: figure 6E). In cross-section the pores display ragged borders, rather than the smooth, globular cross-sections of the bacterially degraded bones from terrestrial burial sites, and branch more frequently. The tunnels attributed to cyanobacteria do not appear to respect the natural micro-architecture of the bone in the same way that terrestrial soil bacteria do, and there is no evidence for local demineralization or reprecipitation of HAP.

Diagenetic Pathways

There are two predominant mechanisms of bone degradation in archaeological soils, which may or may not proceed simultaneously. These are bacterial degradation of the tissues and chemical hydrolysis of bone collagen. In most bones from aerated soils, both mechanisms proceed simultaneously, albeit at different rates, and the net result is a gradual loss of collagen content over time (Figure 1.9). Bacterial degradation is by far the most rapid pathway by several orders of magnitude. Bone buried in tropical countries may be rapidly consumed within a few centuries, whereas bone buried in colder or waterlogged sediments may survive for several thousands of years, or even hundreds of thousands of years. The two mechanisms may be distinguished at the limits of resolution of scanning electron microscopes. Figure 1.10 shows a backscatter image of bone from the medieval site of Wharram Percy in the UK, viewed with a high-resolution field-emission scanning electron microscope. The large circular voids on the left-hand side of the image were created by soil bacteria. The area affected is bounded by a band of dense, reprecipitated HAP that appears bright in backscatter. The countless tiny holes filling the right-hand side of the image are the voids left by the hydrolysis and leaching of the collagen fibres, leaving a negative cast of undissolved HAP. The occasional swirling bands reflect changes in the orientation of collagen fibres in the lamellar structure of the tissue.

The state of preservation of any skeleton, or assemblage of bones, depends upon its early taphonomic history and the particular diagenetic trajectory it follows. The former may be controlled by cultural, economic or social factors, whereas the latter may be controlled solely by geographical location, climatic factors and the character of the burial environment. Thus, preservation may differ markedly depending upon whether a corpse is interred in a stone-lined crypt, a wooden coffin or directly in the soil. The degree of bacterial degradation (or its total absence) may depend upon depth of burial (above or below the water table), the local average soil temperature and the dissolved oxygen content of the soil waters. Graves cut into alkaline chalky soils, particularly those in hot countries, show evidence of collagen hydrolysis and shrinkage cracks, as well as some limited bacterial attack (Jans *et al.*, 2004). Waterlogged acidic or neutral soils produce bones with little or no bacterial attack and a high residual collagen content, but with some potential for demineralization and permineralization with exogenous mineral species. Bodies that find their way into sphagnum bogs, either by accident or as some ritual sacrifice, exhibit another type of preservation in which the collagen matrix survives as part of a ‘tanned’ bog body, but where the mineral may have been completely lost.

It is worth noting here that the relationships between bone preservation at the microscopic level (as revealed by histological studies) and gross preservation as perceived by visual

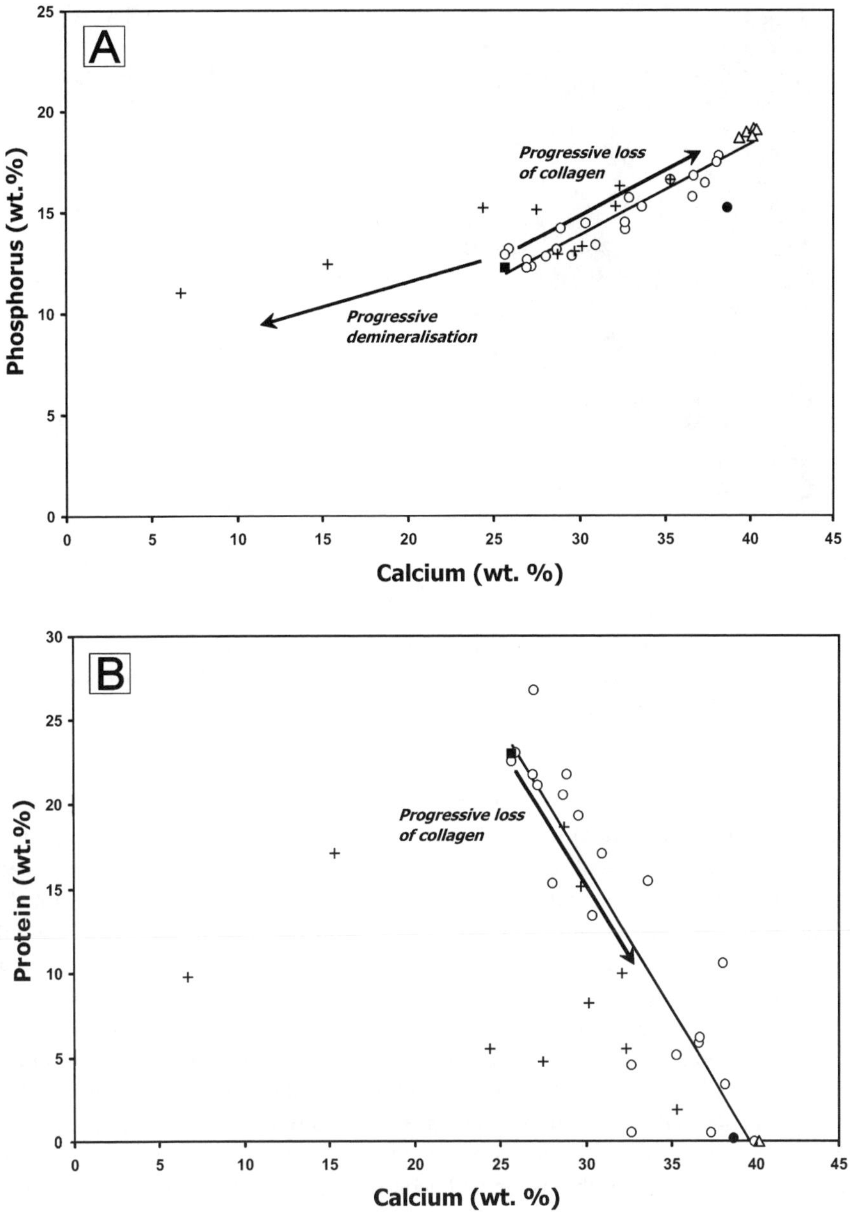

Figure 1.9 (a) Graph of calcium versus phosphorus (in weight percent) for archaeological bones from different sites. The open circles represent bones excavated from aerated soils. The filled square is modern sheep bone. The open triangles represent chemically deproteinated modern bone (i.e. pure bone mineral). The crosses represent bones excavated from a waterlogged site lying beneath peat deposits. It is clear that some bones suffered bacterial degradation (movement along collagen loss trajectory) prior to demineralisation in the peat deposits. The single filled circle is bone from a very alkaline soil that contained diagenetic calcite. (b) Graph of calcium versus residual protein for bones from different sites. Symbols are as described for (a). Unpublished data from Turner-Walker (1993)

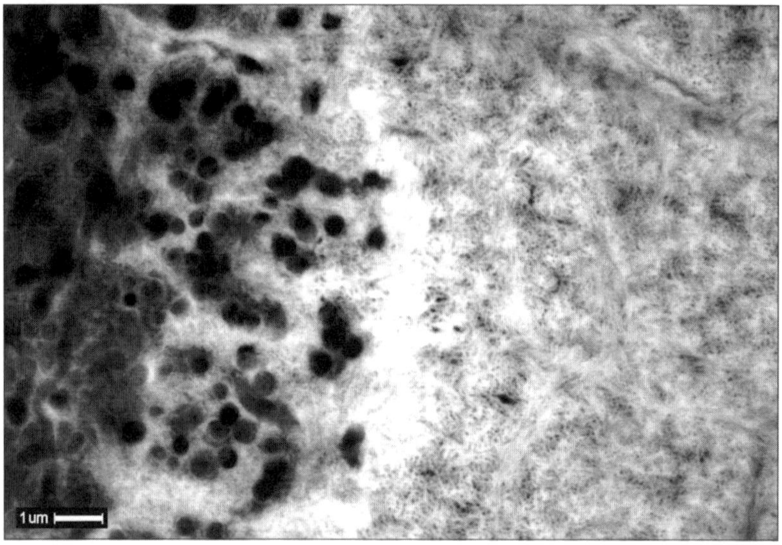

Figure 1.10 High-resolution field-emission scanning electron microscope backscatter image of medieval human femur from Wharram Percy, UK. The circular holes (approximately 500 nm in diameter) on the left side of the image were created by soil bacteria. Hydrolysis and leaching of the collagen fibres is revealed as the numerous small holes (40–50 nm in diameter) filling the rest of the image. The central band that appears bright in backscatter represents HAP that has been dissolved by bacteria and reprecipitated at a higher density

inspection of archaeological skeletons are far from clear. Some skeletal assemblages (particularly those from alkaline soils) that are grossly very well preserved may exhibit very poor preservation histologically, with considerable destruction of cortical bone by soil bacteria. Often the outer millimetre of bone tissue where it has been in direct contact with the soil is preferentially preserved (Figure 1.6a) compared with the interior. Bacterial degradation may have to be sufficiently advanced to render archaeological bone prone to fragmentation and dissolution before it substantially reduces the ability of osteoarchaeologists to extract useful data from an assemblage. Conversely, skeletons that exhibit especially good histological preservation, such as those from medieval Trondheim, can sometimes show very variable gross preservation, including erosion of bone surfaces. This observation has obvious implications for the palaeopathological diagnosis of many diseases that affect the surface texture or morphology of bones and raises the question as to what extent palaeohistological work can aid the diagnosis of disease in archaeological skeletons (Turner-Walker and Mays, Chapter 7).

What it is important to appreciate is that, over archaeological time-scales, burial environments can change, either as a direct result of human activities (drainage of wetlands, build up of deep cultural layers in urban contexts, etc.) or by natural geological or climatic processes. The soil conditions from which bones are excavated may not necessarily represent those of antiquity; therefore, the preservation state of the bones may not reflect the preserving qualities of the surrounding soil. Bones, therefore, travel through time along different diagenetic trajectories or pathways, which in some cases may be approximately linear but in other cases may make abrupt changes in course according to the evolution of the burial environment (Figure 1.9a). It is possible that a history of the burial environment is preserved in the histology of the bones themselves, and in certain circumstances that is clearly the case (Turner-Walker, 1998b; Turner-Walker and Jans, in press).

Despite the considerable advances made over the past decade in the understanding of diagenetic alteration of archaeological bones, much still remains to be done. There is still the unresolved question of which microorganism (or group of microorganisms) is responsible for the tunnelling that is such a common feature in ancient bone. It would also be interesting to trace the antiquity of the organism(s) responsible by tracking this tunnelling through the fossil record. The exact relationships between microbial destruction of mineralized tissues and the survival of biomolecular evidence are also unresolved, although all evidence to date points to a regular relationship between histological preservation and the ability to extract intact biomolecules. In particular, the precise location of preserved DNA in archaeological bones and teeth, and how this influences the success or failure of decontamination and extraction protocols, would seem a pressing question. In conclusion, it may be expected that the coming decade will see the answers to these and many other questions relating to the potential data locked away in archaeological and fossil bones.

REFERENCES

Agarwal SC, Grynpas MD. 1996. Bone quantity and quality in past populations. *Anat Rec* **246**: 423–432.

Badone E, Farquhar RM. 1982. Application of neutron activation analysis to the study of element concentration and exchange in fossil bones. *J Radioanalytic Chem* **69**: 291–311.

Baron H, Hummel S, Herrmann, B. 1996. *Mycobacterium tuberculosis* complex DNA in ancient human bones. *J Archaeol Sci* **23**: 667–671.

Bartsiokas A, Middleton AP. 1992. Characterization and dating of recent and fossil bone by X-ray diffraction. *J Archaeol Sci* **19**: 63–72.

Behrensmeyer AK, Hill AP. 1980. *Fossils in the Making: Vertebrate Taphonomy and Palaeoecology*. University of Chicago Press: Chicago.

Bell LS. 1990. Palaeopathology and diagenesis: an SEM evaluation of structural changes using backscattered electron imaging. *J Archaeol Sci* **17**: 85–102.

Bell LS, Boyd A, JonesSJ. 1991. Diagenetic alteration to teeth in situ illustrated by backscattered electron imaging. *Scanning* **13**(2): 173–183.

Bell LS, Skinner MF, Jones SJ. 1996. The speed of post mortem change to the human skeleton and its taphonomic significance. *Forensic Sci Int* **82**(2): 129–140.

Bethel PH, Carver MOH. 1987. Detection and enhancement of decayed inhumations at Sutton Hoo. In *Death Decay and Reconstruction*, Boddington A, Garland, AN, Janaway RC (eds). Manchester University Press: Manchester; 10–21.

Bond MD, van Wart HE 1984. Characterisation of the individual collagenases from *Clostridium histoliticum*. *Biochemistry* **23**: 3085–3091.

Braun M, Cook DC, Pfeiffer S. 1998. DNA from *Mycobacterium tuberculosis* complex identified in North American, pre-Columbian human skeletal remains. *J Archaeol Sci* **25**: 271–277.

Briggs DEG. 1992. Conodonts: a major extinct group added to the vertebrates. *Science* **256**: 1285–1286.

Buikstra JE, Frankenberg S, Lambert JB, Xue L. 1989. Multiple elements: multiple expectations. In *The Chemistry of Prehistoric Human Bone*, Price TD (ed.). Cambridge University Press: Cambridge; 155–210.

Carver M. 2005. *Sutton Hoo: A Seventh-Century Princely Burial Ground and its Context*. British Museum Press: London.

Cattaneo C, Gelsthorpe K, Phillips P, Sokol RJ. 1995. Differential survival of albumin in ancient bone. *J Archaeol Sci* **22**: 271–276.

Child AM. 1995. Microbial taphonomy of archaeological bone. *Stud Conserv* **40**: 19–30.

Collins MJ, Riley MS, Child A, Turner-Walker G. 1995. A basic mathematical simulation of the chemical degradation of ancient collagen. *J Archaeol Sci* **22**: 175–183.

Davis PG. 1997. The bioerosion of bird bones. *Int J Osteoarchaeol* **7**: 388–401.

De Cupere VM, Van Wetter J, Rouxhet PG. 2003. Nanoscale organization of collagen and mixed collagen – pluronic adsorbed layers. *Langmuir* **19**: 6957–6967.

De Vries H, Oakley KP. 1959. Radiocarbon dating of the Piltdown skull and jaw. *Nature* **184**: 224–226.

Dupras TL, Schwarcz HP. 2001. Strangers in a strange land: stable isotope evidence for human migration in the Dakhleh Oasis, Egypt. *J Archaeol Sci* **28**: 1199–1208.

Eastoe JE, Eastoe B. 1954. The organic constituents of mammalian compact bone. *Biochem J* **57**: 453–459.

Efremov IA. 1940. Taphonomy: a new branch of palaeontology. *Pan-Am Geol* **74**: 81–93.

El-Kammar A, Hancock RGV, Allen RO. 1989. Human bones as archaeological samples: changes due to contamination and diagenesis. In *Archaeolaeolgical Chemistry IV*, Allen RO (ed.). American Chemical Society: Washington, DC.

Evershed RP, Turner-Walker G, Hedges REM, Tuross N, Leyden A. 1995. Preliminary results for the analysis of lipids in ancient bone. *J Archaeol Sci* **22**: 277–290.

Geigl E-M. 2002. On the circumstances surrounding the preservation and analysis of very old DNA. *Archaeometry* **44**(3): 337–342.

Gernaey AM, Waite ER, Collins MJ, Craig OE, Sokol RJ. 2001. Survival and interpretation of archaeological proteins. In *Handbook of Archaeological Science*, Brothwell DR, Pollard AM (eds). Wiley: Chichester; 323–329.

Gilbert MTP, Rudbeck L, Willerslev E, Hansen AJ, Smith C, Penkman K, Prangenberg, K, Nielsen-Marsh C, Jans M, Arthur P, Lynnerup N, Turner-Walker G, Biddle M, Kjølbye-Biddle B, Cooper A, Collins MJ. 2005. Biochemical and physical correlates of DNA contamination in archaeological bones and teeth excavated at Matera, Italy. *J Archaeol Sci* **32**: 785–793.

Gilbert MTP, Hansen AJ, Willerslev E, Turner-Walker G, Collins MJ. 2006. Insights into the processes behind the contamination of degraded human teeth and bone samples with exogenous sources of DNA. *Int J Osteoarchaeol* **15**: 1–9.

Gill-King H. 1997. Chemical and ultrastructural aspects of decomposition. In Forensic Taphonomy: *The Postmortem Fate of Human Remains*, Haglund WD, Sorg MH (eds). CRC Press: Boca Raton; 93–108.

Giraud-Guille MM. 1988. Twisted plywood architecture of collagen fibrils in human compact bone osteons. *Calcified Tissue Int* **42**: 167–180.

Götherström A, Collins MJ, Angerbjörn A, Liden K. 2002. Bone preservation and DNA amplification. *Archaeometry* **44**: 395–404.

Grupe G, Piepenbrink H. 1989. Impact of microbial activity on trace element concentrations in excavated bones. *Appl Geochem* **4**: 293–298.

Grupe G, Dreses-Werringloer U, Parsche F. 1993. Initial stages of bone decomposition: causes and consequences in prehistoric human bone. In *Archaeology at the Molecular Level*, Lamberts JB, Grupe G (eds). Springer: Berlin; 257–274.

Hackett CJ. 1981. Microscopical focal destruction (tunnels) in exhumed human bones. *Med Sci Law* **21**: 243–265.

Hassan AA, Ortner DJ. 1977. Inclusions in bone material as a source of error in radiocarbon dating. *Archaeometry* **19**: 131–135.

Hedges REM. 2002. Bone diagenesis: an overview of processes. *Archaeometry* **44**: 319–328.

Hedges REM, Millard AR. 1995. Bones and groundwater towards the modelling of diagenetic processes. *J Archaeol Sci* **22**: 155–165.

Hedges REM, Millard AR, Pike AWG. 1995. Measurements and relationships of diagenetic alteration of bone from three archaeological sites. *J Archaeol Sci* **22**: 201–211.

Horowitz LK, Goldberg P. 1989. A study of Pleistocene and Holocene hyena coprolites. *J Archaeol Sci* **16**: 71–94.

Jans MME, Nielsen-Marsh CM, Smith CI, Collins MJ, Kars H. 2004. Characterisation of microbial attack on archaeological bone. *J Archaeol Sci* **31**: 87–95.

Kaehler S. 1999. Incidence and distribution of phototrophic shell-degrading endoliths of the brown mussel *Perna perna*. *Mar Biol* **135**: 505–514

Keeley HCM, Hudson GE, Evans J. 1977. Trace element contents of human bones in various states of preservation: 1. The soil silhouette. *J Archaeol Sci* **4**: 19–24.

Koon HEC. 2006. Detecting cooked bone in the archaeological record: a study of the thermal stability and deterioration of bone collagen. Unpublished doctoral thesis, University of York.

Lambert JB, Szpunar CB, Buikstra JE. 1979. Chemical analysis of excavated human bone from middle and late Woodland sites. *Archaeometry* **21**: 115–129.

Lambert JB, Vlask SM, Thometz AC, Buikstra JE. 1982. A comparative study of the chemical analysis of ribs and femurs in Woodland populations. *Am J Phys Anthropol* **59**: 289–294.

Lambert JB, Simpsom SV, Buikstra JE, Hanson D. 1983. Electron microprobe analysis of elemental distribution in excavated human femurs. *Am J Phys Anthropol* **62**: 409–423.

Lambert JB, Simpson SV, Szpunar CB, Buikstra JE. 1985a. Bone diagenesis and dietary analysis. *J Hum Evol* **14**: 477–482.

Lambert JB, Simpson SV, Weiner SG, Buikstra JE. 1985b. Induced metal ion exchange in excavated human bone *J Archaeol Sci* **12**: 85–92.

Lee-Thorpe JA, van der Merwe NJ. 1987. Carbon isotope analysis of fossil bone apatite. *S Afr J Sci* **83**: 712–715.

Lee-Thorpe J, van der Merwe NJ. 1991 Aspects of the chemistry of modern and fossil biological apatites. *J Archaeol Sci* **18**: 343–354.

Lowenstam HA, Weiner S. 1989. *On Biomineralization*. Oxford University Press: Oxford.

Mann RW, Feather ME, Tumosa CS, Holland TD, Schneider KN. 1998. A blue encrustation found on skeletal remains of Americans missing in action in Vietnam. *Forensic Sci Int* **97**: 79–86.

Marchiafava V, Bonucci E, Ascenzi A. 1974. Fungal osteoclasia: a model of dead bone resorption. *Calcified Tissue Res* **14**: 195–210.

Maritan L, Mazzoli C. 2004. Phosphates in archaeological finds: implications for environmental conditions of burial. *Archaeometry* **46**: 673–683.

Mays S. 1998. *The Archaeology of Human Bones*. Routledge: London.

Mays S. 1999. Osteoporosis in earlier human populations. *J Clin Densitom* **2**: 71–78.

Mays S. 2000. New directions in the analysis of stable isotopes in excavated bones and teeth. In *Human Osteology in Archaeology and Forensic Science*, Cox M, Mays S (eds). Greenwich Medical Media: London; 425–438.

Mays S, Turner-Walker G, Syversen U. 2006. Osteoporosis in a population from medieval Norway. *Am J Phys Anthropol* **116**: 34–44.

Muyzer G, Sandberg P, Knapen MHJ, Vermeer C, Collins M, Westbroek P. 1992. Preservation of the bone protein osteocalcin in dinosaurs. *Geology* **20**: 871–874.

Newesely H. 1989. Fossil bone apatite. *Appl Geochem* **4**: 233–245.

Nielsen-Marsh CM. 1997. Studies in archaeological bone diagenesis. Unpublished doctoral thesis, University of Oxford.

Nielsen-Marsh CM, Hedges REM. 1999. Bone porosity and the use of mercury intrusion porosimetry in bone diagenesis studies. *Archaeometry* **41**: 165–174.

Nielsen-Marsh CM, Hedges REM. 2000a. Patterns of diagenesis in bone I: the effects of site environments. *J Archaeol Sci* **27**: 1139–1150.

Nielsen-Marsh CM, Hedges REM. 2000b. Patterns of diagenesis in bone II: effects of acetic acid treatment and the removal of diagenetic CO_3^{2-}. *J Archaeol Sci* **27**: 1151–1159.

Nielsen-Marsh CM, Gernaey AM, Turner-Walker G, Hedges REM, Pike AWG, Collins MJ. 2000. The chemical degradation of bone. In *Human Osteology in Archaeology and Forensic Science*, Cox M, Mays S (eds). Greenwich Medical Media: London; 439–454.

Olsen EC. 1980. Taphonomy: its history and role in community evolution. In *Fossils in the Making: Vertebrate Taphonomy and Palaeoecology*, Behrensmeyer AK, Hill AP (eds). University of Chicago Press: Chicago; 5–19.

Ortner DJ, Turner-Walker G. 2003. The biology of skeletal tissues. In *Identification of Pathological Conditions in Human Skeletal Remains* (2nd edition), Ortner DJ (ed.). Academic Press: New York; 11–35.

Painter TJ. 1983. Residues of D-lyxo-5-hexosulopyranuronic acid in sphagnum holocellulose, and their role in cross-linking. *Carbohydr Res* **124**: C18–C21.

Painter TJ. 1991a. Lindow Man, Tollund Man and other peat-bog bodies: the preservation and antimicrobial action of sphagnan, a reactive glycuronoglycan with tanning and sequestering properties. *Carbohydr Polym* **15**: 123–142.

Painter TJ. 1991b. Preservation in peat. *Chem Industr* **17**: 421–424.

Painter TJ. 1995. Chemical and microbiological aspects of the preservation process in sphagnum peat. In *Bog Bodies: New Discoveries and New Perspectives*, Turner RC, Scaife RG (eds). British Museum Press: London; 88–99.

Painter TJ. 1998. Carbohydrate polymers in food preservation: an integrated view of the Maillard reaction with special reference to discoveries of preserved foods in sphagnum-dominated peat bogs. *Carbohydr Polym* **36**: 335–347.

Pate FD, Hutton JT, Norrish K. 1989. Ionic exchange between soil solution and bone: towards a predictive model. *Appl Geochem* **4**: 303–316.

Petchley F, Higham T. 2000. Bone diagenesis and radiocarbon dating of fish bones at the Shag River mouth site, New Zealand. *J Archaeol Sci* **27**: 135–150.

Piepenbrink H. 1986. Two examples of biogenous dead bone decomposition and their consequences for taphonomic interpretation. *J Archaeol Sci* **13**: 417–430.

Piepenbrink H. 1989. Examples of chemical changes during fossilization. *Appl Geochem* **4**: 273–280.

Pike AWG. 1993. Bone porosity, water and diagenesis: towards a grand unified theory of bone diagenesis. Unpublished undergraduate dissertation, University of Bradford.

Pike AWG, Nielsen-Marsh CM, Hedges REM. 2001. Modelling bone dissolution under different hydrological regimes. In *Archaeological Sciences '97*, Millard AR (ed.). BAR International Series 939. Archaeopress: Oxford; 127–132.

Posner AS. 1985. The structure of bone apatite surfaces. *J Biomed Mater Res* **19**: 241–250.

Price TD (ed.). 1989. *The Chemistry of Prehistoric Human Bone*. Cambridge University Press: Cambridge.

Price TD, Schoeninger MJ, Armelagos GJ. 1985. Bone chemistry and past behaviour: an overview. *J Hum Evol* **14**: 419–448.

Roux W. 1887. Über eine Knochen lebende Gruppe von Fadenpilzen (*Mycelites ossifragus*). *Zeitschr Wiss Zool* **45**: 227–254.

Sansom IJ, Smith MP, Armstrong HA, Smith MM. 1992. Presence of the earliest vertebrate hard tissues in conodonts. *Science* **256**: 1308–1311.

Sarathchandra P, Pope F, Ali S. 1999. Morphometric analysis of type I collagen fibrils in the osteoid of osteogenesis imperfecta. *Calcif Tissue Int* **65**: 390–395.

Schultze H-P. 1996. Conodont histology: an indicator of vertebrate relationship? *Mod Geol* **20**: 275–286.

Smith CI, Craig OE, Prigodich RV, Nielsen-Marsh CM, Jans MME, Vermeer C, Collins MJ. 2005. Diagenesis and survival of osteocalcin in archaeological bone. *J Archaeol Sci.* **32**: 105–113.

Stott AW, Evershed RP, Jim S, Jones V, Rogers JM, Tuross N, Ambrose S. 1999. Cholesterol as a new source of palaeodietary information: experimental approaches and archaeological applications. *J Archaeol Sci* **26**: 705–716.

Taylor GM, Crossey M, Saldanha J, Waldron T. 1996. DNA from *Mycobacterium tuberculosis* identified in mediaeval human skeletal remains using polymerase chain reaction. *J Archaeol Sci* **23**: 789–798.

Trueman CN, Martill DM. 2002. The long term preservation of bone: the role of bioerosion. *Archaeometry* **44**: 371–382.

Turner-Walker G. 1993. The characterisation of fossil bone. Unpublished PhD thesis, Department of Archaeology, University of Durham.

Turner-Walker G. 1998a. The West Runton fossil elephant: a pre-conservation evaluation of its condition, chemistry and burial environment. *Conservator* **22**: 26–35.

Turner-Walker G. 1998b. Pyrite and bone diagenesis in terrestrial sediments: evidence from the West Runton fresh water bed. *Bull Geol Soc Norfolk* **48**: 3–26.

Turner-Walker G, Jans MME. in press. Reconstructing taphonomic histories using histological analysis. *Palaeogeog Palaeoclimatol Palaeoecol.*

Turner-Walker G, Parry TV. 1995. The tensile strength of archaeological bone. *J Archaeol Sci* **22**: 185–191.

Turner-Walker G, Peacock EE. in press. Preliminary results of bone diagenesis in Scandinavian bogs. *Palaeogeog Palaeoclimatol Palaeoecol.*

Turner-Walker G, Syversen U. 2002. Quantifying histological changes in archaeological bones using BSE-SEM image analysis. *Archaeometry* **44**: 161–168.

Turner-Walker G, Nielsen-Marsh CM, Syversen U, Kars H, Collins MJ. 2002. Sub-micron spongiform porosity is the major ultra-structural alteration occurring in archaeological bone. *Int J Osteoarchaeol* **12**: 407–414.

Tuross N. 2003. Recent advances in bone, dentin and enamel biochemistry. In *Identification of Pathological Conditions in Human Skeletal Remains*, 2nd edition, Ortner DJ (ed.). Academic Press: New York; 65–72.

Van Klinken GJ. 1999. Bone collagen quality indicators for palaeodietary and radiocarbon measurements. *J Archaeol Sci* **26**: 687–695.

Waldron T. 1987. The relative survival of the human skeleton: implications for palaeopathology. In *Death Decay and Reconstruction*, Boddington A, Garland AN, Janaway RC (eds). Manchester University Press: Manchester; 55–64.

Wandeler P, Smith S, Morin PA, Pettifor RA, Funk SM. 2003. Patterns of nuclear DNA degeneration over time a case study in historic teeth samples. *Mol Ecol* **12**: 1087–1093

Wedl C. 1864. Über einen im Zahnbein und Knochen keimenden Pilz. *Mineral Biol Erd* **50**: 171–193.

Weiner S, Bar-Yosef O. 1990. States of preservation of bones from prehistoric sites in the near east. *J Archaeol Sci* **17**: 187–196.

Weiner S, Traub W. 1992. Bone structure; from ångströms to microns. *FASEB J* **6**: 879–885.

Weser U, Kaup Y, Etspüler H, Kenward N, Hedges REM. 1996. Biochemically and immunologically active alkaline phosphatase in archaeologically important bone samples. *J Archaeol Sci* **23**: 723–730.

Yamamoto S, Nakamura F, Hitomi J, Shigeno M, Sawaguchi S, Abe H, Ushiki T. 2000. Atomic force microscopy of intact and digested collagen molecules. *J Electron Microsc* **49**: 423–427.

2

How Representative Are Human Skeletal Assemblages for Population Analysis?

Ron Pinhasi[1] and Chryssi Bourbou[2]

[1] *Department of Archaeology, University College Cork, Cork, Ireland*
[2] *28th Ephorate of Byzantine Antiquities, Chania, Crete, Greece*

INTRODUCTION

In recent decades the palaeopathological study of health and disease patterns has demonstrated a shift from the descriptive realm of case studies to the more theoretical realm of palaeoepidemiology and cross-population analysis. Bioarchaeologists can address research questions regarding diachronic changes in health status in past populations, changes in morbidity and mortality profiles, and the prevalence of certain pathological conditions. However, any population-level analysis is affected by a set of biases and limitations that typify archaeological skeletal samples. These include methodological aspects, such as age and sex estimation, excavation and recovery methods, and survival biases of skeletal samples.

The aim of this chapter is to discuss some of the sources of bias that potentially exert significant effects on population-based palaeoepidemiological assessments. The next section of the chapter addresses biases in the representativeness of archaeological samples. A short review of aspects of the interpretation of morbidity status in palaeopopulations and the effects of some demographic parameters on the interpretation of palaeopathological data are then provided. The chapter concludes with a demonstration of an age-correction epidemiological method to archaeological material.

Advances in Human Palaeopathology Edited by Ron Pinhasi and Simon Mays

REPRESENTATIVENESS OF ARCHAEOLOGICAL SAMPLES

Under ideal conditions, the skeletal sample studied represents the total deaths in the population during the time period under investigation (Alesan *et al.*, 1999). However, this is hardly ever the case, as some of the skeletons, particularly those of foetuses and infants, do not survive or were buried elsewhere. Moreover, the archaeologist usually does not excavate the total cemetery due to various budgetary, time and logistical constraints. In most cases, lack of written documentation makes it impossible to ascertain whether the excavated sample is a true random sample of the cemetery skeletal population. Therefore, we do not know how well the skeletal sample represents the living population and we have no means to resolve this issue, as various non-random taphonomic, cultural, and demographic factors may have played a major part in the sampling process (Waldron, 1994; Hoppa, 2002). The sections below discuss some of the more important influences on the degree to which a cemetery sample is representative of the original population.

Bone Survival and Recovery

A number of physical, chemical and biological factors play a significant role in determining the preservation of human remains (Waldron 1987, 1994; Mays, 1998; Bourbou, 1999; Caffel *et al.*, 2001; Turner-Walker, Chapter 1). A buried body has a better chance of surviving than an exposed one, the latter being more vulnerable to environmental degradation and animal attacks (Henderson, 1987). Different components of the skeleton exhibit different preservation patterns, as they vary in their physical strength and structure. Trabecular bone seems to decay more rapidly in the soil than cortical bone, partially due to the larger surface area of trabecular bone favouring chemical exchange between bone and soil (Lambert *et al.*, 1982; Grupe, 1988) and partially due to its mechanical fragility. Studies on the relative survival and recovery of bones of adults from the Romano-British site at West Tenter Street, London (Waldron, 1987), and on medieval skeletons from the Blackfriars site in Ipswich, England (Mays, 1992), confirm that small bones and those with high trabecular and low cortical bone contents are poorly represented. The low mineralization of subadult bones and their fragile nature contribute to their poor preservation (Currey and Butler, 1975; Gordon and Buikstra, 1981; Specker *et al.*, 1987; Nicholson, 2001).

The underrepresentation of specific skeletal parts is of great relevance to palaeopathologists. For example, in leprosy, peripheral nerve damage causes loss of sensation and eventually ulceration and secondary infections in both hand and foot (Steinbock, 1976; Aufderheide and Rodríguez-Martín, 1998). Consequently, some of the hand and foot phalanges are expected to be missing from the archaeological assemblage owing to disease. A systematic analysis of the prevalence of leprosy in a skeletal population must take into consideration whether missing hand and foot phalanges should be possibly attributed to both leprosy and bone survival and recovery aspects. Similarly, some of the main diagnostic criteria for leprosy involve resorption of the anterior nasal spine, rounding of the margins of the nasal aperture, perforation of the hard palate and ante-mortem loss of the central upper incisors (Møller-Christensen, 1961). In archaeological assemblages, one or more of these indicative features is frequently damaged by taphonomic processes.

The survival of bones with pathological lesions will vary depending on the nature and severity of the specific condition. Bone is more susceptible to decomposition and post-mortem damage in the case of a predominantly lytic process (e.g. osteomalacia, Paget's

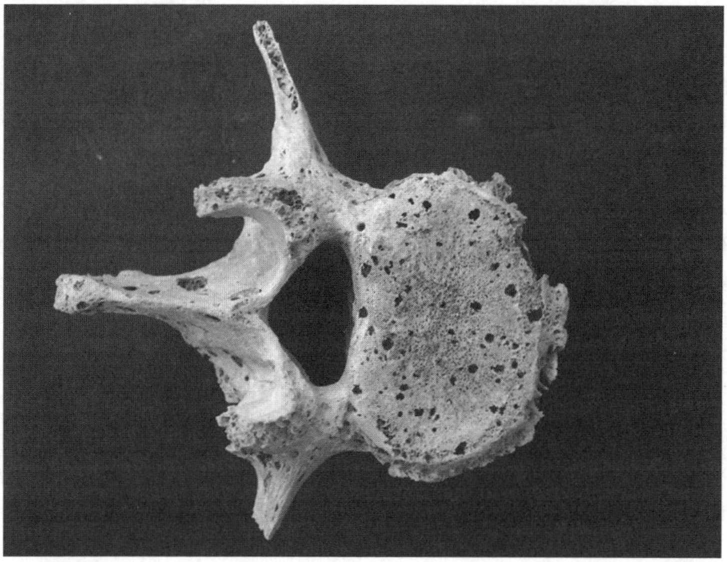

Figure 2.1 Lumbar vertebrae of a 10th-century AD adult male from Turkey with numerous non-sclerotic punched-out lytic lesions caused by multiple myeloma

disease) than in a predominantly blastic process (e.g. osteophytosis, DISH). An example of the effect of metastatic lytic bone lesions on bone completeness and preservation is illustrated in Figure 2.1.

In recent years, more researchers have considered taphonomic processes as an important aspect of archaeological and forensic anthropological analysis (see contributions in Haglund and Sorg (1997, 2002), Nicholson (2001) and Hunter and Cox (2005). Dry and wet sieving of soil remaining in the grave after the skeleton has been lifted is now standard practice. It facilitates recovery of small bones and other remains, such as calcified cysts, kidney stones, calcified blood-vessel walls, lymph nodes, pleural plaques, loose teeth or infant bones (Wells and Dallas, 1976; Morris and Rodgers, 1989; Baud and Kramer, 1991; Baker *et al.*, 2005).

Subadult Underrepresentation

A major concern in cemetery excavations is the possible underrepresentation of young individuals (Angel, 1969; Acsádi and Nemeskéri, 1970; Weiss, 1973a; Johnston and Zimmer, 1989). Guy *et al.* (1997) compared the number of infants aged 0–1 years from Hungarian archaeological cemeteries (10th–12th centuries AD), as a proportion of every 1000 burials, to figures from 18th–20th century burial registers from Europe. They noted an underrepresentation of infants in the archaeological samples and argue that this cannot be explained by taphonomic factors, but is more likely due to a mixture of factors, including type of burial and associated burial practices, and archaeological recovery strategies.

The underrepresentation of infants or any subadult age cohort should be assessed on a case-by-case basis. Acsádi and Nemeskéri (1970) noted that, in Hungarian medieval material, infants were underrepresented. This skeletal material was hand recovered, so underrepresentation of infants might be due to poor recovery of perinatal/infant bones rather than taphonomic factors. Saunders *et al.* (2002) examined age distributions of subadults and adults of the 19th century AD skeletal population from the St Thomas Anglican church, Belleville, Ontario, and assessed the skeletal age-at-death pattern against available parish records. They reported a general excess of infants and a deficiency of elderly adults in the skeletal sample when compared with the registers. The difference between the parish records and the skeletal representation is particularly pronounced for infants aged 0–1 years and for adults aged >60 years. The underrepresentation of older adults may be because these are often classified osteologically as adults of indeterminate age and, hence, are excluded from palaeodemographic analysis. The high proportion of infants in the skeletal sample might be due to the use of sieving to recover small bones (Saunders *et al.*, 2002); and that the numbers were higher than expected from parish records may indicate that burials of infants were sometimes not recorded in the burial registers. Mays (1993) examined infant age at death in several Romano-British sites over a wide variety of soil conditions and found that skeletons of perinatal age are often better preserved than those of older infants. Several studies of prehistoric skeletal samples from Italy indicate that subadult remains are underrepresented in cemeteries associated with settlements, whereas they are overrepresented in mortuary deposits in caves (Skeates, 1991, 1997; Robb, 1994).

Pseudopathology

Pseudopathology refers to post-mortem skeletal changes that may be mistakenly diagnosed as ante-mortem pathological conditions. Soil may be responsible for warping of bones and erosion of the bone cortex, which can be mistaken for specific pathological conditions, like rickets or trauma, and cribra orbitalia or periostitis respectively (Wells, 1967, Ubelaker, 1991; Buikstra and Ubelaker, 1994). However, the most prevalent type of pseudopathology is post-mortem fractures. These may be caused by damage from excavation tools, pressure from overlying soil, and rough handling of excavated remains (Aufderheide and Rodríguez-Martín, 1998). Perimortem fractures need to be distinguished from post-mortem damage. The best way to differentiate these is to examine variations in bone colour at the fracture sites. Lighter 'fresh' colour indicates recent post-mortem trauma, whereas regions of perimortem trauma should be of the same colour as the rest of the bones (Ubelaker and Adams, 1995; Bennike, Chapter 14). However, post-depositional breaks occurring during the interval between deposition and recovery may also leave weathered edges; hence, differentiation between perimortem and post-mortem fractures is not always straightforward. It is also important to investigate the direction of the force that operated on the bone in relation to the anatomical position of the skeleton in order to assess whether the given blows could have been applied during life. For example, in a study of a skeleton from Georgia, USA (Ubelaker and Adams, 1995), it was noted that whereas fractures on the left tibia and fibula were in a similar location on the diaphysis, they displayed fracture patterns which suggest the forces that produced them came from nearly opposite directions. Had the trauma occurred perimortem, with the bones still in articulation, the pattern would imply the occurrence of two separate traumatic events originating from opposite directions, but involving the same area of the lower leg. This is very unlikely; hence, it seemed more likely that this trauma pattern was sustained post-mortem, a suggestion supported by the clean edges of the breaks.

Rounded perforations of bones caused by various post-mortem processes may be mistaken for pathological or traumatic lesions. For example, two nearly symmetric perforations observed on the fifth metatarsals of a female skeleton from a Byzantine tomb from Ramat

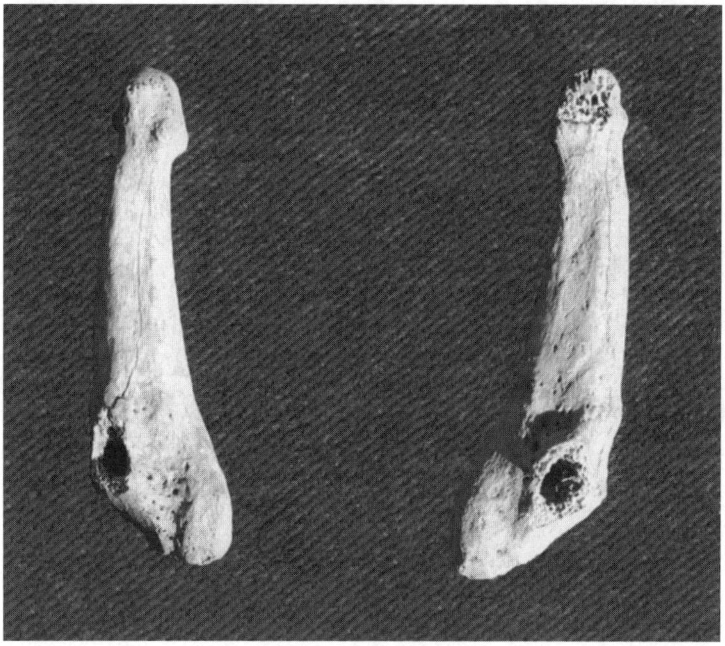

Figure 2.2 Fifth metatarsals showing post-depositional perforations from the Byzantine site of Ramat Handiv, Israel

Hanadiv, Israel, could be mistakenly attributed to ante-mortem trauma (Figure 2.2), and lesions of this type have in the past been misdiagnosed as pathology (Elliot-Smith, 1908). Rodents and carnivores often cause gnawing damage to bones (Figure 2.3) that may potentially be mistaken for trauma or cannibalism (Haglund and Sorg, 2002). Bone gnawing by rodents can be distinguished from that of carnivores by the characteristic parallel series of furrows created by the incisors (Haglund, 1992).

Burial Practices

Individuals in cemeteries are often buried in specific zones according to age, social status, sex, and other aspects. Sex- and age-specific zones are known from excavations of Swedish medieval Christian cemeteries (Gejvall, 1960; Kjellström *et al.*, 2005). At some Christian cemeteries in England, spatial clusters of infant burials have been identified; for example, at medieval Wharram Percy, England, a concentration of infant burials was found close by the north wall of the church (Mays, 1997). Similarly, Boddington (1987) reports age-specific spatial clusters in the early medieval English cemetery at Raunds.

Segments of the population, such as social outcasts or specific age groups, may be accorded different burial rites from the rest of the population (Barley, 1995; Parker-Pearson, 1999) and, therefore, may be missing or underrepresented in skeletal samples. An example is the Irish *cillini.* This was a special resting place (e.g. deserted churches and graveyards, ancient megalithic tombs and secular earthworks, sea or lake shores) used from early Christian times to the 20th century for stillborn babies, unbaptized infants and some other members of society

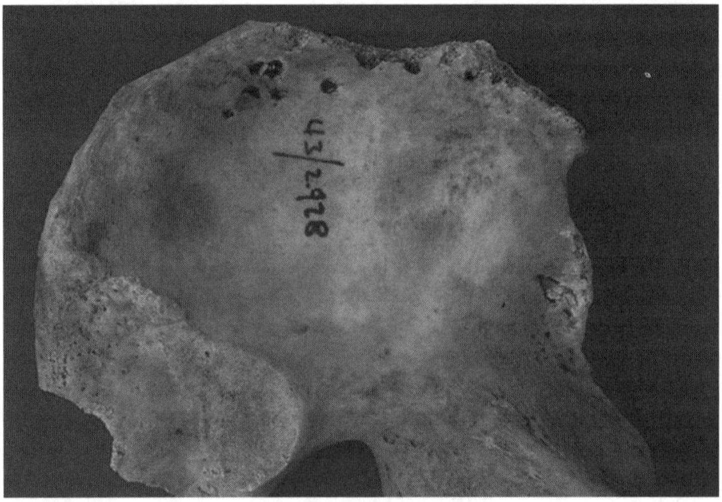

Figure 2.3 Carnivore tooth puncture marks on ilium. Reproduced with permission from Dr. C. Rodríguez–Martín, Instituto Canario de Bioantropalogía

(e.g. mentally retarded, strangers, the shipwrecked, criminals, famine victims, suicides), who were considered unsuitable for burial in consecrated ground (Murphy and McNeil, 1993; Hurl and Murphy, 1996; Donnelly *et al.*, 1999). An interesting case of social exclusion is an adult skeleton recovered from an Iron Age well in Athens (Little and Papadopoulos, 1998). The middle-aged male has multiple healed fractures to the skull and the vertebral column. The authors suggest that the head injuries probably caused permanent neurological defects, and that this may be the explanation for the unusual burial treatment of this individual.

Exclusion from the communal burial ground is also known in the case of individuals who suffered from leprosy. Written sources indicate that, from the 13th century AD, lepers in Europe were generally buried in cemeteries associated with leprosaria rather than being taken back to their place of origin (Roberts *et al.*, 2002). By contrast, leprosy cases in the Avar period (8th–9th centuries AD) in Hungary and Lower Austria were buried in the same cemetery as the rest of the local population (Pinhasi *et al.*, unpublished data).

Analysis of the age and sex distribution of a skeletal sample may provide indirect evidence for a catastrophic (hence unnatural) assemblage that may be the outcome of a single event. The age distribution of skeletal assemblages from catastrophic episodes often resembles the living age distribution, with greater numbers of older children, adolescents, and young adults than in mortality profiles typical of most archaeological cemeteries (Paine, 2000). Hence, a mortality profile of this type suggests deaths resulted from a catastrophe of some sort, perhaps due to violence, natural disaster or epidemic.

MORBIDITY STATUS

Over the last 15 years there has been much debate regarding the interpretation of prevalence data in palaeopathology. In particular, a concept termed the 'osteological paradox' by Wood *et al.* (1992) has been influential. Wood *et al.* (1992) point out that a morbidity profile reconstructed from a skeletal sample, since it represents non-survivors, may not properly reflect the morbidity profile of the once living population from which it derived. The reason is the possibility that selective mortality and heterogeneity in risk of death may confound the interpretation of skeletal disease prevalence. A high prevalence of skeletal lesions may indicate not an unhealthy population but one in which disease was regularly survived for long enough to cause bone changes. By contrast, a population showing few skeletal lesions may do so not because they were generally healthy, but because their resistance to disease was too poor for them regularly to survive it long enough for bone lesions to develop. Such interpretations were held by Wood *et al.* (1992) to be rather counter-intuitive and to run counter to 'conventional' palaeopathological interpretations, which equate low skeletal lesion prevalence with good health and high bone lesion prevalence with poor health – hence the osteological 'paradox'.

Since the Wood *et al.* (1992) article, the concept of the osteological paradox and its implications have been hotly debated (references in Milner *et al.* (2000)), but the extent to which it actually undermines conventional palaeopathological interpretation remains unclear. For example, a recent examination of stress in relation to growth of subadults from two medieval skeletal assemblages from Denmark – a disadvantaged group from a leprosarium in Næstved and a relatively privileged medieval sample from Æbelholt – showed that the subadults from Næstved displayed higher frequency and severity of nutritional and disease stress (Bennike *et al.*, 2005). This study, and others like it (e.g. Cohen, 1997), suggest that in some cases the morbidity status of an archaeological sample is a valid indication of the actual health profile of the original population. Currently, the debate regarding the osteological paradox has yet to be resolved. Perhaps the best we can do is to be conscious that different interpretations of prevalence data are possible, and to attempt to use supporting data in order to advance some interpretations at the expense of others, rather than attempting to interpret prevalence data in isolation from other evidence. For example, analysis of the palaeodemographic profile of the population(s) under study and, in the case of infectious lesions, recording whether lesions were active at death or healed (Mays, 1997) may help to elucidate the extent to which the osteological paradox is likely to apply to the particular material under study.

PALAEODEMOGRAPHY

Palaeodemographic research published during the 1970s compared the age-at-death profiles and life tables of prehistoric skeletal samples in order to reconstruct the profiles in the original populations from which they were the physical remains (Acsádi and Nemeskéri, 1970; Weiss, 1973a,b; Klepinger, 1979). However, it soon became evident that there are difficult problems of ascertainment of age of specimens and there is bias in the skeletal deposition of certain age groups (Weiss, 1976). During the next 10–15 years it became evident that these major methodological concerns could not be ignored (Bocquet-Appel and Masset, 1982) and, in fact, in most cases they have not yet been overcome. In the following sections, we selectively focus only on the aspects that have a direct effect on palaeopathological analysis rather than on current issues in the field of palaeodemography (for the latter, see various contributions in Hoppa and Vaupel (2002)).

Age-at-Death Tables and the Health Status of Past Populations

Calculation of age-at-death mortality tables of past populations is of particular interest to any population-based palaeopathological analysis. These require the calculation of crude mortality rates for skeletal populations (Jackes, 1992). The mortality profile of a past population can be interpreted in relation to the prevalence of various specific and non-specific stress indicators in order to derive a 'health index' for the specific population (Steckel *et al.*, 2002). A major source of bias in the calculation of age-at-death categories is the poor accuracy and high error ranges associated with many ageing methods, and these are particularly severe in the case of older adults (Cox, 2000). In the last two decades, palaeodemographic research has moved from the calculation of mortality tables and life tables to the use of hazards models and maximum likelihood estimators (Wright and Yoder, 2003). Such studies assess the nature of the demographic structure of the skeletal sample (Milner *et al.*, 2000; Hoppa and Vaupel, 2002). However, none of the statistical manipulations can overcome the inherent problems with adult skeletal age indicators.

The Application of Age Adjustment

Age at death is clearly important for bioarchaeologists who wish to assess the age-specific health profile of the population from which the study sample is drawn. However, many bioarchaeologists are only concerned with the analysis of the general prevalence of given pathological condition in skeletal samples. Even in this latter case, it nevertheless remains important to control for the effects of age: different diseases strike different age groups with different frequencies; and bony pathologies generally accumulate with age, so that, in skeletal series, older age cohorts tend to show greater lesion prevalences. Direct age adjustment involves the use of prevalence parameters obtained from a reference or hypothetical 'standard' population that should be divided into the same age categories as the archaeological population under study. The researcher may use clinical epidemiological data on a population that they believe to be the most appropriate standard for their archaeological material. Alternatively, they may combine age-specific prevalence taken from previous bioarchaeological studies and use that as their standard. By using a single standard population one eliminates the possibility that observed differences between the archaeological samples are the result of age differences in the populations (Gordis, 2000). By applying each age-specific prevalence rate to the population in each age group of the standard population it is then possible to calculate the expected rate for each age group and to standardize the rates in each new archaeological population according to these rates (Gordis, 2000: 52–54).

Application of Standardized Mortality Ratio to the Study of Gout: A Case Study

Another approach is indirect age adjustment by applying standardized mortality ratios (SMRs; Gordis, 2000: 54). The SMR is the observed prevalence of a disease in the archaeological sample divided by the expected prevalence of the disease in the base population. SMRs are then calculated for each age group and can be compared for other archaeological samples. An $SMR = 1$ will indicate that the prevalence in the archaeological population is identical to the one of the standard population, whereas $SMR > 1$ indicates a higher prevalence. Again, by addressing prevalence by age group it is possible to avoid at least some of the more crude palaeodemographic biases and to derive comparable figures.

An unpublished study of joint disease in an early medieval skeletal sample from the site of Zwölfaxing, Lower Austria, revealed a high prevalence of gout. Of a total of 130 skeletons over 16 years of age (77 males and 53 females), 14.2 % were diagnosed with gout (using a combination of macroscopic, microscopic and radiological methods). The sex-specific prevalence was 15.6 % for males and 5.7 % for females.

The prevalence of gout in modern populations is shown in Table 2.1. The highest modern prevalence reported is 10.2 % among Maori males (Prior *et al.*, 1966). The prevalence in the archaeological material from Zwölfaxing exceeds even this, but it is possible that the Zwölfaxing prevalence was confounded by the age structure of the sample. In order to investigate this, the SMR was calculated for the aged and sexed samples. The total population was divided to broad age categories: 16–30, 31–50, and > 50 years. Next, a standard reference population was selected from the literature. A standard reference population should closely resemble genetically the archaeological sample under investigation. In the current case, no adequate data on gout prevalence are available for modern Austria. In addition, the choice of standard was constrained by the fact that many epidemiological studies do not present their data in ways that allow the necessary comparisons. The standard reference data chosen to calculate the SMRs were from modern England (Harris *et al.*, 1995). They comprise 300 376 individuals (146 973 males, 153 403 females) of which 2865 were diagnosed of having gout. The prevalence in Zwölfaxing was standardized following the above-described method. The results are presented in Table 2.2. None of the archaeological cases diagnosed with gout was younger than 30 years of age; hence, the SMR value of both males and females is nil for this age category. With the age category of 31–50 years, the Zwölfaxing SMR is 10-fold higher than in the reference British population for males and 66-fold higher for females. For the age category of >50 years the SMR is 6-fold higher for males and 20-fold higher for females.

The results confirm the high prevalence figures obtained before age adjustment. However, owing to the small sample size, the SMR for Zwölfaxing females should be interpreted

Table 2.1 A summary table of the prevalence of gout among various modern populations

Sample/location	Period of study	Sex	Sample size	Age	Gout prevalence (%)	Reference
Framingham, USA	1950s–1960s	Both	5 127	30–59	0.25 (at age 44)	Hall *et al.* (1967)
England	1991	Both	300 376	–	1.64 (♂) 0.95 (♀)	Harris *et al.* (1995)
Paris, France	1980	Both	4 663	35–44	1–2	Zalokar *et al.* (1981)
Greece	1996–1999	Both	8 547	≥19	0.47	Andrianakos *et al.* (2003)
New Zealand, Maori	1960s	Both	370 (♂) 385 (♂)	20–>70	10.2 (♀) 1.8 (♀)	Prior *et al.* (1966)
Cook Islands, Rarotonga and Pukapuka	1964	♂	431	Adult	2.4 (Rarotogan) 5.3 (Pukapuka)	Prior *et al.* (1966)
China, Shanghai	1997–1998	Both	3 190 (♂) 3 394 (♀)	≥15	0.63 (♀) 0.059 (♀)	Dai *et al.* (2003)

Table 2.2 Computation of SMR for gout in a medieval Austrian population from Zwölfaxing using indirect age adjustment based on 1991 epidemiological data on the age- and sex-specific prevalence in England[a]

	Modern England		Zwölfaxing					
	Prevalence rate (per 1000)		Total		Observed cases		SMR	
Age (years)	Male	Female	Male	Female	Male	Female	Male	Female
16–30	0.96	0.09	15	13	0	0	0.00	0.00
31–50	18.14	1.12	44	27	8	2	10.02	66.14
>50	35.48	3.84	18	13	4	1	6.26	20.03

[a] Modern data from Harris *et al.* (1995).

with caution; the more reliable figures are those obtained for the males. Of course, the methods used for the diagnosis of gout in skeletal samples differ to a fair extent from those used to diagnose the disease in modern clinical studies. The application of rigorous palaeopathological methodology for differential diagnosis, together with taphonomic aspects of bone representation and degree of preservation, will tend to result in underestimation of the true prevalence of the condition in the archaeological population, supporting the idea that the greater frequency of gout in the Zwölfaxing population than the modern reference population is a robust result. All archaeological specimens with gouty lesions were at the chronic tophaceous disease stage, which in modern clinical cases develops, on average, 10 or more years after the onset of acute intercritical gout (Edwards, 2001). Therefore, it seems that the onset of gout in some of the specimens studied from the necropolis population was during (or prior to) the third decade of life. The age distribution suggests that the age of onset of gout in this population is, on average, much earlier than in the modern reference population, although some cases of early onset (early 20s) of chronic gout in males are reported in the medical literature (Resnick *et al.*, 1975).

CONCLUSIONS

Factors that may affect the nature of an archaeological sample can be broadly grouped in two categories: the first, uncontrolled by the researcher, includes biases caused by burial practices, differential bone survival, and age- and sex-specific biases in the representativeness of the excavated sample; the second, controlled by the researcher, includes the excavation methods and recovery strategies, the application of specific age-correction algorithms, and the like. Palaeopathologists should not be discouraged by these biases. The future lies in a paradigmatic shift towards a population-based approach in palaeopathology. As Wright and Yoder (2003: 49) assert:

> Accurate age-at-death estimates are critical for interpreting the impact of pathological lesions on well-being at the population level. Analysis of pathological lesion abundance by age-at-death cohorts may be a useful approach for evaluating the significance of lesions in terms of morbidity and mortality.

Future work on palaeopathological and palaeoepidemiological aspects should, therefore, incorporate and synthesize palaeodemographic factors. Such an approach will potentially lead to a better understanding of the complex relationship between mortality and morbidity profiles in skeletal samples of past populations.

ACKNOWLEDGEMENTS

We thank Professor Yizhar Hirschfeld, Department of Archeology, Hebrew University, Jerusalem, Israel, for the permission to publish Figure 2.2.

REFERENCES

Acsádi G, Nemeskéri J. 1970. *History of Human Life Span and Mortality*. Akadémiai Kiadó: Budapest.

Alesan A, Malgosa A, Simo C. 1999. Looking into the demography of an Iron Age population in the western Mediterranean (1): mortality. *Am J Phys Anthropol* **110**: 285–301.

Andrianakos A, Trontzas P, Christoyannis F, Dantis P, Voudouris C, Georgountzos A, Kaziolas G, Vafiadou E, Pantelidou K, Karamitsos D, Kontelis L, Krachtis P, Nikolia Z, Kaskani E, Tavaniotou E, Antoniades C, Karanikolas G, Kontoyanni A. 2003. Prevalence of rheumatic disease in Greece: a cross-sectional population based epidemiological study. *J Rheumatol* **30**: 1589–1601.

Angel JL. 1969. The bases of paleodemography. *Am J Phys Anthropol* **30**: 427–438.

Aufderheide AC, Rodríguez-Martín RM. 1998. *The Cambridge Encyclopaedia of Human Palaeopathology*. Cambridge University Press: Cambridge.

Baker BJ, Dupras TL, Tocheri MW. 2005. *The Osteology of Infants and Children*. Texas A&M University Press: College Station, TX.

Barley N. 1995. *Dancing on the Grave. Encounters with the Dead*. John Murray: London.

Baud CA, Kramer C. 1991. Soft tissue calcifications in paleopathology. In *Human Paleopathology. Current Syntheses and Future Options*, Ortner DJ, Aufderheide AC (eds). Smithsonian Institution Press: Washington, DC; 87–89.

Bennike P, Lewis ME, Schutkowski H, Valentin F. 2005. Comparison of child morbidity in two contrasting medieval cemeteries from Denmark. *Am J Phys Anthropol* **128**: 734–746

Bocquet-Appel JP, Masset C. 1982. Farewell to paleodemography. *J Hum Evol* **11**: 321–333.

Boddington A. 1987. Raunds, Northamptonshire: analysis of a country churchyard. *World Archaeol* **18**: 411–425.

Bourbou C. 1999. Digging up burials: excavation techniques, handling and transportation of bones. *Archaeol Art* **67**: 55–61.

Buikstra JE, Ubelaker, DH. 1994. *Standards for Data Collection from Human Skeletal Remains*. Arkansas Archaeological Survey, Research Seminar Series 44. Arkansas Archaeological Survey: Fayetteville,AR.

Caffell A, Roberts CA, Janaway RC, Wilson A. 2001. Pressures on osteological collections-the importance of damage limitation. In *Human Remains: Conservation, Retrieval and Analysis*, Williams E (ed.). British Archaeological Reports International Series 934. Archaeopress: Oxford; 187–197.

Cohen MN. 1997. Does paleopathology measure community health? A rebuttal of 'the osteological paradox' and its implication for world history. In *Integrating Archaeological Demography: Multidisciplinary Approaches to Prehistoric Population*, Paine RR (ed.). Centre for Archaeological Investigations Occasional Paper No. 24. Southern Illinois University Press: Carbondale, IL; 242–260.

Cox M. 2000. Ageing adults from the skeleton. In *Human Osteology in Archaeology and Forensic Science*, Cox M, Mays S (eds). Greenwich Medical Media: London; 61–82.

Currey JD, Butler G. 1975. The mechanical properties of bone tissue in children. *J Bone Jt Surg* **57**: 810–814.

Dai SM, Han XH, Zhao DB, Shi YQ, Meng JM. 2003. Prevalence of rheumatic symptoms, rheumatoid arthritis, ankylosing spondylitis, and gout in Shanghai, China: a COPCORD study. *J Rheumatol* **30**: 2245–2251.

Donnelly S, Donnelly C, Murphy EM. 1999. The forgotten dead: the cilliní and disused burial grounds of Ballintoy, County Antrim. *Ulster J Archaeol* **58**: 109–113.

Edwards NL. 2001. Gout. A: Epidemiology, pathology and pathogenesis. In *Primer on the Rheumatic Diseases*, 12th edition, Klipper JH (ed.). Arthritis Foundation: Altanta, GA; 307–312.

Elliot-Smith G. 1908. The alleged discovery of syphilis in prehistoric Europeans. *Lancet* **2**: 521–524.

Gejvall N-G. 1960. *Westerhus. Mediaeval Population and Church in the Light of the Skeletal Remains.* Vitterhets Historie och Antikvitets Akademien: Lund.

Gordis L. 2000. *Epidemiology*, 2nd edition. WB Saunders: Philadelphia.

Gordon CG, Buikstra JE. 1981. Soil pH, bone preservation and sample bias at mortuary sites. *Am Antiquity* **46**: 566–571.

Grupe G. 1988. Impact of the choice of bone samples on trace element data in excavated human skeletons. *J Archaeol Sci* **15**: 123–129.

Guy H, Masset C, Baud CA. 1997. Infant taphonomy. *Int J Osteoarchaeol* **7**: 221–229.

Haglund WD. 1992. Contribution of rodents to postmortem artifacts of bone and soft tissue. *J Forensic Sci* **37**: 1459–1465.

Haglund WD, Sorg MH (eds). 1997. *Forensic Taphonomy. The Postmortem Fate of Human Remains.* CRC Press: Boca Raton, FL.

Haglund WD, Sorg MH (eds). 2002. *Advances in Forensic Taphonomy. Method, Theory and Archaeological Perspectives.* CRC Press: Boca Raton, FL.

Hall AP, Barry PE, Dawber TR, McNamara PM. 1967. Epidemiology of gout and hyperuricemia: a long term population study. *Am J Med* **42**: 27–37.

Harris CM, Lloyd DCEA, Lewis J. 1995. The prevalence and prophylaxis of gout in England. *J Clin. Epidemiol* **48**: 1153–1158.

Henderson J. 1987. Factors determing the state preservation of human remains. In *Death, Decay and Reconstruction. Approaches to Archaeology and Forensic Science*, Boddington A, Garland AN, Janaway RC (eds). Manchester University Press: Manchester; 43–53.

Hoppa RD. 2002. Paleodemography: looking back and thinking ahead. In *Paleodemography: Age Distributions from Skeletal Samples*, Hoppa RD, Vaupel JW, (eds). Cambridge University Press: Cambridge; 9–23.

Hoppa RD, Vaupel JW (eds). 2002. *Paleodemography: Age Distributions from Skeletal Samples.* Cambridge University Press: Cambridge.

Hunter J, Cox M. 2005. *Forensic Archaeology. Advances in Theory and Practice.* Routledge: London.

Hurl DP, Murphy EM. 1996. Life and death in a County Antrim tower house. *Archaeol Ireland* **10**: 20–23.

Jackes M. 1992. Paleodemography: problems and techniques. In *The Skeletal Biology of Past Peoples*, Saunders S, Katzenberg MA (eds). Wiley-Liss: New York; 189–224.

Johnston FE, Zimmer LO. 1989. Assessment of growth and age in the immature skeleton. In *Reconstruction of Life from the Skeleton*, İşcan MY, Kennedy KAR (eds). Wiley-Liss: New York; 11–22.

Kjellström A, Tesch S, Wikström A. 2005. Inhabitants of a sacred townscape: an archaeological and osteological analysis of skeletal remains from late Viking Age and medieval Sigtuna, Sweden. *Acta Archaeol* **76**: 87–110.

Klepinger LL. 1979. Paleodemography of the Vadivia III phase at Real Alto, Ecuador. *Am Antiquity* **44**: 305–309.

Lambert JB, Vlasak SM, Tometz AC, Buikstra JE. 1982. A comparative study of the chemical analysis of ribs and femurs in Woodland populations. *Am J Phys Anthropol* **59**: 284–294.

Little L, Papadopoulos K. 1998. A social outcast in early Iron Age Athens. *Hesperia* **67**: 375–404.

Mays SA. 1992. Taphonomic factors in a human skeletal assemblage. *Circaea* **9**: 54–58.

Mays SA. 1993. Infanticide in Roman Britain. *Antiquity* **67**: 883–888.

Mays SA. 1997. Life and death in a mediaeval village. In *Death and Burial in Mediaeval Europe: Papers of the 'Mediaeval Europe Brugge 1997' Conference*, volume 2, De Boe G, Verhaege F (eds). Instituut voor het Archeologisch Patrimonium Rapporten 2. Instituut voor het Archeologisch Patrimonium: Zellik; 121–125.

Mays SA. 1998. *The Archaeology of Human Bones*. Routledge: London.

Milner GR, Wood JW, Boldsen JL. 2000. Paleodemography. In *Biological Anthropology of the Human Skeleton*, Katzenberg MA, Saunders SR (eds). Wiley-Liss: New York; 467–497.

Møller-Christensen V. 1961. *Bone Changes in Leprosy*. Munksgaard: Copenhagen.

Morris AG, Rodgers AL. 1989. A probable case of prehistoric kidney stone disease from the Northern Cape Province, South Africa. *Am J Phys Anthropol* **79**: 521–527.

Murphy EM, McNeil TE. 1993. Human remains excavated at Doonbought Fort, County Antrim, 1969. *Ulster J Archaeol* **56**: 120–138.

Nicholson RA. 2001. Taphonomic investigations. In *Handbook of Archaeological Science*, Pollard MA, Brothwell D (eds). Wiley: Chichester; 179–190.

Paine RP. 2000. If a population crashes in prehistory, and there is no paleoedemographer there to hear it, does it make a sound? *Am J Phys Anthropol* **112**: 181–190.

Parker-Pearson M. 1999. *The Archaeology of Death and Burial*. Sutton: Stroud.

Prior IAM, Rose BS, Harvey HPB, Davidson F. 1966. Hyperuricemia, gout, and diabetic abnormality in Polynesian people. *Lancet* **1**: 333–338.

Resnick D, Reinke RT, Taketa RM. 1975. Early-onset gouty arthritis. *Radiology* **114**: 67–73.

Robb J. 1994. Burial and social reproduction in the peninsular Italian Neolithic. *J Mediterr Archaeol* **7**: 27–71.

Roberts CA, Lewis ME, Manchester K (eds). 2002. *The Past and Present of Leprosy*. British Archaeological Reports International Series S1054. Archaeopress: Oxford.

Saunders SR, Herring A, Sawchuk L, Boyce G, Hoppa R, Klepp S. 2002. The health of the middle class: the St. Thomas Anglican church cemetery project. In *The Backbone of History: Health and Nutrition in the Western Hemisphere*, Steckel RH, Rose JC (eds). Cambridge University Press: Cambridge; 130–161.

Skeates R. 1991. Caves, cult and children in Neolithic Abruzzo, central Italy. In *Sacred and Profane*, Garwood P, Jennings D, Skeates R, Toms J (eds). Oxford University Centre for Archaeology: Oxford; 122–134.

Skeates R. 1997. The human use of caves in east-central Italy during the Mesolithic, Neolithic and Copper Age. In *The Human Use of Caves*, Bonsall C, Tolan-Smith C (eds). British Archaeological Reports International Series 667. Archaeopress: Oxford; 79–86.

Specker BL, Brazerol W, Tsang RC, Levin R, Searcy J, Stechen J. 1987. Bone mineral content in children 1 to 6 years of age. *Am J Dis Child* **141**: 343–344.

Steckel RH, Sciulli PW, Rose JC. 2003. A health index from skeletal remains. In *The Backbone of History: Health and Nutrition in the Western Hemisphere*, Steckel RH, Rose JC (eds). Cambridge University Press: Cambridge; 61–93.

Steinbock RT. 1976. *Paleopathological Diagnosis and Interpretation*. Charles C Thomas: Springfield.

Ubelaker DH. 1991. Perimortem and postmortem modification of human bone. Lessons from forensic anthropology. *Anthropologie* **29**: 171–174.

Ubelaker DH, Adams BJ. 1995. Differentiation of perimortem and postmortem trauma using taphonomic indicators. *J Forensic Sci* **40**: 509–512.

Waldron T. 1987. The relative survival of the human skeleton: implications for paleopathology. In *Death, Decay and Reconstruction. Approaches to Archaeology and Forensic Science*, Boddington A, Garland AN, Janaway RC (eds). Manchester University Press: Manchester; 55–64.

Waldron T. 1994. *Counting the Dead. The Epidemiology of Skeletal Populations*. Wiley: Chichester.

Weiss KM. 1973a. *Demographic Models for Anthropology*. Memoirs of the Society for American Archaeology No. 27. Society of American Archaeology: Washington, DC.

Weiss KM. 1973b. A method for approximating age-specific fertility in the construction of life tables for anthropological populations. *Hum Biol* **45**: 195–210.

Weiss KM. 1976. Demographic theory and anthropological inference. *Ann Rev Anthropol* **5**: 351–381.

Wells C. 1967. Pseudopathology. In *Diseases in Antiquity*, Brothwell DR, Sandinson AT (eds). Charles C. Thomas: Springfield; 5–19.

Wells C, Dallas C. 1976. Romano-British pathology. *Antiquity* **50**: 53–55.

Wood JW, Milner GR, Harpending HC, Weiss KM. 1992. The osteological paradox. Problems of inferring health from skeletal samples. *Curr Anthropol* **33**: 343–370.

Wright LE, Yoder CJ. 2003. Recent progress in bioarchaeology: approaches to the osteological paradox. *J Archaeol Res* **11**: 43–70.

Zalokar J, Lellouch J, Claude JR. 1981. Serum urate and gout in 4663 young male workers. *Sem Hop* **57**: 664–670 (in French).

3

Epidemiological Approaches in Palaeopathology

Ron Pinhasi[1] and Katy Turner[2]

[1] *Department of Archaeology, University College Cork, Cork Ireland*
[2] *Division of Epidemiology, Public Health and Primary Care, Imperial College, Praed Street, St Mary's Campus, London W2 1PG, UK*

INTRODUCTION

Epidemiological investigation of disease involves the elucidation of the aetiology of a specific disease or group of diseases by combining epidemiological data with information from other sources (genetics, biochemistry, microbiology), and the evaluation of consistency of epidemiological data with aetiological hypotheses developed either clinically or experimentally (Lilienfeld and Stolley, 1994). Palaeoepidemiology can be broadly defined as '... an interdisciplinary area that aims to develop more suitable epidemiological methods, and to apply those in current use, to the study of disease determinants in human populations in the past' (de Souza *et al.*, 2003: 21). What is the relationship between palaeoepidemiology and medical epidemiology? Or, in other words, what concepts and methods of medical epidemiology are of use to palaeopathologists working on skeletal remains? In an attempt toward addressing these issues, some of the basic concepts of epidemiology will be briefly reviewed, as these are also central to palaeoepidemiological studies.

Traditionally, epidemiology was a discipline inferring causal relationships between risk factors and disease. Pearce (2005: 9) points out that '... the key feature of epidemiological studies is that they are quantitative (rather than qualitative) observational (rather than experimental) studies of the determinants of disease in human populations (rather than individuals)'. Epidemiology has a descriptive dimension that involves the identification and documentation of disease trends, differential diagnosis of disease and injury and other related phenomena (Rockett, 1999). Epidemiology is a multi-step process that involves what Gordis (2000) defines as 'epidemiologic reasoning', a process that begins with descriptive analysis

Advances in Human Palaeopathology Edited by Ron Pinhasi and Simon Mays

but which should also proceed to address casual relationships between disease, demographic, social and cultural factors.

Both the descriptive and analytical sides of epidemiology are of relevance to palaeopathologists; as yet, however, most palaeopathological studies do not incorporate epidemiological methods in their analyses. To some extent, the limited use of epidemiological methods by palaeopathologists reflects a general scepticism as to the degree to which a given skeletal assemblage is a reliable representative sample of the parent population. Additionally, epidemiological theory and methods were developed for medical rather than palaeopathological research, so they require some modification to make them suitable for research using skeletal material.

RESEARCH DIRECTIONS IN THE STUDY OF DISEASE AT A POPULATION LEVEL IN PALAEOPATHOLOGY

A significant body of research exists that involves assessing overall health status in a skeletal population using a series of disease markers. In order to produce workable quantities of data, markers are usually chosen that generally have reasonably high rates of occurrence in skeletal series, e.g. dental enamel hypoplasias, cribra orbitalia, dental caries, ante-mortem tooth loss, lesions associated with skeletal infections, degenerative joint disease and trauma. For example, in the 1980s, indicators such as these were assessed in agricultural and pre-agricultural populations from various parts of the world in order to assess the effects of the transition to agriculture on the health status of earlier human groups (Cohen and Armelagos, 1984; Cohen, 1989). Building on this work, Steckel *et al.* (2002, 2003) combined data on different pathological features in order to produce overall 'health indices' for palaeopopulations from various geographical and temporal contexts in the Americas in order to assess their health profiles.

Although the 'health index' approach is potentially a useful one, there are several issues that are of concern. An assumption in calculating a health index is that the palaeopathological conditions that comprise it can be combined and compared between skeletal populations. This approach is complicated by the fact that the occurrences of the different palaeopathological conditions are, to varying extents, correlated with one another, as some factors may be involved in the causes of more than one condition. In order to control for this from an epidemiological perspective, it is essential to use methods such as logistic regression in order to detect the main uncorrelated variables, rather than simply combining pathological features to form a single index. Many of the skeletal pathological features used to derive a health index have multiple and incompletely understood aetiologies. They are not diseases, but simply lesions that can be identified on bone. Therefore, the health index is a palaeopathological construct that is not directly comparable with any medical epidemiological index. Although the health index may provide a measure of morbidity in a past population, it cannot shed light on patterns of particular diseases. By contrast, medical epidemiological reasoning focuses on the study of associations between social and environmental factors and specific diseases, rather than on health status, which is in any event a rather nebulous concept. It may be more useful for palaeopathological study to focus, where possible, on specific diseases in order to maintain the link between the study of disease in the past and medical research on disease in present-day populations. By doing so, palaeoepidemiological research will not become a detached subdiscipline with ill-defined palaeopathological 'conditions' and will, instead, continue to benefit from an ongoing discourse with medical epidemiological research.

The study of ancient DNA of pathogens is beginning to make an impact in palaeoepidemiology. Whilst early palaeopathological studies of pathogen DNA focused on single cases or small numbers of individuals, more recent work has begun to analyse for pathogen DNA in large numbers of skeletons, opening the way toward biomolecular-based palaeoepidemiological work. For example, Aufderheide *et al.* (2003) analysed ancient DNA of *Trypanosoma cruzi* from soft tissue samples taken from 283 naturally desiccated mummies from Chile and Peru. The specimens ranged in date from 7050 BC to approximately the time of the Spanish conquest, AD 1500. Of the 283 mummies, 115 (40.6%) were positive for *T. cruzi*. No statistically significant differences in the prevalence of the pathogen (indicated by positive test results) were noted between any of the cultural groups. There was no sex difference, but analysis of prevalence rates by age indicated that the prevalence of the disease was significantly higher among infants of 0–2 years of age. The transmission of Chagas' disease depends on the ability of the insect vector to infest the wild animals' nests or lairs, providing opportunities for the insect's blood meal and transmission of the infectious agent (*T. cruzi*). Aufderheide *et al.* (2003) assert that the lack of a significant diachronic trend in the prevalence rates of Chagas' disease among the human populations studied suggests that the earliest human groups that colonized the Andean coast offered the Chagas vector a physical environment for access to a blood meal that was equivalent to the nests and lairs of various indigenous feral (host) animals.

Palaeomicrobiology, the study of the antiquity and molecular evolution of pathogens, most usually involves the study of DNA from modern pathogens rather than ancient DNA. However, it is likely that, as work on large samples of skeletons becomes more common and techniques for amplifying and studying ancient DNA improve, the study of pathogen DNA from ancient skeletons will begin to make a significant contribution to the understanding of the evolution and spread of microbial human pathogens (Chapter 8).

Another related research direction involves the study of major epidemiological transitions associated with particular cultural–historical changes in human history. Barrett *et al.* (1998) provide a detailed account of major epidemiological changes in human host–pathogen systems that are associated with cultural/evolutionary changes during the Palaeolithic, Neolithic and the Industrial Revolution periods. They adopt an evolutionary historical perspective, using an expanded framework of epidemiologic transition theory that views major changes in host–pathogen systems as being directly related to corresponding changes in human modes of subsistence and social organization. Barrett *et al.* (1998) suggest that the first epidemiological transition occurred about 10 000 years ago, when the first agricultural settlements emerged in the Near East. The transition involved a drastic increase in infectious disease and mortality associated with changes in aggregation, social organization, domestication of animals (and the emergence of zoonotic infections), diet and other socio-cultural aspects of the Neolithic lifestyle. The second epidemiologic transition roughly coincided with the Industrial Revolution in mid-19th century Europe and North America. This period involved a marked decline in infectious disease mortality within developed countries, the disappearance of infectious diseases pandemics and a rise in chronic and degenerative non-infectious diseases. The third epidemiologic transition occurs during the last 25 years in the context of globalization, global trade, changes in disease ecology and mass migrations and a drastic increase world travel. It is associated with the appearance of numerous new diseases, an increase in the incidence and prevalence of pre-existing infectious diseases, and re-emerging drug-resistant strains of pathogens, such those responsible for tuberculosis and syphilis.

The study of the evolution and spread of disease requires interdisciplinary collaboration between epidemiologists, anthropologists and geneticists. The contribution of palaeopathology to such research involves the following. First, the detection and analysis of certain infectious diseases that leave diagnostic lesions on bone. The identification of early cases of leprosy, brucellosis, syphilis and other diseases in antiquity will at least provide geneticists with a preliminary age for the antiquity of diseases that leave traces on bone and also potentially indicate the geographic regions in which a disease existed during past epochs. Second, a systematic analysis of large archaeological skeletal collections from various time periods can give information concerning changes in disease prevalence over time. A good example is the evidence for a sharp increase in prevalence rates for leprosy during late medieval times, which was calculated from an observed increase in the number of skeletons with palaeopathological lesions pathognomonic of leprosy (Roberts, 2002). Third, palaeopathologists should oversee the taking of bone samples for ancient DNA to ensure that they are from archaeological specimens with unambiguous archaeological context. Fourth, the study of bone degradation (Chapter 1) provides information about chances of recovery of ancient DNA sequences from a given archaeological bone sample. Fifth, sampling for the pathogen DNA requires palaeopathological knowledge about the disease at hand. For example, preferential sites for sampling for *Mycobacterium leprae* are within in the nasal cavity, rather than from lesions in the hand and foot bones, which are principally due to secondary infections.

EPIDEMIOLOGICAL STUDIES OF SKELETAL SAMPLES: RELEVANT CONCEPTS AND LIMITATIONS

An approach of clear value in palaeoepidemiology simply involves the application of medical epidemiological methods to the investigation of disease in archaeological skeletal samples (Waldron, 1991a–c, 1994). Fundamental observations in epidemiology are measures of the occurrence of disease; the main measurements are those of risk, incidence and prevalence (Rothman, 2002). In a population with N individuals and in which A individuals developed the disease of interest during a period of time, risk is calculated (Rothman, 2002: 24) as

$$\text{Risk} = \frac{A}{N} = \frac{\text{Number of subjects developing disease during a time period}}{\text{Number of subjects followed for the same period}}$$

The average risk in the population is also known as the ‘incidence proportion’. In most cases risk is used in reference to a single person’s risk of developing the disease, whereas incidence proportion refers to groups of people (Rothman, 2002). Incidence rate is similar to incidence proportion, but instead of measuring the number of subjects with the disease as a proportion of the number of subjects that were initially followed, cases A are divided by a specific period of time T, which is the summation, across all individuals, of the time experiences by the population being followed (Rothman, 2002). Palaeopathological studies are cross-sectional in nature and, therefore, cannot measure incidence (Waldron, 1994). It is best, instead, to focus on the assessment of prevalence (Baker and Pearson, 2006). Prevalence is the number of people P in a population of N individuals who have specific disease (Rothman, 2002). The prevalence is P/N and is often multiplied by 1000 and reported in epidemiological studies as a rate per 1000 (Gordis, 2000). Prevalence may be measured as point prevalence, i.e. over a short period of time, or as a period prevalence, in which the time period is longer (Gordis, 2000). It is important to note that prevalence does not take into consideration aspects such as when the disease developed and its duration.

Researchers who intend to apply a palaeoepidemiological approach to the study of disease in archaeological skeletal samples must take into consideration certain problems and limitations. First, it is necessary to assess the degree to which a sample is truly representative of the population (Waldron, 1994; Chapter 2). In the case of palaeoepidemiological studies, representativeness does not necessarily apply to the true biological parent population, but rather to the group of interest to the epidemiological analysis. In fact, the sample used by palaeopathologists will almost never be random in the epidemiological sense (Waldron, 1994). Nonetheless, this does not prevent the palaeopathologist from deriving a random subsample from the parent skeletal 'population'.

A second concern is the state of preservation of specimens in a given skeletal sample. Palaeopathologists must make decisions concerning what to do with skeletons in which some skeletal parts are damaged or missing, particularly in cases in which pathognomonic features of a given disease require the preservation of specific skeletal features. In studies of archaeological populations, bone preservation may vary not only between archaeological sites, but also between and within cultural layers in the excavated area. The researcher, therefore, should assess the preservation of specimens from the various archaeologically defined areas or strata in order to decide whether differential preservation may bias prevalence estimates.

A third aspect is the time-scale involved. Most modern epidemiological studies focus on the time interval of years or decades (Waldron, 1994). Palaeoepidemiology cannot usually assess prevalence and other epidemiological aspects in this temporal resolution. Mean prevalence of a disease over a time span of several hundred years may obscure variations in disease prevalence during the time interval (Waldron, 1994). This, however, is more of an issue in the study infectious diseases with characteristic episodic peaks and troughs than it is for non-infectious conditions such as osteoarthritis. Because the palaeopathologists cannot usually address palaeoepidemiological aspects in refined temporal resolution, the focus is more often on broad chronological phases or culturally defined subdivisions of the skeletal sample, or on aspects such as gender or social status.

A fourth aspect is the assessment of error in the diagnosis of conditions. Only a small number of palaeopathological studies (e.g. Waldron and Rogers, 1990) involve a systematic assessment of observer errors and error in the diagnosis of disease from lesions on bone. This would involve: evaluation of repeatability of diagnosis of specific conditions by the same observer and by other observers; assessment of the accuracy of the method, in terms of the degree to which it does not exclude cases with the condition or include those without; the skill required to assess the condition on the basis of the specific set of criteria; and the investment of time that is required in order to evaluate the condition. Clearly, a more comprehensive approach to the diagnosis of a specific condition may entail the recording of an extensive set of diagnostic features. However, multiple features may be highly correlated, so it is sufficient to include the minimal number of criteria that are pathognomonic. Recording of fewer skeletal features means that fewer specimens need be excluded from study due to incompleteness or time constraints on the work.

Disease Prevalence in Past Populations: Case Studies

In the following sections, two hypothetical data sets are provided in order to demonstrate new methods for the calculation of prevalence rates in skeletal populations which take into account missing data, differential diagnostic criteria and undiagnosed specimens

Calculation of Prevalence Rates Based on Differentially Weighted Criteria

Table 3.1 is a hypothetical study of prevalence of leprosy based on a set of morphological criteria and on the methodology described by Law (2005). The columns represent a series of observed pathological features. All conditions are recorded on a scale of '0-3', where '0' indicates the trait is absent and '1–3' indicate the escalating scale of degree of severity for the presence of the feature. Features that cannot be recorded because parts are missing or damaged are marked with 'D'. An 'if' condition was then applied so that a case was diagnosed with leprosy if at least one of the rhinomaxillary criteria was given a score of '3', or one of the changes on the fibula/tibia and one or more of the changes affecting the joints of the hand and/or feet are present. This can be reduced to the following logical expression:

$C_i = 1$ if {(RM1 or RM2 or RM3 or RM4 = 3) OR [(VG or SPE $\geq$ 3) AND (NBD or CDR $\geq$ 3)]}; else $C_i = 0$

where C_i is case i in a skeletal population and a value of '1' denotes a specimen diagnosed with leprosy and '0' a specimen not diagnosed with leprosy.

The drawback of this system is that in some specimens it may be impossible to record one or more of the rhinomaxillary traits due to post-mortem damage. It is difficult to decide whether a skeleton should be included in which, for example, the anterior nasal spine and/or the alveolar process of the premaxilla are damaged. Clearly, no high score for the rhinomaxillary syndrome may be obtained in such instances simply because of post-depositional damage. Nevertheless, it is possible with some slight modifications to use the above methodology to calculate prevalence rates of both infectious and non-infectious conditions. Pathognomonic aspects may be included with a logical condition so that a score of '1' is only obtained once these are present. Moreover, data can then be used by other researchers who can modify the conditional phrase in order to compare their study with others. Alternatively, a probabilistic approach may be adopted by applying a dichotomous (absent, present) assessment to a set of features rather than giving particular weight to

Table 3.1 Hypothetical study of the prevalence of leprosy by a set of morphological criteria[a]

	Rhinomaxillary				Fibula and tibia		Hands and feet			
Case no.	RM1	RM2	RM3	RM4	VG	SPE	NBD	CDR	N (complete)	C_i
1	0	1	3	D	1	2	3	2		1
2	2	2	2	2	2	2	0	0	1	0
3	0	1	0	0	0	0	3	3	1	0
4	3	0	D	D	0	0	1	1		1
5	D	D	D	0	0	1	3	2		0
5	D	0	0	D	1	2	D	D		0
6	0	3	3	2	1	2	3	3	1	1

[a] N (complete) denotes the total number of specimens that minimally have one complete tibia, fibula, hand and foot bones and preserved facial morphology allowing the diagnosis of rhinomaxillary features. RM1–RM4 are the four rhinomaxillary changes described by Møller-Christensen (1961) on the scale of 0–3, where 0 denotes the normal non-pathological condition; D: parts are missing or damaged. VG: vascular grooves; SPE: subperiosteal exostoses; NBD: new bone deposition; CDR concentric diaphyseal resorption.

pathognomonic features (Boldsen, 2001). The drawback of the probabilistic approach is that it gives equal weight to each feature and eliminates any scoring scale for the manifestation of a given feature.

Age- and Sex-Specific Disease Prevalence in Skeletal Samples

The following is a hypothetical example of the calculation of prevalence of gout in a skeletal population of 1100 individuals (550 males and 550 females). The sample is then stratified by age and sex (Table 3.2).

Data in Table 3.2 are based on the assumption that it was possible to age and sex all of the specimens in the skeletal population. The 'unknown' specimens are those in which no pathognomonic gouty lesions were observed but in which a key diagnostic skeletal part or parts were missing or damaged (e.g. the halluces in the case of gout).

The following values were calculated: total unknown $U_t = 140$; average unknown per subgroup $U_a = 140/6 = 23$; and the proportion of unknown cases is $L_i = U_i/N_i$. The researcher may decide not to calculate prevalence when the proportion of skeletons whose diagnostic status is unknown exceeds a specific value for a given subgroup or for the whole sample as this may suggest that the overall preservation of the sample is too poor to allow a reliable palaeoepidemiological investigation.

The prevalence of the disease P in the stratified cells excludes all unknown specimens. It is impossible to assess what proportion of the unknown specimens had the disease. However, it is possible to estimate minimum and maximum values by assuming that either all or none of the unknown specimens had the disease.

Next, we set the null hypothesis that there is no significant difference in L_i for the subcategories. Simple chi-square analysis of the unknown skeletons by each subcategory indicates that we should reject the null hypothesis that there is no significant age or sex bias in the distribution of unknown cases ($\chi^2 = 12.73$, $p > 0.05$, 4 d.f.). This may indicate that there are problems with the analysis of prevalence rates in this population. There are, in fact, no simple solutions to this scenario, as pooling the subcategories and calculating the crude prevalence figure will not resolve this bias.

Next, we examine the null hypothesis of no significant difference in prevalence rates for the subcategories. A chi-square analysis of the prevalence of gout in each subcategory

Table 3.2 Age and sex specific prevalence of gout in a hypothetical skeletal population

	18–30 years		30–50 years		>50 years	
	Males	Females	Males	Females	Males	Females
Gout	10	15	30	35	35	45
No gout	95	95	150	150	150	150
Unknown U	20	20	30	30	30	10
N (gout + no gout))	105	110	180	185	185	195
N (gout+no gout+unknown)	125	130	210	215	215	205
Proportion unknown L	0.16	0.15	0.14	0.14	0.14	0.05
Prevalence P	0.10	0.14	0.17	0.19	0.19	0.23

yielded a non-significant value ($\chi^2 = 8.59$, $p > 0.05$, 4 d.f.), so there is no evidence for a difference in the prevalence rates of the various subcategories.

Comparing Prevalence Rates Between Skeletal Populations and Confidence Intervals of Prevalence

The comparison of the prevalence rates in several populations requires either direct or indirect standardization of the age- and sex-specific rates (Waldron, 1994; Chapter 2). A standard population can be either another skeletal population with comparable age and sex categories (and for which prevalence rates of the same disease are available), epidemiological data on a modern population, or an entirely artificial population (as in Table 3.2). The standardization of prevalence rates requires that both populations are subdivided according to the same age and sex categories and that the same methods are applied for the diagnosis of the disease of interest. It calls, therefore, for the use of standardized ageing and sexing methods in population-based studies in palaeopathology, and preferably for the use of well-defined age intervals. At present, there are no standard prevalence data that take into consideration the effect of specimens where diagnostic parts are missing or damaged. Nonetheless, researchers that wish to apply such a method to the study of several populations can combine these into a total sample, and use the prevalence rates obtained to derive standardized rates for each of the populations. Alternatively, it is possible to standardize the rates using medical epidemiological prevalence figures and then compare the observed rates with the expected rates using indirect standardization. The outcome of such a standardization procedure is the derivation of palaeoepidemiological data that are comparable and, hence, is a step towards meta-analysis of palaeopathological studies and pooling and/or comparing data from different skeletal samples.

Palaeopathologists usually calculate the prevalence of a disease in a skeletal sample without calculating its associated confidence interval. Consequently, they do not take into account the range of possible values for the calculated prevalence in the population. The 95 % confidence interval provides a range of values within which there is a 95 % chance that the true figure lies. An estimate of the 95 % confidence interval for the true population mean is provided by

$$\mathrm{CI}_{95} = P \pm 1.96 \times \mathrm{SE}(P)$$

where P is the prevalence rate of the disease, 1.96 is the z value for a 95 % confidence interval of a normal distribution, and SE(P) is the standard error of the prevalence under the binomial model:

$$\mathrm{SE}(P) = \sqrt{\frac{P(1-P)}{N}}$$

where N is the total cases excluding those with diagnostic parts missing or damaged.

The bottom three rows in Table 3.3 provide standard errors of the prevalence from the above data set (Table 3.2) and the minimum and maximum values of the 95 % confidence intervals. It is evident that there are no great differences in standard error for the various columns. The interval obtained (ranging between the minimum and maximum prevalence figures in Table 3.2) and the standard error of the prevalence rate provide the researcher with additional information about the prevalence of disease in each age category.

Table 3.3 Standard errors of the prevalences in Table 3.2 and the associated minimum and maximum values of the 95 % confidence interval

	18–30 years		30–50 years		>50 years	
	Males	Females	Males	Females	Males	Females
Prevalence *P*	0.095	0.136	0.167	0.189	0.189	0.231
SE	0.029	0.033	0.028	0.029	0.029	0.030
CI_{MIN}	0.039	0.072	0.112	0.133	0.133	0.172
CI_{MAX}	0.151	0.200	0.221	0.246	0.246	0.290

Analytical Palaeoepidemiology: Case-Control Studies

A case-control study is based on a non-random sampling from the source population and, hence, can begin with people with a disease (the cases) and compare them with people without the disease (the controls) (Gordis, 2000). The case-control study is suitable for palaeoepidemiological investigations that start from the identification of disease in bone or other tissue (Waldron, 1994).

Matching cases to controls may be done in two ways, group matching and individual matching. Group matching involves selecting the controls in such a manner that the proportion of certain characteristics, such as sex and age, are identical to the proportions among the cases. Individual matching involves selecting, for each case, a control that matches its specific variables of concern, such as age and sex (Gordis, 2000). The researcher may apply a 1:1 ratio of cases and controls or alternatively use multiple controls, applying a ratio of 1:2, 1:3 or even 1:4 cases to controls. Multiple controls are used in medical epidemiology when the researcher wishes to increase the overall sample size without having to increase the number of cases (which may be difficult with, for example, rare conditions).

An example of a palaeoepidemiological case-control study is the analysis of the association between osteoarthritis of the hands and knee. This association exists in modern populations; a case-control study was used to investigate whether it also extended back into the past. One hundred and fifteen skeletons with hand osteoarthritis from 18th–19th century AD cemeteries in London were examined (Waldron, 1997). These cases with hand osteoarthritis were selected based on the condition that knee joints were also present for observation. The 115 cases were individually matched for sex and age with 115 controls that did not have osteoarthritis of the hands and which had knee joints preserved to allow the assessment of osteoarthritis. Eight cases had osteoarthritis of the knee in comparison with only two controls. The results appear to confirm that an association between osteoarthritis of the hand and osteoarthritis of the knee already prevailed in 18th–19th century AD British populations.

FUTURE DIRECTIONS

Both medical epidemiology and palaeopathology share a common interest in disease patterns and change over time, and a focus on processes that concern populations rather than individuals. However, palaeopathological research, which in most instances is based on the analysis of archaeological skeletal samples, calls for the further development of new epidemiologically based methods and techniques (de Souza *et al.*, 2003). It calls for a focus on biocultural approaches that form a sound link between the biological and socio-cultural aspects that affected the health status of past societies. These should be addressed using a population-based approach which can, on the one hand, apply the causal analytical approach of epidemiological reasoning (and, hence, focus on the assessment of specific disease factors) and, on the other hand, anchor these to the specific archaeological contexts of the skeletal samples analysed. The growing trend in palaeopathology to shift away from the descriptive realm to the analytical realm requires the application of analytical quantitative methods and a focus on hypothesis-driven research.

Prevalence rates of disease in skeletal samples are probably the most useful parameter in palaeoepidemiological investigations based on the analysis of archaeological skeletal samples. Prevalence figures for archaeological skeletal series must take into consideration issues of preservation and representativeness of specimens in the sample, and particularly in cases when poor preservation or missing skeletal features preclude the diagnosis of a given disease.

Biomolecular methods potentially allow the derivation of prevalence rates of infectious diseases from bone/tissue samples from past populations. Unlike palaeopathological studies, in which prevalence rates are assessed from lesions, these rates are based on identification of pathogen DNA fragments. Palaeoepidemiological investigations that apply ancient DNA analysis of pathogens may, therefore, open a new window to our understanding of the antiquity of various diseases and cultural – historical changes in prevalence rates. However, such studies must take into consideration the factor of latency; that is, the detection of a pathogen, such as *Mycobacterium tuberculosis*, in a skeleton does not necessarily indicate active disease. The calculation of prevalence rates of infectious diseases using ancient DNA must also take into consideration the bias of false negatives due to non-survival of ancient DNA.

In future it may be possible to derive inferences about the evolution, spread and natural history of a given disease from the use of epidemiologic mathematical models. Modern infectious disease epidemiology is underpinned by a theoretical framework based on the reproductive number R_0 (Anderson and May, 1991). The reproductive number R_0 is defined as the average number of secondary cases caused by one infectious case in a fully susceptible population. Therefore, if $R_0 > 1$, then the infection is able to spread; and if $R_0 < 1$, then the epidemic dies out. An important influence on R_0 is contact patterns between humans and disease-causing pathogens. This contact may be a result of contact with infected humans or with other reservoirs of disease. To our knowledge, the concept of R_0 has yet to be used as an analytical tool in palaeoepidemiology, but it could be potentially applicable to such work. Contact patterns in past populations could be estimated from archaeological data, such as the study of the internal architecture of houses, location of refuse pits, kitchens, distance between animal pens and human living areas, and other aspects of settlement morphology. Population size also plays a role in determining the persistence of disease, by modifying the number of available susceptibles and, hence, the number of effective contacts. Estimates for population size in different periods/regions can be obtained from archaeological studies of average size and settlement pattern analysis.

The discipline of palaeopathology may need to evolve and develop a better cross-disciplinary dialogue which is grounded in a firmer theoretical basis. So far, palaeopathologists have managed successfully to incorporate medical, demographic and archaeological concepts in their analysis and interpretation for disease in past populations, but much more work is needed on disease aetiology, causation and spread in past populations. The time is ripe to broaden the current scope and to work towards a better dialogue between palaeopathology and other disciplines, such as disease ecology and molecular biology and the development of a palaeoepidemiological perspective.

REFERENCES

Anderson R, May R. 1991. *Infectious Diseases of Humans: Dynamics and Control*. Oxford University Press: Oxford.

Aufderheide AC, Salo W, Madden M, Streitz J, Buikstra J, Guhl F, Arriaza B, Renier C, Wittmers Jr LE, Fornaciari G, Allison M. 2003. A 9,000 years old record of Chaga's disease. *Proc Natl Acad Sci U S A* **17**: 2034–2039.

Baker J, Pearson OM. 2006. Statistical methods for bioarchaeology: applications of age-adjustment and logistic regression to comparisons of skeletal populations with differing age-structures. *J Arch Sci* **33**: 218–226.

Barrett R, Kuzawa XW, McDade T, Armelagos GJ. 1998. Emerging and re-emerging infectious diseases: the third epidemiologic transition. *Ann Rev Anthropol* **27**: 247–271.

Boldsen JL. 2001. Epidemiological approach to the palaeopathological diagnosis of leprosy. *Am J Phys Anthropol* **115**: 380–387.

Cohen MN. 1989. *Health and the Rise of Civilization*. Yale University Press: New Haven, CT.

Cohen MN, Armelagos GJ. 1984. *Paleopathology at the Origins of Agriculture*. Academic Press: Orlando, FL.

De Souza SM, de Carvalho DM, Lessa A. 2003. Paleoepidemiology: is there a case to answer? *Mem Inst Oswaldo Cruz* **98**(Suppl 1): 21–27.

Gordis L. 2000. *Epidemiology* (2nd edition). WB Saunders: Philadelphia.

Law A. 2005. A simple method for calculating the prevalence of disease in a past human population. *Int J Osteoarchaeol* **15**: 146–147.

Lilienfeld DE, Stolley PD. 1994. *Foundations of Epidemiology* (3rd edition). Oxford University Press: New York.

Møller-Christensen V. 1961. *Bone Changes in Leprosy*. Munksgaard: Copenhagen.

Pearce N. 2005. *A Short Introduction to Epidemiology* (2nd edition). Occasional Report Series No. 2, Centre for Public Health Research. Massey University: Wellington.

Roberts CA. 2002. The antiquity of leprosy in Britain: the skeletal evidence. In *The Past and Present of Leprosy*, Roberts CA, Lewis ME, Manchester K (eds). British Archaeological Reports, International Series 1054. Archaeopress: Oxford: 213–222.

Rockett IR. 1999. Population and health: an introduction to epidemiology. *Popul Bull* **54**: 1–44.

Rothman KJ. 2002 *Epidemiology: An Introduction*. Oxford University Press: New York.

Steckel RH, Rose JC, Larsen CS, Walker PL. 2002. Skeletal health in the western hemisphere from 4000 B.C. to the present. *Evol Anthropol* **11**: 142–155.

Steckel RH, Sciulli PW, Rose JC. 2003. A health index from skeletal remains. In *The Backbone of History: Health and Nutrition in the Western Hemisphere*, Steckel RH, Rose JC (eds). Cambridge University Press: Cambridge; 61–93.

Waldron T. 1991a. Rates for the job. Measures of disease frequency in palaeopathology. *Int J Osteoarchaeol* **1**: 17–25.

Waldron T. 1991b. Variations in the rates of spondylolysis in early populations. *Int J Osteoarchaeol* **1**: 63–65.

Waldron T. 1991c.The prevalence of, and the relationship between some spinal diseases in a human skeletal population from London. *Int J Osteoarchaeol* **1**: 103–110.

Waldron T. 1994. *Counting the Dead: The Epidemiology of Skeletal Populations*. Wiley: Chichester.

Waldron T. 1997. Association between osteoarthritis of the hand and knee in a population of skeletons from London. *Ann Rheum Dis* **56**: 116–118.

Waldron T, Rogers J. 1990. Inter-observer variation in coding osteoarthritis in human skeletal remains. *Int J Osteoarcheol* **1**: 49–56.

4

Macroscopic Analysis and Data Collection in Palaeopathology

Anne L. Grauer

Department of Anthropology, Loyola University of Chicago, 6525 N. Sheridan Road, Chicago, IL 60626, USA

INTRODUCTION

Palaeopathology has embraced technology. The incorporation of precision instrumentation, along with the application of new biochemical and biomedical techniques such as DNA analysis, radiographic imaging and trace element analysis, has provided researchers with new insights into the past and a promise of new theoretical directions to follow. Ironically, however, the foundation of palaeopathological investigation, i.e. macroscopic analysis and data collection, is rarely the focus of discussion or debate regardless of the fact that it serves as the starting point for virtually all methodological approaches.

The lack of discussion towards improving macroscopic techniques and data collection is not surprising given the fact that, at its most cursory level, the researcher needs only good (or slightly magnified) vision, a reasonable light source, a sound measuring device, and pen and paper. This low-tech tool kit hardly sparks the imagination of the young, nor does it garner the excitement that genetic research, biochemical analyses, and spectacular visuals can achieve in the public eye. Even within our discipline, discussion and focus on the recognition, assessment, and recording of lesions noticeable to the naked eye remain negligible, as new techniques overshadow old arguments.

Macroscopic analysis, however, stands as the primary and most pervasive means by which researchers worldwide begin skeletal assessment. As recognized by Lovell (2000: 219):

> Visual observation is generally the first method employed when examining archaeological remains for pathological lesions. In many cases it may be the only method required, while in some circumstances it may be the only method available.

Advances in Human Palaeopathology Edited by Ron Pinhasi and Simon Mays

Arguably, without macroscopic evaluation of human skeletal material even the most technologically sophisticated research would be hindered. For instance, is every recovered or curated skeleton to be earmarked for DNA, histological, or biochemical analysis in our effort to answer our questions about the past? No. We have neither the time nor resources to tackle this endeavour. The unrealistic initiative would fall short in answering many of our questions. We would still find ourselves relying on our visual assessments of skeletal remains to select new datasets for evaluation, to highlight human variation, and to isolate variables.

This being the case, macroscopic evaluation of skeletal material ought to warrant as much of our time, concern and effort as any new technique. However, in spite of the significance of macroscopic analysis, it seems that macroscopic work has become a poor cousin to the technologically innovative, cutting-edge means by which new information is acquired. This must change.

HISTORICAL BACKGROUND

In order to understand the problems and promise of macroscopic analysis and data collection in palaeopathology it is essential to understand the road that palaeopathology has taken. Palaeopathology is a relatively new field with a long history. While curiosity about the antiquity of human life developed amidst Renaissance Europe's fascination with the 'ancient world', specific focus on the history of diseases through the examination of human remains (both skeletal and mummified) only emerged during the 20th century. Numerous workers have explored the origins of palaeopathology and have chronicled its growth and development (e.g. Moodie, 1923; Williams, 1929; Wells, 1964; Janssens, 1970; Jarcho, 1966a; Brothwell and Sandison, 1967; Buikstra and Cook, 1980; Buikstra and Beck, 2007). All find a similar trajectory: that scientific understanding of the history of human diseases began as a physician's 'hobby' and developed as an anthropological endeavour much later in the 20th century.

Indeed, a review of publications produced in the late 19th and early 20th centuries clearly highlights this trend. Rudolf Virchow (1821–1902), a renowned physician, Frederic Wood Jones (1879–1954), an anatomist by training, Grafton Elliot Smith (1871–1937), chair and professor of anatomy at the Governmental School of Medicine in Cairo, and Sir Marc Armand Ruffer (1859–1917), a professor of bacteriology and the first director of the British Institute of Preventative Medicine (heralded by Sandison (1967) as the 'pioneer of palaeopathology'), to name a few, focused much of their skill and attention on the recognition of specific pathological lesions in human bone and the presence of anatomical variation between human groups. Relying on their diagnostic skills and familiarity with human anatomy developed during their medical training, and utilizing the current and burgeoning body of medical terminology, the presence of particular diseases witnessed in particular skeletal specimens became the focal point of interest. To the credit of these early researchers, their works contributed greatly to macroscopic analysis. The diagnosis and recognition of skeletal lesions became more firmly based on known clinical manifestations of the diseases. Suggested diagnoses were increasingly corroborated by histological and radiographic investigation, and the analysis of mummified remains allowed these early researchers to link soft-tissue conditions more firmly with skeletal alterations.

By the close of the first quarter of the 20th century a new approach towards understanding the antiquity of disease began to emerge. Exemplified by Hooton (1930), a substantial redirection takes hold within palaeopathology. Researchers now focused on disease processes in humans, which were seen as complexly intertwined with ecological variables, human behaviour and human culture (Kerley and Bass, 1967). Calvin Wells, for instance, asserted that 'the pattern of disease of injury that affects any group of people is never a matter of chance. It is invariably the expression of stresses and strains to which they were exposed, a response to everything in their environment and behaviour' (Wells, 1964: 17). Similarly, Brothwell's broader and bolder perspective on human culture, environmental conditions and disease explored prehistoric populations up to and including the Iron Age in Europe and the Middle East (Brothwell, 1967).

Palaeopathological research also began to adopt greater diagnostic rigour. Møller-Christensen, for instance, in his studies of the presence of leprosy in human populations offered clear criteria for diagnosis (Møller-Christensen, 1967), meticulously described pathological lesions (Møller-Christensen, 1961, 1978), offered differential diagnoses, and used known clinical manifestations and measurements of the disease to aid in the diagnoses of archaeological specimens (Møller-Christensen, 1974). In spite of these efforts, and those of others, concerns over the uneven adoption of diagnostic rigour within palaeopathology were voiced (Jarcho, 1966b; Brothwell and Sandison, 1967: xii).

In response, a number of important efforts were launched to deal with methodological needs. Brothwell and Sandison's seminal work, for instance, which stemmed 'from a feeling among students of early disease, that the time has come for some form of palaeopathological stock-taking and pooling of recently collected data' (Brothwell and Sandison, 1967: xiv), provided the field with a compendium of skeletally recognized conditions and diseases. Here, insights into the antiquity and geographical scope of skeletal lesions, as well as radiographic, histological, and photographic documentation, were provided. Later, Steinbock (1976: ix), in an effort to 'provide a basic framework for those interested in diagnosing and interpreting bone lesions', led the way by closely synthesizing clinical data with archaeological specimens, providing information on the pathogenesis of conditions, gross morphology, radiographic appearance, differential diagnosis and photographic examples of the conditions in archaeological samples. More recent efforts to meet palaeopathologists' need for diagnosis and interpretation of lesions by Ortner and Putschar (1981), Manchester (1983), Aufderheide and Rodriguez-Martin (1998), and Ortner (2003), along with the extensive database of images from Ortner's collection, currently residing on the Ohio State University server (http://global.sbs.ohio-state.edu/cd-contents/Ortner-slides), have provided researchers bases upon which careful visual examination of bone lesions can lead to diagnoses. For macroscopic analysis, this has meant that a compendium of clinically and archaeologically derived information, along with visual images created by palaeopathologists, is available for researchers seeking to compare and understand specimens under their investigation.

Alongside the new emphases on cultural/environmental contexts of diseases and new rigorous means of providing diagnoses in the 1960s came changes in the understanding of the nature and scope of human disease. Derived in part from Audy's (1967) expanded and holistic definition of health and disease, along with Selye's (1973) development of the general adaptation syndrome, which explored the impact of unspecified physiological disruption on the human body, palaeopathologists began to avert their attention from lesions solely associated with singular pathogens and to direct it toward a broad spectrum of abnormal bone. 'Multiple stress indicators' and 'non-specific indicators of stress' (Huss-Ashmore *et al.*, 1981; Armelagos, 1997; Goodman and Martin, 2002) included conditions such as aetiologically non-specific periosteal reaction and indicators of growth disruption.

The new theoretical approaches launched many methodological changes within palaeopathology. For one, they initiated the need to incorporate many historically ignored variables into skeletal analyses. The age at death and sex of the individual, the nutritional and social environment, and developmental and genetic factors became important considerations in recognizing 'disease clusters' and interpreting the presence of 'stressors' (Armelagos *et al.*, 1982; Armelagos and van Gerven, 2003). Researchers placed an emphasis on population analysis rather than on individuals and required contextual analyses, both archaeological and cultural. They demanded continued improvement in the recording and diagnosis of disease through stringent comparison with clinical data and with the support of technological tools.

The impact of these changes on data collection was immense. With earlier focus on individual specimens and the diagnosis of specific conditions there was little need for systematic means of data collection. In fact, data collection often consisted of gathering other researchers' diagnosed specimens. Case studies, which dominate the early literature, dealt with sample sizes of one. The new osteoarchaeological and biocultural approaches, however, required the development of more standardized recording of skeletal variables to meet the needs for population-based approaches and cross-cultural comparison. In 1992, responding to continued dissatisfaction with terminology and classification of skeletal lesions, the newsletter of the Paleopathology Association (Ragsdale, 1992) published a series of nouns and modifiers to be used or avoided when describing bone lesions. A similar effort to standardize terminology followed with Thillaud's (1992: 4) suggestions for macroscopic terminology and Lovell's (2000) synthesis of previous work. The success of these endeavours was limited. While they drew attention to the need for careful, well-formulated descriptions, palaeopathologists made independent decisions on whether to adopt the suggestions or ignore them. Hence, suggested 'terms to avoid' can still be found in the literature.

Efforts to improve the standardization of data recording in order to facilitate data sharing and population-based study were also initiated. The Paleopathology Association began to tackle this challenge in 1988, with discussion of the need and direction that standardization might take (*Paleopathology Newsletter*, September 1988). In 1989, Jonathan Haas, then the Vice President for Collections and Research at the Field Museum of Natural History in Chicago, convened a workshop to develop standards for the collection of osteological data. These efforts were further spurred in 1989 by the enactment of United States Public Law 101-185 (The National Museum of the American Indian Act) and in 1990 by the United States Public Law 101-601 (The Native American Graves Protection and Repatriation Act), which created the immediate need for massive and careful data collection, and for collaboration between researchers (Rose *et al.*, 1996).

A number of contributions to palaeopathology stemmed from these efforts. One was the Paleopathology Association Skeletal Database Committee Recommendations (Rose *et al.*, 1991), which sought to provide guidance in what types of data to record and methods to use (Rose *et al.*, 1991: 1). Another contribution, *Standards for Data Collection From Human Skeletal Remains* (Buikstra and Ubelaker, 1994), provided guidelines for terminology (which were supported and exemplified by photographs and description), data collection (including, for instance, inventory, taphonomy, metric and non-metric analyses, dentition, and pathology), and created specific codes for recording variables in order for data to be more fully comparable between researchers. Focus was turned sharply away from determining aetiology of conditions and offering diagnoses, towards the codification of skeletal remains.

Arising from the recognition that vast amounts of data would be generated and that the computerization of collected data would enhance researchers' ability to maintain and collect information, a free-standing relational database called the Standardized Osteological Database (SOD) was created and made available without cost to interested parties (this database remains available as a downloadable application at http://www.cast.uark.edu/cast/sod). More recently, building on their familiarity with the strengths and weaknesses of entering alphanumeric codes into text-based screens, as designed and required by Buikstra and Ubelaker (1994) and SOD, researchers at the Smithsonian Institution Repatriation Osteology Laboratory have developed a relational database with an enhanced graphical user interface. The strengths of this new system rest on the researcher's ability to link different data tables together through key fields, to prevent the duplication of data or records, to compare many records simultaneously, and to manipulate large amounts of data (Ousley *et al.*, 2006). For further discussion of databases see Chapter 9.

Efforts to tackle issues inherent to data collection were also recognized and initiated in Britain. Focusing on current research method and theory, the British Association of Biological Anthropology and Osteoarchaeology (BABAO) developed a guideline for data collection that emphasized utility and function over detail and strict replicability (Brickley and McKinley, 2004). These efforts, culminating with the production of the *Guidelines to the Standards for Recording Human Remains* (Brickley and McKinley, 2004), provide researchers with a well-built foundation upon which individual recording procedures should be developed and shared.

METHODOLOGIES

As the historical background to macroscopic analysis and data collection illustrates, changes in the theoretical foci within palaeopathology have influenced what alterations in the human skeleton warrant analysis, and discussions of how alterations are recorded and interpreted. There is no single method by which macroscopic diagnoses and data collection is accomplished. However, guidelines developed by Buikstra and Ubelaker (1994), Ortner (1991, 1992, 1994, 2003), and Brickley and McKinley (2004) indicate that a standard of rigour can be created and adopted. Synthesizing the work of these contributors, six suggestions toward rigorous macroscopic diagnosis are set out in the following sections. All or most of them ought to be adopted.

Assessment and Recording of Processes and Variables

Ortner and Putschar (1981: 36) argue that:

> in a descriptive system for abnormal bone conditions, there are several essential elements. These include (1) an unambiguous terminology, (2) precise identification of the location and distribution of abnormal bone, and (3) a descriptive summary of the morphology of the abnormal bone.

Unambiguous Terminology

Although attempts to standardize terminology within palaeopathology have met with limited success (see Ragsdale (1992), Buikstra and Ubelaker (1994) and Lovell (2000) for suggestions and guidance), unambiguous terminology is rooted in established physiological processes, and clinical derivation. The term 'periostitis', for instance, while commonly used in the palaeopathological literature to indicate the presence of fusiform bone hypertrophy involving the periosteum, carries with it the clinical suggestion that an inflammatory process (involving vascular changes and phagocytic activity) is involved. As proliferative bone reactions of the periosteum can be triggered by a variety of conditions (other than inflammation), and macroscopic analysis of dry bone may mask changes to the cortex and/or endosteum, 'periostosis' is a less ambiguous and more precise term to adopt, as it suggests the presence of hypertrophy of the periosteum without inserting an unsubstantiated cause. Researchers must choose their terminology carefully.

Identification and Recording

Precise recording of anatomical location and distribution of abnormal bone is also integral to macroscopic analysis. This is because different pathogens, along with physiological changes in body functions and abilities, differentially affect areas and/or groups of bones. To this end, Ortner (2003: 49–50) offers guidance in recording the distribution of lesions within the skeleton, as do Buikstra and Ubelaker (1994). The critical aspect of recording anatomical location lies in the need for details that include not only the precise bone affected, but also the component of the bone involved (e.g. the epiphysis or proximal third of the diaphysis), the aspect of the bone (e.g. anterior/lateral or posterior/medial), and affected features (e.g. being circumscribed by the presence of vascular channels, affecting foramina, or traversing sutures).

Descriptive Summaries

Providing a detailed and descriptive summary of the morphology of the abnormal bone is key to macroscopic evaluation. Since variables associated with bone change are essential components leading toward possible diagnosis, recognition and recording of all known variables is mandatory. As an example, abnormal bone formation is recognizably complex. New bone tissue might be composed of woven or compact bone, requiring the type of deposited bone to be differentiated and recorded. The pace of formation also varies, resulting in the development of different organizational structures (e.g. plaques of bone tissue deposited over well-organized compact bone, or rapidly formed spicules of bone). The point of origin of bone formation serves as another variable, as endosteal surfaces, cortical structures, and/or periosteal surfaces can be impacted in isolation or serve as the starting point of diffuse change.

Creating a Detailed Bone Inventory

Recognizing the presence of pathological lesions in bone and understanding disease processes requires documenting the presence and condition of all bones and bone fragments in the skeletal sample under investigation. Reports of the absence or low frequency of a particular pathological condition within an individual or group is only meaningful if the investigator has shown that requisite bones, or aspects of requisite bones, were available for examination in all members of the population. Low frequency rates within a population may simply be due to the repeated failure of the anatomical feature to have survived or to have been recovered. As another example, recording that a left humerus was present for evaluation is meaningless to an investigator seeking to explore the frequency of degenerative joint disease unless the presence and condition of all articular facets was also recorded. Buikstra and Ubelaker (1994) and Brickley and McKinley (2004) offer ways in which components of bone can be evaluated and recorded, allowing researchers to evaluate the frequency of conditions within a population and/or allowing comparisons between archaeological populations to be made.

Inclusion of Demographic Information

An increased understanding of the role of sex and gender in disease processes (Grauer and Stuart-Macadam, 1998), along with the development of population approaches within palaeopathology, have necessitated our inclusion of age at death and sex determination in skeletal analysis. Conditions such as iron-deficiency anaemia have been recognized to affect skeletal tissue differently depending on age of onset (Stuart-Macadam, 1985), and reproductive capacities can directly (as in cases of malignancy of reproductive organs) and indirectly (as in the predilection of gout in males or osteoporosis in women) affect the presence of diseases. It is essential, therefore, to collect and incorporate demographic information into the macroscopic evaluation, as it serves as a vital component for differential diagnosis and cross-population comparisons. The assessment of age at death and sex, however, must include the use of as many techniques as possible and address, when appropriate, the potential effects of disease process on growth and development.

Appreciation of Multiple Conditions

The tendency to link the presence of macroscopic lesions with singular causes or pathogens is common within palaeopathology. However, both the limited pathognomonic response of bone tissue and the synergistic interactions between diseases often render this goal simplistic and misleading. As an example, many diseases/conditions provoke osteoblastic activity, obscuring our ability to determine the specific pathogen or condition responsible for the bone change. Equally confounding are the synergistic effects between disease processes. Parasitic or chronic infection may trigger iron-deficiency anaemia, whereas the presence of iron-deficiency anaemia (or low serum iron levels) might protect individuals from other pathogenic infections. There is also the potential for a single individual to have several concurrent diseases or conditions. An appreciation of possible complexities lessens the tendency to be oversimplistic and increases the power of differential diagnosis.

Use of Differential Diagnosis

An important component of macroscopic analysis is the use of differential diagnosis. Creating an exhaustive list of potential causes of a lesion is a reasonable starting point. Included on the list must be taphonomic processes that potentially mimic *in vivo* bone change. Adding demographic data, as well as the archaeological and environmental context of the specimen, provide further evaluative tools. The first step toward differential diagnosis is effectively to argue that the bone alteration under investigation is not due to post-mortem processes (Chapter 2). This may require radiographic and/or microscopic investigation, as well as biochemical tests for diagenesis. Conservatively, only after post-mortem changes have been ruled out should further steps towards differential diagnosis be taken. The next step requires adopting clinically created and palaeopathologically supported criteria for the presence of disease. The former, based upon a wide range of test results performed on the genetic, cellular, tissue and system levels, serves as a base line for diagnosis in clinical settings. With only bone tissue available for evaluation in most archaeological samples, the adoption of clinical criteria is often impossible. Palaeopathologically supported criteria of disease seek to unite clinical and palaeopathological research by finding common ground through histological evaluation, microscopy, and a range of imaging techniques (Aufderheide and Rodriguez-Martin, 1998; Ortner, 2003; Schultz, 2001, 2003). While incorporating these approaches can lead to a narrowing of possible aetiologies, they often leave the researcher with more than a single possible cause. Waldron (1994) and Ortner (2003) aptly suggest that palaeopathologists relinquish the tendency to assert the presence of specific diseases derived from macroscopic analysis and to utilize broader disease categories (e.g. metabolic disorder or arthropathy) when warranted.

Use of Multiple Lines of Inquiry

Similar to the diagnosis of disease in clinical settings where multiple tests are conducted and a variety of data are accumulated prior to diagnosis, palaeopathological analysis of skeletal lesions on dry bone must adopt multiple lines of inquiry whenever possible. To begin this process, determine the feasibility or appropriateness of different approaches. For instance, is the preservation of the skeletal material adequate for analysis? The use of some techniques may require that minimal post-mortem damage has occurred to the bone. Imaging techniques, along with the analysis of isotopes, trace elements, and/or DNA from bone tissue, can be compromised by diagenesis and contamination (Chapters 1, 5 and 8). Next, weigh the promise of the techniques adopted against the costs (both financial and material). Running technologically complex tests can cost a considerable sum. Is the information sought worthy of the expense? Perhaps more controversial is to pose the question of whether the information sought is worth causing destruction to the sample. While there is no absolute answer to these questions, it is the investigator's responsibility to stand accountable for decisions made. Last, use strong inference, since hasty associations between macroscopic lesions and variation in bone microstructure, chemistry, or genetics can be misleading. Have all possible explanations for the association or variation been explored? Does only one explanation stand as a possibility? With the adoption of these steps, the incorporation of multiple lines of inquiry, alongside macroscopic analysis, can provide new and robust directions for palaeopathological diagnosis and interpretation.

APPLYING TECHNIQUES

A number of examples of new methodological directions in macroscopic analysis can found in the palaeopathology literature. Many of these stem from repeated efforts to improve the diagnosis of macroscopically recognizable lesions (Mays, 2001, 2005; Santos and Roberts, 2001; Brickley *et al.*, 2005; Matos and Santos, 2006; Mays *et al.*, 2006). They result in nuanced understandings of disease processes in humans and aid in our attempts to understand the presence and evolution of human disease.

New methodological directions are also evident in studies adopting decidedly biocultural and/or evolutionary perspectives. Palfi *et al.* (1999), for instance, based on the contributions of many researchers, offer a compendium of work focusing on the multifaceted approaches towards understanding tuberculosis. Here, the authors compile research on radiological, biomolecular, histological, epidemiological, historical, and palaeopathological accounts of the disease. More recently, and more synthetically, Roberts and Buikstra (2003) offer an integration and interpretation of archaeological, historical, and clinical data on tuberculosis. Successfully integrating detailed descriptions of skeletal lesions with discussion of the systemic effects of the disease, and incorporating demographic data with differential diagnosis, the authors exemplify how multiple lines of inquiry can begin to provide a 'global view' of a disease.

Efforts to understand and interpret the presence of porotic lesions of the cranium (frequently referred to as porotic hyperostosis) have also provided strength to the macroscopic analysis of bone. Schultz (2001, 2003), utilizing histological examination and clinical data alongside macroscopic analysis, elucidated a wide range of medical conditions capable of creating porotic lesions of the skull and offered new means for differential diagnoses. These insights, coupled with work specifically focusing on macroscopic, radiographic and microscopic (scanning electron microscope) skeletal manifestations of scurvy found in the Americas (Ortner and Ericksen, 1997; Ortner *et al.*, 1999, 2001), have led to the recognition of this condition and, equally important, its ramifications in the Old World (Roberts and Cox, 2003; Brickley and Ives, 2006).

The growing emphasis on global and evolutionary perspectives within palaeopathology has led to an increased need and more common adoption of standardized recording measures. Two efforts in particular illustrate the potential of these initiatives. First, the 'Western Hemisphere Project', in its efforts to integrate environmental, socio-economic, and biological variables towards an understanding of human health and nutrition over the past 7000 years, argued for the circumscribed recording of seven skeletal conditions, which were then used to construct a health index (Steckel and Rose, 2002). Deliberate measures to train participants in the recognition and recording of the skeletal conditions were undertaken to ensure uniform data collection. Although the limitations and oversimplifications have been a source of concern to many, including to those involved in the project, the potential for collaboration, the development of a substantial database, along with opportunities and commitments to refine and re-evaluate the methods used for interpretation renders this project monumental in the development of bioarchaeological and palaeopathological method and theory.

Another ambitious effort was undertaken by Roberts and Cox (2003). Here, the authors chronicle health and disease in Britain over a 10 000-year span. Using the work of many researchers over many years, they recoded and re-evaluated the presence of macroscopic lesions. Although issues of data compatibility were undeniable obstacles, the authors' determination to develop a new synthesis of previously published and unpublished data allowed for an unprecedented examination of the relationships between human disease and biosocial environments.

PROBLEMS AND ISSUES

Have we solved all our problems? Do we stand as a unified body of researchers capable of effectively and efficiently recognizing, diagnosing, interpreting, and communicating our ideas to others? Can we regularly and successfully share our data in an effort to answer new questions and adopt new approaches? Unfortunately, the answer is 'No'.

Differing 'Environments'

The environment within which palaeopathological analysis takes place today may affect macroscopic evaluation and data collection. No longer a physician's hobby, skeletal analysis in some countries is conducted by researchers working in two arenas: academia/medicine, and 'contract' work or 'applied palaeopathology'. In the first instance, researchers in palaeopathology are driven to create research agendas which are fundable, publishable, and problem oriented. That is, the researcher chooses a pertinent problem or question and seeks to find the answer. In the second instance, skeletal analysis is conducted under contractual obligation, with success measured by the collection of data and the creation of a database. As seen in Table 4.1, a number of differences between palaeopathological work completed within academia and/or medicine and 'under contract' or 'applied palaeopathology' exist.

The repercussions of these differences have noticeable impacts on macroscopic analyses and data collection. Within academe, the researcher must be careful to focus closely on a stated problem/question and streamline the data presented (and collected). Reviewing peers must agree that the data and analyses provided to the reader substantially support the conclusion, and 'extraneous' information is uniformly excluded. As noted by Wright and Yoder (2003: 56):

Table 4.1 Differences between palaeopathological work in academia and applied palaeopathology

Palaeopathology within academia/medicine	Palaeopathology 'under contract' (applied palaeopathology)
Greater emphasis on problem-oriented research and seeking answers to set questions	Emphasis on thorough data collection
Skeletal variables investigated are defined by researcher and limited by the questions being posed	As many skeletal variables as possible are examined and recorded
Time-line of project usually created by researcher	Time restraints often guided by external contractual obligations
Successful in gaining permission and acquiring funds to engage in invasive and/or technologically complex techniques	Rarely successful in gaining permission to alter skeletal material and/or to use technologically complex techniques

> Many palaeopathological studies focus on a single indicator of health, such as enamel defects, stature, or porotic hyperostosis. In part, this narrow focus is facilitated by the constraints of journal publishing formats, which encourage us to carve up large projects into manageable units.

Hence, problem-oriented studies, which often limit the variables under investigation at the onset of the project, might provide even less information to the reader due to publication restraints. Conversely, applied palaeopathological work may appear truncated and/or appended to archaeological site reports; and when it is allowed to be synthesized with archaeological and cultural information, it can remain unpublished or published without peer review. Arguably, each difference impacts macroscopic analysis and data collection. Importantly, the dichotomy is not necessarily created by work conducted by two separate groups of individuals, but rather is created by the impetus for the study.

Changing Criteria

As if the tensions and limitations created by research conducted within two different professional arenas are not enough of a challenge to palaeopathology, macroscopic analysis and data collection is also strapped by its greatest asset: progress. The analysis of degenerative joint disease (DJD) serves as a good example of this conundrum.

The recognition of DJD in archaeological human populations is almost a century old (e.g. Ruffer and Rietti, 1912). Years later, assisted by the use of clinical manifestations of the condition, Jurmain (1975, 1977) and Steinbock (1976) offered means of recording and identifying DJD by creating ordinal scales, or recognizable 'degrees' of severity. In 1981, Ortner and Putschar (1981: 420), while being careful to assert that 'marginal lipping can represent a feature of joint remodeling without degenerative cartilage changes', declared that 'for our purpose in the study of dry bone, it [marginal lipping] has to be included in the manifestations of degenerative joint disease'. Hence, DJD in skeletal populations became identified by marginal lipping (and/or other bony changes, such as eburnation) on one or more articular surfaces of the skeleton (as undertaken by Owsley *et al.* (1987) and Bridges (1991)).

More recently, Waldron and Rogers (1991) and Rogers and Waldron (1995) assert that:

> the palaeopathological diagnosis of OA [osteoarthritis, which is frequently used as a synonym for DJD] should be simple and straightforward; it depends first and foremost on demonstrating the presence of eburnation. Where eburnation is absent, then we suggest that it should be diagnosed only when at least two of the following are present: marginal osteophyte and/or new bone on the joint surface; pitting on the joint surface; or alteration in the bony contour of the joint (Rogers and Waldron, 1995: 43–44).

Importantly, they warn that 'OA must never be diagnosed if marginal osteophytosis is the only abnormality...' (Rogers and Waldron, 1995: 44). Thus, researchers who might, after careful consideration and research, have recorded prevalences of marginal lipping on multiple articular surfaces, and who have used an ordinal scale of severity based upon prior published methods, might find now that, at best, they will need to re-evaluate their data in order to determine the frequency and pattern of DJD in their populations or that, at worst, they may find they failed to collect the requisite information to make any assertions.

Conflation and Confusion

Yet another issue rendering macroscopic evaluation and data collection problematic is the tendency within palaeopathology to conflate and confuse description and diagnosis. Wells (1964: 36) was quick to recognize these problems and suggested that 'to reduce this chaotic rabble of pathology into an orderly scheme of disease some form of classification must be devised'. He provided two proposals: to group conditions based upon anatomy and physiology, which creates rubrics 'diseases of the skin, diseases of bone, and cardiovascular, ophthalmic or neurological lesions'; or 'to group diseases according to their cause: congenital abnormalities, infection, allergy, and other basic principles' (Wells, 1964). While both systems, he found, have strengths and weaknesses, he chose to organize his book based on the underlying cause of a disease (i.e. congenital abnormalities, injury, infection, degenerative conditions, etc.). Clearly, the strength of this approach is in allowing researchers to evaluate and compare frequencies of conditions within and between populations, and allowing new questions about human disease to be posed. The drawback to this system is that, along with the assessment of the presence of abnormal bone tissue, an integrated judgment involving the cause of the lesion is asserted. Data collected using this approach record the presence/absence, severity, and location of the already-diagnosed lesion. If the criteria upon which we make our diagnoses change (as discussed above), then the recorded data may prove to be useless to future researchers.

Recognizing this dilemma, Buikstra and Cook (1980: 442) suggest that:

> Data collection strategies should be based upon careful and objective description of abnormal bony remodelling, avoiding diagnostic labels. Although the investigator may emphasize certain conditions, such as periostitis and osteomyelitis, in most cases more generalised description of pathologic change is preferable.

Staunchly agreeing with this approach, and offering considerable guidance and discussion on the subject, Ortner and co-workers (Ortner and Putschar, 1981; Ortner, 1990, 2003; Ortner and Aufderheide, 1991) provide researchers with substantial bases upon which macroscopic evaluation and data collection can be built.

However, even guidelines created for objective description of bone change are limited and may lead to conflation and confusion. As an example, skeletal lesions categorized as porotic hyperostosis in Buikstra and Ubelaker (1994: 115) are included in their skeletal pathology code key as a separate category from all other bone processes and conditions. Limiting the lesion to the frontal, parietal, and occipital bones, they ask the researcher to record the 'degree' of the lesion (barely discernable, porosity only, porosity with coalescing foramina, and coalescing with diploic thickening), as well as whether the lesions are active or healed. These specific variables were based upon strong evidence that the presence of 'porotic hyperostosis', a term offered by Angel (1966), was associated with hereditary haemolytic anaemia (Angel, 1966, 1967, 1978), as well as with acquired iron-deficiency anaemia (Carlson *et al.*, 1974; El-Najjar, 1976; Stuart-Macadam, 1991, 1992).

The development of recordable variables based upon limited diagnostic criteria has led inadvertently to the conflation of diagnosis and description in the palaeopathological literature. The descriptive term 'porotic hyperostosis' can be found used *de facto* to indicate the presence of iron-deficiency anaemia, as offered by Larsen (1997: 30): 'The skeletal changes associated with iron deficiency anaemia are part of a generalised syndrome called porotic hyperostosis'. While Larsen is careful to insert the caveat 'are part of a generalised syndrome', to many researchers who have presented their work at meetings, and have published in site reports and papers, the term 'porotic hyperostosis' has become unequivocally associated with iron-deficiency anaemia. However, more recent work, indicating that porous lesions of the ectocrania can be caused by infection, metabolic disorders, tumours, haemorrhagic processes, and cranial deformation (Ortner and Putschar, 1981; Aufderheide and Rodriguez-Martin, 1998; Schultz, 2001, 2003; Ortner, 2003), requires a re-examination of the variables and diagnostic criteria to be adopted by researchers prior to interpretation.

Clearly, the conflation of descriptive terminology with absolute cause is dangerous. It fails to follow the careful processes needed for differential diagnosis (in fact, it skips over differential diagnosis completely). The end result might be an accumulating body of research with limited comparative ability and questionable diagnoses.

Static Models and Evolutionary Processes

Germane to palaeopathology is the assumption that macroscopic skeletal manifestations of today's diseases are similar to those in antiquity. Diagnostic sophistication within palaeopathology has been achieved through careful correlation of bone change in clinical populations with similar recognizable change in archaeological populations. However, if we model the pathogenesis of particular diseases on modern population studies, we ignore the large body of work within medicine, immunology, parasitology, epidemiology, and genetics which indicate that pathogens, along with host-pathogen interactions, have profoundly changed over time. Although circumventing this issue might be impossible, acknowledging that environmental, immunological, genetic, and nutritional differences between past and present populations might influence pathogenesis can serve as new directions for research into the presence and influence of past diseases.

Thoroughness Has Its Limitations

As aptly explained in this volume, there are a growing number of methodological considerations in palaeopathological investigation. The impact on the field is great. For macroscopic analysis, the inclusion of new imaging techniques along with biochemical and genetic analyses has allowed researchers to diagnose more robustly the presence of particular diseases and to become more sensitive to disease – environment interactions. However, there are substantial limitations to the widespread adoption of these techniques within palaeopathology (Roberts, 2002). First, multiple imaging techniques and chemical and/or genetic study of bone tissue are expensive analytical methods. Palaeopathologists' budgets rarely support the implementation of these methods on more than a few individuals within a population. Second, a number of the new methodological approaches require substantial time for completion to ensure that variables are adequately controlled and replicability is possible. Third, many of the new methodologies require specialized technology and expertise in preparing specimens, running protocols, and interpreting results. Rarely will a palaeopathologist have the budget, time or expertise to implement many different methods on all individuals within the population under investigation.

LOOKING TOWARDS THE FUTURE

In spite of the limitations and frustration surrounding efforts to describe clearly, carefully record, and accurately diagnose skeletal lesions, palaeopathology has taken great strides in bringing to light the lives of our ancestors. As macroscopic analysis will likely continue to serve as the premier means by which palaeopathological research is conducted, it is essential that efforts to improve macroscopic evaluation are equal to our efforts to refine, improve, and develop new technologically complex techniques. Towards this goal, a few suggestions are offered here.

Plan to utilize multiple lines of evidence. Balancing the promise of including new techniques and methods into macroscopic analysis with recognized budget and time limitations is difficult. It is not impossible. As practitioners within our field more repeatedly and effectively argue that incorporating multiple methodologies alongside macroscopic analysis is essential (not tangential) in palaeopathological analysis, the more likely it is to become routine. An *a priori* argument for the need of multiple imaging techniques, trace element analysis or ancient DNA, for example, based on a preliminary understanding of the population to be excavated, might be more influential when seeking funding for skeletal analysis than asking for financial support after macroscopic analysis has been completed and questions arise from it.

Focus on standardized data collection. The contributions of Buikstra and Ubelaker (1994), Rose *et al.* (1991), Brickley and McKinley (2004), and Ousley *et al.* (2006) toward developing standards by which skeletal data can be collected have deeply impacted the discipline. Not only have the studies offered means for structure and consistency in data collecting within the field, they have initiated a dialogue between researchers that continues to shape the field today. The overwhelming amount of data to be collected, as well as the concrete suggestions guiding researchers in what and how data might be collected, are current topics of discussion, sources of concern, and focal points of debate. The contributions, therefore, can be measured not only by the comprehensive nature of the endeavours, but also by the continuing dialogue that they have ignited. Clearly, efforts to facilitate discussion and revision of techniques used within the field must include the voices of colleagues worldwide, and must strive to keep our field dynamic. Perhaps, most importantly, the development of standards for data collection is only useful if all skeletal analysts have access to the methods and techniques.

We need to do our homework. Most researchers would agree that a first step toward answering a question is to explore previously published material on the subject. All PhD dissertations and most published journal articles place the investigator's question into context by reviewing the work that has been conducted in the past. A skeletal report ought, likewise, to contain information on the origin of selected techniques adopted by the investigator. While researchers will often explore the frequency and patterns of the disease under investigation cited in publications, or will adopt techniques used by other researchers, rarely will an investigator seek to find out how data were recorded and evaluated. That is, they fail to seek out how the previous researchers recorded lesions, what scores and scales were created and implemented, what criteria were used, what variables were recognized, and what level of reproducibility can be met.

For example, in their attempt to offer concrete guidelines for the collection of human skeletal data, Mann and Murphy (1990: 13) make the following suggestion:

> If, after examining ten or fifteen skeletons, you find that you have been scoring a lesion as moderate in severity and feel that it should only be scored as slight, then you should modify the criteria for the trait or lesion in question.

I would argue that if an investigator has to change the criteria of data collection midway through analysis, then the investigator did not do their homework. Developing criteria in a vacuum, let alone changing criteria mid project, easily renders the data and subsequent conclusions unusable to the investigator and to any other interested researchers. Ortner and Aufderheide (1991), Grauer (2002), and Roberts and Cox (2003) recount these issues in their efforts to compare populations across time and space, and as they attempt to correlate the work of many researchers.

Efforts to ensure that comparisons between studies and between populations are possible in the future ought to include detailed descriptions or photographing examples of each level of lesion severity denoted, or each type of lesion classified, or each variable delineated in the investigator's macroscopic analysis. As offered by Robb (2000: 479):

> In designing a skeletal research project, to avoid finishing empty handed or having to repeat data collection, data analysis must be built into research designs from the start. This is all the more so since skeletal research almost always requires striking a practical balance between the number of specimens studied, the amount of detail recorded, and research time and money.

Err on the side of caution. The tendency among researchers to label or diagnose skeletal lesions prematurely or inaccurately is unfortunately common (Ortner and Aufderheide, 1991; Waldron, 1994; Miller *et al.*, 1996; Lovell, 2000). However, with palaeopathologists' increasing emphasis on non-specific conditions which impact human health, interests have shifted away from documenting the antiquity of specific diseases towards understanding disease processes, the evolution of disease in human groups, and the impact of human culture on human biology. Within these research agendas all abnormal skeletal changes (even those not labelled as pathological by the medical community) have the potential, eventually, to provide insight into the past.

Hence, for the palaeopathologist, carefully recognizing and recording even slight bone changes and abnormalities becomes an essential component of evaluation. This means that palaeopathologists cannot simply record conditions with which they are familiar, or focus on conditions that they can allegedly quickly diagnose. They cannot choose to minimize the presence of variation within their population by creating categories of lesions that are anatomically, physiologically or pathogenically illogical. Rather they must get in the habit of carefully describing what they see. With amazing data collection and data storage technology at our fingertips, we need not fear collecting too much information. Ortner and Aufderheide (1991: 8) assert that a goal is to:

> develop and apply a descriptive methodology to the analysis and publication of pathological specimens that does not necessarily preclude classification or diagnosis but, at the very least, permits the reader to reach his or her own conclusion regarding the nature of the pathology, without having to accept the diagnostic opinion of the author.

Consider collaboration. Collaboration in the sciences is not novel, it is *de rigueur.* As robust macroscopic evaluation of skeletal remains benefits from employing multiple lines of evidence, so too will palaeopathological research and publication benefit from the inclusion of many specialists contributing throughout the analysis and publication process. Individuals with a wide range of specializations must be routinely asked to participate in palaeopathological projects, and not be called upon as an occasional resource under particular circumstances. Research in South America showcases the promise of collaboration, as scientists across the globe, specializing in many different fields, work to synthesize data in an effort to create a holistic, nuanced and evolutionary perspective of life in Peru, Chile, and Brazil.

Publish with an eye for comparison. The success of our discipline relies on our ability to communicate and to provide a hefty foundation of data and ideas for future palaeopathological research. This is only accomplished if the ideas contained within our missives are thoroughly researched, robustly argued, and based upon sound, reproducible data. In spite of the tendency for skeletal reports to be truncated and/or placed into appendices, and in spite of the tendency for journal publishing formats to encourage us to carve up large projects into manageable units, it remains our responsibility to get the message across. Seeking to synthesize the maximum archaeological and cultural data with pathological analysis must become a greater priority for palaeopathologists asked to analyse skeletal collections. Similarly, including more detail and a broader scope must become a greater priority in our journal submissions, as this will allow investigators worldwide to develop cross-cultural comparisons.

A recent analysis conducted by the Paleopathology Association (Stodder *et al.*, 2006) highlights the controversy over the utility and need for the publication of descriptive case studies (Armelagos and van Gerven, 2003; Larsen 2005; Stojanowski and Buikstra, 2005) and brings to light the scope and frequency of palaeopathological research published in journal format. Palaeopathological research, they found, appeared in 286 different journals from 1996 to 2005, with 15 journals publishing 10 or more articles on the subject (Stodder *et al.*, 2006: 9). Clearly, palaeopathological research can be interpreted from these findings as contributing to a variety of disciplines and scientific fields.

CONCLUSION

There is no easy fix to rid palaeopathology of its difficulties and stumbling blocks. Advancements in technology will not supersede the need to inspect human remains macroscopically, record findings, and communicate conclusions. It is equally unlikely that the problems inherent in the visual examination of skeletons, common in data collection, and impacting communication and cooperation, will disappear if ignored. This being the case, the onus is on us, the palaeopathologists. We must focus on clear description, develop and use standardized recording techniques, share our information, familiarize ourselves with recording techniques used in comparative samples, employ many analytical techniques, and maintain a dialogue focused on the importance of macroscopic evaluation and data collection within our discipline.

REFERENCES

Angel JL. 1966. Porotic hyperostosis, anemias, malarias, and marshes in prehistoric eastern Mediterranean. *Science* **153**: 760–763.

Angel JL. 1967. Porotic hyperostosis or osteoporosis symetrica. In *Diseases in Antiquity*, Brothwell D, Sandison AT (eds). Charles C. Thomas: Springfield, IL; 378–389.

Angel JL. 1978. Porotic hyperostosis in the eastern Mediterranean. *Med Coll Virginia Q* **14**: 10–16.

Armelagos GJ. 1997. Palaeopathology. In *History of Physical Anthropology*, Vol. 2, Spencer F (ed.). Garland: New York; 790–796.

Armelagos GJ, van Gerven DP. 2003. A century of skeletal biology and palaeopathology: contrasts, contradictions, and conflicts. *Am Anthropol* **105**: 53–64.

Armelagos GJ, Carlson DS, van Gerven DP. 1982. The theoretical foundation and development of skeletal biology. In *A History of American Physical Anthropology, 1930–1980*, Spencer F (ed.). Academic Press: New York; 305–328.

Audy R. 1967. Measurement and diagnosis of health. In *Environ/Mental: Essays on the Planet as a Home*, Shepard P, McKinley D (eds). Houghton Mifflin: Boston, MA.

Aufderheide AC, Rodriguez-Martin C. 1998. *The Cambridge Encyclopedia of Human Paleopathology*. Cambridge University Press: Cambridge.

Brickley M, Ives R. 2006. Skeletal manifestations of infantile scurvy. *Am J Phys Anthropol* **129**: 163–172.

Brickley M, McKinley JI. 2004. *Guidelines to the Standards for Recording Human Remains*. IFA Paper Number 7. BABAO: Southampton.

Brickley M, Mays S, Ives R. 2005. Skeletal manifestations of vitamin D deficiency osteomalacia in documented historical collections. *Int J Osteoarchaeol* **15**: 389–403.

Bridges PS. 1991. Degenerative joint disease in hunter-gatherers and agriculturalists from southeastern United States. *Am J Phys Anthropol* **85**: 379–391.

Brothwell D. 1967. The bio-cultural background to disease. In *Diseases in Antiquity*, Brothwell D, Sandison AT (eds). Charles C. Thomas: Springfield, IL; 56–68.

Brothwell D, Sandison AT. 1967. Editorial prolegomenon: the present and future. In *Diseases in Antiquity*, Brothwell D, Sandison AT (eds). Charles C. Thomas: Springfield, IL; xi–xiv.

Buikstra JE, Beck L. 2007. *Bioarchaeology: The Contextual Analysis of Human Remains*. Elsevier: San Diego, CA.

Buikstra JE, Cook DC. 1980. Palaeopathology: an American account. *Ann Rev Anthropol* **9**: 433–470.

Buikstra JE, Ubelaker DH. 1994. *Standards for Data Collection from Human Skeletal Remains*. Arkansas Archaeological Survey Research Series No. 44. Arkansas Archaeological Survey: Fayetteville, AR.

Carlson DS, Armelagos GJ, van Gerven DP. 1974. Factors influencing the etiology of cribra orbitalia in prehistoric Nubia. *J Hum Evol* **3**: 405–419.

El-Najjar MY. 1976. Maize, malaria and the anemias in the Pre-Columbian New World. *Yrbk Phys Anthropol* **20**: 329–337.

Goodman AH, Martin DL. 2002. Reconstructing health profiles from skeletal remains. In *The Backbone of History: Health and Nutrition in the Western Hemisphere*, Steckel RH, Rose JC (eds). Cambridge University Press: Cambridge; 11–60.

Grauer AL. 2002. Taking the trauma out of trauma: a callous approach. In *Annual Meeting of the Midwest Bioarchaeology and Forensic Anthropology Association*, Indianapolis, IN.

Grauer AL, Stuart-Macadam P. 1998. *Sex and Gender in Paleopathological Perspective*. Cambridge University Press: Cambridge.

Hooton EA. 1930. *The Indians of Pecos Pueblo: A Study of their Skeletal Remains*. Yale University Press: New Haven, CT.

Huss-Ashmore R, Goodman A, Armelagos GJ. 1981. Nutritional inference from palaeopathology. Adv *Archaeol Method Theory* **5**: 395–474.

Janssens PA. 1970. *Palaeopathology: Diseases and Injuries of Prehistoric Man*. Humanities Press: New York.

Jarcho S. 1966a. *Human Paleopathology*. Yale University Press: New Haven, CT.

Jarcho S. 1966b. The development and present condition of human palaeopathology in the United States. In *Human Palaeopathology*, Jarcho S (ed.). Yale University Press: New Haven, CT; 3–30.

Jurmain RD. 1975. *Distribution of degenerative joint disease in skeletal populations*. PhD dissertation, Harvard University, Cambridge, MA.

Jurmain RD. 1977. Stress and the etiology of osteoarthritis. *Am J Phys Anthropol* **46**: 353–366.

Kerley ER, Bass WM. 1967. Paleopathology: meeting ground for many disciplines. *Science* **157**: 638–644.

Larsen CS. 1997. *Bioarchaeology: Interpreting Behavior from the Human Skeleton*. Cambridge University Press: Cambridge.

Larsen CS. 2005. Description, hypothesis testing, and conceptual advances in physical anthropology: have we moved on? In *Annual Meeting of the American Association of Physical Anthropologists*, Milwaukee, WI.

Lovell NC. 2000. Paleopathological description and diagnosis. In *Biological Anthropology of the Human Skeleton*, Katzenberg MA, Saunders SR (eds). Wiley-Liss: New York; 217–248.

Manchester K. 1983. *The Archaeology of Disease*. Bradford University Press: Bradford.

Mann RW, Murphy SP. 1990. *Regional Atlas of Bone Disease*. Charles C. Thomas: Springfield, IL.

Matos V, Santos AL. 2006. On the trail of pulmonary tuberculosis based on rib lesions: results from the human identified skeletal collection from the Museu Bocage (Lisbon, Portugal). *Am J Phys Anthropol* 130: 190–200.

Mays SA. 2001. Paleopathological and biomolecular study of tuberculosis in a medieval skeletal collection from England. *Am J Phys Anthropol* **114**: 298–311.

Mays SA. 2005. Paleopathological study of hallux valgus. *Am J Phys Anthropol* **126**: 139–149.

Mays SA, Brickley M, Ives R. 2006. Skeletal manifestations of rickets in infants and young children in a historical population from England. *Am J Phys Anthropol* **129**: 362–374.

Miller E, Ragsdale B, Ortner D. 1996. Accuracy in dry bone diagnosis: a comment on paleopathological methodology. *Int J Osteoarchaeol* **6**: 221–229.

Møller-Christensen V. 1961. *Bone Changes in Leprosy*. Munksgaard: Copenhagen.

Møller-Christensen V. 1967. Evidence of leprosy in earlier peoples. In *Diseases in Antiquity*, Brothwell D, Sandison AT (eds). Charles C. Thomas: Springfield, IL; 295–306.

Møller-Christensen V. 1974. Changes in the anterior nasal spine and the alveolar process of the maxillae in leprosy. *Int J Leprosy* **42**: 431–435.

Møller-Christensen V. 1978. *Leprosy Changes of the Skull*. Odense University Press: Odense.

Moodie RL. 1923. *Paleopathology: An Introduction to the Study of Ancient Evidences of Disease*. University of Illinois Press: Urbana, IL.

Ortner DJ. 1990. Introduction. In *Regional Atlas of Bone Disease*, Mann RW, Murphy SP, (eds). Charles C. Thomas: Springfield, IL; ix–xii.

Ortner DJ. 1991. Theoretical and methodological issues in paleopathology. In *Human Paleopathology: Current Syntheses and Future Options*, Ortner DJ, Aufderheide AC (eds). Smithsonian Institution Press: Washington, DC; 5–12.

Ortner DJ. 1992. Skeletal palaeopathology: probabilities, possibilities, and impossibilities. In *Disease and Demography in the Americas*, Verano JW, Ubelaker DH (eds). Smithsonian Institution Press: Washington, DC; 5–15.

Ortner DJ. 1994. Descriptive methodology in paleopathology. In *Skeletal Biology in the Great Plains*, Owsley DW, Jantz RJ (eds). Smithsonian Institution Press: Washington, DC; 73–80.

Ortner DJ. 2003. *Identification of Pathological Conditions in Human Skeletal Remains*, second edition. Academic Press: San Diego, CA.

Ortner DJ, Aufderheide AC (eds). 1991. *Human Paleopathology: Current Syntheses and Future Options*. Smithsonian Institution Press: Washington, DC.

Ortner D, Ericksen M. 1997. Bone changes in the human skull probably resulting from scurvy in infancy and childhood. *Int J Osteoarchaeol* **7**: 212–220.

Ortner DJ, Putschar WGJ. 1981. *Identification of Pathological Conditions in Human Skeletal Remains*. Smithsonian Contributions to Anthropology, No. 28. Smithsonian Institution Press: Washington, DC.

Ortner D, Kimmerle E, Diez M. 1999. Probable evidence of scurvy in subadults from archaeological sites in Peru. *Am J Phys Anthropol* **108**: 321–331.

Ortner D, Butler W, Cafarella J, Milligan L. 2001. Evidence of probable scurvy in subadults from archaeological sites in North America. *Am J Phys Anthropol* **114**: 343–351.

Ousely S, Dudar JC, Jones E, London M, Madden G, Mulhern D, Wilczak C. 2006. Standardizing the standards: computerized documentation of skeletal pathology at the Smithsonian Institution. In *Workshop presented at the Annual Meeting of the Palaeopathology Association*, Anchorage, Alaska.

Owsley DW, Orser CE, Mann RW, Moore-Jansen PH, Montgomery RL. 1987. Demography and pathology of an urban slave population from New Orleans. *Am J Phys Anthropol* **74**: 185–197.

Palfi G, Dutour O, Deak J, Hutas I. 1999. *Tuberculosis Past and Present*. Golden Books: Szeged.

Ragsdale B. 1992. Task force on terminology: provisional word list. *Palaeopathol Newslett* **78**: 7–8.

Robb J. 2000. Analyzing human skeletal data. In *Human Osteology in Archaeology and Forensic Science*, Cox M, Mays S (eds). Greenwich Medical Media: London; 475–490.

Roberts C. 2002. Palaeopathology and archaeology: the current state of play. In *The Archaeology of Medicine*, Arnott R (ed.). British Archaeological Reports, International Series 1046. Archaeopress: Oxford; 1–20.

Roberts C, Buikstra JE. 2003. *The Bioarchaeology of Tuberculosis*. University Press of Florida: Gainesville, FL.

Roberts C, Cox M. 2003. *Health and Disease in Britain*. Sutton: Stroud.

Rogers J, Waldron T. 1995. *A Field Guide to Joint Disease in Archaeology*. Wiley: Chichester.

Rose J, Anton SC, Aufderheide AC, Eisenberg L., Gregg JB, Neiburger EJ, Rothschild B. 1991. *Skeletal Database Committee Recommendations*. Paleopathology Association, Detroit, MI.

Rose JC, Green TJ, Green VD. 1996. NAGPRA is forever: osteology and the repatriation of skeletons. *Ann Rev Anthropol* **25**: 81–103.

Ruffer MA, Rietti A. 1912. On the osseous lesions in ancient Egyptians. *J Pathol Bacteriol* **16**: 439–465.

Sandison AT. 1967. Sir Marc Armand Ruffer (1859–1917) pioneer of palaeopathology. *Med Hist* **11**: 150–156.

Santos AL, Roberts CA. 2001. A picture of tuberculosis in young Portuguese people in the early 20th century: a multidisciplinary study of the skeletal and historical evidence. *Am J Phys Anthropol* **115**: 38–49.

Schultz M. 2001. Paleohistopathology of bone: a new approach to the study of ancient diseases. *Yrbk Phys Anthropol* **44**: 106–147.

Schultz M. 2003. Light microscopic analysis in skeletal palaeopathology. In *Identification of Pathological Conditions in Human Skeletal Remains*, second edition, Ortner DJ (ed.). Academic Press: San Diego, CA; 73–108.

Selye H. 1973. The evolution of the stress concept. *Am Sci* **61**: 692–699.

Steckel RH, Rose JC. 2002. *The Backbone of History: Health and Nutrition in the Western Hemisphere*. Cambridge University Press: Cambridge.

Steinbock RT. 1976. *Paleopathological Diagnosis and Interpretation*. Charles C. Thomas: Springfield, IL.

Stodder ALW, Johnson K, Chan A, Handwerk E, Rudolph KZ. 2006. Publishing patterns in paleopathology: findings of the Publications Explorations Committee. *Paleopathol Newslett* **134**: 6–13.

Stojanowski CM, Buikstra JE. 2005. Research trends in human osteology: a content analysis of papers published in the American Journal of Physical Anthropology. *Am J Phys Anthropol* **128**: 98–109.

Stuart-Macadam P. 1985. Porotic hyperostosis: representative of a childhood condition. *Am J Phys Anthropol* **66**: 391–398.

Stuart-Macadam P. 1991. Anaemia in Roman Britain: Poundbury Camp. In *Health in Past Societies. Biocultural Interpretations of Human Skeletal Remains in Archaeological Contexts*, Bush H, Zvelebil M (eds). British Archaeological Reports, International Series 567. Tempus Reparatum: Oxford; 101–113.

Stuart-Macadam P. 1992. Anemia in past human populations. In *Diet, Demography, and Disease: Changing Perspectives on Anemia*, Stuart-Macadam P, Kent S (eds). Aldine de Gruyter: New York; 151–170.

Thillaud PL. 1992. Retrospective diagnosis in paleopathology. *Palaeopathol Newslett* **80**(suppl): 1–4.

Waldron T, Rogers J. 1991. Interobserver variation in coding osteoarthritis in human skeletal remains. *Int J Osteoarchaeol* **1**: 49–56.

Waldron T. 1994. *Counting the Dead*. Wiley: Chichester.

Wells C. 1964. *Bones, Bodies and Diseases*. Thames and Hudson: London.

Williams HU. 1929. Human paleopathology, with some original observations on symmetrical osteoporosis of the skull. *Arch Pathol* **7**: 839–902.

Wright LE, Yoder CJ. 2003. Recent progress in bioarchaeology: approaches to the osteological paradox. *J Archaeol Res* **11**: 43–70.

5

Radiography and Allied Techniques in the Palaeopathology of Skeletal Remains

Simon Mays

English Heritage Centre for Archaeology, Fort Cumberland, Eastney, Portsmouth PO4 9LD, UK

INTRODUCTION

Imaging of skeletal lesions using plain-film radiography has been used on ancient human remains for over 100 years, and it remains the most important augment to visual examination of specimens in the description and diagnosis of disease in palaeopathology. Since the discovery of radiographic imaging in the late 19th century, many allied techniques using X-rays have been developed and used in clinical medicine. The most important development in radiographic imaging is computed tomography (CT), which is covered elsewhere in this book (Lynnerup, Chapter 6). However, as well as imaging techniques, methods involving the use of X-rays yielding quantitative data have also been developed. Measurements of cortical thickness, taken from radiographs, either manually using rulers or callipers or using image analysis with computer software, provide quantitative data on cortical bone. Taking measurements of cortical bone from radiographs is known as radiogrammetry. Bone density can also be measured radiologically. Historically, a variety of techniques have been used to do this, but currently one method, dual X-ray absorptiometry (DXA), dominates both clinical and palaeopathological bone densitometry.

This chapter discusses radiographic imaging of palaeopathological lesions and the use of radiogrammetry and bone densitometry (particularly DXA) in palaeopathology. In each case, the technique and its development in clinical science is described. Its areas of application in palaeopathology are discussed, and the issues raised by using methods and techniques developed for living patients on excavated skeletal remains are considered. Finally, some possible future directions in radiography and allied techniques in palaeopathology are discussed.

Advances in Human Palaeopathology Edited by Ron Pinhasi and Simon Mays

RADIOGRAPHIC IMAGING

Production of a radiograph requires the passing of X-rays through a specimen and the capture of the negative image that results from the attenuation of the X-rays (for details, see Lynnerup, Chapter 6). Traditionally, images have been captured on radiographic film, which is then processed chemically in a dark room to develop and fix the image. Digital image capture, which may include the facility to view the radiographic image in real time, is becoming increasingly available (Lang *et al.*, 2005). As with photography, digital image capture will doubtless in time oust film, but the substantial investment that has been made by institutions in film-based methods means that this will probably be a slower process than has been the case for photography.

Plain-film radiography produces a two-dimensional image. CT is required to generate a full three-dimensional image; but in order to at least partially reconstruct the three-dimensional appearance of a lesion using radiography, two views of the specimen are normally obtained, classically medio-lateral and antero-posterior. For further details of practical aspects of skeletal radiography of palaeopathological specimens the reader is referred to Ortner (2003: 57–62).

History of Radiography in Palaeopathology

X-rays were discovered by Roentgen in November 1895. It was clear from his initial experiments that the new rays were an effective means of visualizing bone (Fiori and Nunzi, 1995), and barely 2 months later the first clinical radiographs were being made (Spiegel, 1995). Archaeologists were quick to grasp the potential of the new rays for studying ancient human remains. As early as March 1896, radiographic images were being produced of Egyptian mummies (Böni *et al.*, 2004). Most of the early applications of radiography in archaeology were carried out on mummified remains, chiefly from Egypt, simply to discern what lay within the wrappings (Böni *et al.*, 2004). Perhaps the first publication devoted solely to a radiographic skeletal abnormality in human remains came in 1898, with a report of an abnormal number of sesamoids in the hand of an Egyptian mummy (Clendinnen, 1898). Radiography soon began to be applied not only to mummified material, but also to the study of dry bones, with the specific purpose of visualizing pathological changes (Eaton, 1916; Means, 1925; Williams, 1929). However, radiography failed to become routine in skeletal palaeopathology, and as late as 1963 Calvin Wells (1963: 401) bemoaned that it was still 'largely neglected'. In part, the neglect to which Wells refers may stem from the opinion, promulgated particularly by palaeopathologists with medical backgrounds, that radiography of specimens, and in particular diagnostic radiography, was beyond the grasp of the palaeopathologist. Wells (1963: 410) himself stated that: 'It cannot be too strongly emphasised that the interpretation of the films must invariably be the work of a professionally trained radiologist'. This view was also echoed by others during the 1960s (e.g. Sandison, 1968). This attitude reflects the status of palaeopathology at that time as a still nascent discipline 'borrowing' the skills of others, rather than a fully fledged discipline with its own academic traditions, training and skills base. This began to change from the 1970s. The Palaeopathology Association was set up as the professional organization for practitioners, and this was a vital step in the development of palaeopathology as a discipline in its own right. This period also saw the publication of the first textbooks aimed specifically at collating diagnostic criteria for the purpose of the identification of disease in ancient skeletal remains (Steinbock, 1976; Zimmerman and Kelley, 1982; Ortner and Putschar, 1985). With this avowed aim, radiography could not be ignored, nor could it be dismissed as an arcane set of skills which a palaeopathologist could not hope to acquire. These handbooks included not only illustrations of dry bone specimens, but also radiographic views of clinical cases of bone disease, of pathology museum specimens with known conditions, and of diseased archaeological specimens. Palaeopathological diagnostic criteria were formulated from a combination of both gross and radiographic features. These and subsequent texts (Aufderheide and Rodríguez-Martin, 1998; Ortner, 2003) have helped to ensure that radiography became a routine aid to description and diagnosis, and a fully integrated core skill in palaeopathology.

Methodological Issues in Radiographic Imaging in Palaeopathology

In palaeopathology, lesions need first to be adequately described, both in terms of their morphology and in terms of their distribution in the skeleton. Then, generally using recent cases of known disease as a baseline, competing diagnoses can be considered and, in some cases, a most likely cause determined. Radiography is of value in palaeopathology both in description and in diagnosis.

For all but the most superficial lesions, radiography aids the elucidation of their morphology by revealing those parts hidden by overlying bone (Figure 5.1). Lesions completely confined to the bone interior will, of course, be invisible on gross examination in intact

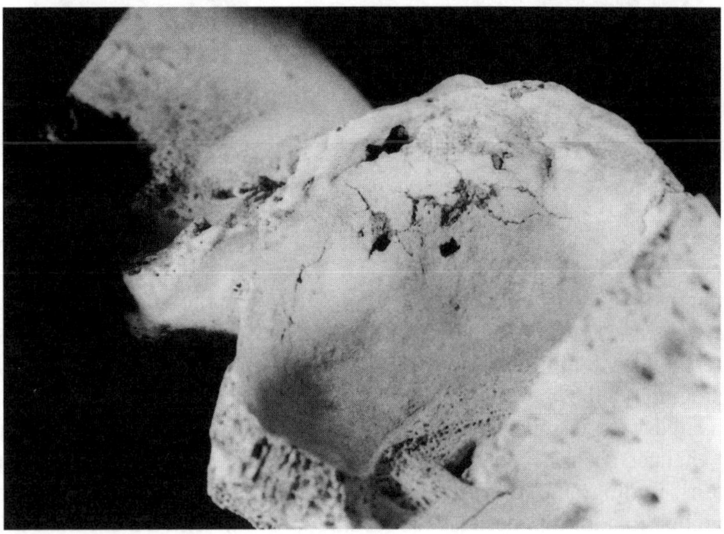

Figure 5.1 (a) The acetabular roof in this specimen from Wharram Percy shows united fissure fractures and several holes in the joint surface which penetrate the subchondral bone. (b) Radiograph of the specimen in (a), revealing that the holes in the joint surface communicate with a large lytic area within the subchondral bone. This lesion represents a supra-acetabular cyst, most probably due to passage of synovial fluid into the bone as a result of the hip injury

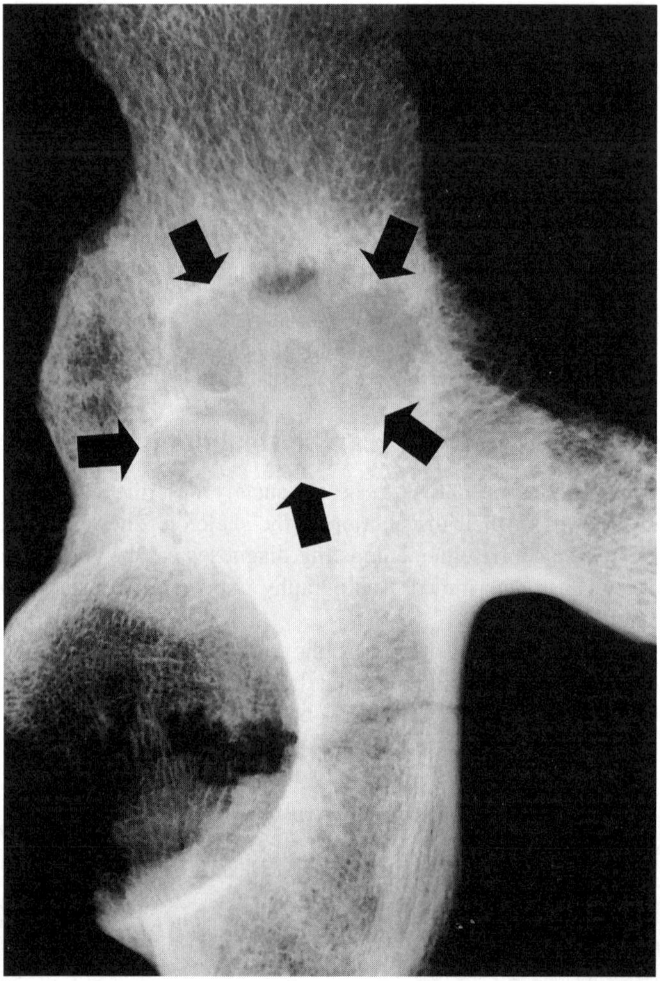

Figure 5.1 (Continued)

specimens, but may be identified on X-ray. Therefore, skeletons suspected from gross examination as having some systemic bone disease should be X-rayed in their entirety so that a full assessment of the distribution of lesions in surviving elements can be obtained (Figure 5.2).

Using radiography, the appearance of lesions seen in archaeological skeletons can be directly compared with lesion morphology in patients. Radiography provides a direct link between the changes seen in palaeopathological specimens and those in living people with known diseases. Radiographs from modern patients are useful in aiding palaeopathological interpretation; but, in addition, because the technique was invented more than 100 years ago, a substantial radiographic record of disease exists in the medical literature from the first half of the 20th century. This means that, for plain-film radiography, unlike more recently developed imaging modalities such as CT, there is a substantial published corpus depicting cases of disease whose progress was unhindered by modern treatment, and such cases more closely resemble those which we might expect to encounter in ancient skeletons. These cases

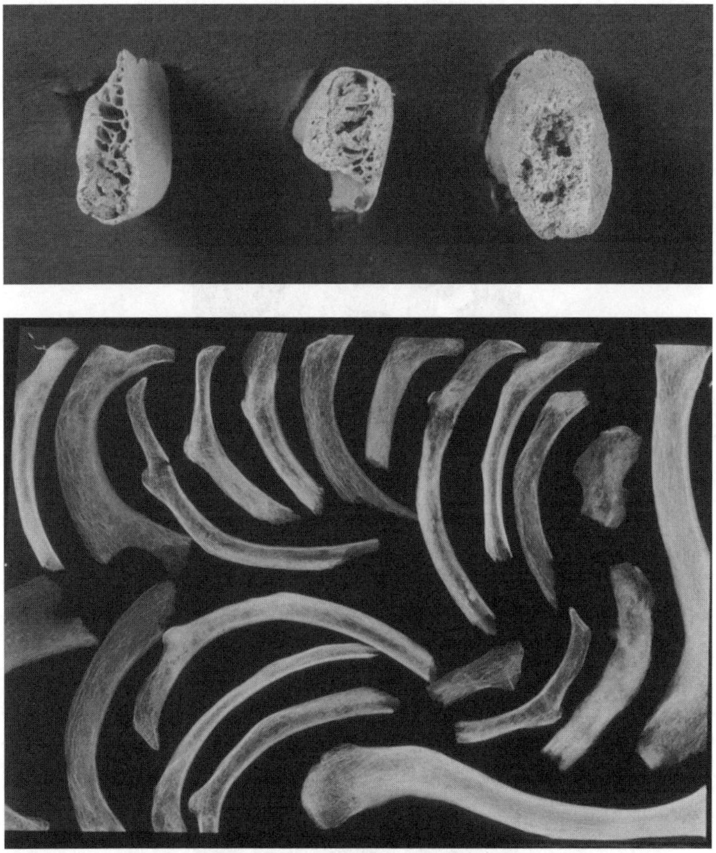

Figure 5.2 A case of metastatic carcinoma from Wharram Percy. (a) Endosteal new bone formation is visible in some rib elements via post-depositional breaks. (b) Radiography reveals much more widespread changes, with areas of patchy sclerosis in most ribs and both clavicles

are a particularly important augment to our baseline for the diagnosis of palaeopathological specimens.

Post-depositional changes in skeletal remains may cause problems in the interpretation of radiographs. Severe soil infiltration may prevent the production of diagnostically useful radiographs (e.g. Ortner and Mays, 1998). In other cases it may produce radiographic artefacts (consisting of rather 'fluffy', irregular areas of radiodensity) even if it does not obscure the whole picture (Figure 5.3). Localized soil erosion of cortical or cancellous bone may produce radiolucencies which mimic the effects of disease. Post-depositional uptake of lead may alter radiographic images, simulating diseases such as osteopetrosis, but noticeable effects on plain-film radiographs only occur when lead concentrations exceed about 1.5 %, so that this effect is only likely when lead sources lie close to the body, as, for example, in individuals interred in lead coffins (Molleson *et al.*, 1998) or buried with lead objects as grave goods. The possibility of post-depositional artefacts in palaeopathological radiographs means that they should always be examined in conjunction with the specimen itself (Wells, 1967), and the palaeopathologist should be aware of the archaeological context of the specimen.

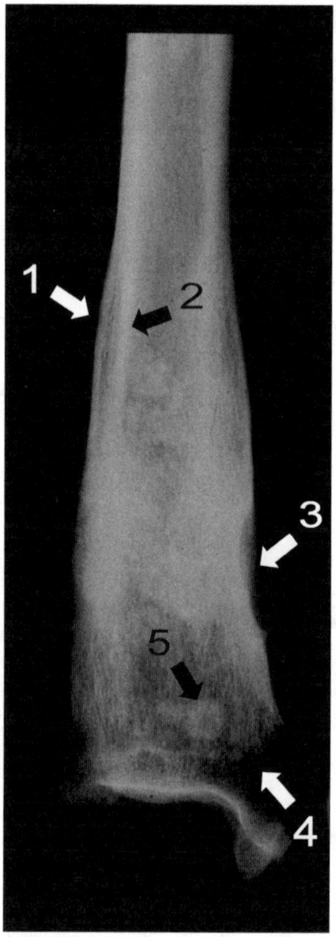

Figure 5.3 Radiograph of the distal parts of a tibia from School Street, Ipswich. There is a shell of new bone (involucrum) (1) surrounding the original cortex (sequestrum) (2), which has been partially resorbed. The radiolucency (3) is a healed cloaca. These are typical features of chronic osteomyelitis. The radiolucency (4) is a post-depositional artefact. The radio-opacities at (5) and elsewhere in the metaphysial interior and in the medullary cavity are post-depositional ingress of soil

Wells (1963: 403) stated that 'it is difficult to think of any pathological condition which is not made more intelligible when the accompanying internal bony changes are revealed'. The veracity of this statement is evident in palaeopathology textbooks, which present radiographic images illustrating changes in most major classes of disease. However, it is fair to say that radiography plays more of a role in the study of some conditions than in others. Radiography and allied techniques are clearly vital to the study of conditions such as osteoporosis (Mays, Chapter 11), which involve loss of bone mineral without change to the normal anatomical shape of the bone. Radiography is also necessary for the study of lesions that are wholly hidden within the bone, for example Harris lines (Mays, 1995; Ameen *et al.*, 2005). It is also important for the identification of diseases for which gross bony changes are obvious but non-specific. An example is Paget's disease of bone, where the gross appearance of the thickened, pitted bones can resemble a variety of other conditions (e.g. treponematosis). Although the gross specimen is difficult to interpret, the radiographic appearance is highly characteristic (Mays, Chapter 11).

Even in diseases where dry bone changes are more helpful in diagnosis, radiography, in combination with careful assessment of dry bone morphology, is usually of value. This may particularly be the case in disease conditions where clinical diagnosis is heavily based on radiographic features. Radiography has, since its introduction into clinical medicine, played a major role in the identification of the different arthropathies (Rogers and Waldron, 1995: 3), and it continues to be a major diagnostic tool in rheumatology today. It is also important in description and differential diagnosis of many types of arthritis in palaeopathology. For example, various arthropathies may cause joint ankylosis, but the type of bony union differs in different conditions. For example, both DISH and sero-negative spondylolarthropathies may cause ankylosis of the sacro-iliac joints. In DISH, bony union takes place via ligamentous ossification so that union is confined to the margins and the joint space remains intact (Rogers and Waldron, 1995: 50). By contrast, in sero-negative spondyloarthropathies, union may occur across the whole joint surface with eventual trabecular continuity between sacrum and ilium (Ortner, 2003: 572). Radiography may aid in distinguishing different types of joint ankylosis.

In the radiographic study of lytic lesions, it is important to note not only lesion morphology, but also the nature of lesion margins. Radiographically, a slowly developing lytic lesion tends to have a margin that is well defined and often shows some sclerosis. In more rapidly expanding lesions, the margin may be well defined but lack sclerosis, and very aggressive lesions may be poorly circumscribed with a gradient of radiolucency rather than a well-defined edge (Ortner, 2003: 52). Thus, for example, slow-growing urate crystal deposits (tophi) in gout may lead to lytic lesions at the joints. These effectively represent pressure defects of bone (Watt, 1989), and have well-defined, frequently sclerotic margins (Figure 5.4). Similarly, benign tumours, which generally grow rather slowly, tend, when they lead to bony destruction, to produce well-circumscribed lesions with sclerotic margins. By contrast, a malignant osteolytic lesion, such as may occur in metastatic carcinoma, may, if the cancer is aggressive, produce defects which have poorly defined margins. Because it reflects to a great extent the rate at which a lesion was expanding at time of death, the radiographic appearance of lytic lesion margins may also vary with the phase of the disease when the individual died. For example, in rheumatoid arthritis, margins of erosions are often fuzzy in acute-phase disease but are more circumscribed during regressive phases (Jensen and Steinbach, 1977).

Turning to blastic lesions, the distribution of endosteal sclerosis can only be adequately evaluated radiographically. As well as aiding the documentation of changes in a specimen, the distribution of any endosteal sclerosis also often informs diagnosis (Figure 5.5). The bone in blastic lesions tends, if they are slow growing, to be dense and well corticated, whereas more rapidly deposited bone tends, if deposited immediately prior to death, to consist of poorly structured woven bone. Thus, for example, sub-periosteal bone deposits may occur in metastatic carcinoma, particularly of the prostate. Since this is a lethal condition, lesions are generally active at time of death and consist of diffuse woven bone of low radiodensity. Conversely, sub-periosteal bone in conditions that are not rapidly fatal may be well consolidated and of similar radiodensity to normal bone tissue. In instances where blastic lesions consist of circumscribed bony overgrowth(s) rather than diffuse periostitis,

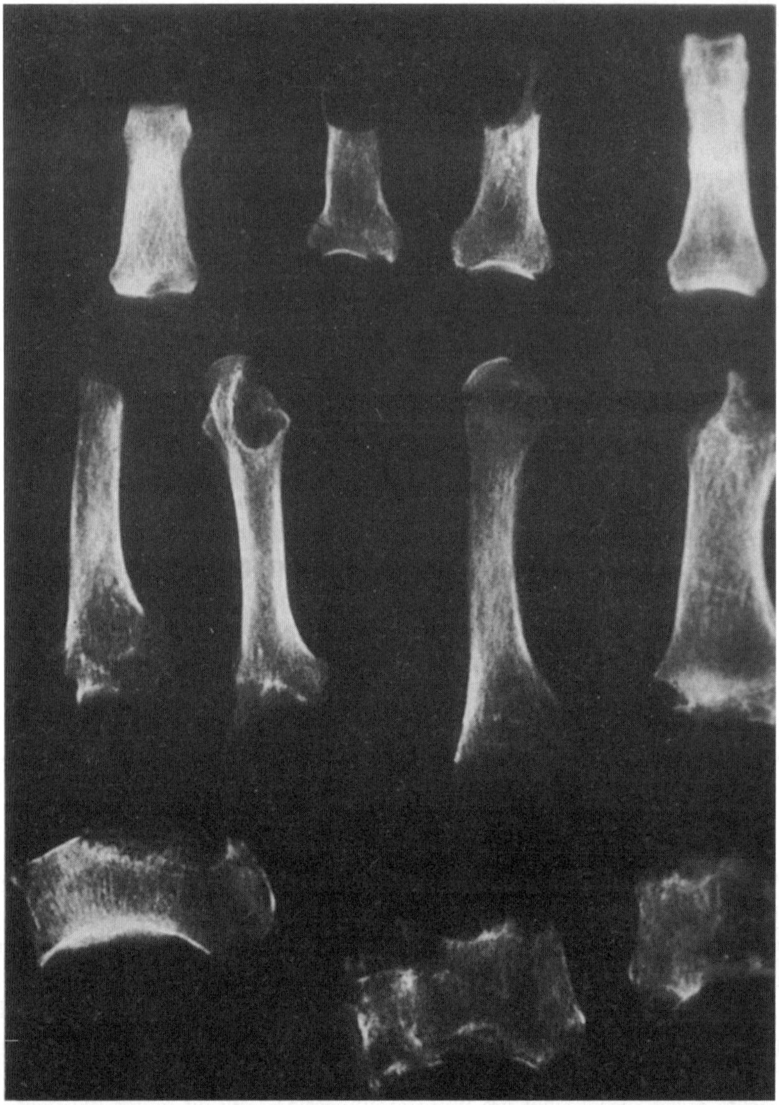

Figure 5.4 A case of gout from Barton-on-Humber. There are erosions, some of which show sclerotic margins. Reprinted from *Journal of Archaeological Science* Vol. 14, 1987, Rogers *et al.*, 'Arthropathies in palaeopathology: the basis of classification according to most probable cause', p. 191. Copyright (1987), with permission from Elsevier

visualization of the internal structure of the 'lump' or 'bump' by radiography aids diagnosis (Figure 5.6).

For primary bone diseases, clinical radiographic diagnostic criteria can often be directly applied to palaeopathological cases. This is the case, for example, with Paget's disease of bone (Mays, Chapter 11) and with skeletal neoplasms. However, for most conditions, the application of clinical radiographic diagnostic criteria is less straightforward. For diseases that affect soft tissue as well as bone, clinical radiographic diagnostic criteria frequently include

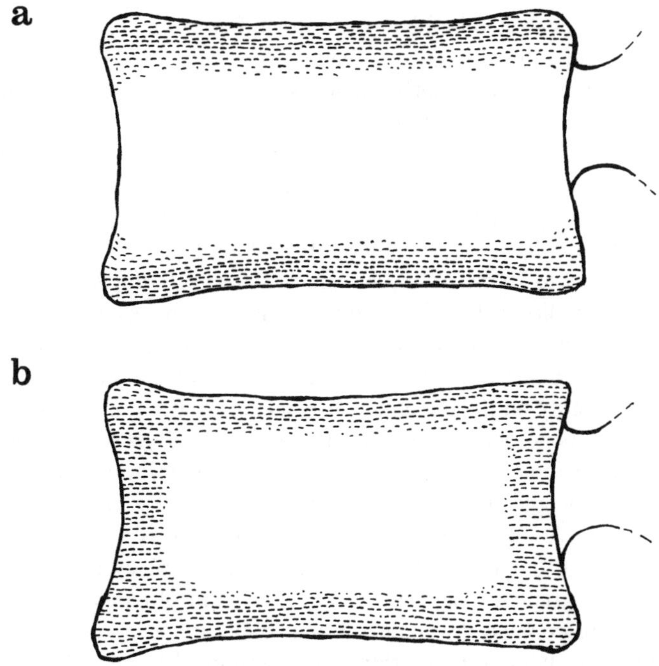

Figure 5.5 (a) Vertebral body osteosclerosis (stippled area) in renal osteodystrophy. The band-like sclerosis of the inferior and superior parts is termed 'rugger-jersey spine'. (b) In Paget's disease of bone, sclerosis may occur around the entire periphery of the vertebral body; an appearance termed 'picture-frame' sclerosis. After Resnick and Niwayama (1995: Fig. 64–22)

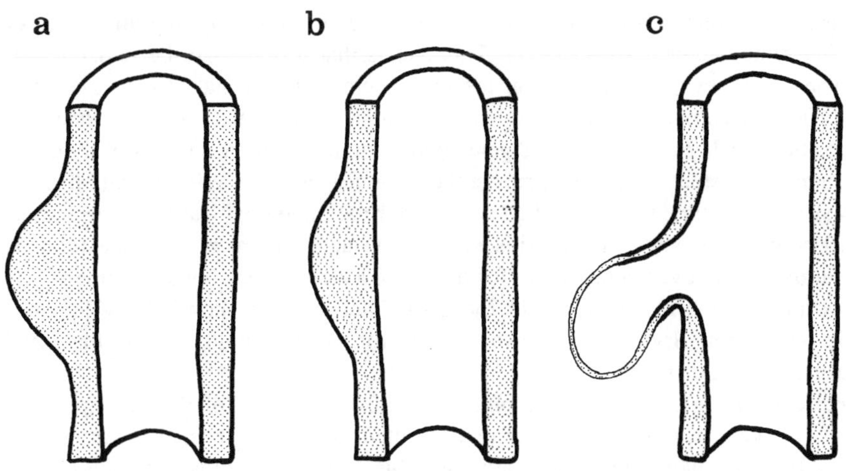

Figure 5.6 Internal structure of some circumscribed blastic lesions due to differing causes. (a) Bulging of cortical contour due to well-consolidated ossified haematoma. (b) Osteoid osteoma: bulging of cortical contour with radiolucent central focus. (c) Osteochondroma: the cancellous and cortical bone of the exostosis is continuous with that of the underlying normal bone

changes that cannot be evaluated in the dry specimen. For example, joint-space narrowing is a cardinal clinical radiographic diagnostic feature in osteoarthritis (Rogers *et al.*, 1990). In dry bones, joint-space narrowing clearly cannot be directly evaluated. In some diseases, the clinical radiographic literature emphasizes alterations that can be recorded in palaeopathology but for which radiography is redundant as they can be readily seen with the naked eye in dry specimens. For example, in rickets, most radiographic clinical diagnostic criteria relate to bending deformities, metaphysial broadening, concavity of metaphysial subchondral bone and 'fraying' of bone beneath the epiphysial plate (Thacher *et al.*, 2000; Pettifor, 2003: 555–557). These changes are obvious in the dry specimen, so palaeopathologists are selective in their application of clinical radiographic diagnostic criteria to dry bones. Radiographic diagnostic criteria for rickets in palaeopathology emphasize alterations to the internal bone structure (e.g. coarsening/thinning of trabecular structure and loss of cortico-medullary distinction) that are not visible grossly on the undamaged specimen (Mays *et al.*, 2006a).

For diseases where bone changes are superficial and easily seen on gross examination, radiography of palaeopathological specimens may be of rather limited value. For some such conditions, dry bone diagnostic criteria have been specifically developed with reference to radiographic criteria used by clinicians. For example, in osteoarthritis, in addition to joint-space narrowing, osteophyte formation, and subchondral cyst formation and sclerosis are important radiographic diagnostic features for clinical cases (Rogers *et al.*, 1990). In palaeopathology, osteophyte development and joint surface porosis and eburnation are important dry bone indicators of osteoarthritis (Rogers and Waldron, 1995: 44). Joint surface porosis and eburnation correspond to subchondral cyst formation and subchondral sclerosis identified in clinical radiographs: joint surface pores often communicate with subchondral cysts, and eburnated bone is sclerotic on X-ray (Rogers and Waldron, 1995: 44). For other conditions, palaeopathological diagnostic criteria have been developed based more on dry bone cases than on clinical radiographic features. This is the case, for example, with leprosy and scurvy (Ortner, 2003: 263–271, 383–393); radiography plays only a minor role in their palaeopathological study.

A change in bone density of about 40 % is needed before any alteration is visible on plain-film radiography (Ortner, 1991). This means that lesions may be invisible on X-ray even when they are quite obvious on the dry specimen to the naked eye. Therefore, it is often difficult to compare disease frequencies obtained from skeletal remains using dry bone palaeopathological criteria with those generated from radiographic surveys of living patients. For example, in a study of osteoarthritis at the knee, Rogers *et al.* (1990) found that only 2 of the 24 skeletal specimens they studied were abnormal radiographically, whereas there were obvious bony changes of osteoarthritis in 16 on visual assessment. The greater sensitivity of visual inspection over radiography for detecting lesions in dry bone specimens (provided changes are not entirely confined to the bone interior) also shows the value of, where possible, developing dry bone diagnostic criteria for palaeopathology rather than relying on clinical radiographic criteria.

In palaeopathology, most radiographic work is in the study of disease; applications to the study of skeletal trauma are rather more limited. In cases of healed fractures, a fracture line may be discernable radiographically, and this may aid diagnosis in such cases, but in many instances the value of radiography is quite limited as far as diagnosis is concerned. Whether a thickened cuff of bone represents a fracture callus is usually obvious to the naked eye, particularly where, as is generally the case with fracture union in antiquity, there is shortening or abnormal angulation of the bone. In cases where the diagnosis is unclear, this is generally because it may be a fracture that occurred long before death, so that remodelling of the callus is very thorough (Figure 5.7). In instances where remodelling is advanced it is not generally possible to visualize the fracture line on X-ray either; hence, the diagnosis cannot be confirmed. Although it is often not very useful for fracture diagnosis, radiography may, in cases where remodelling of the lesion is not too thorough, enable visualization of the direction of the fracture and the disposition of the broken ends. In this way, the degree of displacement on healing and, hence, the efficacy of any splinting or other treatment received may be assessed (Grauer and Roberts, 1996). Radiography may also of value for visualization of weapon points and other foreign objects embedded in bone. Corrosion often means that it is difficult to discern the shape of embedded iron objects, but this may be readily revealed on radiography (Figure 5.8).

RADIOGRAMMETRY

Measurement of cortical thickness using radiogrammetry was originated over 40 years ago as a means of studying bone loss in clinical work on osteoporosis (Barnett and Nordin, 1960). Measurements are generally taken, traditionally using callipers or rulers, on radiographs of the second metacarpal. This bone is selected for a number of reasons (Garn, 1970: 5–8). Adequate views can be obtained from standard postero-anterior hand radiographs. The bone section approaches circularity, which means that the method is robust to minor discrepancies in bone orientation, and, if desired, cortical area and other biomechanical parameters can readily be estimated. Measurements are generally made at the midshaft (Figure 5.9). Total bone width T and medullary width M are taken. Cortical thickness is simply given by $T - M$, but for cross-sectional, population studies the cortical index $\mathrm{CI} = 100 \times (T - M)/T$ is preferable as it is a measure of cortical thickness standardized for bone size.

Major radiogrammetric studies of cortical bone in modern populations were undertaken during the 1960s. Garn and co-workers conducted a monumental series of studies on the second metacarpal. Looking at various world populations, they investigated both appositional growth in cortical thickness in childhood and loss of cortical bone with advancing age in adults. This work, summarized in Garn (1970), showed that bone is added to the outer surface beneath the periosteum throughout the growth period. During infancy and childhood, bone is simultaneously resorbed from the endosteal surface, enlarging the medullary cavity. Endosteal resorption generally occurs at a slower rate than periosteal apposition. During adolescence, bone is added at both the periosteal and endosteal surfaces. The upshot is that, as well as an increase in overall bone width, there is an increase in thickness of cortical bone throughout the growth period. Peak cortical thickness is normally attained during early adult life. Slight periosteal bone apposition continues throughout adulthood, but, from about middle age, bone begins once more to be resorbed from the endosteal surface at a rate that outstrips periosteal apposition, leading to thinning of cortical bone. There is greater endosteal loss in females, principally as a response to the hormonal changes that accompany menopause (Garn, 1970).

In addition to Garn and colleagues' work, a number of other studies have been done on specific populations to obtain reference values for cortical bone. The largest of these is by Virtama and Helelä (1969), who conducted a radiogrammetric study of most of the tubular bones of the body in individuals from a Finnish population aged 1–90 years. Studies

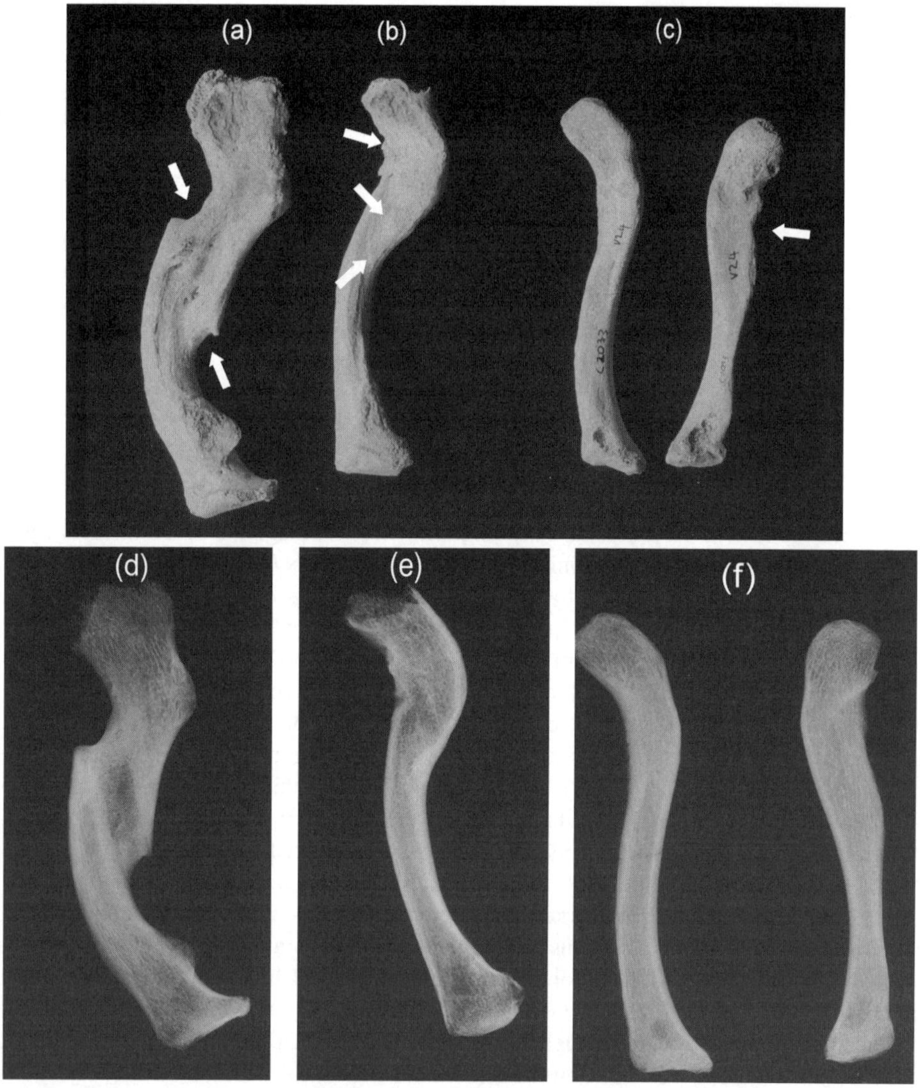

Figure 5.7 (a)–(c) Inferior view of some clavicles, showing probable healed fractures. (a) Specimen from a burial from Launceston Castle. The fracture line, at midshaft, is clearly evident (arrows). (b) Specimen from burial V33 from Wharram Percy. Although less obvious than in the Launceston Castle specimen, a fracture line can be discerned (arrows). (c) Specimens from burial V24 from Wharram Percy. The right bone is shorter than its antimere and is somewhat thickened in its lateral third, suggesting a united fracture in this region (arrow). (d)–(f) Infero-superior radiographs of the specimens in (a)–(c). (d) Launceston Castle: as in the gross specimen, the fracture line is clearly evident. (e) Wharram Percy, V33. Although the course of the fracture could be traced with careful examination of the gross specimen, it is difficult to discern in the radiograph. (f) Wharram Percy, V24. No fracture line can be seen. This sequence illustrates the limited value of radiography in the identification of united fractures in skeletal remains. In no case did the radiograph add significantly to diagnostic interpretation based on gross examination

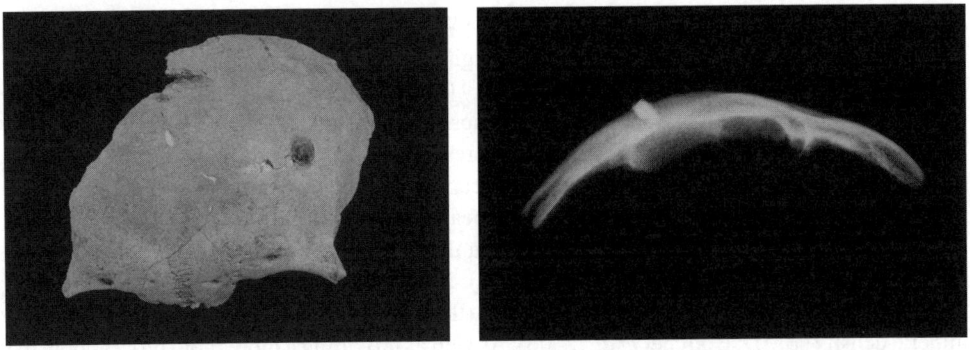

Figure 5.8 (a) Frontal bone from a burial from Ipswich Blackfriars. There is an iron fragment embedded in the left side. (b) Radiograph (superior view) showing the morphology of the embedded fragment. It appears to be the tip of a projectile point, implement or weapon that has broken off in the bone. It fails to penetrate the full thickness of the frontal bone

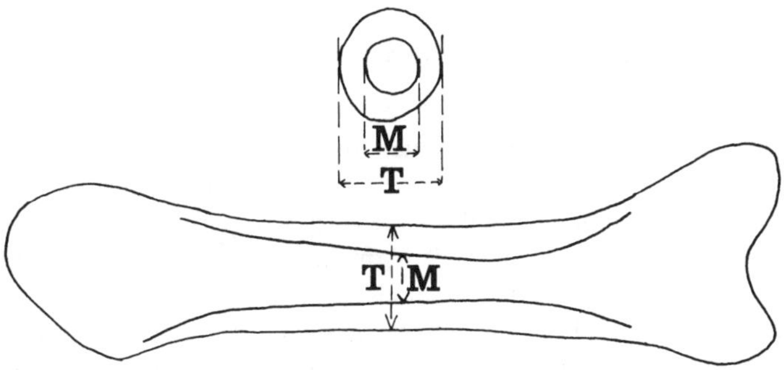

Figure 5.9 Schematic diagram of a second metacarpal, showing the methodology for measuring total bone width T and medullary width M at the midshaft from postero-anterior radiographs for the purpose of calculating cortical index

specifically of the second metacarpal have also been done on other populations, notably by Bugyi (1965) on Hungarians and by Dequeker *et al.* (1971) on Belgians.

Measurements of metacarpal cortical bone are of value in investigating osteoporosis, as they are a guide to the status of those elements (hip, spine and wrist) most at risk of osteoporotic fracture: metacarpal cortical bone is a useful indicator both of bone density (Adami *et al.*, 1996; Dey *et al.*, 2000) and fracture risk (Crespo *et al.*, 1998; Haara *et al.*, 2006) at these skeletal sites. Although successful in a research setting, the manual measurement technique meant that radiogrammetric quantification of cortical bone was labour intensive and time consuming, and these were significant disadvantages for its use for routine clinical monitoring of osteoporosis. With the commercial availability of automated photon absorptiometric methods for measuring bone density in the 1970s, and DXA in the 1980s, the use of radiogrammetry declined, although occasional research studies based on its use continued to appear in the clinical literature (e.g. Dequeker, 1976; Meema and Meema, 1987; van Hemmert *et al.*, 1993; Derisquebourg *et al.*, 1994; Maggio *et al.*, 1997).

Recently, there has been a revival of interest in radiogrammetry as a diagnostic tool in osteoporosis, due largely to the development of automated computerized versions of the traditional manual measurement technique (e.g. Dey *et al.*, 2000; Hyldstrup and Nielsen, 2001; Montalban *et al.*, 2001; Nielsen, 2001; Rosholm *et al.*, 2001; Ward *et al.*, 2003; Reed *et al.*, 2004; Boonen *et al.*, 2005; Haara *et al.*, 2006). Computerized digital image analysis versions of radiogrammetry are referred to as digital X-ray radiogrammetry (DXR) in the clinical literature. A frequently used DXR method in clinical research is the Pronosco system, originated in Denmark. The software uses an algorithm to locate 'regions of interest' (ROIs) around the narrowest parts of the diaphyses of the second, third and fourth metacarpals from a standard postero-anterior hand radiograph. In each metacarpal, more than 100 measurements are performed of T and M within the ROI, and a weighted average of results from the three bones is used to calculate an overall cortical index (Hyldstrup and Nielsen, 2001). If desired, the system can combine cortical thickness with analyses of porosity of cortex from the radiograph to produce an estimate of bone mineral density (BMD; Böttcher *et al.*, 2006). As well as providing a more rapid process than the manual measurement technique, this helps to minimize method error (Hyldstrup and Nielsen, 2001). Owing to the methodological differences, DXR results are not directly comparable to those generated from traditional, manual radiogrammetry. To my knowledge, DXR has yet to be applied to archaeological human remains

Radiogrammetry in Palaeopathology

Population studies on cortical bone thickness began to be undertaken in palaeopathology soon after the pioneering studies of Garn and others on modern populations (e.g. Dewey *et al.*, 1969; van Gerven *et al.*, 1969; Armelagos *et al.*, 1972). Most palaeopathological studies used radiogrammetry, but some, particularly the early work, used direct measurement of cut sections of bone (e.g. Dewey *et al.*, 1969; Armelagos *et al.*, 1972; Hummert, 1983; van Gerven *et al.*, 1985). Bone sectioning methods have the advantage that they allow direct measurement of cortical areas, but the results using such methods are not directly comparable to those generated by radiogrammetry on living subjects, and in any event cause damage to collections that is unacceptable in most curatorial situations.

There have been two principal foci for studies of cortical thickness in palaeopathology. First, echoing the work done on living subjects, studies in adults have been used to investigate osteoporosis, and both peak cortical bone thickness and age-related patterns of its loss have been investigated (Mays, Chapter 11). Second, the quantity of cortical bone accumulated during growth has been used as a stress indicator in skeletal populations. Although, other than *in extremis*, there is no relationship in adults between skeletal maintenance and diet either in terms of protein or caloric intake (Garn, 1970), a consistent link has been found between poor childhood nutrition and deficient appositional bone growth (Adams and Berridge, 1969; Garn *et al.*, 1969; Barr *et al.*, 1972; Himes *et al.*, 1975). Most workers studying cortical bone as a stress indicator have focused on patterns of increase in cortical thickness with age in juveniles (e.g. Cook, 1979; Huss-Ashmore *et al.*, 1982; Hummert, 1983; van Gerven *et al.*, 1985; Mays, 1985, 1995, 1999), but some studies taking this perspective have compared peak cortical bone levels in adults from different populations (Hatch *et al.*, 1983; Pfeiffer and King, 1983; Owsley, 1991; Rewekant, 2001).

Most palaeopathological osteoporosis work using radiogrammetry has used the second metacarpal, reflecting this bone's pre-eminence in monitoring cortical bone loss in osteoporosis in modern populations. A few studies have investigated loss of cortical bone in other elements, principally the femur (e.g. Ekenman *et al.*, 1995; Mays *et al.*, 1998). By contrast, most of the studies using cortical thickness as an indicator of non-specific stress during the growth period have studied long-bones, most often the femur. That major long-bones rather

Table 5.1 Repeatability of cortical index taken at the midshaft of the second metacarpal and femur from antero-posterior radiographs, and of DXA measurements of BMD in the proximal femur, in adult skeletons from Wharram Percy, England. Data from Mays (1996) and Mays *et al.* (1998)[a]

Variable	N	S_m	S	R
MCCI	10	2.270	11.52	0.040
FEMCI	9	1.280	9.180	0.019
BMDN	144	0.017	0.192	0.008
BMDW	144	0.026	0.239	0.012

[a]MCCI: metacarpal cortical index; FEMCI: femur cortical index; BMDN; BMD at femur neck, BMDW: BMD at Ward's triangle; BMD values in g cm^{-2}. N: number of repeat measurements, S_m: standard deviation of the measurement (method error statistic); S: sample standard deviation; R: variance of measurement/sample variance, i.e. S_m^2/S.

than the metacarpal have been used at least in part probably reflects difficulties in taking reliable measurements on the small metacarpals of young individuals, measurement error being proportionately greater for smaller measurements.

In skeletal studies, provided that specimens showing soil erosion are excluded, accurate measurements of cortical thickness are readily obtained using radiogrammetry. Metacarpal radiogrammetry on archaeological specimens produces results which are closely comparable to those obtained clinically. The metacarpal can be readily positioned on the film so that its orientation mimics that of the bone in hand radiographs of living subjects, and given the fairly small soft tissue thicknesses, differential radiographic enlargement of T and M between clinical and dry bone studies is likely minor; the effect of radiographic enlargement on CI is likely to be negligible.

The repeatability of radiogrammetric measurements on palaeopathological specimens appears good. Replicability data for metacarpal and femoral cortical index in adult skeletons from a large archaeological assemblage (from Wharram Percy, England), with re-radiography of specimens, and analysed using the method error statistic, are shown in Table 5.1. The results indicate that measurement error made up approximately 2 % and 4 % of sample variance for femoral and metacarpal cortical indices respectively. That intra-observer error is low has been confirmed by other workers on archaeological material (e.g. Lazenby, 2002; Ives and Brickley, 2004). However, inter-observer error may be a problem if workers are using different criteria for defining the endosteal border in their measurement of medullary width (Ives and Brickley, 2004).

BONE DENSITOMETRY

The oldest technique for measuring bone density from radiographs is photodensitometry (Mack *et al.*, 1939). A standard of known density, usually an aluminium step-wedge, is exposed in the radiograph alongside the bone, and the standard used to estimate bone density using an optical densitometer (Lees *et al.*, 1998). This method has been little used in palaeopathology (although see Ives and Brickley (2005)).

Single, and later dual, photon absorptiometry were introduced commercially to measure bone density in osteoporosis in a clinical setting in the 1970s. These techniques used a radionuclide source to generate photons, the attenuation of which by bone was used to assess density (Lees *et al.*, 1998). Some early palaeopathological work on osteoporosis used photon absorptiometry (e.g. Perzigian, 1973; Laughlin *et al.*, 1979), but the technology is now obsolete.

DXA has replaced photon densitometry as a means of measuring bone density, and it is currently the 'gold standard' by which bone loss is assessed clinically in osteoporosis (Kanis and Glüer, 2000). DXA uses an X-ray source that emits beams at two different energy levels. In the living patient this enables differing attenuation due to bone and soft tissue to be calculated. The areas scanned are generally those sites which are most vulnerable to osteoporotic fracture, namely the hip, spine and forearm. For the site scanned, the computer software in the machine selects a standard ROI within which bone density is measured. For example, for the hip the standard ROIs are the femoral neck, Ward's triangle and the greater trochanter. The bone mineral content in the ROI is measured, and the result divided by the area scanned. This gives an 'areal' density (grams per square centimetre) rather than a true volumetric density (Lees *et al.*, 1998). Because of this, DXA BMD results are not fully normalized for bone size. Larger bones will give greater apparent densities than smaller ones simply because the X-rays have passed through a greater thickness of bone. This will merely result in a minor degree of random noise in cross-sectional studies in adult populations, but it is more of a difficulty when studying age-related change in BMD in children (Nelson and Koo, 1999). To deal with this problem, formulae have been derived to estimate volumetric densities taking into account bone dimensions (e.g. Kröger *et al.*, 1992; Boot *et al.*, 1997).

Dual X-Ray Absorptiometry in Palaeopathology

DXA scanners became commercially available to clinicians in 1987 (Banks, 2001), and within a few years the first DXA work on archaeological bone, to investigate osteoporosis, was being undertaken (Hammerl *et al.*, 1990). Since then, DXA has been quite widely used to study osteoporosis in ancient populations (Mays, Chapter 11). The value of BMD measured using DXA as a stress indicator in juvenile skeletons has also recently begun to be investigated (McEwan *et al.*, 2005).

In general, palaeopathological studies of BMD in osteoporosis using DXA are conducted with the aim of comparing patterns seen in the study population with those in other groups, either a modern reference population or some other ancient population (Mays, Chapter 11). A general problem with making comparisons between different studies is that, due to hardware and software differences between different DXA scanners, absolute BMD values from different machines cannot be directly compared (Genant *et al.*, 1994; Hui *et al.*, 1997; Boonen *et al.*, 2003). In order to do such comparisons, a cross-calibration between machines needs to be carried out. In a clinical setting, this may be accomplished using a bone phantom, a calibration standard containing inserts of different density (Lees *et al.*, 1998). Alternatively, cross-calibration data may be derived by scanning the same set of subjects in different machines (e.g. Hui *et al.*, 1997). Similarly, in a palaeopathological setting, a subset of the specimens under study can be scanned on different machines and cross-calibration equations obtained (Mays *et al.*, 2006b).

In DXA scanners, attenuation of X-rays by soft tissue is taken into account by the computer software when BMD is calculated. This means that, for scanning, archaeological specimens generally need to be placed in a material whose density approximates to that of soft tissue, such as water (Kneissel *et al.*, 1994) or dry rice (Mays *et al.*, 1998). Nevertheless, because archaeological specimens lack marrow and soft tissue, absolute BMD values cannot be compared with those on living subjects (Lees *et al.*, 1993; Chappard *et al.*, 2004). This means that peak BMD cannot be compared between ancient and modern subjects. However, provided that significant diagenetic changes in bone density in archaeological specimens can be excluded, valid comparisons of age-related patterns in BMD between ancient and modern data can be made and peak BMD can be compared between different skeletal populations.

The possibility of diagenetic change in density of buried bone, either due to physical contamination with extraneous matter from the burial environment or due to chemical or microstructural alteration, is perhaps the most significant potential problem in bone densitometry studies in palaeopathology using DXA or other methods. One way of evaluating the possibility of diagenetic change in BMD is with reference to the results themselves. For example, in cases where patterning in BMD is as expected on physiological grounds (e.g. decline in BMD with age in both sexes but greater in females, greater and earlier age-related loss of BMD at Ward's triangle than at the femur neck) then this suggests that any post-depositional change in BMD is minor on the basis that it is highly unlikely that some complex pattern of differential diagenesis can have fortuitously reproduced such patterns (e.g. Mays *et al.*, 1998). However, even in cases such as this it is desirable that there be some independent evidence that diagenesis in BMD has not occurred to any great extent. To check for gross physical contamination, specimens should be screened prior to scanning for the presence of internal soil infiltration using plain-film radiography (e.g. Mays *et al.*, 1998). The possibility of minor soil infiltration, too subtle to be evident on plain-film radiographs, can be assessed using microscopic study. Chemical analysis or spectroscopic study of bone samples is also of value to detect intrusive minerals (Mays *et al.*, 2006b).

Microstructural diagenetic deterioration of bone, of the type described by Hackett (1981) and Hedges *et al.* (1995), has the potential to alter bulk bone density. Relatively little work has been done to determine whether it actually does, although a study has recently been undertaken involving comparison of two medieval populations, from Norway and England (Mays *et al.*, 2006b). The bones from Norway (Trondheim) showed excellent microstructural preservation of bone. Those from England (Wharram Percy) showed severe microstructural diagenesis. Nevertheless, DXA revealed similar absolute BMD values, and similar age-related patterns in BMD, in the two groups. Were microstructural diagenesis an important influence on BMD, then some difference between the two groups, at opposite ends of the spectrum of diagenetic alteration, would have been expected. Consistent with the notion that the microstructural diagenesis in the Wharram Percy bones had not caused appreciable change in bulk density, examination of histological sections (Turner-Walker and Syversen, 2002) showed that alterations had resulted in discrete areas of hyper- and hypo-mineralization, visible as areas of different tonal values under scanning electron microscopy. Elemental analysis indicated that, although diagenesis caused dissolution and re-precipitation of bone mineral, movement of bone mineral in this process was a highly localized phenomenon and bulk calcium content was little changed (Turner-Walker and Syversen, 2002; Mays, 2003). In summary, existing work clearly demonstrates that valid bone densitometry studies can be carried out on ancient skeletal remains. However, the possibility of diagenetic change in BMD should always be evaluated in the skeletal population under study.

The replicability of DXA results on ancient remains appears good. For example, repeat scannings, with repositioning of specimens, on the Wharram Percy adults suggests that about 1–2 % of sample variance consisted of measurement error (Table 5.1). That the precision of

DXA on ancient remains is good is also supported by studies on other assemblages (Lees *et al.*, 1993; Poulsen *et al.*, 2001).

Other Densitometric Techniques in Palaeopathology

Some more recently developed techniques can also produce BMD estimates from bone. Energy-dispersive low-angle X-ray scattering (EDLAXS) is one such technique. This technique has some advantages over DXA. It produces an estimate of true (volumetric) BMD, rather than the areal BMD of DXA, so that results are truly size-standardized. By adjusting the measurement area, BMD purely for trabecular bone can be obtained if desired. In addition, the spectrum generated by EDLAXS of different minerals is unique. This means that the presence of different minerals in a bone sample can be recognized and their quantities calculated. It is, therefore, another way of investigating diagenesis. The technique has been trialled using archaeological bone, and a good correlation with physical measures of density has been reported (Farquharson *et al.*, 1997; Farquharson and Brickley, 1997). However, EDLAXS has no clinical application, precluding comparisons with living subjects, and the equipment is not commercially available. For these reasons, future applications in palaeopathology are likely to be limited.

Quantitative CT (qCT) is a method that has become widely used for measuring BMD in clinical research. Like EDLAXS, it provides an estimate of volumetric BMD and permits the isolation of a specific volume of bone so that exclusively trabecular bone density can be measured if desired. However, as with DXA, the fact that skeletal remains lack marrow and soft tissue means that bones need to be placed in soft-tissue equivalents (e.g. water) for scanning, and absolute BMD values are not directly comparable between skeletal remains and living subjects. The high cost and higher radiation dose than for DXA have restricted the clinical application of qCT (Lees *et al.*, 1998; Banks, 2001; Kanis, 2002; Moyad, 2003). A small and rather inconclusive pilot study (Gonzalez-Reimers *et al.*, 2007) has been conducted on qCT in ancient bones, but further work is needed to assess adequately its value in palaeopathology.

CONCLUSIONS

Plain-film radiography has traditionally been the most important augment to gross examination for identifying and interpreting skeletal lesions in palaeopathology. It seems likely that this will continue to be the case for the foreseeable future. CT scanning may increase in importance, but cost considerations mean that its use in palaeopathology is unlikely to become routine. In time, digital image capture will doubtless oust film-based radiography, as has been the case with photography. Provided that resolution is adequate, real-time radiography provides the potential for more rapid screening of remains for pathological lesions. The increasing availability of portable digital radiographic equipment facilitates radiographic work at locations that lack radiographic facilities, such as field situations.

Radiogrammetry offers a straightforward technique by which cortical bone can be quantified. It provides data that are closely comparable to those obtained on living subjects, facilitating comparison between ancient and modern populations. It uses facilities (plain-film radiography, callipers) readily available to most palaeopathologists. In a clinical setting, the rather time-consuming nature of traditional manual radiogrammetry was a significant disadvantage, and the revival of clinical radiogrammetry was heralded by the advent of automated DXR techniques. However, it remains to be seen whether currently available DXR systems, such as Pronosco, can be applied to dry bones. In some ways DXR would seem to offer few advantages to palaeopathologists over manual radiogrammetry. Rapidity, which is the great advantage of DXR in a clinical context, is less critical in a research setting. Clinically, the decreased method error of DXR is important, but this is less so palaeopathologically, where the study is of patterning in cross-sectional population samples rather than, for example, attempting to detect change in sequential radiographs of individual patients. In addition, method errors introduced by imprecision in manual radiogrammetry are often minor compared with those introduced into work on palaeopopulations by imprecision in other methods; for example, limitations in skeletal age determination, rather than method error in radiogrammetry, are the major limitation on the resolution with which we can study age-related patterns of decline in cortical bone in osteoporosis. The Pronosco system of DXR relies on measurements from three metacarpals, a requirement that will reduce sample size in fragmentary archaeological remains. Despite these disadvantages, it would be useful to investigate the applicability of DXR to skeletal remains, if only so that archaeological data can be compared with data currently being generated on modern populations.

Turning to the assessment of bone density, advantages of using DXA in the study of osteoporosis in archaeological populations are that it is the clinical 'gold standard' by which the BMD in osteoporosis is assessed, and it can be used to estimate bone density specifically at the skeletal sites vulnerable to osteoporotic fracture. A disadvantage in its use in skeletal remains is that, because absolute BMD values are not directly comparable to those obtained from living patients, only patterns of age-related bone loss rather than peak bone density can be compared between living and skeletal populations. An additional problem (also shared by other densitometric methods) is the potential for diagenesis to alter the density of buried bone. Although review of the published literature reveals little evidence that diagenetic change in BMD is a pervasive problem in palaeopathological studies, it is a possibility that should be investigated in each archaeological population under study.

REFERENCES

Adami S, Zamberlan N, Gatti G, Zanfisi C, Braga V, Broggini M. 1996. Computed radiographic absorptiometry and morphometry in the assessment of post-menopausal bone loss. *Osteoporos Int* **6**: 8–13.

Adams P, Berridge SR. 1969. Effects of kwashiorkor on cortical and trabecular bone. *Arch Dis Child* **44**: 705–709.

Ameen S, Staub L, Ulrich S, Vock P, Ballmer F, Anderson SE. 2005. Harris lines of the tibia across two centuries: a comparison of two populations, medieval and contemporary in central Europe. *Skel Radiol* **34**: 279–284.

Armelagos GJ, Mielke JH, Owen KH, van Gerven DP, Dewey JR, Mahler PE. 1972. Bone growth and development in prehistoric populations from Sudanese Nubia. *J Hum Evol* **1**: 89–119.

Aufderheide AC, Rodríguez-Martin RM. 1998. *The Cambridge Encyclopaedia of Paleopathology*. Cambridge University Press: Cambridge.

Banks LM. 2001. Dual energy X-ray absorptiometry (DXA). In *Diagnostic Radiology: A Textbook of Medical Imaging* (4th edn), vol. 1, Grainger RG, Allison DJ, Adam A, Dixon AK (eds). Churchill and Livingstone: London; 185–195.

Barnett E, Nordin BEC. 1960. The radiological diagnosis of osteoporosis: a new approach. *Clin Radiol* **11**: 166–174.

Barr DGD, Shmerling DH, Prader A. 1972. Catch-up growth in malnutrition studied in celiac disease after institution of a gluten-free diet. *Pediatr Res* **6**: 521–527.

Böni T, Rühli FJ, Chhem RK. 2004. History of paleoradiology: early published literature, 1896–1921. *Can Assoc Radiol J* **55**: 203–210.

Boonen S, Kaufman J-M, Reginster J-Y, Devogelaer J-P. 2003. Patient assessment using standardized bone mineral density values and a national reference database: implementing uniform thresholds for the reimbursement of osteoporosis treatments in Belgium. *Osteoporos Int* **14**: 110–115.

Boonen S, Nijs J, Borghs H, Peeters H, Vanderschueren D, Luyten FP. 2005. Identifying postmenopausal women with osteoporosis by calcaneal ultrasound, metacarpal digital radiogrammetry and phalangeal radiographic absorbtiometry: a comparative study. *Osteoporos Int* **16**: 93–100.

Boot AM, de Ridder MAJ, Pols HAP, Krenning EP, Keizer-Schrama SMPFdeM. 1997. Bone mineral density in children and adolescents: relation to puberty, calcium intake and physical activity. *J Clin Endocrinol Metab* **82**: 57–62.

Böttcher J, Pfeil A, Rosholm A, Schäfer M-L, Malich A, Petrovich A, Seidl B, Lehman G, Mentzel H-J, Hein G, Wolf G, Kaiser WA. 2006. Computerized digital imaging techniques provided by digital X-ray radiogrammetry as new diagnostic tool in rheumatoid arthritis. *J Digital Imag* **19**: 279–288.

Bugyi B. 1965. Über Alternsabhängigkeit des Kortikalen Index des II Metakarpalknochens. *Anat Anz* **116**: 378–383.

Chappard D, Moquereau M, Mercier P, Gallois Y, Legrand E, Baslé MF, Audran M. 2004. *Ex vivo* bone mineral density of the wrist: influence of medullar fat. *Bone* **34**: 1023–1028.

Clendinnen FJ. 1898. Skiagram of hand of Egyptian mummy, showing abnormal number of sesamoid bones. *Intercolon Med J Austr* **3**: 106.

Cook DC. 1979. Subsistence base and health in prehistoric Illinois Valley: evidence from the human skeleton. *Med Anthropol* **4**: 109–124.

Crespo R, Revilla M, Usabiago J, Crespo E, Garcia-Areno J, Villa LF, Rico H. 1998. Metacarpal radiogrammetry by computed radiography in postmenopausal women with Colles' fracture and vertebral crush fracture syndrome. *Calcif Tissue Int* **62**: 470–473.

Dequeker J. 1976. Quantitative radiology: radiogrammetry of cortical bone. *Br J Radiol* **49**: 912–920.

Dequeker J, Remans J, Franssen R, Waes W. 1971. Ageing patterns of trabecular and cortical bone and their relationship. *Calcif Tissue Res* **7**: 23–30.

Derisquebourg T, Dubois B, Devogelaer JP, Meys E, Duquesnoy B, Nagant de Deuxchaisnes C, Delacambre B, Marchandise X. 1994. Automated computerised radiogrammetry on the second metacarpal and its correlation with absorptiometry of the forearm and spine. *Calcif Tissue Int* **54**: 461–465.

Dewey JR, Bartley MH, Armelagos GJ. 1969. Rates of femoral cortical bone loss in two Nubian populations. *Clin Orthop* **65**: 61–66.

Dey A, McCloskey EV, Taube T, Cox R, Pande KC, Ashford RU, Forster M, de Takats D, Kanis JA. 2000. Metacarpal morphometry using a semi-automated technique in the assessment of osteoporosis and vertebral fracture risk. *Osteoporos Int* **11**: 953–958.

Eaton GF. 1916. The collection of osteological material from Machu Picchu (Peru). *Mem Conn Acad Arts Sci* **5**: 1–96.

Ekenman I, Eriksson SAV, Lindgren JU. 1995. Bone density in mediaeval skeletons. *Calcif Tissue Int* **56**: 355–358.

Farquharson MJ, Brickley M. 1997. Determination of mineral composition of archaeological bone using energy-dispersive low-angle X-ray scattering. *Int J Osteoarchaeol* **7**: 95–99.

Farquharson MJ, Speller RD, Brickley M. 1997. Measuring bone mineral density in archaeological bone using energy dispersive low angle X-ray scattering techniques. *J Archaeol Sci* **24**: 765–772.

Fiori MG, Nunzi MG. 1995. The earliest documented application of X-rays to the examination of mummified remains and archaeological materials. *J R Soc Med* **88**: 76–69.

Garn SM. 1970. *The Earlier Gain and Later Loss of Cortical Bone*. Charles C Thomas: Springfield, IL.

Garn SM, Guzman MA, Wagner B. 1969. Subperiosteal gain and endosteal loss in protein – calorie malnutrition. *Am J Phys Anthropol* **30**: 153–156.

Genant HK, Grampp S, Glüer CC, Faulkner KG, Jergas M, Engelke K, Hagiwara S, van Kuijk C. 1994. Universal standardization for dual X-ray absorptiometry: patient and phantom cross-calibration results. *J Bone Miner Res* **9**: 1503–1514.

Gonzalez-Reimers E, Velasco-Vázquez J, Arnay de la Rosa M, Machado-Calvo M. 2007. Quantitative computerized tomography for the diagnosis of osteopenia in prehistoric skeletal remains. *J Archaeol Sci* **34**: 554–561.

Grauer AL, Roberts CA. 1996. Palaeoepidemiology, healing and possible treatment of trauma in the mediaeval cemetery population of St Helen-on-the-Walls, York, England. *Am J Phys Anthropol* **100**: 531–544.

Haara M, Heliövaara M, Impivaara O, Arokoski JPA, Manninen P, Knekt P, Kärkkäinen A, Reunanen A, Aromaa A, Kröger H. 2006. Low metacarpal index predicts hip fracture. *Acta Orthop* **77**: 9–14.

Hackett CJ. 1981. Microscopic focal destruction (tunnels) in exhumed bone. *Med Sci Law* **21**: 243–265.

Hammerl J, Protsch R, Happ J, Frohn J, Hor G. 1990. Osteodensitometrie des Femurhalses an historischen Skeletten. In *Osteologie-Interdisziplinar*, Werner R, Mattiass HH (eds). Springer: Berlin; 139–142.

Hatch JW, Willey PS, Hunt EE. 1983. Indicators of status-related stress in Dallas society: transverse lines and cortical thickness in long bones. *Midcontinent J Archaeol* **8**: 49–71.

Hedges REM, Millard AR, Pike AWG. 1995. Measurements and relationships of diagenetic alteration of bone from three archaeological sites. *J Archaeol Sci* **22**: 201–209.

Himes JH, Matorell R, Habicht J-P, Yarbrough C, Malina RM, Klein RE. 1975. Patterns of cortical bone growth in moderately malnourished preschool children. *Hum Biol* **47**: 337–350.

Hui SL, Gao S, Zhou X-H, Johnson CC, Lu Y, Glüer CC, Genant HK. 1997. Universal standardisation of bone density measurements: a method with optimal properties for calibration among several instruments. *J Bone Miner Res* **12**: 1463–1470.

Hummert JR. 1983. Cortical bone growth and dietary stress among subadults from Nubia's Batn el Hajar. *Am J Phys Anthropol* **62**: 167–176.

Huss-Ashmore R, Goodman AH, Armelagos GJ. 1982. Nutritional inference from paleopathology. *Adv Archaeol Meth Theor* **5**: 395–474.

Hyldstrup L, Nielsen SP. 2001. Metacarpal index by digital X-ray radiogrammetry. *J Clin Densitom* **4**: 299–306.

Ives R, Brickley MB. 2004. A procedural guide to metacarpal radiogrammetry in archaeology. *Int J Osteoarchaeol* **14**: 7–17.

Ives R, Brickley M. 2005. Metacarpal radiogrammetry: a useful indicator of bone loss throughout the skeleton? *J Archaeol Sci* **32**: 1552–1559.

Jensen PS, Steinbach HL. 1977. Roentgen features of the rheumatic diseases. *Med Clin North Am* **61**: 389–404.

Kanis JA. 2002. Diagnosis of osteoporosis and assessment of fracture risk. *Lancet* **359**: 1929–1936.

Kanis JA, Glüer CC. 2000. An update on the diagnosis and assessment of osteoporosis with densitometry. *Osteoporos Int* **11**: 192–202.

Kneissel M, Boyde A, Hahn M, Teschler-Nicola M, Kalchhauser G, Plenk H. 1994. Age- and sex-dependent cancellous bone changes in a 4000y BP population. *Bone* **15**: 539–545.

Kröger H, Kotaniemi A, Vainio P, Alhava E. 1992. Bone densitometry of the spine and femur in children by dual-energy X-ray absorptiometry. *Bone Miner* **17**: 75–82.

Lang J, Middleton A, Ambers J, Higgins T. 2005. Radiographic images. In *Radiography of Cultural Material* (2nd edition), Lang J, Middleton A (eds). Elsevier: Oxford; 20–48.

Laughlin WS, Harper AB, Thompson DD. 1979. New approaches to the pre- and post-contact history of Arctic peoples. *Am J Phys Anthropol* **51**: 579–587.

Lazenby RA. 2002. Circumferential variation in human second metacarpal cortical thickness: sex, age and mechanical factors. *Anat Rec* **267**: 154–158.

Lees B, Molleson T, Arnett TR, Stevenson JC. 1993. Differences in proximal femur bone density over two centuries. *Lancet* **341**: 673–675.

Lees B, Banks LM, Stevenson JC. 1998. Bone mass measurements. In *Osteoporosis*, Stevenson JC, Lindsay R (eds). Chapman and Hall: London; 137–160.

Mack PB, O'Brien AT, Smith JM, Bauman AW. 1939. A method for estimating the degree of mineralisation of bones from tracings of roentgenograms. *Science* **89**: 467.

Maggio D, Pacifici R, Cherubini A, Simonelli G, Luchetti M, Aisa MC, Cucinotta D, Adami S, Senin U. 1997. Age-related cortical bone loss at the metacarpal. *Calcif Tissue Int* **60**: 94–97.

Mays SA. 1985. The relationship between Harris line formation and bone growth and development. *J Archaeol Sci* **12**: 207–220.

Mays SA. 1995. The relationship between Harris lines and other aspects of skeletal development in adults and juveniles. *J Archaeol Sci* **22**: 511–520.

Mays S. 1996. Age-dependent cortical bone loss in a medieval population. *Int J Osteoarchaeol* **6**: 144–154.

Mays S. 1999. Linear and appositional long bone growth in earlier human populations: a case study from mediaeval England. In *Human Growth in the Past: Studies Using Bones and Teeth*, Hoppa RD, Fitzgerald CM (eds). Cambridge University Press: Cambridge; 290–312.

Mays SA 2003. Bone strontium:calcium ratios and duration of breastfeeding in a mediaeval skeletal population. *J Archaeol Sci* **30**: 731–741.

Mays S, Lees B, Stevenson JC. 1998. Age-dependent cortical bone loss in the femur in a medieval population. *Int J Osteoarchaeol* **8**: 97–106.

Mays S, Brickley M, Ives R. 2006a. Skeletal manifestations of rickets in infants and young children in an historic population from England. *Am J Phys Anthropol* **129**: 362–374.

Mays S, Turner-Walker G, Syversen U. 2006b. Osteoporosis in a population from mediaeval Norway. *Am J Phys Anthropol* **131**: 343–351.

McEwan J, Mays S, Blake G. 2005. The relationship of bone mineral density and other growth parameters to stress indicators in a mediaeval juvenile population. *Int J Osteoarchaeol* **15**: 155–163.

Means JH. 1925. A roentgenological study of skeletal remains of the prehistoric Mound Builder Indians of Ohio. *Am J Roentgenol* **13**: 359–367.

Meema HE, Meema S. 1987. Postmenopausal osteoporosis: simple screening method for diagnosis before structural failure. *Radiology* **164**: 405–410.

Molleson TI, Williams CT, Cressey G, Din VK. 1998. Radiographically opaque bones from lead-lined coffins at Christ Church, Spitalfields, London – an extreme example of bone diagenesis. *Bull Soc Geol France* **169**: 425–432.

Montalban SJ, Rico LH, Cortes PJ, Pedrera ZJD. 2001. Cortical bone mass and risk factors for osteoporosis among postmenopausal women in our environment. *Rev Clin Esp* **201**: 16–20.

Moyad MA. 2003. Osteoporosis: a rapid review of risk factors and screening methods. *Urol Oncol* **21**: 375–379.

Nelson DA, Koo WWK. 1999. Interpretation of absorptiometric bone mass measurements in the growing skeleton: issues and limitations. *Calcif Tissue Int* **65**: 1–3.

Nielsen SP. 2001. The metacarpal index revisited: a brief overview. *J Clin Densitom* **4**: 199–207.

Ortner DJ. 1991. Theoretical and methodological issues in paleopathology. In *Human Paleopathology: Current Syntheses and Future Options*, Ortner DJ, Aufderheide AC (eds). Smithsonian Institution Press: Washington; 5–11.

Ortner DJ. 2003. *Identification of Pathological Conditions in Human Skeletal Remains* (2nd edition). Academic Press: New York.

Ortner DJ, Mays S. 1998. Dry-bone manifestations of rickets in infancy and early childhood. *Int J Osteoarchaeol* **8**: 45–55.

Ortner DJ, Putschar WGJ. 1985. *Identification of Pathological Conditions in Human Skeletal Remains*. Smithsonian Institution Press: Washington, DC.

Owsley DW. 1991. Temporal variation in femoral cortical thickness of North American Plains Indians. In *Human Paleopathology: Current Syntheses and Future Options*, Ortner DJ, Aufderheide AC (eds). Smithsonian Institution Press: Washington, DC; 105–110.

Pettifor JM. 2003. Nutritional rickets. In *Pediatric bone. Biology and Diseases*, Glorieux F, Pettifor J, Jüppner H (eds). Academic Press: New York; 541–565.

Pfeiffer S, King P. 1983. Cortical bone formation and diet among protohistoric Iroquoians. *Am J Phys Anthropol* **60**: 23–28.

Perzigian AJ. 1973. Osteoporotic bone loss in two prehistoric Indian populations. *Am J Phys Anthropol* **39**: 87–96.

Poulsen LW, Qvesel D, Brixen K, Versterby A, Boldsen JL. 2001. Low bone mineral density in the femoral neck of medieval women: a result of multiparity. *Bone* **28**: 454–458.

Reed MR, Murray JRD, Abdy SE, Francis RM, McCaskie AW. 2004. The use of digital X-ray radiogrammetry and peripheral dual X-ray absorptiometry in patients attending fracture clinic after distal forearm fracture. *Bone* **34**: 716–719.

Resnick D, Niwayama G. 1995. *Diagnosis of Bone and Joint Disorders* (3rd edition). WB Saunders: London.

Rewekant A. 2001. Do environmental disturbances of an individual's growth and development influence the later bone involution process? A study of two medieval populations. *Int J Osteoarchaeol* **11**: 433–443.

Rogers J, Waldron T. 1995. *A Field Guide to Joint Disease in Archaeology*. Wiley: Chichester.

Rogers J, Watt I, Dieppe P. 1990. Comparison of visual and radiographic detection of bony changes at the knee joint. *Br Med J* **300**: 367–368.

Rosholm A, Hyldstrup L, Baeksgaard L, Grunkin M, Thodberg HH. 2001. Estimation of bone mineral density by digital X-ray radiogrammetry: theoretical background and clinical testing. *Osteoporos Int* **12**: 961–969.

Sandison AT. 1968. Pathological changes in the skeletons of earlier populations due to acquired disease, and difficulties in their interpretation. In *The Skeletal Biology of Earlier Human Populations*, Brothwell DR (ed.). Pergamon: Oxford; 205–243.

Spiegel PK. 1995. The first clinical X-ray made in America. *Am J Roent* **164**: 241–243.

Steinbock RT. 1976. *Paleopathological Diagnosis and Interpretation*. Charles C Thomas: Springfield, IL.

Thacher TD, Fischer PR, Pettifor JM, Lawson JO, Manaster BJ, Reading JC. 2000. Radiographic scoring method for the assessment of the severity of nutritional rickets. *J Trop Pediatr* **46**: 132–139.

Turner-Walker G, Syversen U. 2002. Quantifying histological changes in archaeological bones using BSE-SEM image analysis. *Archaeometry* **44**: 461–468.

Van Gerven DP, Armelagos GJ, Bartley MH. 1969. Roentgenographic and direct measurement of femoral cortical involution in a prehistoric Mississippian population. *Am J Phys Anthropol* **31**: 23–38.

Van Gerven DP, Hummert JR, Burr DP. 1985. Cortical bone maintenance and geometry of the tibia in prehistoric children from Nubia's Batn el Hajar. *Am J Phys Anthropol* **66**: 275–280.

Van Hemert AM, Vanderbroucke JP, Hofman H, Valkenberg HA. 1993. Metacarpal bone loss in middle-aged women: "horse racing" in a 9-year population based follow up study. *J Clin Epidemiol* **43**: 579–588.

Virtama P, Helelä T. 1969. Radiographic measurements of cortical bone. Variation in a normal population between 1 and 90 years of age. *Acta Radiol* **293**(Suppl): 1–268.

Ward KA, Cotton J, Adams JE. 2003. A technical and clinical evaluation of digital X-ray radiogrammetry. *Osteoporos Int* **14**: 389–395.

Watt I. 1989. The radiology of erosions. In *The Antiquity of the Erosive Arthropathies*, Dieppe P, Rogers J (eds). Arthritis and Rheumatism Council for Research, Conference Proceedings No. 5. Arthritis and Rheumatism Council: London; 18–20.

Wells C. 1963. The radiological examination of human remains. In *Science in Archaeology*, Brothwell D, Higgs E (eds). Thames and Hudson: London; 401–412.

Wells C. 1967. Pseudopathology. In *Diseases in Antiquity*, Brothwell DR, Sandison AT (eds). Charles C Thomas: Springfield, IL; 5–19.

Williams HU. 1929. Human paleopathology with some original observations of symmetrical osteoporosis of the skull. *Arch Path* **7**: 839–902.

Zimmerman MR, Kelley MA. 1982. *Atlas of Human Paleopathology*. Praeger: New York.

6

Computed Tomography Scanning and Three-Dimensional Visualization of Mummies and Bog Bodies

Niels Lynnerup

Laboratory of Biological Anthropology, The Panum Institute, Blegdamsvej 3, DK-2200 Copenhagen, Denmark

INTRODUCTION

Mummies and bog bodies present unique opportunities for the palaeopathologist. While many pathological processes may be identified in the skeleton, bones are overall a 'slow-reacting' tissue, so that pathological changes mainly reflect chronic diseases (Ortner, 2003). The presence of soft tissues may expand the scope of pathological studies, so that more acute diseases and diseases that do not affect bone tissue may be identified (Cockburn and Cockburn, 1980). The mummification and preservation of the various soft tissues, though, is very variable. Interior organs, particularly of the digestive system, are often completely decomposed, and organs may be very shrunken and difficult to identify morphologically due to desiccation. Furthermore, various funerary rites that include embalming may entail the complete removal of internal organs, including the brain (as seen for Egyptian mummies). Generally, the most commonly preserved soft tissues are those with a high content of collagen, such as the dermis, muscle fasciae and tendons. Still, even just the presence of skin may give important clues to pathology and trauma; for example, penetrating wounds and cuts, scars and even warts (Lowenstein, 2004).

However, the very presence of soft tissue, especially the skin, makes it difficult to examine the body. One may conduct an autopsy, but while such mummy autopsies have been carried out often (David, 1979; Aufderheide, 2003), most archaeologists, conservators, physical anthropologists and pathologists are reluctant to permit this. An autopsy is an invasive

Advances in Human Palaeopathology Edited by Ron Pinhasi and Simon Mays

and destructive procedure; the integrity of the mummy or bog body as an archaeological specimen may be destroyed. Attitudes have perhaps shifted from an earlier, more clinical, medical approach to mummies and bog bodies (in a certain sense seeing the body as any other unidentified body that must be properly examined in a forensic way) to one that views a mummy or a bog body as an invaluable archaeological artefact.

X-RAYS AND MUMMIES

The most well-known non-invasive technique for visualizing the internal structures of mummies and bog bodies employs X-rays. Indeed, the first use of X-radiography in mummy research occurred only 1 year after William Röntgen discovered X-rays in 1895 (Koenig, 1896). X-rays are electromagnetic waves. Because they are generated by the conversion of the energy acquired by electrons accelerated through an electrical field gradient, they are more often characterized by their photon energies. The electrical field is produced by an anode and cathode in the X-ray tube. The cathode is made of a tungsten filament, and this emits the electrons. When the electrons strike the anode, a small fraction of the energy is emitted as X-rays. The X-rays then pass out of the tube, through adjustable diaphragms in order to limit the beam to the minimally required area one wishes to examine (collimation). The intensity of the X-rays produced at a given voltage is determined by the number of electrons impacting the anode, and is expressed in milliamperes. The X-ray dose is proportional to the time during which the beam current flows. Intensity and dosage are major parameters in clinical use, as X-rays are potentially hazardous due to their ionizing properties. As the X-rays pass through the object to be examined, they interact with the object, being either scattered or absorbed. The combined effect of this is expressed as attenuation. After passing through the object, the X-rays then strike a photographic plate. Since the X-rays are thus attenuated differentially by different tissues and materials, they end up striking the photographic plate at different intensities, resulting in differing grey-values. After development, these are rendered from white (tissues or materials with high attenuation, e.g. bone) to black (e.g. air in body cavities). Modern clinical radiographic equipment is now fully digitized, i.e. the X-rays do not end on a photographic plate but rather in special sensors which translate the attenuation to direct pixel-based images. The digital equipment has the advantage of being able to control the beam modalities and the image building more directly, enabling sharper pictures (Fleckenstein and Tranum-Jensen, 1993; Carlton and Adler, 2001).

Radiography has the advantage of being almost universally available and easily performed. As mentioned above, radiography has a long track record when it comes to mummies and bog bodies, in effect having been the only method available for 'looking inside'. Flinders Petrie used X-rays in the studies of mummies in 1897 (Petrie and Griffith, 1898). Moodie published the findings of his analyses of 17 Egyptian mummies at the Field Museum in Chicago in 1931; this probably constitutes the first systematic radiographic analysis of a major mummy collection (Moodie, 1931). All the royal mummies housed at the Cairo Museum were radiographed in 1967 (Harris and Weeks, 1973; Harris and Wente, 1980). Radiography of mummies has also been performed on site in some very remote locations (Notman *et al.*, 1987; Notman and Beattie, 1995).

The primary purpose of radiographing mummies has often been archaeological as well as medical; for example, searching for amulets in Egyptian mummy wrappings (Christensen, 1969). However, the determination of sex and age, based on skeletal traits, has usually been carried out whenever possible (e.g. Fawcitt *et al.*, 1984). Grafton Elliot Smith evaluated the epiphyseal union of the mummy of Tuthmosis IV in order to ascertain age at death (Smith, 1912). Cranial traits, and cranial morphometry based on radiographs, have been studied in order to try to establish kinship (Harris and Wente, 1980). This pre-empted one of the uses of computed tomography (CT) scanning: the production of three-dimensional solid models of skulls for facial reconstruction purposes (see below). Neave (1979) managed to make a facial reconstruction of an Egyptian mummy based on radiographs.

The pathological changes observed by X-raying mummies include arthritis, atheroma, healed fractures, and parasite-induced changes (Brothwell and Sandison, 1967; Christensen, 1969; David 1979; Harris and Wente, 1980; Bloomfield, 1985). Most of the pathological processes observed reflect diseases affecting bone or calcified structures, such as a radiodense structure in an Egyptian mummy that was revealed to be a guinea worm parasite at a later autopsy (David, 1979). This is a reflection of the limitations of radiography. Skeletal structures are usually easily identifiable due to the high attenuation, but it may be very difficult to discriminate between various soft-tissue remains, especially as these will be superimposed upon another on the radiograph. In radiography, a three-dimensional structure is presented in two dimensions, so the skin and fasciae of the ventral aspect of a mummy will be superimposed over the remains of both the interior organs, membranes and fasciae, and the skin and fasciae of the dorsal aspect. Since the soft tissues already may have near equal attenuation coefficients, this inhibits greatly the possibilities for investigating soft-tissue pathology.

COMPUTED TOMOGRAPHY SCANNING

A further development of radiographic imaging came with the advent of CT scanners (computer-assisted tomography). Unlike conventional X-ray images, the regions of interest in CT scannings are presented without superimpositions of juxtaposed structures. The X-ray tube revolves around the item being analysed (Figure 6.1). Coupled with a specific array for reading out the X-ray attenuation, this means that a CT scanner will generate an image that represents a slice of an item (Figure 6.2). The item is moved through the CT scanner, generating hundreds or thousands of slices, which may be viewed as serial, two-dimensional slice images (Hsieh, 2002).

The development of the first clinical CT scanners began in 1967 with Godfrey Hounsfield (1976), and the first clinical CT scanner was installed in 1971. The first patient scanned was diagnosed with a large cyst (Ambrose, 1975). Hounsfield was awarded the Nobel Prize for his work, and he shared it with Allan M. Cormack, who had undertaken pioneering work on the mathematical basis for image reconstruction. Since the introduction of the first scanners, many technical advances have been made. These have produced much finer images and much more rapid image acquisition times (from more than 1 min to less than 0.1 s per slice) (Hsieh, 2002). The advances in CT scanners are generally recorded as 'generations'. First-generation scanners had only one beam being recorded at a time, whereas second-generation scanners had multiple beams, enabling a faster recording time. Third-generation scanners were able

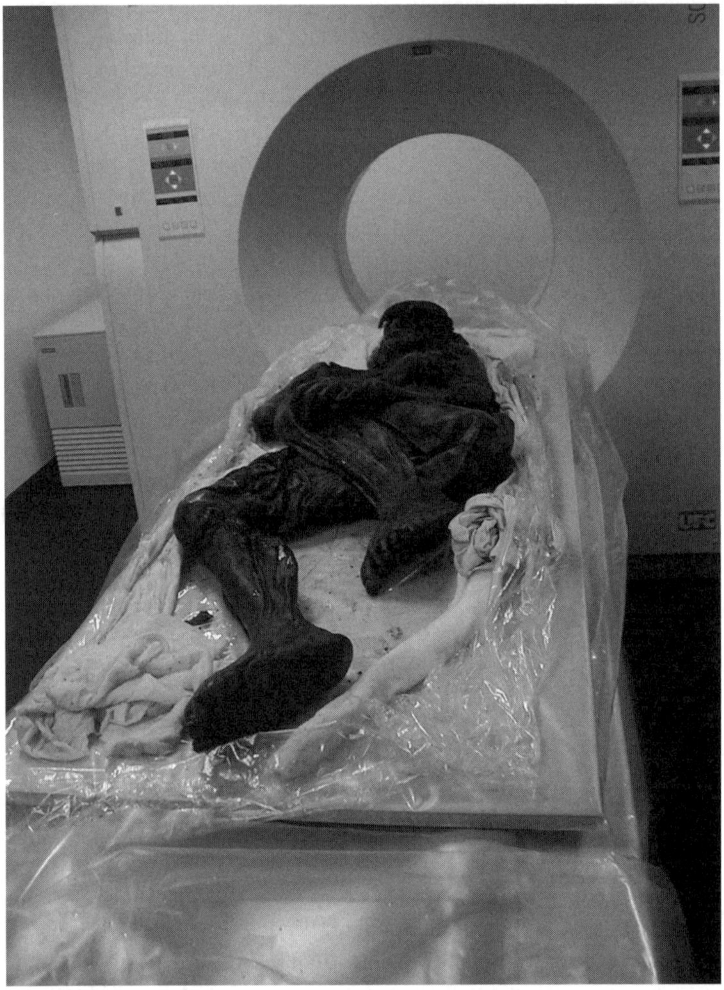

Figure 6.1 CT scanning the Borremose Woman (photograph: N. Lynnerup)

to irradiate the entire object, and with the introduction of multislice scanners achieved better volume rendering (Hsieh, 2002).

The CT scanner generates the slice images as an array of pixels (usually 512 × 512), with each pixel having a value depending on the attenuation of the X-rays as they pass through the object being scanned (Carlton and Adler, 2001). The attenuation is represented by Hounsfield units (HU), which are scaled and calibrated arbitrarily according to the attenuation of water, so that water has an HU of 1000 and air will have an HU of −1000 on the Hounsfield scale. The Hounsfield units are then converted to a grey scale potentially covering 256 shades of grey. This is more than the human eye can discern, so, consequently, the attenuation is remapped to about 20 grey-scale shades (image reconstruction). This minimizing of data is offset by adjusting the window width (i.e. the number of HUs differentiated on the grey-scale

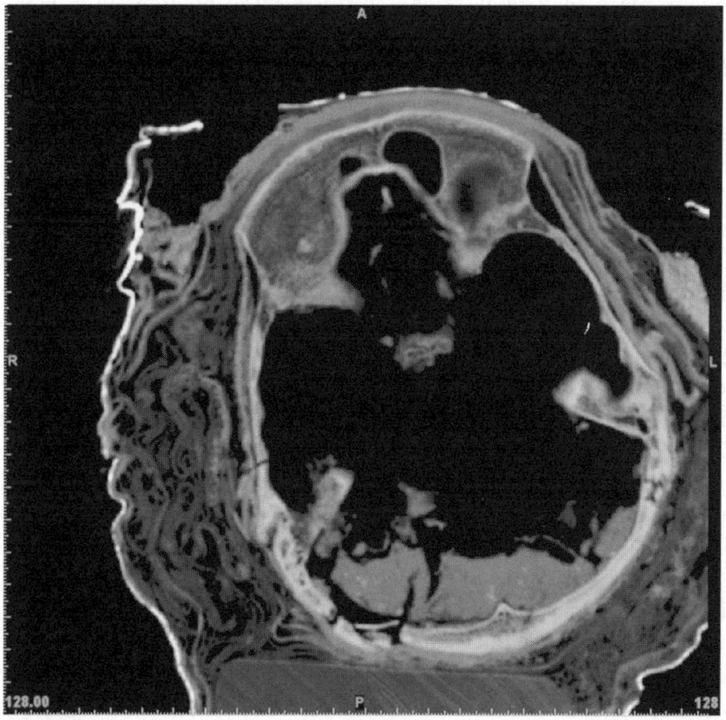

Figure 6.2 A single slice through a Ptolemaic Period Egyptian mummy (Carlsberg Gyptotek Museum, Copenhagen, Denmark). The many layers of wrapping are clearly seen. The scanning is of the head, and the cranial vault is clearly visible. The brain has been removed, as was customary for the embalming techniques (although there are some remains in the back of the vault). This has been done through the nasal aperture and ethmoid, and other bony structures in this area are indeed fractured. Since the brain has been removed, the CT scanner renders the cranial cavity as black. The skull fracture seen in the back of the vault is post-mortem

shades in the reconstructed remap) and the window level (the midpoint value of the window). Adjusting window width and window level enables viewing the structures of interest at the best possible resolution (Fleckenstein and Tranum-Jensen, 1993).

CT scanning of mummies was first carried out in 1977 (Harwood-Nash, 1977; Lewin and Harwood-Nash, 1977). Eleven mummies at the Museum of Fine Arts in Boston were CT scanned by Marx and D'Auria (1986) in what was then the most systematic CT scanning of a collection of mummies. Increasing computing power enabled imaging with an ever-finer resolution and rapid three-dimensional visualizations (e.g. Marx and D'Auria, 1988; Pickering *et al.*, 1990). Basically, the CT-scanner computer 'restacks' the single slices and, based on different algorithms, connects object boundaries between the slices (Zollikofer and Ponce de León, 2005). CT scanners used in hospitals come equipped with various computer programs that allow rapid image building and three-dimensional visualization, but it must be noted that the software is developed for medical purposes, and tuned to tissues and organ systems of the living. For example, the Hounsfield units associated with living bone do not vary much (pathology aside), so it is easy to preprogram a three-dimensional rendering of the skeletal tissues in a CT scanner, so that, for example, the skull of a scanned patient can be visualized just by a click on the appropriate menu. Mummies, and especially bog bodies may require much more work at this stage (see below). While the earlier CT scanners operated with a slice thickness of several millimetres, new CT scanners can achieve slice thicknesses of less than 1 mm, as well as the capacity for 'overlapping'. This is an important parameter for three-dimensional visualization, since the slice thickness will affect the quality of the three-dimensional image. If the slice thickness is several millimetres or more, then the resulting image will have a 'terraced' appearance. Obviously, this may also lead to misinterpretations of a three-dimensional structure, so it is always important to note the slice thickness when appraising CT scans (Hsieh, 2002; Zollikofer and Ponce de León, 2005). Large, detailed studies on embalming techniques and mummy wrappings of Egyptian mummies, employing three-dimensional imagery, have lately been performed by Hoffman *et al.* (2002) and Jansen *et al.* (2002). Recently, three Inca mummies were found on top of Mount Llullaillaco in Argentina. Owing to the extreme altitude, the 500-year old mummies were frozen. This meant that the internal organs were exceptionally well preserved and easily visible upon CT scanning (Previgliano *et al.*, 2003).

Three-dimensional visualization may be an important tool for the physical anthropologist. Since CT scanning is based on X-rays, bone is generally easy to visualize. This means that skeletal structures may be viewed so that they can be assessed morphologically, as well as measured. Visualizing the skull and pelvis makes it possible to assess the sex-specific traits virtually and use many of the methods described for 'real' bones (e.g. Buikstra and Ubelaker, 1994; Mays, 1998; White and Folkens, 2005). Virtual ageing may also be performed by focusing on dental development and epiphyseal morphology (Buikstra and Ubelaker, 1994; Mays, 1998; White and Folkens, 2005).

The benefits of three-dimensional visualization also apply to the assessment of possible pathological changes. For example, a mummy that had already been scanned in 1983 (Vahey and Brown, 1984) was re-scanned in 1990 in order to make three-dimensional visualizations of the skull and pelvis to confirm a possible fracture (Pickering *et al.*, 1990). Dental disease has also been visualized in this way (Melcher *et al.*, 1997), as has an Egyptian mummy where an intravital, foreign object (presumably a toe prosthesis) was found (Nerlich *et al.*, 2000). However, given all the mummies that have been CT scanned, pathological or traumatic finds have been rather sparse. Probably one of the more well-known cases involved the CT scanning of the Iceman, a glacier mummy found in Italy in 1992. The mummy had been X-rayed and CT scanned after the find (zur Nedden and Wicke, 1992). However, not until a renewed scanning 10 years later was an arrowhead identified in the left shoulder region (Gostner and Vigl, 2002). Rheumatoid arthritis has been diagnosed based on CT-visualized bone erosions and joint subluxation (Ciranni *et al.*, 2002). A CT scan of a natural mummy from the 19th century AD from a friary in Italy revealed a distended bladder and a ring of dense tissue at the site of the prostate, indicative of prostatic hyperplasia (Fornaciari *et al.*, 2001). Bone tumours have been identified in two Egyptian mummies (Taconis and Maat, 2005). Bone pathology as observed by CT scanning, indicative of tuberculosis, has been correlated with ancient-DNA analyses (Pap, personal communication). It should be mentioned that CT scanning has also been used to locate pathological processes or specific organs, in order to make precise incisions to perform biopsies (Brothwell *et al.*, 1990; Bennike, 2003), or to guide endoscopic examinations (Rühli *et al.*, 2002).

Problems in Mummy and Bog Body Computed Tomography Scanning and Three-Dimensional Visualization

The presence of many layers of mummy wrappings, especially those closely adhering to the skin, as well as organ removal, etc., may impede the interpretation of the body structures of a mummy. Diagenetic changes, mainly desiccation, may also have an impact. However, the skeletal structures are usually intact in both natural and artificial mummies. This means that structures useful for ageing and sexing may still be visualized. Availability of the skeletal images also results in having many reference points, which makes it easier to assess the remains of internal organs and structures.

Special problems of radiography and CT scanning arise when the diagenetic changes are so massive that the remaining tissues, including bone, are severely degraded. This is perhaps best shown by bog bodies. Owing to the acidic bog environment, calcium is leached from the bones, causing demineralization of the bony tissues, which consequently lose their hardness and become pliable (van der Sanden, 1996). When a bog body is X-rayed, bones are often very badly visualized. The bones have an appearance as if they were made of glass. This may be demonstrated by the radiographs taken of the Grauballe Man, a Danish bog body from the Iron Age, excavated in 1950 and X-rayed in 1955 (Figure 6.3). Consequently, when CT scanning a bog body, applying the same range of Hounsfield units for bones as in clinical work may result in the bones not being visualized at all. Furthermore, the demineralization is not necessarily uniform, but may differ within the skeletal system or a single bone due to the diagenetic microenvironment. This may generate a patchy appearance of the bone, even though it is intact morphologically. The second

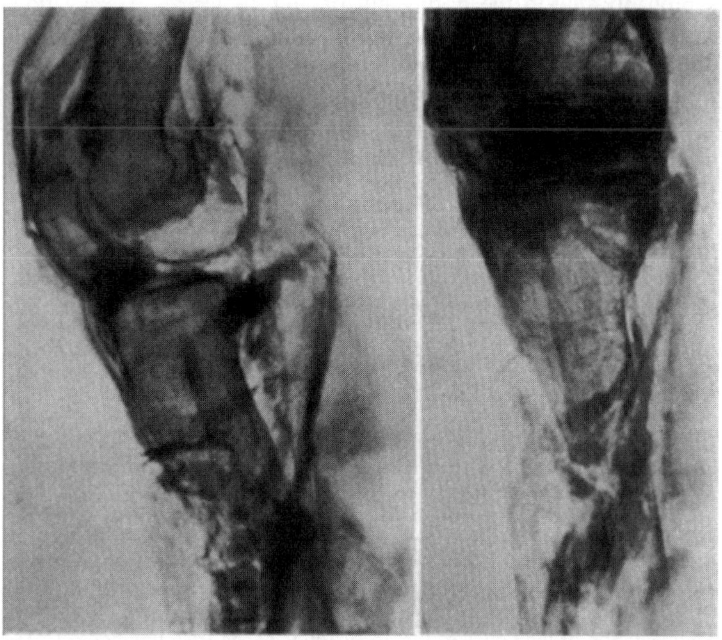

Figure 6.3 X-ray from 1955 of the left leg of the Grauballe Man (from Munck (1956))

difference is the fact that other tissues seem to acquire a more radiodense structure (i.e. the attenuation of the X-ray beams is increased). This is especially seen in some of the connective tissues, e.g. ligaments, fasciae and the subcutis. The reason for this is probably a deposition of mineral salts (containing metals such as iron) from the soil in collagenous tissues.

In the Iron Age, bogs in northern Europe had a special significance. Not only could they be difficult (even dangerous) to pass, but they were also used for rituals and sacrifices. The people of the Iron Age probably sacrificed weapons, boats, chariots, animals, and even humans in the bogs. They knew that people thrown in would be 'swallowed' up by the wet and cold bogs. They probably even knew that special preservation might take place in the bogs, so that textiles, leather and even human bodies might be preserved for a long time. Today, many bogs have disappeared, having been drained and then made into arable land. One other use of the bogs was for peat. Most northern European finds of bog bodies were discovered as a result of peat cutting (van der Sanden, 1996). In Denmark, peat was excavated throughout the last few centuries, but it was especially in the 1930s–1950s that peat was excavated on a large scale. It is from this period that most bog bodies were found, as accidental finds. Peat is still excavated in Ireland, and the extent to which the bog environment may have an effect on bog bodies can be illustrated by our work on three such recently found bog bodies from Ireland ('Clony Cavan Man', 'Old Man Croghan' and 'Derry Cashel'). These bog bodies have been subjected to a series of scientific studies, including CT scanning. The head of Clony Cavan Man shows extreme lateral flattening, with a facial breadth of only approximately 2–3 cm. The humeri of Old Man Croghan are an example of differential preservation: the right humerus is intact enough to be readily recognizable, whereas the left humerus is very difficult to visualize (Figure 6.4).

Many bog bodies are also shrunken. Whereas some shrinkage may take place in the bog, most is probably due to the drying out of bog bodies after excavation. Drying out was previously the only preservation and conservation method for 'wet' bog bodies. In the mid-20th century, more targeted methods were used, such as attempts to substitute the water with alcohols and tannic oils, bark extracts and wax, while today freeze-drying is used.

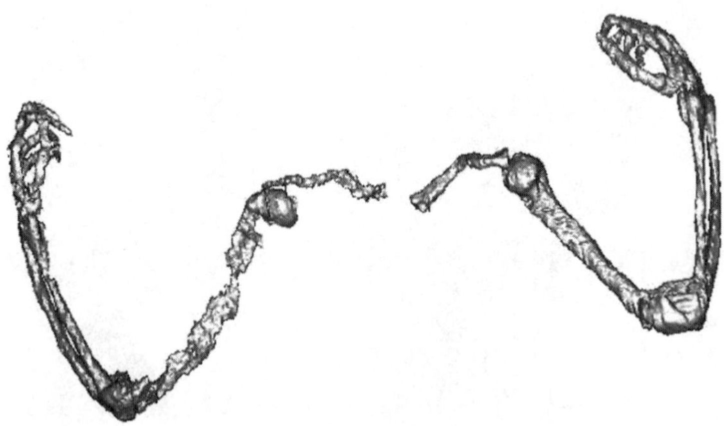

Figure 6.4 Segmentation of the bones of the upper limbs of the Irish bog body 'Old Man Croghan'. Note the difference in preservation of the humeri

Whatever the method, some shrinkage is probably inevitable; and it may be pronounced, especially for the older bog finds.

Thus, diagenetically modified tissues and organs may have to be delineated (or segmented) manually on almost every single slice. At present, features allowing manual slice-by-slice image editing are seldom available on the computer programs of the (clinical) CT scanners. The images, therefore, need to be transferred to another program that allows editing (post-image-capture processing) of the CT data.

Post-Capture Processing of Computed Tomography Scan Images

Several computer programs exist that allow post-image-capture processing of the CT data, especially the very important ability to edit the single image-slices manually. At our laboratory, we currently use a program package (MIMICS®) from Materialise®, Belgium (www.materialise.be). After import of the CT-scanner data file, the single-slice images may be shown. The program also allows for immediate multiplanar reformatting (Figure 6.5). The single image elements may then be edited. In order to visualize, for example, the head of the Grauballe Man, a colour coding is applied 'over' the single grey-scale pixels. Different tissues and organs may be given specific colours (Figure 6.5). This usually involves a process of identifying the single structures on the single-slice images and then following these structures through on the adjacent slices. For example, the bones of the skull may

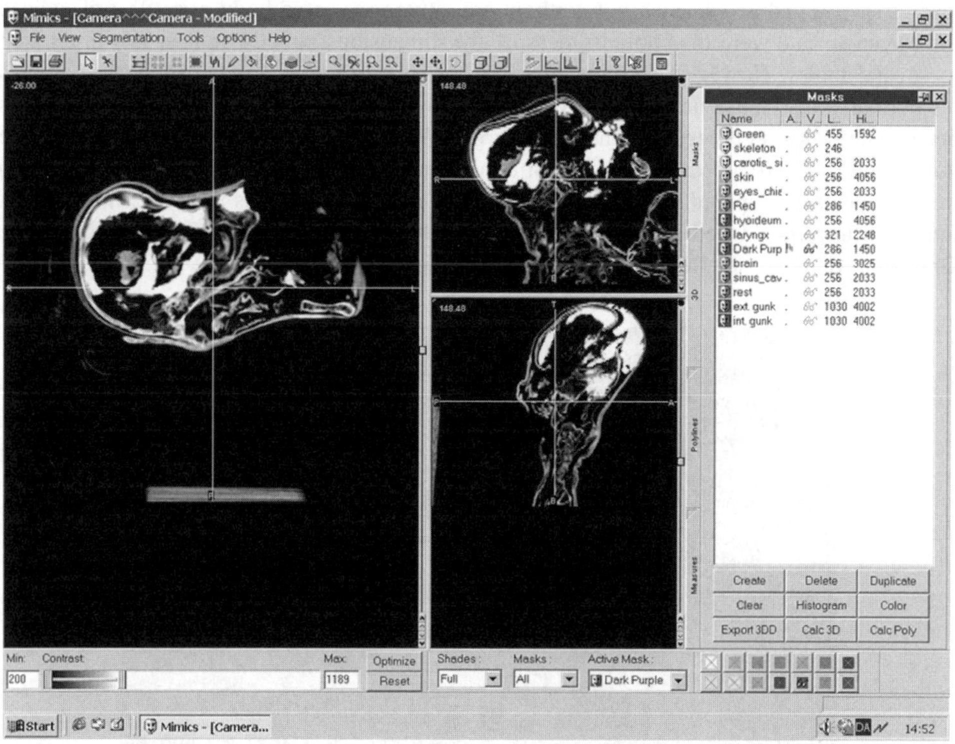

Figure 6.5 Program interface of MIMICS® showing a CT slice image being segmented by application of 'masks' (colour coding) based on thresholding of the grey-scale values

first be picked out, then the brain, etc. All the single slices have to be edited, so that that ultimately the individual structures are completely delineated (segmented). The colour-coded pixels associated with the various structures may then be extracted, and used as a basis for three-dimensional rendering.

The manual process of delineating and extracting the relevant items (tissues, organs, etc.) necessitates a certain anatomical knowledge, the more so as the bones often are bent and deformed and organs are often shrunken. There is some room for subjectivity when performing the segmentation. Along with the previously stated caveat on slice thickness (and indeed on the inherent properties of X-ray-based image acquisition), this means that CT-scanning images and three-dimensional renderings should not be viewed as a totally objective and 'true' representation of internal structures and tissues. Some pathological processes may be falsely ascribed to diagenetic processes, and vice versa.

Once tissues and organs have been segmented and visualized, they can then be measured and assessed morphologically. An initial benefit is the visualization of bone structures that may be used for sexing and ageing. We segmented the Grauballe Man pelvis into the single constituent bones and were thus able to display the auricular surface (Figure 6.6). Although the resolution does not, for example, allow a direct application of the auricular ageing technique (Lovejoy *et al.*, 1985), some information may be salvaged.

Although some bog bodies have preservation of soft tissues that allow direct sexing, other bog bodies, especially body parts, rely on identifying certain features associated with the male and female skeleton. This also includes measurements. The Frer foot is a foot in a leather shoe found in a Danish bog without any other associated body parts. We visualized the calcaneus and were then able to measure the dimensions directly and set this in relation to several tables. However, applying the raw measures in tables and formulae developed on normal bones must be done with some caution. As noted above, not only do bones become deformed in a bog, they may also shrink. When we analysed the Grauballe Man, we noted an overall 10 % decrease in several bone measurements (both cranial and post-cranial measurements), compared with the mean values of non-bog skeletons from the same period. It is then uncertain whether this 10 % decrease reflects diagenetic change, or whether the Grauballe Man was smaller than other people of his day and age.

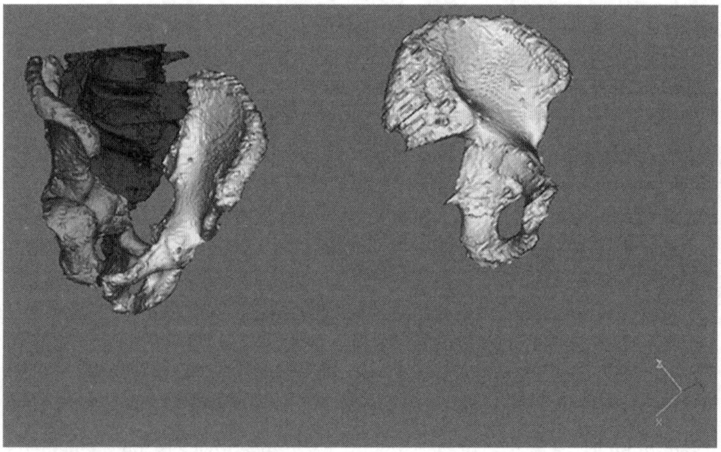

Figure 6.6 The pelvis and an isolated innominate showing the auricular surface: Grauballe Man

Bog Body Lesions and Trauma

Grauballe Man was first examined in 1950. Forensic pathological and radiographic analyses at that time showed that he had had his throat cut, and probably had sustained a cranial fracture and a tibial fracture perimortally. This was interpreted by the archaeologists as evidence of a sacrificial execution, whereby he had his leg broken in order to incapacitate him, then a blow to his head and finally a sharp, mortal wound to his throat (Munck, 1956; Glob, 1965).

We CT scanned the body in 2001, and after image segmentation it was possible to visualize the throat organs (Figure 6.7). The hyoid bone and the larynx were not fractured. This indicates that when the Grauballe Man had his throat slit, that this was done with his head held back. This supports the theory of execution: by holding his head back, his throat was exposed, and his throat was then slit from ear to ear, probably by a person also standing at his back. A direct blow frontally to his throat, for example by a sword or axe, would more probably have fractured or lacerated the larynx or hyoid.

On the other hand, the previously noted cranial fracture did not seem as certain. When visualizing the cranium, it was clearly seen how the appearance was more suggestive of post-mortem influences (Figure 6.8). Unlike the Clony Cavan Man, the head had not been crushed completely, but had bilateral impressions. On the right side, the bone was also

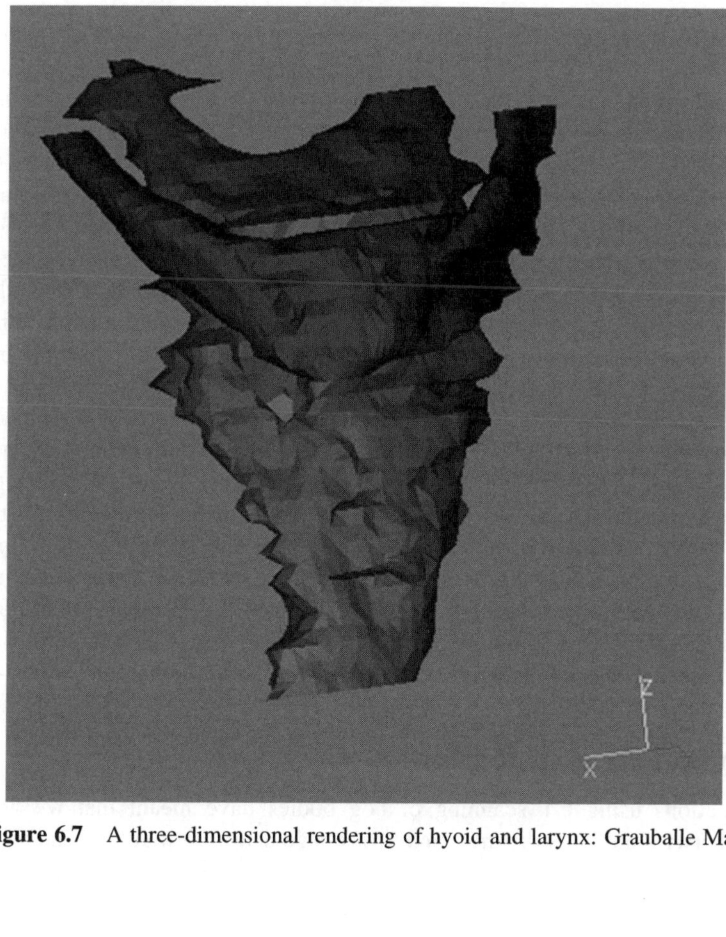

Figure 6.7 A three-dimensional rendering of hyoid and larynx: Grauballe Man

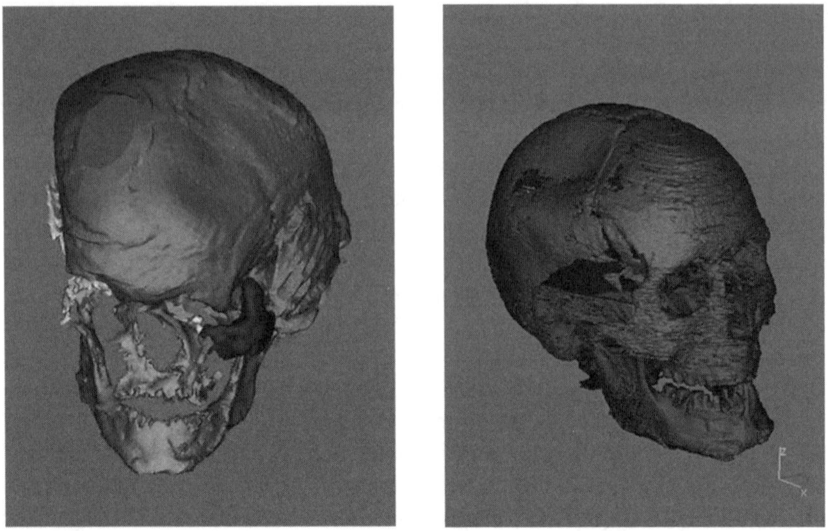

Figure 6.8 The skull of the Tollund Man (right) and the Grauballe Man (left). Note the impressions in the parietal regions of both crania

fractured and bent inward. The fracture did not resemble a peri-mortem fracture as seen in modern clinical cases. We felt this could be substantiated when we studied another bog body. The Tollund Man is from the same time period as Grauballe Man, and found in a bog near that of the Grauballe Man. Tollund Man's head is extremely well preserved; yet, when visualizing his skull, the same kind of concave impression may be seen (Figure 6.8).

The Grauballe Man's leg fracture at first did seem to resemble a peri-mortem traumatic event. Indeed, the initial appraisal was that this tibia fracture did look somewhat like present-day traffic accidents, where a person is hit by the fender of a car. However, the fibula had not fractured, and the leg had a rather pronounced rotation (Figure 6.9). This could be a case of a peri-mortem fracture, followed by post-mortem degradation and rotation. When we first analysed the Grauballe Man, we could not reach a decision as to whether the fracture was peri- or post-mortem. However, a few years later, in 2004, we CT scanned the Borremose Woman (an Iron Age bog body from northern Jutland) and we noted that she had a femoral fracture. Due to the way her legs lay, it seemed clear to us that this fracture was post-mortem due to soil pressure, the fractured femur lying across the other leg. Also, the fracture did not present itself as identical to modern clinical cases. When perusing the single CT-images of the Borremose Woman's thigh, we noted that the fracture not only showed itself as lines between bone segments, but also as a curious separation between trabecular bone and compact bone. This, in fact, was the same kind of fracture also seen in the Grauballe Man tibia fracture. We find that since the Borremose Woman most probably had a post-mortem fracture with these characteristics, then probably the Grauballe Man tibia fracture is also post-mortem.

Pathology and Pseudopathology

Our examinations using CT scanning of bog bodies have meant that we were able to visualize otherwise hidden structures, and post-image-acquisition editing further allowed us

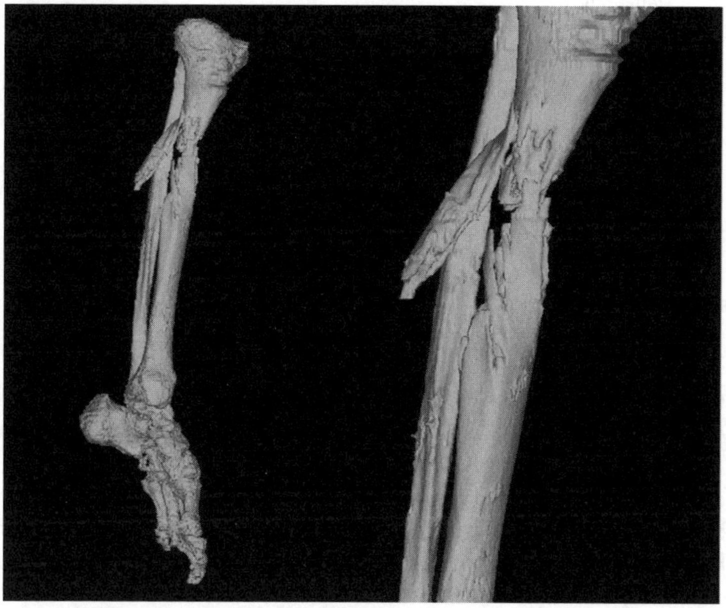

Figure 6.9 Tibia fracture: Grauballe Man

to visualize demineralized bone. This, in turn, enabled us to reappraise some previously described lesions. The acid bog diagenetics mean that bone will be demineralized, become pliable and, upon subsequent excavation and drying out, also shrink and warp. This means that, under these conditions, aetiological attribution of pathology and trauma lack certainty. It is our contention that some previously described lesions associated with the bog bodies are more an effect of post-mortem factors than trauma. This is, of course, complicated by the fact that a perimortal lesion may also be diagenetically affected.

The aforementioned Irish bog body finds may also illustrate damage due to excavation, aside from the post-mortem degradation due to the bog. Unlike earlier finds in the 19th century and up until the 1950s, peat is now excavated from bogs by machine. Peat-cutting machinery may sever body parts, and subsequent transport on conveyor belts may further damage the body. Clony Cavan Man had lost his forearms and lower abdomen on this account.

Based on our analyses, therefore, we suggest that bog body lesions and pathology should be viewed in a two-axis continuum: one axis covering pre-, peri-, and post-mortem time periods, and the other axis ranging from 'true' lesions to clearly pseudopathology, i.e. lesions or pathology due to diagenetic change (Figure 6.10).

For example, the Grauballe Man throat wound is without doubt a peri-mortal lesion. Also, it clearly presents itself as a lesion, with only little post-mortem change (the skin has shrunk from the cut surfaces). This would yield a clear point in the graph (number 1 on Figure 6.10). The Tollund Man head impression is, on the other hand, clearly post-mortem, due to diagenetic change (number 2 on Figure 6.10). The Grauballe Man cranial lesion is most probably quite like the Tollund Man lesion; but, since there is some fracturing and

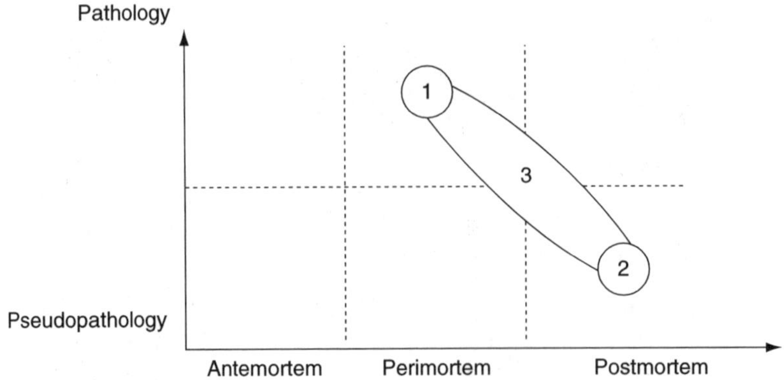

Figure 6.10 Pathology and pseudopathology versus time-period when lesion was inflicted

bone loss, it cannot be definitively ruled out that a peri-mortem lesion was there originally, which has since been heavily degraded and warped post-mortem. This corresponds to point 3 on Figure 6.10.

As a final case in point, we looked at the Borremose Woman's head. Her skull was found to be heavily fractured, and from the archaeological viewpoint this was seen as perhaps how she had been executed: by a blow or multiple blows to the face (Ry Andersen and Geertinger, 1984). From a forensic viewpoint this would be somewhat unusual: craniofacial fracturing of that magnitude is mainly found in high-energy impacts, e.g. traffic accidents. Inflicting these fractures would be difficult using a blunt weapon even if she was lying on her back on the ground. The Borremose Woman was found lying face down in the bog. Owing to the diagenetic effects, the fracturing of the facial skeleton might be more indicative of sutures coming apart and soil pressure 'flattening the face' in an antero-posterior direction (unlike the more lateral compression seen for Grauballe Man and Tollund Man, who both lay with their head on the side). The three-dimensional visualization of her skull shows that the cranial bones show major dislocation, with some bones separating at the sutures, while some bones do present fracture lines. This means that a peri-mortem lesion cannot be ruled out, perhaps only with minor fractures, but which has since been heavily degraded post-mortem, resulting in the total separation of the cranial and facial bones. Indeed, we feel that she might even have been dealt a single blow to the back of the head, resulting in an impact fracture to the posterior cranium, with fracture lines extending forward to the facial bones and cranial base (as, indeed, is seen in modern forensic pathology of head trauma). Further post-mortem degradation, especially due to the soil pressure, could then explain the findings.

Computed Tomography Scanning and the Biocultural Perspective

As with other investigative techniques and methods employed for the study of mummies, the data generated from CT scanning and three-dimensional visualization must ultimately be put in a biocultural context. In this respect, the differentiation between natural and artificial mummies is important. What can be seen in artificial mummies will require a basic knowledge of the embalming techniques employed, which may alter or remove organs. On the other hand, natural mummies reflect the diagenetic changes to which the body has been subjected. At one end of the spectrum, the natural mummies from Mount Llullaillaco, preserved in permafrost, are examples of an exceptional degree of preservation of internal organs. Bog bodies stand at the other end, being very much degraded due to the acidic bogs. This establishes limitations in terms of what can be visualized by CT scanning. Overall, perusing the results of CT scanning of mummies and bog bodies, it seems to this author that the major contribution of CT scanning to the biocultural interpretations of mummy finds is in the identification of mummification methods for artificial mummies and of trauma in natural mummies. Palaeopathological lesions are generally rare, especially in non-mineralized tissues. For those pathological conditions identified, it may perhaps be discussed to what extent minor pathologies have a major impact on biocultural understanding; trauma, however, is clearly indicative of an active, maybe even sacrificial, cultural setting. The Tyrolean Iceman had an arrow point lodged in his shoulder (Gostner and Vigl, 2002), while nearly all the bog bodies have definite signs of having been put to death (van der Sanden, 1996).

One should, perhaps, focus just as much on the physical anthropological data derived from CT scannings. For example, three-dimensional visualizations make it possible to ascertain age and sex, and these data may in themselves be important when placing mummies in a biocultural perspective. The Mount Llullaillaco mummies were all determined to be children or juveniles, which is highly informative of Inca human sacrificial customs (Previgliano *et al.*, 2003). Likewise, the Greenland mummies from Qilakitsoq were all children (two) and women (six), again attesting to Greenlandic Thule culture social structures. Very few male skeletons are found, presumably because the males have a higher risk of death while out hunting. On the other hand, if the male hunter dies, the rest of the family may die from starvation (Hart Hansen, 1989).

CONCLUSION

The methods of CT scanning and subsequent three-dimensional visualization are powerful analytical tools for palaeopathological and physical anthropological analyses of mummies and bog bodies. The technique allows the visualization of internal structures, and especially of bones and teeth. However, post-image-capture processing is often necessary to extract the full information from the CT data. This is especially the case with bog bodies, where there is much taphonomic alteration of the skeletal structures. We have CT scanned several Danish bog bodies, and based on our findings we feel that some lesions previously claimed as signs of peri-mortem trauma may have to be re-evaluated. Owing to the sometimes extreme post-mortem, diagenetic influences on the bog bodies, it may be impossible ever to be certain of the exact nature of the observed lesions. We propose to view the lesions as a continuum, covering both pseudopathology and 'true' lesions, over pre- to post-mortem time periods.

Finally, we wish to draw attention to the fact that CT scanning may not only be a valuable analytical tool, but also an extraordinarily precise tool for documenting bog bodies and mummies. The CT scanning process generates data, literally millimetre by millimetre and inside out, in a digital format that may be accessed and shared with other scientists or museums. Our cooperation with Argentine scientists in the investigation of the Mount

Llullaillaco mummies is an example of such a data sharing. The CT scanning data may also be important in terms of future assessments of preservational status of the bog bodies. Using so-called stereolithography, it is possible to produce 1:1 models of CT-scanned structures directly from the computer visualizations (zur Nedden and Wicke, 1992; Hjalgrim *et al.*, 1996; Cesarani *et al.*, 2004), detailing structures at a 1 mm resolution (Figure 6.11). This has some very evident uses for exhibits, and has been used as basis for facial reconstruction of two bog bodies (Figure 6.12). Aside from advances in resolution and faster CT scanners, future prospects will probably include more CT-scanned bog bodies and mummies that will lead to better comparative studies. These can be expected to result in a better understanding of mummification, taphonomical changes, pseudopathologies, and the nature of observed lesions and tissue preservation. In this respect, use of CT scanning in mummy studies as a 'screening tool' for locating pathological changes for subsequent precise sampling or for guiding endoscopic procedures will probably become more common.

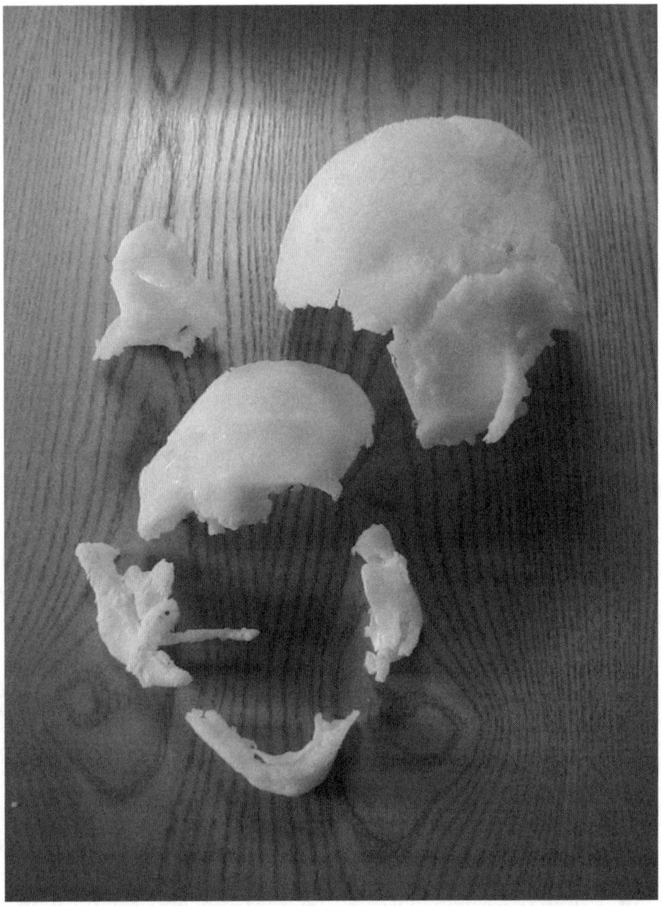

Figure 6.11 Stereolithographic models of the Borremose Woman cranial bones

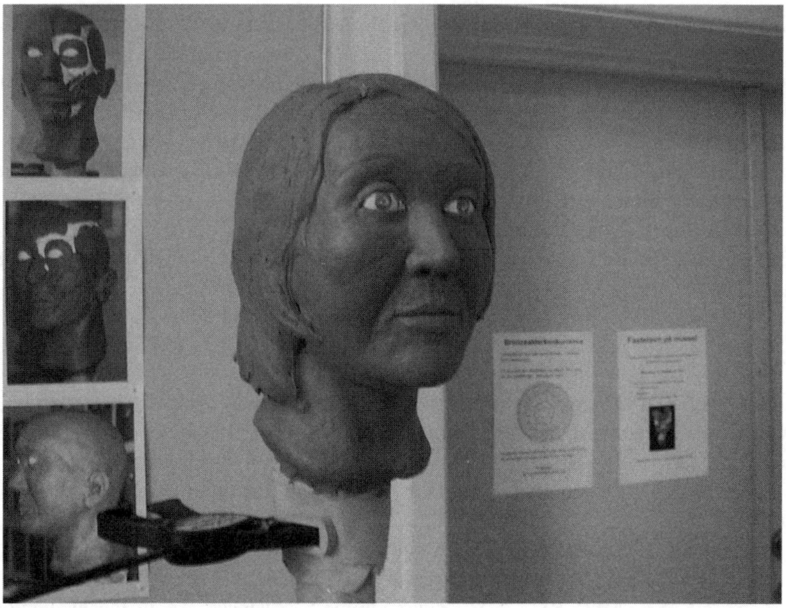

Figure 6.12 Facial reconstruction of the Borremose Woman (photo: N. Lynnerup)

REFERENCES

Ambrose J. 1975. A brief review of the EMI scanner. *Proc Brit Inst Radiol* **48**: 605–606.

Aufderheide A. 2003. *The Scientific Study of Mummies*. Cambridge University Press: Cambridge.

Bennike P. 2003. Bodies and Skeletons from Danish Bogs. Do they tell the same story? In *Mummies in a New Millennium, Proceedings of the 4th World Congress on Mummy Studies*, Lynnerup N, Andreasen C, Berglund J (eds). Danish Polar Centre, Publication No. 11. Danish Polar Centre: Copenhagen; 39–42.

Bloomfield JA. 1985. Radiology of Egyptian mummy. *Austr Radiol* **29**: 64–66.

Brothwell D, Sandison AT (eds). 1967. *Diseases in Antiquity*. Charles. C. Thomas: Springfield.

Brothwell D, Liversage D, Gottlieb B. 1990. Radiographic and forensic aspects of the female Huldremose body. *J Danish Archaeol* **9**: 157–178.

Buikstra JE, Ubelaker DH. 1994. *Standards for Data Collection from Human Skeletal Remains*. Arkansas Archaeological Survey Research Series, 44. Arkansas Archaeological Survey Fayetteville, AR.

Carlton RR, Adler AM. 2001. *Principles of Radiographic Imaging. An Art and a Science*. Delmar: Albany, NY.

Cesarani F, Martina MC, Grilletto R, Boano R, Roveri AMD, Capussotto V, Giuliano A, Celia M, Gandini G. 2004. Facial reconstruction of a wrapped Egyptian mummy using MDCT. *Am J Radiol* **183**: 755–758.

Christensen OE. 1969. Un examen radiologique des mommies Egyptienne des musées Danois. *La Semaine Hôpit* **45**: 1990–1998.

Ciranni R, Garbini F, Neri E, Melai L, Giusti L, Fornaciari G. 2002. The "Braids Lady" of Arezzo: a case of rheumatoid arthritis in a 16th century mummy. *Clin Exp Rheumatol* **20**: 745–752.

Cockburn A, Cockburn E. 1980. *Mummies, Disease and Ancient Cultures*. Cambridge University Press: Cambridge.

David RA. 1979. *The Manchester Museum Mummy Project*. Manchester University Press: Manchester.

Fawcitt RA, Jarvis H, Isherwood I. 1984. X-raying the Manchester mummies. In *Evidence Embalmed. Modern Medicine and the Mummies of Ancient Egypt*, David R, Tapp E, editors. Manchester University Press: Manchester.

Fleckenstein P, Tranum-Jensen J. 1993. *Anatomy in Diagnostic Imaging*. Munksgaard: Copenhagen.

Fornaciari G, Ciranni R, Ventura L. 2001. Paleoandrology and prostatic hyperplasia in Italian mummies (XV–XIX century). *Med Secoli* **13**: 269–284.

Glob PV. 1965. *Mosefolket*. Gyldendal: Copenhagen.

Gostner P, Vigl E. 2002. Report of radiological-forensic findings on the Iceman. *J Archaeol Sci* **29**: 323–326.

Harris JE, Weeks KR. 1973. *X-Raying the Pharaohs*. Schribners and Sons: Chicago, IL.

Harris JE, Wente EF. 1980. *An X-Ray Atlas of the Royal Mummies*. Chicago University Press: Chicago, IL.

Hart Hansen JP. 1989. The mummies from Qilakitsoq – paleopathological aspects. In *The Mummies from Qilakitsoq – Eskimos in the 15th Century*, Hart Hansen JP, Gulløv HC (eds). Meddelelser om Grønland, Man and Society Series, 12. Danish Polar Centre: Copenhagen; 69–82.

Harwood-Nash DC. 1979. Computed tomography of ancient Egyptian mummies. *J Comput Assist Tomogr* **3**: 768–773.

Hjalgrim H, Lynnerup N, Liversage M, Rosenklint A. 1995. Stereolithography: potential applications in anthropological studies. *Am J Phys Anthropol* **97**: 329–333.

Hoffman H, Torres WE, Ernst RD. 2002. Paleoradiology: advanced CT in the evaluation of nine Egyptian mummies. *RadioGraphics* **22**: 377–385.

Hounsfield GN. 1976. Historical notes on computerized axial tomography. *J Can Assoc Radiol* **27**: 135–142.

Hsieh J. 2002. *Computed Tomography. Principles, Design, Artifacts, and Recent Advances*. SPIE: Bellingham, WA.

Jansen RJ, Poulus M, Taconis W, Stoker J. 2002. High-resolution spiral computed tomography with multiplanar reformatting, 3D surface- and volume rendering: a non-destructive method to visualize ancient Egyptian mummification techniques. *Comput Med Imag Graph* **26**: 211–216.

Koenig W. 1896. *14 Photographien mit Röngten-Strahlen*. Johann Ambrosius Barth: Leipzig.

Lewin PK, Harwood-Nash DC. 1977. Computerized axial tomography in medical archaeology. *Paleopathol Newslett* **17**: 8–9.

Lovejoy CO, Meindl RS, Pryzbeck TR, Mensforth RP. 1985. Chronological metamorphosis of the auricular surface of the illium: a new method for the determination of adult skeletal age at death. *Am J Phys Anthropol* **68**: 15–28.

Lowenstein EJ. 2004. Paleodermatoses: lessons learned from mummies. *J Am Acad Dermatol* **50**: 919–936.

Marx M, D'Auria SH. 1986. CT examination of eleven Egyptian mummies. *RadioGraphics* **6**: 321–330.

Marx M, D'Auria SH. 1988. Three-dimensional CT reconstructions of an ancient human Egyptian mummy. *Am J Radiol* **150**: 147–149.

Mays S. 1998. *The Archaeology of Human Bones*. Routledge: London.

Melcher AH, Holowka S, Pharoah M, Lewin PK. 1997. Non-invasive computed tomography and three-dimensional reconstruction of the dentition of a 2,800-year-old Egyptian mummy exhibiting extensive disease. *Am J Phys Anthropol* **103**: 329–340.

Moodie RL. 1931. *Roentgenologic Studies of Egyptian and Peruvian Mummies*. Field Museum of Chicago: Chicago, IL.

Munck W. 1956. Patologisk-anatomisk og retsmedicinsk undersøgelse af moseliget fra Grauballe. *KUML* (*J Jutland Archaeol Soc*) **1956**: 131–137.

Neave RAH. 1979. The reconstruction of skulls for facial reconstruction using radiographic techniques. In *Science in Egyptology*, David AR (ed.). Manchester University Press: Manchester; 329–336.

Nerlich AG, Zink A, Szeimies U, Hagedorn HG. 2000. Ancient Egyptian prosthesis of the big toe. *Lancet* **356**: 2176–2179.

Notman DNH, Beattie OB. 1995. The paleoimaging and forensic anthropology of frozen sailors from the Franklin Arctic expedition mass disaster (1845–1848): a detailed presentation of two radiological surveys. In *Human Mummies: A Global Survey of their Status and the Technique of Conservation*, Spindler K, Wilfing H, Rastbichler-Zissering E, zur Nedden D, Nothdurfter H (eds). Springer: New York; 3–8.

Notman DNH, Anderson L, Beattie OB, Amy R. 1987. Arctic paleoradiology: portable radiographic examination of two frozen sailors from the Franklin Expedition (1845–1848). *Am J Roent* **149**: 347–350.

Ortner DJ. 2003. *Identification of Pathological Conditions in Human Skeletal Remains*. Academic Press: San Diego, CA.

Petrie WMF, Griifith FLl. 1898. *Deshashesh 1897*. Fifteenth Memoir of the Egypt Exploration Fund. Egypt Exploration Fund: London.

Pickering RB, Conces JF, Braunstein EM, Yurco F. 1990. Three-dimensional computed tomography of the mummy Wenuhotep. *Am J Phys Anthropol* **83**: 49–55.

Previgliano CH, Ceruti C, Reinhard J, Araoz FA, Diez JG. 2003. Radiologic evaluation of the Llullaillaco mummies. *Am J Radiol* **181**: 1473–1479.

Rühli FJ, Hodler J, Böni T. 2002. CT-guided biopsy: a new diagnostic method for paleopathological research. *Am J Phys Anthropol* **117**: 272–275.

Ry Andersen S, Geertinger P. 1984. Bog bodies investigated in the light of forensic medicine. *J Danish Archaeol* **3**: 111–119.

Smith GE. 1912. *The Royal Mummies. Catalogue General des Antiquitees Égyptiennes de Musée du Cairo*. Musée du Cairo: Cairo.

Taconis WK, Maat GJR. 2005. Radiological findings in the human mummies and human heads. In *Egyptian Mummies. Radiological Atlas of the Collections in the National Museum of Antiquities in Leiden*, Raven MJ, Taconis WK (eds). Brepols: Turnhout; 53–80.

Vahey T, Brown D. 1984. Comely Wenuhotep: computed tomography of an Egyptian mummy. *J Comput Assist Tomogr* **8**: 992–997.

Van der Sanden, W. 1996. *Through Nature to Eternity. The Bog People of Northwest Europe*. Drents Museum : Assen.

White TD, Folkens PA. 2005. *The Human Bone Manual*. Academic Press: San Diego, CA.

Zollikofer CPE, Ponce de León MS. 2005. *Virtual Reconstruction. A Primer in Computer-Assisted Paleontology and Biomedicine*. Wiley: New York.

Zur Nedden D, Wicke K. 1992. Der Eisman aus der Sicht der radiologischen und computertomographischen Daten. In *Der Mann in Eis*, Höpfel F, Platzer W, Spindler K (eds). Innsbrück University: Innsbrück; 131–148.

7

Histological Studies on Ancient Bone

Gordon Turner-Walker[1] and Simon Mays[2]

[1] *School of Cultural Heritage Conservation, National Yunlin University of Science and Technology, 123 University Road Sec. 3, Touliou, 640 Yunlin, Taiwan (ROC)*
[2] *English Heritage Centre for Archaeology, Fort Cumberland, Eastney, Portsmouth PO4 9LD, UK*

INTRODUCTION

At a microscopic level, the structure of ancient bone can be examined in a number of different ways. Well-established techniques include optical microscopy, microradiography, scanning electron microscopy (SEM) and transmission electron microscopy. More recent developments in microscopy that see applications in the analysis of archaeological and fossil bones are confocal laser scanning microscopy (CLSM), near-field scanning optical microscopy (NSOM), atomic force microscopy (AFM) and X-ray microtomography, a development of medical computed axial tomography. Some of these latter techniques are highly specialized, requiring large and expensive facilities, and, therefore, are restricted to exceptional specimens or studies that aim to test and refine methodologies. By far the most common techniques applied to archaeological specimens are optical and electron microscopies in their various forms. Both optical and electron microscopy can be applied to untreated specimens to reveal details of three-dimensional architecture, or to embedded and sectioned specimens to explore their morphologies across a two-dimensional plane by preparing polished sections of various thicknesses. Polished thick sections may be examined using reflected light microscopy or backscatter SEM (BSEM). This work will confine itself mainly to the microscopical analysis of planar sections through ancient bone tissues. Bone, and a range of other biological tissues, is routinely examined microscopically in thin sections, the study of which is called histology. The term histology refers to the study of all animal, plant or other biological tissues, which are normally embedded in a substrate such as paraffin wax (or resin) before being cut into fine slices using a microtome. The extreme thinness of the slices means that, generally, only

Advances in Human Palaeopathology Edited by Ron Pinhasi and Simon Mays

one layer of cells is imaged when they are examined using light microscopy or transmission electron microscopy, making the interpretation of structures very much simpler.

A BRIEF HISTORY OF BONE HISTOLOGY

The foundations of histology were laid in the 18th century soon after the invention of the microscope and its application to tissue samples taken from living organisms. Famously, Robert Hooke is credited with the discovery of the 'cell' when examining cork with a microscope of his own design in 1665. However, many historians of science credit the Italian anatomist Marcello Malpighi (1628–1694), as the true 'Father of Histology', although the term histology was not coined until 1819 when the German anatomist Mayer (1819) used the term. The word is derived from two classical Greek root words: *histos* meaning tissue and *logos* meaning word, study or science (Drury and Wallington, 1980). In Greek, the term *histos* can be taken to mean a woven material, web, or loom, but it has been accepted as a general term for the tissues, or layers of cells that constitute the structure of multicellular organisms. The science of histology took a major step forward in 1863 when Wilhelm Waldeyer (more properly Heinrich Wilhelm Gottfried von Waldeyer-Hartz), an influential Berlin anatomist, employed haematoxylin, a natural dye extracted from the logwood tree, to reveal the structures of nerve cells and cell nuclei in histological sections. In that same year, Waldeyer was one of the first histologists to employ Paris blue, one of the newly synthesized aniline textile dyes, to stain tissues.

By the end of the 19th century the microscopic structures of bone had been widely described, techniques of polarizing microscopy developed, and methods for embedding and sectioning (or grinding) of specimens had been perfected. However, the science of archaeological bone histology had not advanced significantly compared with other aspects of archaeology and palaeontology (Garland, 1993; de Ricqlès, 1993). One of the first examples of the successful application of histological techniques to a palaeopathological problem was carried out in 1891 in the USA and concerned the misdiagnosis of syphilis in two tibiae excavated from a Native American burial ground in Colorado. Histological work at the College of Physicians and Surgeons of New York showed that the specimens exhibited only chronic periostitis and osteomyelitis (Garland, 1993). The term palaeohistology was first coined by Moodie (1926) and subsequently formally defined by Graf (1949), who stated 'It would seem rather natural to apply the term palaeohistology to the examination of microscopic sections of human beings and the recognising of tissues and cells in such sections' (quoted from Garland (1993: 2)). Since that date palaeohistology as a discipline has been extended to apply to non-human bones and has most frequently involved examination of skeletonized material in which no soft tissues remain.

Significant work on the palaeohistology of modern and fossil animal bones was published by Enlow and Brown between 1956 and 1958 (Enlow and Brown, 1956, 1957, 1958). These publications give detailed instructions on the preparation and examination of thin sections, together with a definitive classification of the major groups of structural patterns found in bone, covering both recent and fossil genera of fish, mammals and birds. However, this work was primarily concerned with understanding the evolution of bone as a tissue, rather than the modification of bone microstructure either as a result of pathological changes or post-mortem diagenesis. Later studies focused on attempts to elucidate the metabolism of extinct vertebrates. For example, histological studies of juvenile and adult dinosaur bones have shown well-developed vascularity and Haversian systems, providing evidence of high growth rates and high metabolism in some dinosaurs, leading some researchers to conclude that they may have been warm-blooded (de Ricqlès, 1974, 1980). Equivalent studies of the bones of early mammal-like reptiles or therapsids (dinocephalians) indicate that these too may have been endothermic (de Ricqlès, 1974; Bakker, 1975; Kemp, 1982).

In palaeopathology, study of specimens continues to rely chiefly on gross and radiographic examination. Newer techniques, such as biomolecular analyses of ancient DNA and specific lipid markers (Lowenstein and Scheuenstuhl, 1991; Taylor *et al.*, 1996; Faerman *et al.*, 1997, 2000; Gernaey and Minnikin, 2000; Gernaey *et al.*, 2001; Mays *et al.*, 2001; Redman *et al.*, 2002), are also beginning to make an impact. Although it is a technique with a long history, palaeohistology has until recently seen only fairly limited application in palaeopathology. One reason for this may be an understandable reluctance of many museum curators to permit damage or destruction of specimens in order to obtain the necessary samples. This reticence might be expected to decline in future as successful applications of histological methods to problems of diagnosis are published and the rewards of a certain amount of destructive analyses are recognized. Advances made in the extraction and amplification of ancient DNA and other biomolecular evidence from important specimens such as Neanderthal bones (Ovchinnikov *et al.*, 2000; Nielsen-Marsh *et al.*, 2005) have demonstrated that the competing demands of conservation and destructive analysis can be balanced if the resulting damage is minimized and the knowledge gained is valuable and could not have been obtained using a non-destructive approach. Other important factors that may facilitate the greater use of palaeohistology are refinements in sectioning techniques that permit the removal of extremely thin half-sections from bones, thus dramatically reducing the loss of potentially important specimens, and advances in completely non-destructive imaging methods, such as microtomography. These issues will be discussed in greater detail below.

PRACTICAL METHODS AND TECHNIQUES IN PALAEOHISTOLOGY

Techniques used to prepare, section and image specimens of archaeological bones grew out of standard histological methodologies developed for looking at both soft and mineralized modern tissues. When examining samples of bone from biopsies, surgery or autopsy it is still common to decalcify the specimens to leave only the organic (collagen) matrix and its associated cells which can then be fixed, dehydrated, embedded in paraffin wax and sectioned. In fresh bone specimens it is necessary first to stabilize the cells and other soft tissues from putrefaction by 'fixing' them with a cross-linking agent, normally a 10 % solution of formaldehyde in phosphate-buffered saline. The specimens are then dehydrated through successive baths of progressively more concentrated ethanol to dehydrate the tissues. They are then transferred to a clearing solution of an aromatic solvent, such as toluene or xylene, before embedding in hot paraffin wax. This process of decalcification and embedding permits the sectioning of bone tissues into extremely thin sections of 2–8 μm using a microtome. With ancient bone specimens it is impractical and frequently undesirable to decalcify the specimens. Archaeological bones have frequently lost a substantial amount of their organic matrix and are generally too fragile to survive decalcification. Furthermore, decalcification may also remove important evidence of diagenetic alterations that may carry information relating to taphonomic processes or burial environment.

The necessity to work with undecalcified bone specimens means that standard microtomes cannot be used and samples must be sawn or ground to the desired thickness unless the researcher has access to specialized equipment (for example, a Jung K microtome, which is capable of cutting thin sections of only 5 μm). To overcome the tendency for archaeological bone or tooth specimens to crumble or shatter during cutting and grinding they must first be embedded in a penetrating resin with hardness similar to that of the mineralized tissue. Specimens thus embedded in epoxy or other resins can be sectioned by a variety of techniques (using diamond- or glass-blade microtomes, circular diamond saws, etc.) and the resulting thin slices either mounted directly onto glass slides or further reduced in thickness by grinding with abrasives (Caropreso *et al.*, 2000). Hospital pathology departments and larger university laboratories that routinely prepare thin sections of mineralized tissues are generally equipped with sophisticated specialized equipment that is not normally available to researchers working with archaeological material. Nevertheless, with practice and more than a little resourcefulness, remarkably good and consistent results can be achieved using relatively primitive facilities. Frost (1958) advocated a simple and elegant procedure, the 'rapid manual method', for the preparation of thin sections of fresh bone specimens. This protocol is capable of producing high-quality 10–15 μm sections using only a hacksaw, waterproof sandpaper, detergent and tap water. The original method has been adapted by Maat *et al.* (2001, 2006a) to incorporate modifications allowing the preparation of thin sections of archaeological bones and teeth. Once again, preparation is predominantly by hand, with local applications of cyanoacrylate adhesive to reinforce fragile specimens. These publications provide step-by-step instructions that will allow any researcher with modest facilities to prepare thin sections for optical microscopy, provided the specimens retain a modest collagen content and have not suffered appreciable bacterial degradation.

For archaeological specimens where bacterial degradation and other diagenetic processes have left the bone tissue cracked and friable it is necessary to impregnate them with a suitable resin. This, in turn, influences the approach adopted in producing sections, although there is no reason why the rapid manual method should not be applied to embedded specimens. The methodologies described below are suitable for the preparation of both thick sections for electron microscopy, etc., and thin sections for light microscopy. Preparation of undecalcified archaeological specimens of bones or teeth for histological examination is summarized in Figure 7.1.

Obtaining the Sample

Many archaeologists and curators are understandably concerned about clumsy sampling of, and needless damage to, human (or other) remains in their custody. Therefore, the person doing the sampling should aim to be clean and methodical in their approach and strive to minimize damage to the specimen. When sampling specific skeletal elements it is, of course, best to avoid critical features that may be used later by other researchers making morphometric or other measurements; for example, if sampling a long-bone shaft, avoid the midpoint. It is also advisable to sample different skeletons in as near the same anatomical location to simplify comparisons between individuals in, for example, ageing or osteoporosis studies. Where it is possible to obtain the required histological information from a half-section (or two complementary half-sections) this is preferable to cutting the bone into two pieces.

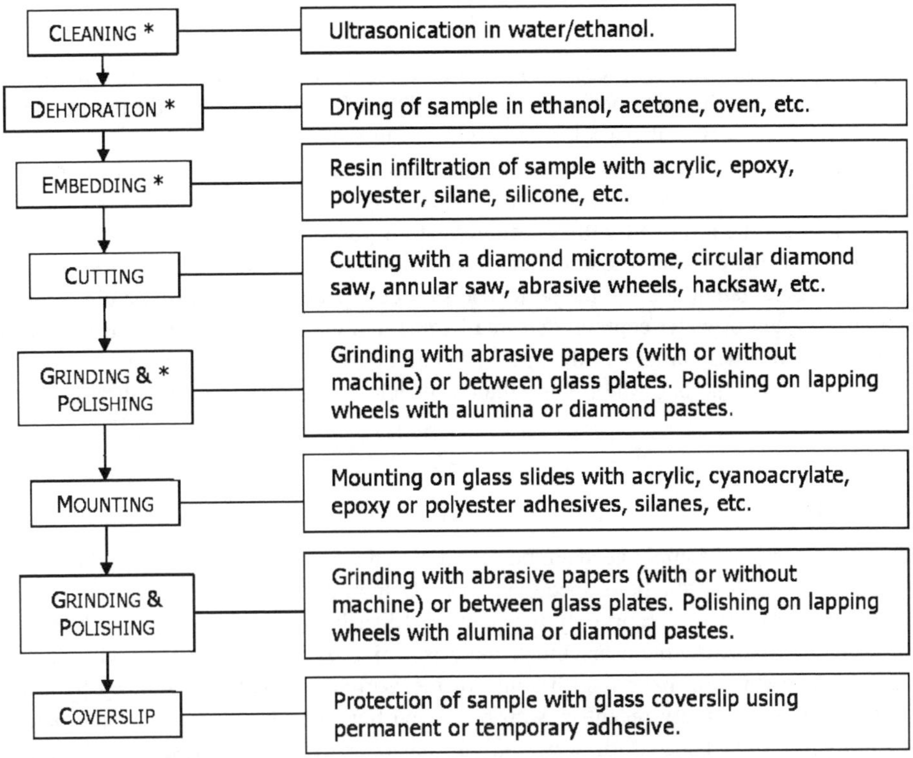

Figure 7.1 Summary of successive processes in the production of histological sections for microscopic analyses. The procedures marked with an asterisk apply to the preparation of thick sections for BSEM imaging. Not all of the successive steps may be required when preparing different specimens for different imaging techniques

Before taking any kind of sample from human remains it is essential to observe relevant health and safety regulations and to protect one's own health by use of fume cupboards, eye and ear protection and dust masks. Mark the location of the cut or cuts in advance. A simple strip of flexible plastic approximately 1 cm wide (such as the plastic binding straps found on packing crates) makes a suitable straightedge for marking the line and extent of the cut. When sampling the shafts of long bones it is useful to mark the centreline with a pencil. For half-sections, rather than making a V-shaped cut, it is better to make two, closely spaced parallel cuts in the diaphysis and then use a thin knife to break the section free. Alternatively, two small holes may be drilled at opposite ends of the proposed cuts and these joined up by two parallel saw-cuts. For thicker half-sections, it is possible to use a small hand-held power drill such as a Dremel® or a flexible-drive drill fitted with a carborundum cutting disc. Make the cuts parallel to the long axis of the bone first and then make the transverse cuts with the same tool or a small hacksaw. When using a powered drill, care must be taken that the chuck does not contact the periosteal surface and cause disfiguring marks. Care should also be taken to avoid overheating the bone tissues, since this can affect the colour of the final thin-section.

Cleaning

It may be advisable in some cases to attempt to remove sediment or other particles of foreign matter from the bone before embedding. The infiltration of resin during the embedding process can carry soil particles deep into the interiors of bones. Macroscopic surface contamination, such as adhering soil, roots or fungal hyphae, may be picked off manually using tweezers under a low-power binocular microscope. Loose particles and bone dust from the sampling can be removed with a compressed air gun or vacuum cleaner, but care must be taken not to dislodge and lose delicate trabeculae from the interior of the specimen or degraded external lamellae from the periosteal surface. Specimens may also be ultrasonicated in deionized water or water containing a small amount of ethanol. This ultrasonication should be brief, and potential damage to the specimen can be reduced by placing it in a small glass beaker or polyethylene bag filled with ethanol, which is then placed in the ultrasonic bath.

Dehydration

The extent to which a sample needs to be dehydrated depends upon a number of factors. The potential water content of a bone specimen depends upon the remaining collagen content and the bone porosity (or more correctly the pore size distribution). Well-preserved bones with high residual collagen content are much more sensitive to moisture than heavily degraded bones or specimens that have lost the bulk of their original organic content by hydrolysis. Thin sections of relatively recent or well-preserved bone can sometimes become detached from the glass slide because of warping, shrinkage or expansion caused by fluctuations in ambient relative humidity. Adequate penetration of the resin into the pore structure of the bone is essential if shrinkage is to be minimized. This is especially true for specimens exposed to the high vacuum of a scanning electron microscope, which frequently develop cracks due to a combination of vacuum drying and localized heating by the electron beam.

If the sample has been ultrasonicated in water/ethanol, then it should be dehydrated rather than allowed to dry in air. It can be dewatered relatively simply in progressive baths of acetone and water. Small wide-mouthed bottles containing different dilutions of acetone which may be reused are useful for this. Acetone has the advantage over ethanol of reducing any remaining lipid that may hinder penetration of the embedding resin. After the last bath in 100 % acetone the sample should be placed into a fume cupboard to evaporate off the solvent and then transferred to a vacuum desiccator until it is required for embedding. Unlike autopsy samples containing soft tissues, archaeological bone specimens can be dried in an oven at 55°C for 120 min (Caropreso *et al.*, 2000). Although this will not remove the tightly bound water adsorbed in the smallest pores it will dry those pores to which even the least viscous resins will penetrate.

Embedding

There is a wide range of resins available, and several methodologies for impregnation are described in the literature. An early, but extremely successful method that still finds widespread application in the medical sciences involves embedding specimens in polymethymethacrylate. Bell *et al.* (1996) used three 24 h changes of 95% (by volume) distilled methymethacrylate with 5% (by volume) styrene and 0.2% (by weight) 2,2′-axo-bis(2-methylpropionitrile). Specimens were then transferred to an oven at 32°C until the resin had fully polymerized.

Polyester resins find widespread application in teaching and research in the biological sciences, where they are used to preserve soft-tissue preparations. Peker *et al.* (2003) describe a three-step embedding process using polyester monomer (Dewester 196 by Dewilux®) with moderate viscosity, but high flowability. Samples were immersed for 24 h in a graded series of acetone–polyester monomer solutions of 25, 50, 75 and 100%. The samples were then transferred to the embedding resin made up of polyester resin (10 ml), benzoyl peroxide accelerator (three drops), and cobalt octoate hardener (three drops). This was polymerized at room temperature for 20 min. Similar methods have been used for embedding undecalcified bone samples (Lataster *et al.*, 1992).

Epoxy resins offer a rapid method for impregnation of bone tissues. Many biological preparation laboratories use specialist embedding epoxies, such as the Spur formulations produced by SPI-Chem™ or some Araldite® formulations, such as Araldite® 753 (known as Araldite® 502 in the USA). These are very low viscosity resins that cure to an inert, hard solid. However, the majority must be cured at elevated temperatures of around 60°C for up to 24 h or at lower temperatures for anything up to 4 days. Curing times can be accelerated by the addition of extra hardener, but this can increase the viscosity and reduce penetration efficiency. In many cases, extreme low viscosity is not necessary and quite acceptable results can be achieved using readily available, non-specialist resins, such as Araldite® 2020 (XW 396/XW 397). Araldite® 2020 is a water-white low-viscosity adhesive that was originally designed for bonding glass, although it may also be used to bond other materials. It is quite widely used in the conservation community and is usually easily obtainable. An additional advantage is a shelf life at room temperature of up to 3 years. Curing time is 16 h at 40°C (Ciba Specialty Chemicals: Publication No. A 282 d GB, July 1998), although it is advised to wait at least 24 h before grinding and polishing the embedded specimens.

Best results are achieved by combining a low-viscosity resin with vacuum impregnation. Purpose-built equipment for the vacuum embedding of biological and metallographic specimens is commercially available, including the Buhler Cast n' Vac systems. The larger version employs a vacuum desiccator with an integral powered sample carrousel and a tiltable cupholder to dispense resin into the sample moulds while the system is under vacuum. The aim is to eliminate the possibility of trapping air bubbles in the pore spaces of porous specimens. For laboratories without access to these facilities it is possible to improvise a simple vacuum-impregnation system using a standard vacuum desiccator fitted with a wire armature passing through a rubber bung. This can be used to push a bone specimen from a shelf into a small cup of premixed resin. After several minutes to allow infiltration of the sample by the resin it can be removed from the desiccator and transferred to a sample mould filled with resin.

Grinding and Polishing

There are a number of commercial systems that can be bought for the grinding and polishing of embedded specimens, and the production of thin sections for optical microscopy. Some systems allow the production of thin sections directly from the embedded block. Jans *et al.* (2002) embedded bone specimens in BioDur® resin and used an annular saw microtome (Leica Microsystems) to cut 30 μm sections. The thin section was stabilized and prevented from crumbling using a coverslip. In the absence of these facilities it is still possible to achieve good, and sometimes better, results with simple equipment and considerable practise. The embedded block can be hand ground on a motorized grinder/polished using successively finer abrasive papers, finishing with a grit size of P1200 or P2000, equivalent to an average particle diameter of 15.3 μm or 10.3 μm respectively. Final polishing is done on a cloth nap with alumina or diamond suspensions or pastes with particle sizes in the range 1–4 μm. If a grinder/polisher is not available, then good results can be achieved using abrasive papers held onto a thick glass or Perspex® sheet with two elastic bands. The block is then placed in a shallow plastic tray filled with water to act as a lubricant and to clear debris from the abrasive paper. Polishing can be done with diamond pastes and a self-adhesive nap cloth stuck onto a thick glass plate.

Polished blocks are suitable for BSEM. For optical microscopy, the upper 3–4 mm of the polished block may be cut off and adhered to a clean glass slide with the same embedding resin, then ground and polished to the appropriate thickness. Often, it is better to adhere the block to the glass slide before cutting of the bulk of the sample to avoid damaging the polished surface when making the saw-cut. Unless a slide holder is available, it can be extremely difficult to hold the thin glass slide during grinding or polishing. This problem can be solved by taping the slide to a Perspex® block cut to the same dimensions. Many adhesive tapes (Sellotape®) have a thickness of about 50 μm. The thickness of the final section can be controlled to some extent by how many turns of the tape are used to secure the slide to the Perspex® block. The time spent in grinding and polishing the section can be reduced by trimming excess resin from the sample block in advance. When grinding or polishing in water-based media it is wise to consider the residual collagen content of the specimen to avoid the section cracking or peeling free from the glass slide. A quick rinse in alcohol will remove much of the excess water and degrease the surface prior to adding a coverslip. Allow the section to air dry before attempting to apply a coverslip with a suitable adhesive. A coverslip will protect the thin section from damage and prevent cracking, but it will then not be possible to use BSEM or microanalytical techniques in combination with optical microscopy.

Obviously, the available equipment and sample preparation techniques have a bearing on the sampling strategy. Preparing sections by hand is not only time consuming, but also consumes more of the bone specimen than using an annular saw to make serial sections. One should always be aware of how much material must be sacrificed during the cutting and polishing procedures when taking the initial sample and take steps to minimize damage to specimens. Several 1–2 mm slices taken from a diaphysis, embedded and ground separately, may be easier than removing one large block and then attempting to cut thinner sections by hand later. It is also pertinent to remember that only a small area can be examined microscopically at any one time; thus, samples taken should be the minimum size required to do the job in hand.

APPLICATIONS OF PALAEOHISTOLOGY

Today, histological studies on ancient human bones may be divided into one of three broad categories: determination of age at death, elucidation of palaeopathological conditions, and the study of diagenesis. In recent years, by far the majority of palaeohistological studies have fallen into this last category, and this work is reviewed elsewhere (Turner-Walker, Chapter 1). Work in the former two categories is discussed below.

Age Determination in Adult Skeletons

The term histomorphometry refers to the measurement and quantitative analysis of physical structures identified in histological sections of biological tissues. For age estimation this usually means the study of the number and distribution of osteons and other remodelling features in histological sections of bone (Boivin and Meunier, 1993). The various parameters recorded and measured when determining remodelling rates include: intact osteon density (number of osteons per square millimetre); fragmentary osteon density (number of osteon fragments per square millimetre); number of remodelling osteons; average osteonal cross-sectional area; average osteonal cross-sectional diameter; average area of Haversian canal; and percentage of osteonal bone. The basis of these analyses in age determination is a comparison between rates of remodelling (derived from the above parameters) of cortical bone tissues in individuals of known age at death with archaeological skeletons or forensic cases of unknown age. The most commonly selected bone elements for histomorphometric analyses are the femoral diaphysis (Kerley and Ubelaker, 1978; Ericksen, 1991, 1997; Chan *et al.*, 2007) the ribs (Stout and Stanley, 1991; Stout and Lueck, 1995; Cho *et al.*, 2006) or both (Pfeiffer *et al.*, 2006).

Even within the same skeletal element there is considerable debate on which features should be counted to arrive at an assessment of age, from which areas, and what proportion of the bone section should be analysed. When evaluating cortical bone remodelling rates for the femoral mid-shaft, Pfeiffer *et al.* (1995) compared histological variables in the anterior, posterior, medial and lateral quadrants, as well as in four zones with different mechanical loadings. They found that there was considerable variation in the measured parameters depending upon sampling site and whether the field analysed lay near the periosteal surface or further towards the interior of the cortex. Similar variations in age estimates depending upon sampled area have been reported by Chan *et al.* (2007), who recommend restricting analysis to the anterior mid-shaft. Some researchers have reported that a significant correlation between osteon counts and age was only observed by analysing the entire cross-section at the femoral mid-shaft (Stout and Stanley, 1991), despite obvious concerns about the unavoidable damage inflicted on skeletal remains sectioned in this way (Kerley and Ubelaker, 1978). More recent work on a contemporary population has demonstrated that satisfactory results can be achieved by removing only a wedge-shaped section of the femoral shaft and examining three, 1 mm^2 regions of the anterior cortex (Maat *et al.*, 2005, 2006b). Rather than count whole and partial osteons, these studies measured the percentage of non-remodelled circumferential lamellar bone close to the periosteal surface. Maat *et al.* (2006b) report that age-related remodelling does not show a linear dependence with age, but they provide quadratic regression equations that can be used to estimate age at death with a standard deviation of the prediction error of ± 11 years. As with other methods for determining age at death of skeletons, the uncertainty increased markedly with advancing age. Maat *et al.* (2005) have compiled a beautifully illustrated catalogue of histological sections taken from a contemporary Dutch population showing sections from individuals with ages from 5 to 90 years.

Many researchers prefer to work on the mid-shafts of ribs, since these bone elements are more readily sacrificed than intact femurs and because the smaller cross-sectional areas of ribs compared with femurs means that a much larger percentage of the sectional area can be realistically analysed, thus giving a more representative and more reliable measure of remodelling rates. Cho *et al.* (2006) measured histomorphometric parameters on approximately 50 % of their rib sections by analysing every alternate microscopic field of view to ensure representative sampling.

Quantitations of histomorphometric parameters are then entered into one of a number of different algorithms to provide an estimate of age at death based on average osteon formation rates measured on historical or modern populations of known age. Any estimation of age at death in an archaeological population, therefore, relies upon a number of assumptions about the reference population and the ancient population under study. Because cortical bone in the limbs is influenced by a number of factors, including genetics, race (Cho *et al.*, 2006), activity patterns (Mays, 2001; Pfeiffer *et al.*, 2006), diet and hormonal influences, one must adopt a modern or historical population that best approximates the lifestyle of the target group (Ericksen, 1997). By choosing to analyse ribs, many researchers hope to avoid the influence of activity patterns, especially lower limb loading on cortical thickness and bone turnover in the femoral diaphysis.

Despite more than 30 years of work in the field of histomorphometric age determinations, the technique does not appear to give more reliable estimates of age at death over other established macroscopic techniques, such as dental attrition (Miles, 1963) or other age-related morphological changes (İşcan and Loth, 1989; Aiello and Molleson, 1993). Histomorphometric measurements are complicated in archaeological skeletal remains by the effects of diagenesis. For example, Ericksen (1997) noted that some of her specimens of prehistoric Chilean femurs suffered exfoliation of the periosteal surface and loss of circumferential lamellae. This loss of un-remodelled bone would skew the measured percentage osteonal bone to higher values and, thus, exaggerate the age of the individual (Ericksen, 1997). On the other hand, microbial alteration of archaeological bone often leads to preferential survival of the outer lamellae and destruction of osteonal bone in the inner cortex (Turner-Walker, Chapter 1). This destruction may lead researchers to examine better-preserved parts of the section rather than the more reliable analysis sites used in the reference population.

Research on thin sections of human permanent single-root teeth suggest that counts of cementum annulations show considerable potential as a measure of age. Cementum is deposited on the surfaces of the roots of teeth in a regular and measurable fashion throughout the life of an individual. Because of differences in collagen fibre orientation, these incremental layers are revealed as light and dark bands when thin sections are viewed in polarized light. A single cycle of one bright and one dark band is thought to represent 1 year of deposition activity (Maat *et al.*, 2006a), although some individuals seem to exhibit multiple annulations in a single year (Grosskopf, 1990). The methodology is comparatively straightforward (Maat *et al.* 2006a), although it necessitates the partial sacrifice of a single-rooted tooth. A recent study of modern teeth from known-age individuals (Wittwer-Backofen *et al.*, 2004) seems to suggest a high accuracy (95 % confidence limits for age estimates did not exceed ±2.5 years). However, other studies have produced less encouraging results (see reviews in Whittaker (2000) and Wittwer-Backofen and Buba (2002)), and systematic testing of the method on archaeological populations of known age at death is needed. Regardless of the outcome of such studies, the time-consuming and destructive nature of analyses will tend to weigh against cementum annulation as an age-estimation method for routine use in archaeological assemblages, given the time and funding constraints under which such research is customarily conducted.

Histomorphometric analyses have also been used to investigate activity patterns in past populations, partly to understand the problems with ageing skeletons better and partly as an insight into behaviour and lifestyle in earlier populations. Pfeiffer *et al.* (2006) examined rib and femur cross-sections from a later Stone Age population from South Africa, post-medieval skeletons from Spitalfields in London (UK) and a 19th century colonial Canadian group from Belleville, Ontario – a sample base covering a date range of 8000 years and three continents. They found that within each group the Haversian canal area did not vary significantly and variation in osteonal area was independent of both age and sex. However, osteonal area was greater in femurs than in ribs. Comparisons between the three groups showed that osteonal areas were smaller in the late Stone Age skeletons than those from Spitalfields and similar to those from Canada. The data appeared to be inconclusive, with there being no obvious causal factors to account for the findings (Pfeiffer *et al.*, 2006). Race and sex differences in cortical bone remodelling rates have been investigated by Cho *et al.* (2006), who measured standard histomorphological parameters in African Americans and Americans of European descent. Once again, the results did not bear out the predictions and expectations of the researchers. The main differences were between European American males and African American females, but the histological data could not strongly support the common observation that African American females are less susceptible to post-menopausal bone loss than women of European descent are. These studies emphasize that, although the relative contributions of each are unknown, a mix of genetic, lifestyle and environmental factors probably influence histomorphological parameters. This also clearly has implications for the reliability of age-at-death estimations for archaeological populations where lifeways were very different from those pertaining to the modern reference populations upon which the various histological methods are based.

In contrast to the work described above, Stout (1983) claimed to have successfully identified consistently higher cortical bone remodelling rates among a New World population whose staple diet was maize, compared with other ancient New World peoples. This higher turnover of bone material may reflect the fact that maize is deficient in calcium and high in phosphorus, a combination that could lead to reduced levels of calcium in blood serum and increases in parathyroid hormone – which is known to stimulate bone remodelling (Stout, 1989). In a separate study, the same author reports fewer osteons in bones from individuals whose mobility was impaired by crippling illness or quadriplegia, with the reduced mechanical stimulation of the bone tissue and its canalicular network resulting in a corresponding slowdown of cortical remodelling (Stout, 1982).

Histological Investigations in Palaeopathology

Diagnosis of disease from histological sections rests on the proper identification of new woven or lamellar bone deposits arising from inflammatory responses, dietary deficiencies, or neoplasms (cancers), lesions caused by disease-related osteolysis, or imbalances in bone remodelling caused by hormonal disturbances or disuse. Although these diagnoses are fairly routine in hospital histopathology laboratories where biopsy or autopsy specimens still retain their cells and unmineralized osteoid, it can be extremely difficult to obtain unambiguous results in archaeological specimens, especially where bone tissue has been obscured by soil staining or destroyed by microbial degradation. It should be noted that almost no pathological changes seen in histological sections are specific to particular diseases. Rather, changes from normal histology arise as a result of inflammation of the periosteum or secondary influences on normal bone metabolism. Thus, palaeohistology should never be used as the sole diagnostic tool, but should be used to support observations of gross anatomy, radiography or other techniques. In seeking to section palaeopathological specimens, the interests of the palaeohistologist and curators of skeletal assemblages frequently clash over the destructive nature of histological analysis and the size of the section removed. Ethical considerations dictate that, in order to preserve the integrity of museum collections, destruction of bone should normally be kept to a minimum. Rather than being routine for every pathological specimen, sampling for palaeohistology should be carefully targeted by restricting it to those skeletons where gross, radiographic or other observations suggest that histological study would help resolve diagnostic dilemmas. In focal disease, it may often be essential to sample from a bone lesion. However, in systemic disease (e.g. many metabolic diseases) this may not be necessary, and in such cases care should be taken to sample from areas that will minimize damage to diagnostically important parts. In general, it is desirable for researchers to return histological slides to the museum or other institution holding the collection once study of them is completed. This helps to minimize damage to the collection by obviating the need for repeat sampling by different researchers over the years.

Histological sections may be examined using a number of imaging techniques. Traditionally, thin sections of un-decalcified bone are ground to 20–100 μm and mounted on microscope slides for optical microscopy. These thin sections may then be stained with standard histological stains; but it is more usual to view ancient bone specimens unstained, since the bulk of organic and cellular matter to which the stains bind are absent and chemical species from the soil can create misleading results. The use of polarized light to view thin sections of bone reveals the orientation of the collagen fibres and closely associated hydroxy-apatite (HAP) crystallites. In the polarizing microscope, light passes through the polarizing filter below the sample and emerges polarized east–west. Some of the light has its plane of polarization rotated when it passes through the collagen and mineral phases (which are both birefringent) in the bone section. This light then proceeds through the analyser, which is oriented to transmit only those components of the light that is polarized north–south. Thus, views of histological sections of archaeological bones under crossed polars display a complex pattern of bright, dark and intermediate zones which carries information about the orientation of collagen/mineral phases and, hence, about the lamellar structure of primary bone and osteonal bone (Figure 7.2a–d). Lamellar bone generally appears as alternating layers of bright and dark stripes, whereas osteons (Haversian systems) appear as characteristic Maltese cross (or Brewster cross) features. Insertion of a quarter lambda ($1/4\lambda$) plate retards some of the wavelengths from a white light source and subsequent interference causes the section to appear vividly coloured. Different colours in the specimen represent specific orientations of birefringent materials in the sample. This, in turn, assists in the interpretation of bone sections by making it easier to see isolated pockets of un-remodelled primary lamellar bone and fragments of osteons (Figure 7.2d). The use of polarized light also provides information about the diagenetic alteration of archaeological bones. Under crossed polars, areas of diagenetically altered bone, which are dark or opaque in plain light, appear as bright, cloudy regions and mineral infiltrations such as calcite appear brilliantly white (Figure 7.3a and b). The birefringence of bone can persist for millions of years (Gillette, 1994) even after all of the original collagen has been lost, but only if the original orientations of the HAP crystallites are preserved.

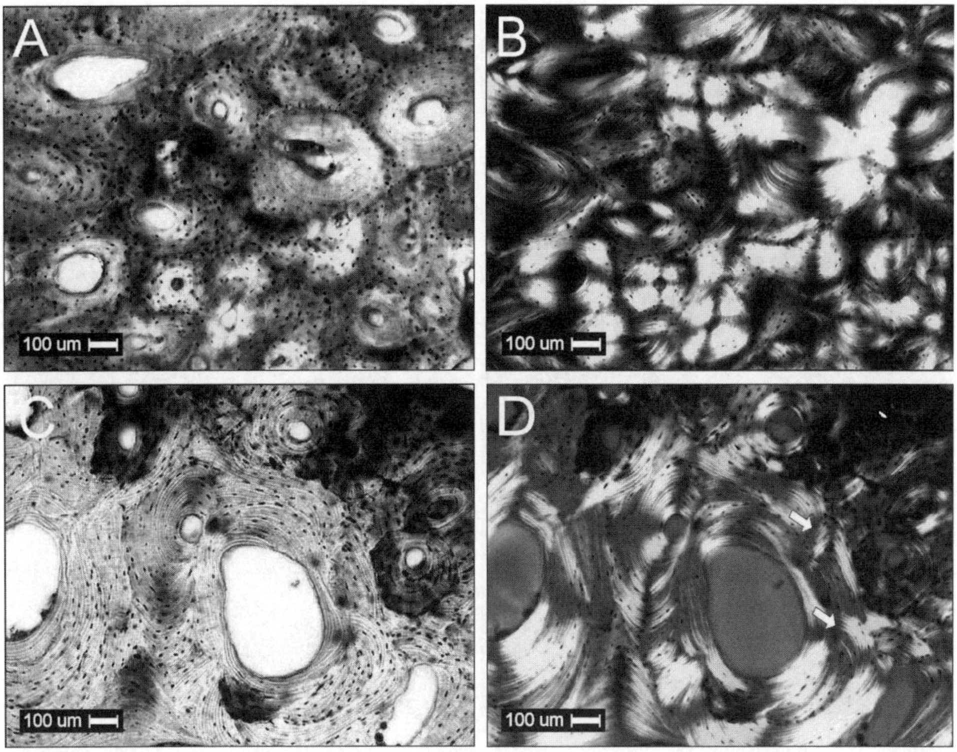

Figure 7.2 Optical microscopy of thin sections of well-preserved archaeological bones. Note the staining in (a) and (c). Staining appears to proceed along the canalicular network. (b) The classic 'Maltese cross' pattern of osteonal bone viewed in crossed polarizers. Note also the isolated pockets of un-remodelled primary lamellar bone and fragments of osteons seen in (d) (arrowed)

One technique that has seen rapid growth is the use of electron microscopy, especially BSEM (Turner-Walker, Chapter 1). Although SEM techniques do not reveal information about the orientation of HAP or collagen, the vastly increased resolution and the sensitivity of backscattered electrons to subtle differences in sample density means that histological structures such as lamellar bone, osteons, cement lines, reversal lines, etc. can be readily identified (Figure 7.3c and d). Detritus from the soil, exogenous minerals and altered bone mineral can be easily distinguished from normal or pathological histology using BSEM imaging techniques (Figure 7.3d). Furthermore, BSEM is not influenced by staining from the soil or many of the optical effects that are responsible for obscuring details in light microscopy. Another advantage over conventional light microscopy is that sample preparation is generally simpler: resin-embedded specimens of arbitrary thickness need only to be polished to a single, optically flat surface, then coated with a conducting layer of carbon.

Microbial degradation of histological structures is general in archaeological bone, but even in bone that has been severely degraded by microbial tunnelling or other diagenetic processes it is still possible to identify the boundaries of osteons. Because bone-destroying bacteria are frequently inhibited from crossing cement lines that are high in non-collagenous proteins, it is

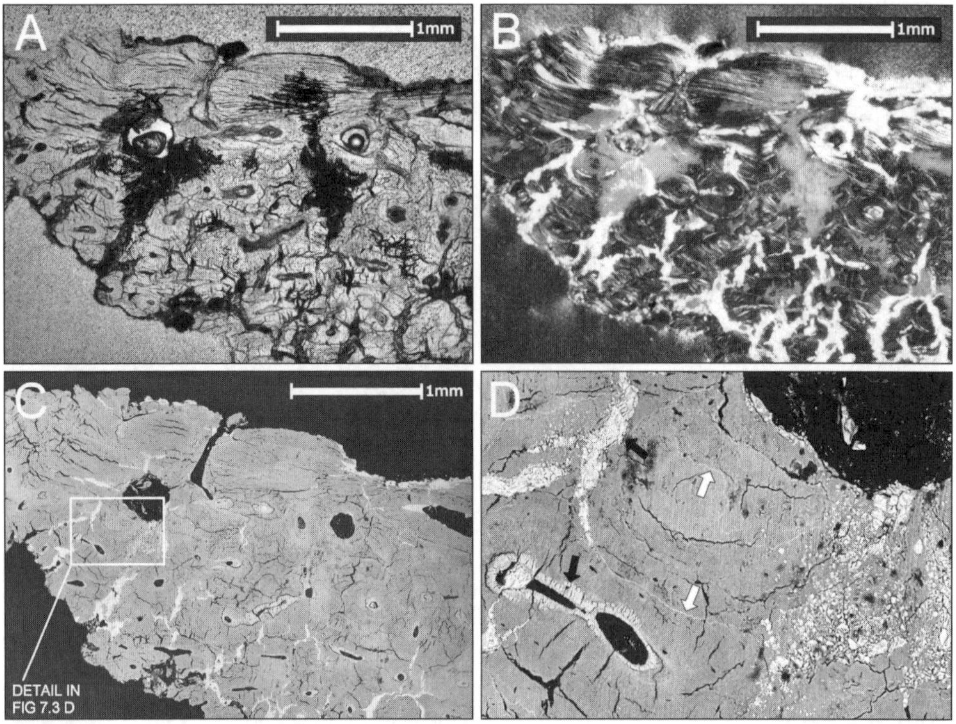

Figure 7.3 Comparison of (a) plain light, (b) polarized light and (c, d) BSEM images of diagenetically altered bone. Cracking and infilling by calcite, visible in places as dark areas under plain light, are dramatically highlighted as bright areas in the polarized light image. This calcite also appears bright in the backscatter images (d, black arrows). In the backscatter images the details of histological structures are still largely visible. Cement lines appear as thin, meandering bright lines (d, white arrows)

still possible to distinguish the outlines of the original osteons. Severely degraded specimens may often retain sufficient histological structure so that diagnostically useful observations can be made. This is illustrated by the following two examples using BSEM imaging of diseased bone.

Figure 7.4 shows BSEM images from a section of a diseased area of a tibia taken from an individual from medieval Ipswich, UK (Mays and Turner-Walker, 1999). Lesions identified during visual and radiographic examination of the skeleton were strongly indicative of Paget's disease of bone. Figure 7.4a shows a lamina of bone tissue where bacterial degradation has been halted by a cement line running longitudinally through the bone. Several more cement lines, representing successive phases of resorption and osteoblastic deposition of new bone, can be seen on the lower right of Figure 7.4a. Figure 7.4b shows similar cement lines exhibiting the characteristic scalloped border that marks the limit of osteoclastic resorption and Howslip's lacunae. The cement lines appear bright because of a higher density and mineral loading compared with mineralized osteoid. Figure 7.4c also shows multiple resorption phases and different degrees of mineralization of the osteoid that suggests rapid bone turnover, i.e. the new bone did not have time to mineralize fully before another resorption episode overtook it. This image also illustrates the enlarged and randomly

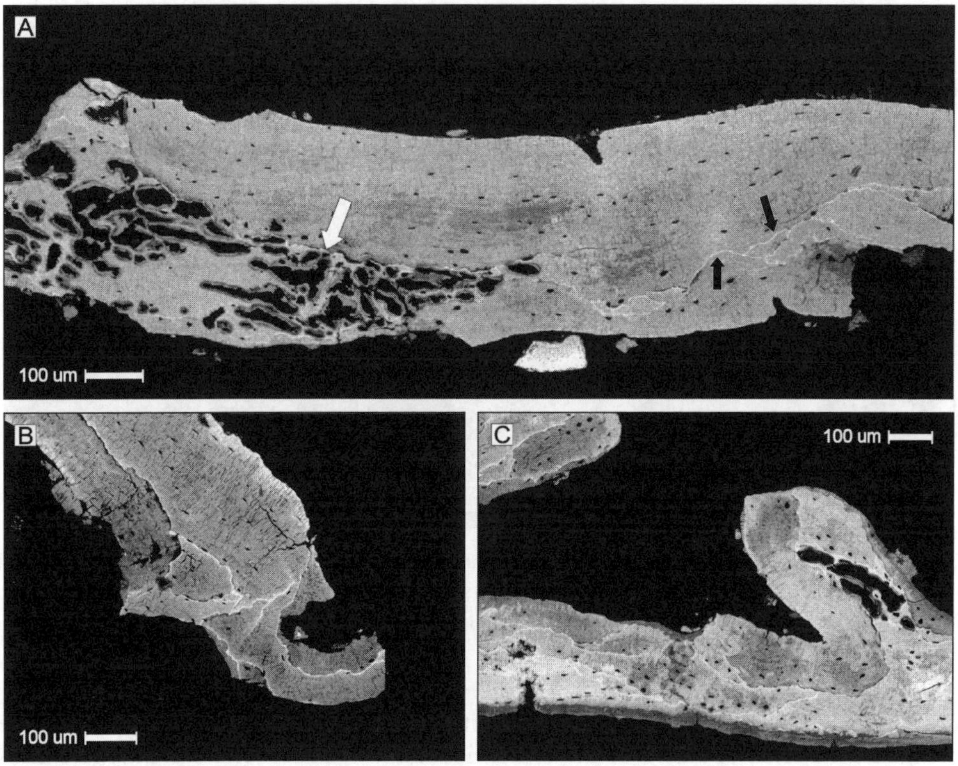

Figure 7.4 BSEM images of bacterially degraded archaeological bone from an individual with suspected Paget's disease. (a) Bacterial degradation has been halted by a cement line running longitudinally through the bone on the left-hand side of the image (white arrow). Several more cement lines, representing successive phases of resorption and osteoblastic deposition of new bone can be seen on the lower right of the image (black arrows). (b) and (c) Multiple cement lines with characteristic scalloped borders are visible representing successive phases of resorption and bone growth. Poorly mineralized zones of osteoid appear darker

distributed osteocyte lacunae typical of Pagetic bone. When viewed in thin section it is clear that the interior of the bone has been severely altered by microbial degradation (Figure 7.5a and b). However, the sub-periosteal bone is better preserved and shows the classic 'mosaic pattern' that so often characterizes the highly remodelled cortical bone in Paget's disease of bone. Compared with Figure 7.2b, which shows a relatively regular arrangement of roughly circular osteons, the detail of the Pagetic bone seen in Figure 7.5c shows a chaotic picture of distorted osteons and osteon fragments with a seemingly random orientation of bone lamellae. Overall, the palaeohistological analyses support a diagnosis of Paget's disease of bone in this skeleton.

A further example of the use of palaeohistological analyses in support of standard palaeopathological studies is given by a possible case of renal osteodystrophy in an adolescent medieval skeleton from Wharram Percy, UK (Mays and Turner-Walker, in press). Gross and radiographic examination of the skeleton (WCO058) revealed deposits of sub-periosteal new bone and extensive cancellous bone growth in the medullary cavities of the left humerus and

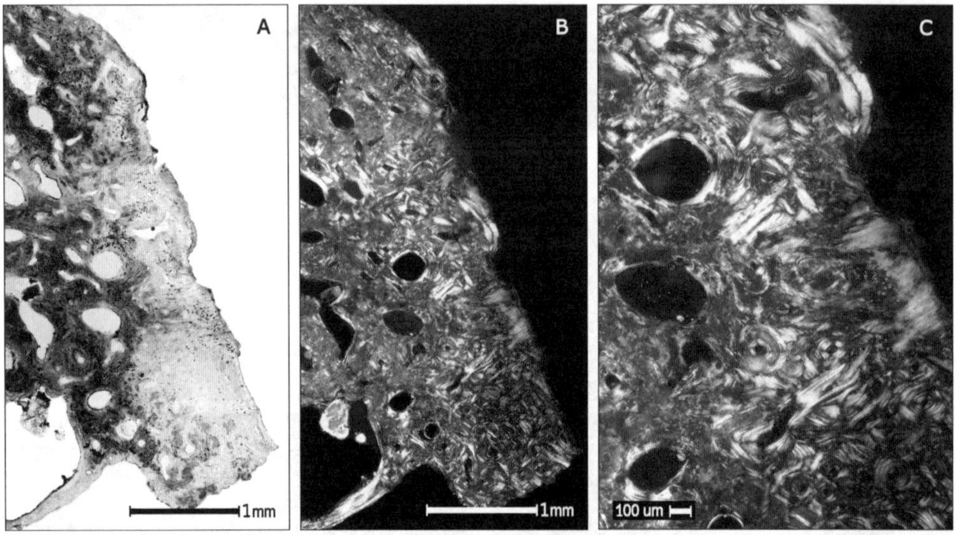

Figure 7.5 Sample from the same skeleton as Figure 7.4 viewed in thin section in (a) plain light and (b, c) polarized light. The sub-periosteal bone shows the characteristic 'mosaic pattern' of incomplete heavily remodelled osteonal bone that is typical of Paget's disease

right radius. There were also new bone deposits within the trabecular bone structure of the proximal metaphysis of the right radius, most ribs and two lumbar vertebrae. Many elements showed sub-periosteal new bone formation, and long-bones showed greater intracortical porosity than was normal for an individual of this age. These observations were suggestive of hyperparathyroidism and increased resorption and turnover of cortical and cancellous bone (Resnick and Niwayama, 1995: 2014–2015). A diagnosis of renal osteodystrophy was considered consistent with the above findings. Unlike primary hyperparathyroidism, in renal osteodystrophy the increased porosity of cortical bone is accompanied by endosteal and sub-periosteal new bone proliferation (Resnick and Niwayama, 1995: 2036, 2943). Histological sections of ribs taken from this individual, and another skeleton of similar age (WCO103), were examined by BSEM and compared with respect to cortical thickness, porosity and cancellous bone development. Despite almost total destruction of original histology by microbial degradation of the tissues, and considerable infiltration of soil into the cancellous bone in the WCO058 specimen, there were clear differences between the two individuals (Figure 7.6). Figure 7.6a shows the normal bone architecture for the individual WCO103 with moderate porosity in the cortical bone and an open, honeycomb structure to the interior trabeculae. The suspected case of renal osteodystrophy in Figure 7.6b shows a thickened but more porous cortex and a chaotic tangle of thin, poorly connected bony spicules in place of the regular trabecular bone. There is no clear demarcation between the cortical bone and the internal spongy bone. In fact, the internal porosity in Figure 7.6b is considerably higher than it appears in the BSEM images due to the presence of soil partly infilling many of the voids. Despite the considerable destruction of original bone tissue by bacteria it is still possible to distinguish the external circumferential lamellae at the periosteal surface in sample WCO103 (Figure 7.7), and some of the boundaries of osteons are hinted at by a change in the pattern of microfocal destruction.

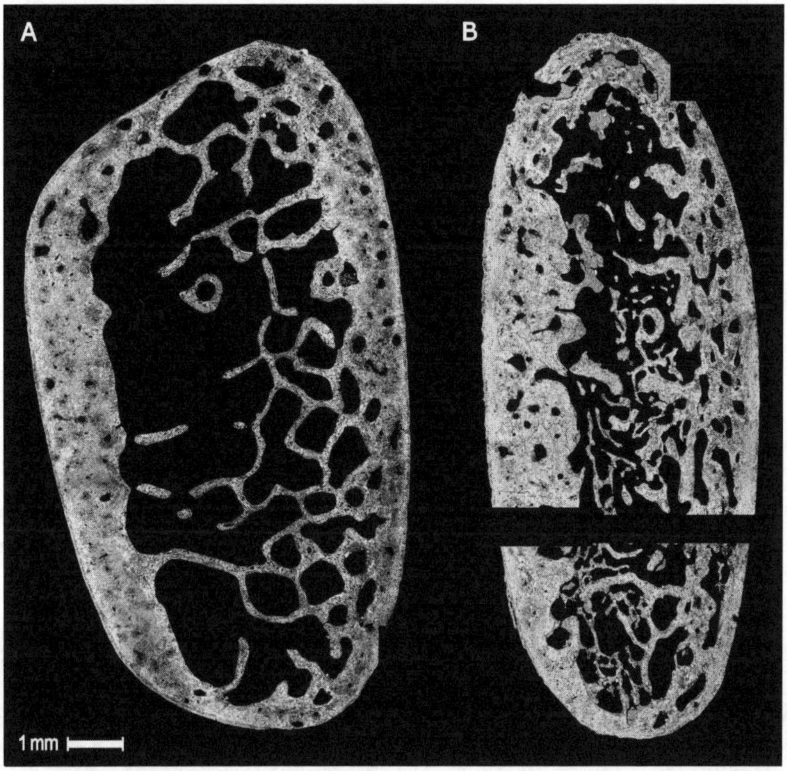

Figure 7.6 BSEM images of rib sections from two inhumations from medieval Wharram Percy, England. (a) Normal cortical bone and trabecular architecture from an undiseased individual (WCO103) aged 13–15 years. (b) Rib section from an individual of similar age with suspected renal osteodystrophy (WCO058). Not only is the cortical bone thickened and more porous, but the normal honeycomb structure of the internal trabeculae has also been replaced by a chaotic tangle of bone spicules that fills the interior of the rib

Although, as demonstrated by the above examples, useful information may be gained from microscopic study of bone showing marked diagenesis, if diagenetic alteration is too advanced then histology may not yield useful diagnostic data. In such specimens, although some histological structures may be discerned, this 'ghost histology' tells us more about diagenetic processes and their relationships with bone microarchitecture and organic and mineral fractions than about the life and health of the individual. Figure 7.8a shows human bone from a Middle Neolithic cemetery site at Ypenburg, The Netherlands. There is massive diagenetic alteration to the bone tissues, but the radial and annular arrangements of the canaliculi have clearly been preserved, presumably by the progressive demineralization of the bone tissues outwards from the natural porosity. Some characteristic focal destruction attributable to bacteria is also visible. Figure 7.8b illustrates human bone from an Iron Age inhumation from the Scilly Isles, UK. Here, only the characteristic focal destruction remains, the rest of the tissue having been leached away by acid soil waters. Once more, the annular structure of the osteons is preserved, since this has clearly influenced the progression of bacterial attack. A fragment of human skull vault from the skeleton of

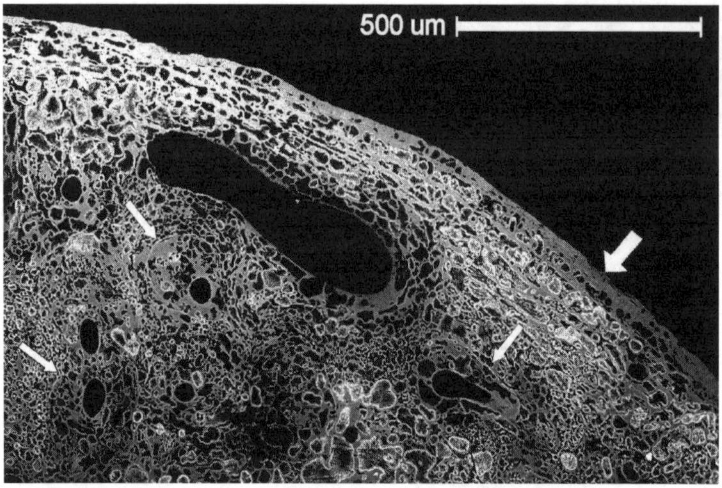

Figure 7.7 BSEM showing detail of sub-periosteal bone in rib of individual WCO103. Despite considerable diagenetic destruction of the bone tissues it is still possible to distinguish circumferential lamellar bone (large arrow) and the boundaries of osteons (small arrows)

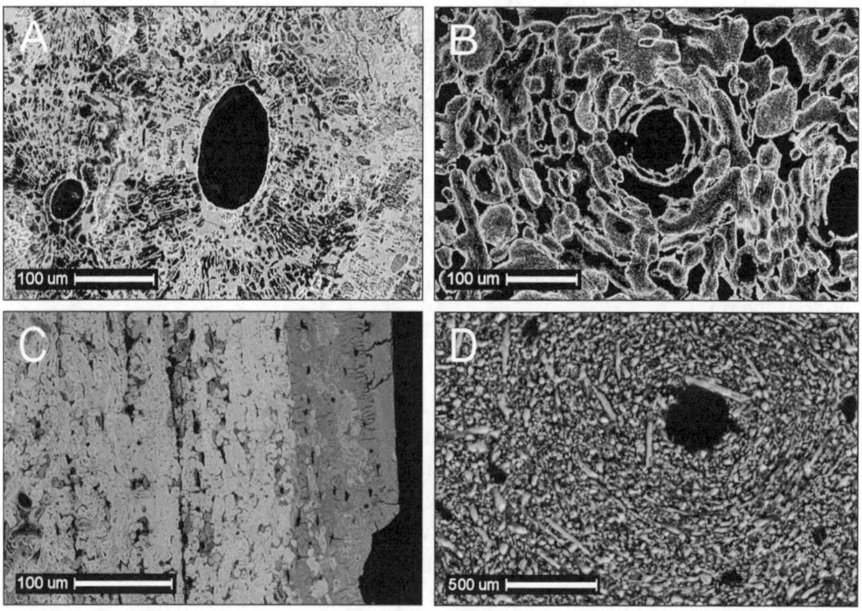

Figure 7.8 Ghost histologies in severely degraded bone samples. (a) Human bone from a Middle Neolithic cemetery site at Ypenburg, The Netherlands. The annular arrangements of the canaliculi have clearly been preserved. (b) Human bone from an Iron Age inhumation from the Scilly Isles, UK. Here, an acid leaching soil has left only the characteristic focal destruction. However, the annular structure of the osteons is preserved. (c) A fragment of human skull vault from a skeleton from Taiwan. The lamellar structure of the tissues is clearly discernable as changes in the density of the focal destruction. (d) Bone from the United Arab Emirates. The normal histology has been replaced by a mass of needle-shaped crystals of HAP. The annular pattern of osteons in preserved

an adolescent from Taiwan is shown in Figure 7.8c. The lamellar structure of the tissues is clearly discernable as changes in the density of the focal destruction. The specimen seen in Figure 7.8d is barely distinguishable as bone. It is from the United Arab Emirates, and for reasons that remain unclear there has been total recrystallization of the HAP into long, thin needle-like crystals. These crystals respect the annular lamellae of the osteon, and the boundaries of the Haversian canal and surrounding osteocyte lacunae are clearly visible.

Palaeohistology may provide visualization of, and quantitative data on, progressive metabolic diseases such as osteoporosis. Osteoporosis is a condition in which there is a general reduction in bone mass, accompanied by microstructural deterioration of bone tissue. Both sexes lose bone mass with increasing age, but skeletal fragility, and thus fracture risk, is greater in elderly women than in elderly men (Chapter 11). Transverse sections of cortical bone show loss of bone tissue both as thinning of cortex and as an increase in intracortical porosity. Figure 7.9 shows part sections from three female medieval skeletons taken from the femoral diaphysis. The section in Figure 7.9a is from a young woman of 21–25 years of age and this skeleton had a bone mineral density at the femoral neck (BMDN), as determined by dual-energy X-ray absorptiometry, of 1.170 g cm^{-2}. The woman in Figure 7.9b was aged over 50 and had a BMDN of 0.618 g cm^{-2}. There is a marked reduction of the cortical thickness (from 4.1 mm to only 2.1 mm) and an increase in porosity due to enlargement of Haversian canals. Figure 7.9c is a section from another woman aged over 50 who obviously suffered from severe osteoporosis. The cortical thickness is reduced to 0.7 mm and this individual had a BMDN of only 0.510 g cm^{-2}. The extreme imbalance between osteoclastic resorption and osteoblastic bone formation that characterizes osteoporosis has reduced cortical bone bordering the endosteal surface to a lace-like tracery of fine trabecular arches. Bone loss in osteoporosis results from disequilibrium between bone formation and resorption in remodelling. Histomorphometry can help identify at a tissue level whether this balance has been

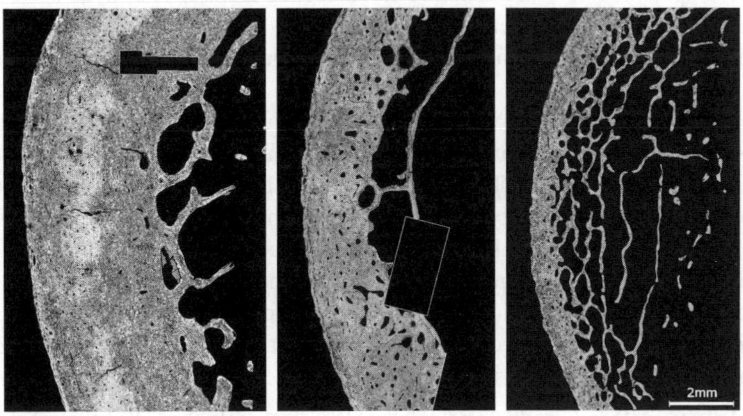

Figure 7.9 BSEM images of transverse sections taken from the femoral shafts of three medieval skeletons from Wharram Percy. (a) Woman aged 21–25 years of age; (b) woman aged >50 years; (c) woman aged >50 whose bone mineral density measurements suggested advanced osteoporosis. Note the marked reduction of the cortical thickness and increasing cortical porosity from specimens (a)–(c)

disrupted. Various parameters can be calculated for cortical bone, including osteon population density, mean osteon size, mean annual activation frequency, mean annual bone formation rate, as well as measures of the amount and porosity of cortical bone (Cho and Stout, 2003). Investigation of changes in these parameters with age is a means by which aspects of osteoporosis may be quantified (see reviews in Cho and Stout (2003) and Brickley and Agarwal (2003)).

FUTURE TRENDS IN PALAEOHISTOLOGY

Recent advances in microscopy offer the potential of improved imaging at higher levels of detail, and hold out the prospect that some of the histological studies which currently involve destructive sampling may be achievable using non-destructive techniques.

Confocal laser Scanning microscopy (CLSM) is able to create crisply focused images of specimens that would appear blurred in a conventional microscope. The principles of the confocal microscope are relatively simple. A narrow beam of light passes through pinhole screens and is scanned in a raster fashion over the sample by way of tilting mirrors. Reflected light passes through a second pinhole to a photomultiplier tube. The arrangement of the pinholes ensures that the amount of scattered light entering the detector is minimal (Sheppard and Shotton, 1997). Because so little light enters the detector, an intense source such as a laser is used to illuminate the sample. The image is created by assembling pixels in the same way that many other scanning microscopy techniques (SEM, AFM, etc.) generate the final picture. The optical geometry of a confocal microscope rejects light that does not derive from the microscope's focal plane. All scattered light and light that originates from above or below the plane of interest is excluded, so that depth of field in a prepared section is no longer a factor in image quality and resolution. By effectively scanning the focal plane through the thickness of a sample it is possible to capture and store serial slices through the specimen and assemble these into a three-dimensional reconstruction. The majority of confocal microscopes create their image either by reflecting light off the surface of the specimen or by stimulating fluorescence from special dyes used to stain the specimen. To date, CLSM has seldom been applied to ancient skeletal remains, although a study of modern and fossil hominid teeth from Israel has been published (Haydenblit *et al.*, 2000).

Near-field scanning optical microscopy (NSOM) is a closely related technique to CLSM, in that it scans the surface with an extremely fine, aluminium-coated fibre optic that shines laser light onto the specimen. The near-field region is defined as the region around the probe encompassed by a distance that is less than the wavelength of the incident light. In NSOM this distance is typically in the order of nanometres. By bringing a fibre-optic probe within this near-field region, NSOM can avoid the limitations arising from diffraction effects in conventional optical microscopy. Light can be shone down the fibre, or light from a second light source can be detected via the fibre. Near-field (or evanescent) light consists of a non-propagating field that is limited to the surface of an object and decays within a few tens of nanometres. The tip is scanned across the surface as in other scanning probe microscopies (see below). NSOM can generate information about the sample's topography and optical properties. The problem with this, and other extreme-resolution scanning probe microscopies, is that the magnifications are so high, and the fields of vision so narrow, that it is frequently impossible to interpret the resulting image when viewing an inhomogeneous

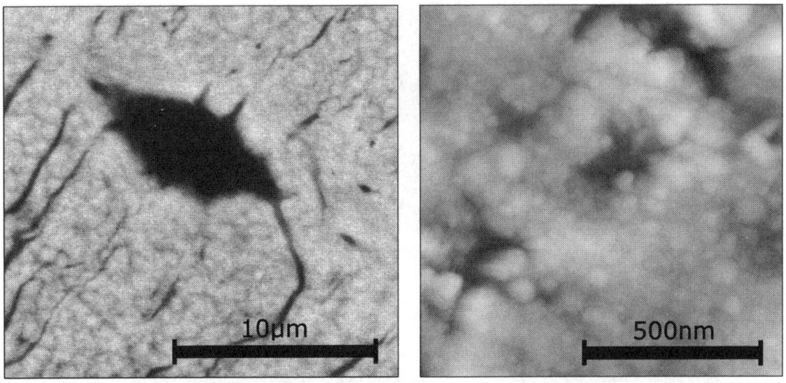

Figure 7.10 Comparison of SEM and NSOM. (a) High-resolution backscatter image of fresh bone from a field-emission scanning electron microscope. The magnification is 3500×. It shows a single osteocyte and associated canaliculi. Note the 'cloudy' texture of the surrounding bone tissue. (b) An equivalent image made with NSOM at a magnification of approximately 70 000×. The 'cloudy' texture is clearly evident, but no other recognizable features are visible in the field of view

or contaminated sample (see Figure 7.10). Although this technique holds great promise for understanding changes in the ultrastructure of skeletal remains, more needs to be learned about sample selection and preparation.

Atomic force microscopy (AFM) is a mature technology widely used in materials research, surface physics and microelectronics. It produces high-resolution, three-dimensional images by scanning a sharp tip over the sample surface. The tip is part of a flexible cantilever mounted on one end of a flexible piezoelectrical tube attached to the top of the microscope. Voltages applied to the X and Y electrodes result in a precise raster scan over the surface, while a voltage applied to the Z electrode controls the vertical height of the tip. As well as precisely measuring the topography of the specimen the tip can also sense subtle differences in the phase or electrical charge of the underlying surface. Figure 7.11 shows AFM images of a thin section of archaeological bone taken with a Dimension 3100 SPM. The three images in the upper part of the figure show the sample imaged in height, phase and amplitude modes. At the top left of the images one can see unaltered bone with several scratches left by the sample preparation. The rest of the field is pitted and represents an area of microbially degraded bone that has not been polished as finely as the unaltered bone. Several pits with diameters in the range 700 nm–1.5 μm can be distinguished which represent the diagenetic porosity seen in BSEM images and mercury porosimetry measurements (Turner-Walker *et al.*, 2002). In the lower part of the image the topographic data are used to reconstruct a three-dimensional perspective view of the specimen surface.

The past decade has seen major advances in micro-computed tomography, particularly microtomography using X-rays from a synchrotron source (SR-μCT). Synchrotron radiation offers an intensely bright source of almost monochromatic and highly collimated X-rays that can be focused using annular diffraction gratings called zone plates (or Fresnel zone plates). The development of finely focused, intense X-ray beams has opened the way for a range of microscopic and analytical techniques at the nanometre level, with resolutions close to 60 nm now possible (Attwood, 2006; Yin *et al.*, 2006). SR-μCT has recently been used to investigate

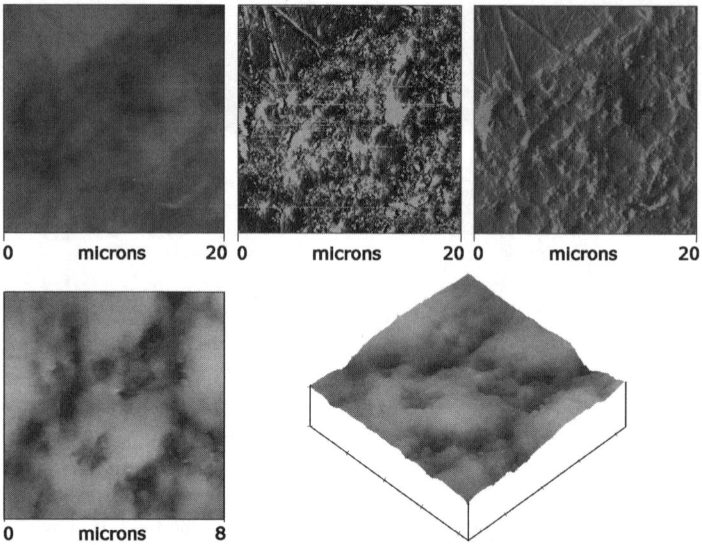

Figure 7.11 AFM images of a thin section of archaeological bone. The three images in the upper part of the figure show the sample imaged in height, phase and amplitude modes (L to R). The lower portion of the field is pitted and represents an area of microbially degraded bone. Several pits with diameters in the range 700 nm–1.5 μm can be seen. In the lower part of the image the topographic data are used to reconstruct a three-dimensional perspective view of the specimen surface

palaeontological and fossil hominid specimens with excellent results. Tafforeau *et al.* (2006) imaged trebacular structures in the tibia of fossil specimens and compared the results of portable μCT equipment with those from SR-μCT obtained at the European Synchrotron Radiation Facility. The latter produced images comparable to BSEM images, but without the necessity of physically sectioning the specimens. Mazurier *et al.* (2006) examined a range of palaeontological specimens, including fossil bones, teeth and an insect embedded in opaque amber. Their published images include both three-dimensional reconstructions and cross-sections. Resolutions of the teeth images are high enough to show clearly the microstructural details of dentine and enamel with voxel sizes (the three-dimensional equivalent of pixels) of 0.7 μm. SR-μCT is capable of producing images that are difficult or impossible to obtain using conventional techniques. Figure 7.12 is a virtual transverse section of the L1 vertebra from a Neanderthal skeleton from Regourdou in Montignac-sur-Vézère, Dordogne, France. The image was created by the SR-μCT laboratory at Grenoble, France. The pixel resolution is 45.5 μm. The section clearly shows the microarchitectural arrangement of trabecular bone in the body of the vertebra and some diagenetic infilling by sediment (small arrow). It is also possible to see changes in mineral density in some of the cortical bone (large arrow) that may reflect diagenetic alteration of the bone tissues. At the time of writing, SR-μCT techniques have not been applied to specific questions relating to palaeopathology or diagenesis studies, but the potential they offer is obvious, particularly in important fossil material where a fully non-destructive approach is essential.

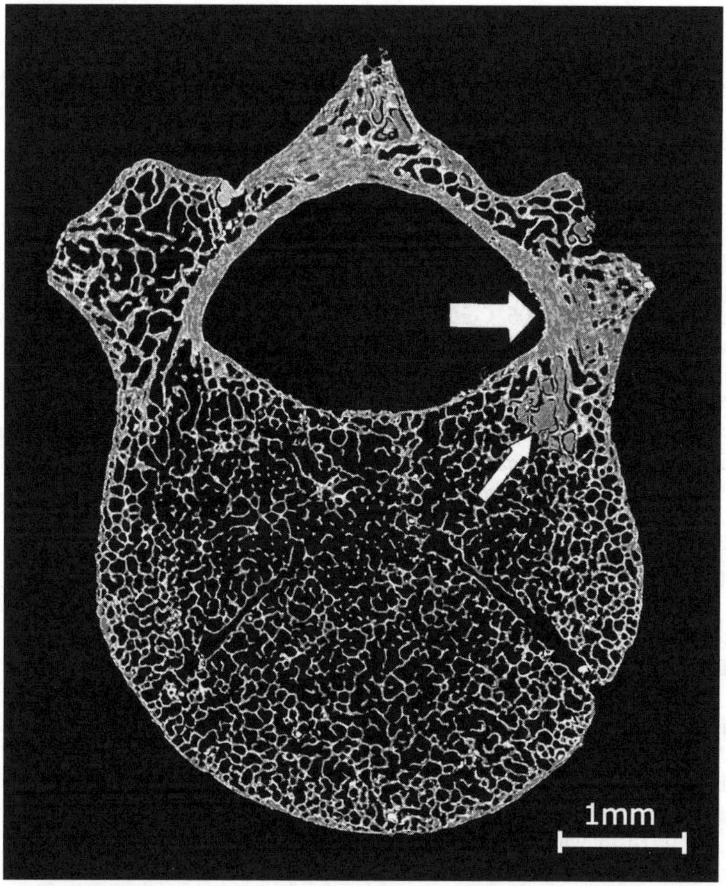

Figure 7.12 Transverse section of the L1 vertebra from a Neanderthal skeleton from Regourdou, a cave site in the Dordogne, France. The image was created by the SR-μCT laboratory at Grenoble, France. The pixel resolution is 45.5 μm. In addition to the trebacular architecture it is possible to identify infiltration by sediments (small arrow) and differences in density in the cortical bone (large arrow). Courtesy of R. Macchiarelli, A. Mazurier and V. Volpato, University of Poitiers

REFERENCES

Aiello L, Molleson T. 1993. Are microscopic ageing techniques more accurate than microscopic ageing techniques? *J Archaeol Sci* **20**: 689–704.

Attwood D. 2006. Nanotomography comes of age. *Nature* **442**: 642–643.

Bakker RT. 1975. Experimental and fossil evidence for the evolution of tetrapod bioenergetics. In *Perspectives in Biophysical Ecology*, Gates D, Shmerl R (eds). Springer: New York; 365–399.

Bell LS, Skinner MF, Jones SJ. 1996. The speed of post mortem change to the human skeleton and its taphonomic significance. *Forensic Sci Int* **82**: 129–140.

Boivin G, Meunier PJ. 1993. Histomorphometric methods applied to bone. In *Histology of Ancient Human Bone: Methods and Diagnosis*, Grupe G, Garland AN (eds). Springer: Berlin; 137–156.

Brickley M, Agarwal S. 2003. Techniques for the investigation of age-related bone loss and osteoporosis in archaeological bone. In *Bone Loss and Osteoporosis: an Anthropological Perspective*, Agarwal SC, Stout SD (eds). Kluwer/Plenum: New York; 157–172.

Caropreso S, Bondioli L, Capannolo D, Cerroni L, Macchiarelli R, Condò SG. 2000. Thin sections for hard tissue histology: a new procedure. *J Microsc* **199**: 244–247.

Chan AHW, Crowder CM, Rogers TL. 2007. Variation in cortical bone histology within the human femur and its impact on estimating age at death. *Am J Phys Anthropol* **132**: 80–88.

Cho H, Stout SD. 2003. Bone remodelling and age-associated bone loss in the past: a histomorphometric analysis of the Imperial Roman skeletal population of Isola Sacra. In *Bone Loss and Osteoporosis: an Anthropological Perspective*, Agarwal SC, Stout SD (eds). Kluwer/Plenum: New York; 207–228.

Cho H, Stout SD, Bishop TA. 2006. Cortical bone remodeling rates in a sample of African American and European American descent groups from the American Midwest: comparisons of age and sex in ribs. *Am J Phys Anthropol* **130**: 214–226.

De Ricqlès AJ. 1974. Evolution of endothermy: histological evidence. *Evol Theory* **1**: 51.

De Ricqlès AJ. 1980. Tissue structure of dinosaur bone: functional significance and possible relation to dinosaur physiology. In *Morphology and Biology of Reptiles*, Bellairs Ad'A, Cox CB (eds). Academic Press: London; 123–150.

De Ricqlès AJ. 1993. Some remarks on palaeohistology from a comparative evolutionary point of view. In *Histology of Ancient Human Bone: Methods and Diagnosis*, Grupe G, Garland AN (eds). Springer: Berlin; 37–77.

Drury RAB, Wallington EA. 1980. *Carleton's Histological Technique*, 5th edition. Oxford University Press.

Enlow DH, Brown SO. 1956. A comparative histological study of fossil and recent bone tissues. Part I. *Texas J Sci* **8**(4): 405–443.

Enlow DH, Brown SO. 1957. A comparative histological study of fossil and recent bone tissues. Part II. *Texas J Sci* **9**(2): 186–214.

Enlow DH, Brown SO. 1958. A comparative histological study of fossil and recent bone tissues. Part III. *Texas J Sci* **10**(2): 187–230.

Ericksen MF. 1991. Histologic estimation of age at death using the anterior cortex of the femur. *Am J Phys Anthropol* **84**: 171–179.

Ericksen MF. 1997. Comparison of two methods of estimating age at death in a Chilean preceramic population. *Int J Osteoarchaeol* **7**: 65–70.

Faerman M, Jankauskas R, Gorski A, Bercovier H, Greenblatt CL. 1997. Prevalence of human tuberculosis in medieval population of Lithuania based on ancient DNA analysis. *Ancient Biomol* **1**: 205–214.

Faerman M, Nebel A, Filon D, Thomas MG, Bradman N, Ragsdale BD, Schultz M, Oppenheim A. 2000. From a dry bone to a genetic portrait: a case study of sickle cell anemia. *Am J Phys Anthropol* **111**: 153–163.

Frost HM. 1958. Preparation of thin undecalcified bone sections by rapid manual method. *Stain Technol* **33**: 271–276.

Garland AN. 1993. An introduction to the histology of exhumed mineralized tissue. In *Histology of Ancient Human Bone: Methods and Diagnosis*, Grupe G, Garland AN (eds). Springer: Berlin; 1–16.

Gernaey AM, Minnikin DE. 2000. Chemical methods in palaeopathology. In *Human Osteology in Archaeology and Forensic Science*, Cox M, Mays S (eds). Greenwich Medical Media: London; 239–254.

Gernaey AM, Minnikin DE, Copley MS, Middleton JC, Dixon RA, Roberts CA. 2001. Mycolic acids and ancient DNA confirm an osteological diagnosis of tuberculosis. *Tuberculosis* **81**: 259–265.

Gillette DD. 1994. *Seismosaurus: The Earth Shaker*. Columbia University Press: New York.

Graf W. 1949. Preserved histological structures in Egyptian and ancient Swedish skeletons. *Acta Anat* **8**: 236–250.

Grosskopf B. 1990. Individualalterbestimmung mit Hilfe von Zuwachsringen im Zement bodengelagerter menschlicher Zähne. *Zähn Rechtsmed* **103**: 351–359.

Haydenblit R, Podbilewicz B, Smith P. 2000. Evolution of human enamel growth analyzed by confocal microscopy. In *Proceedings of the 11th International Symposium on Dental Morphology*, Mayhall J, Heikkinen T (eds). Oulu University Press: Oulu; 228–235.

İşcan MY, Loth SR. 1989. Osteological manifestations of age in the adult. In *Reconstruction of Life from the Skeleton*, İşcan MY, Kennedy KAR (eds). Alan R Liss: New York; 23–40.

Jans MME, Nielsen-Marsh CM, Smith CI, Nord AG, Arthur P, Earl N. 2002. *In situ* preservation of archaeological bone: a histological study within a multidisciplinary approach. *Archaeometry* **44**: 343–352.

Kemp TS. 1982. *Mammal-like Reptiles and the Origin of Mammals*. Academic Press: Oxford.

Kerley ER, Ubelaker DH. 1978. Revisions in the microscopic method of estimating age at death in human cortical bone. *Am J Phys Anthropol* **49**: 545–546.

Lataster LMA, van Mameren H, Terwindt E. 1992. Plastic embedding procedure for 15 μm sections of large undecalcified tissue blocks. *Anat Rec* **233**: 651–656.

Lowenstein JM, Scheuenstuhl G. 1991. Immunological methods in molecular palaeontology. In *Molecules Through Time: Fossil Molecules and Biochemical Systematics*, Eglinton G, Curry GB (eds). The Royal Society: London; 375–380.

Maat GJR, Van den Bos RPM, Aarents MJ. 2001. Manual preparation of ground sections for the microscopy of natural bone tissue: update and modifications of Frost's 'Rapid Manual Method'. *Int J Osteoarchaeol* **11**: 366–374.

Maat GJR, Aarents MJ, Nagelkerke NJD. 2005. *Age Prediction from Bone Replacement: Remodelling of Circumferential Lamellar Bone Tissue in the Anterior Cortex of the Femoral Shaft of the Present Dutch Population.* Barge's Anthropologica 10. Barge's Anthropologica: Leiden.

Maat GJR, Gerretsen RRR, Aarents MJ. 2006a. Improving the visibility of tooth cementum annulations by adjustment of the cutting angle of microscopic sections. *Forensic Sci Int* **159S**: S95–S99.

Maat GJR, Maes A, Aarents MJ, Nagelkerke NJD. 2006b. Histological age prediction from the femur in a contemporary Dutch sample – the decrease of nonremodeled bone in the anterior cortex. *J* Forensic Sci **51**: 230–237.

Mayer AFJK. 1819. *Über Histologie und eine neue Eintheilung der Gewebe des Menschlichen.* Körpers: Bonn.

Mays S. 2001. Effects of age and occupation of cortical bone in a group of 18th–19th century British men. *Am J Phys Anthropol* **116**: 34–44.

Mays S, Turner-Walker G. 1999. A mediaeval case of Paget's disease of bone with complications. *J* Paleopathol **11**: 29–40.

Mays S, Taylor G, Legge A, Shaw D, Turner-Walker G. 2001. Palaeopathological and biomolecular study of tuberculosis in a mediaeval skeletal collection from England. *Am J Phys Anthropol* **114**: 298–311.

Mays S, Turner-Walker G. In press. A possible case of renal osteodystrophy in a skeleton from mediaeval Wharram Percy, England. *Int J Osteoarchaeol.*

Mazurier A, Volpato V, Macchiarelli R. 2006. Improved noninvasive microstructural analysis of fossil tissues by means of SR-microtomography. *Appl Phys A: Mater Sci Process* 83(2): 229–233.

Miles AEW. 1963. The dentition in the assessment of individual age in skeletal material. In *Dental Anthropology*, Brothwell DR (ed.). Pergamon: Oxford; 191–209.

Moodie RL. 1926. Studies in palaeopathology XIII: the elements of the Haversian system in normal and pathological structures among fossil vertebrates. *Biol Generalis* **2**: 63–95.

Nielsen-Marsh CM, Richards MP, Hauschka PV, Thomas-Oates JE, Trinkaus E, Pettit PB, Karavanié I, Poinar H, Collins MJ. 2005. Osteocalcin protein sequences of Neanderthals and modern primates. *Proc Nat Acad Sci U S A* **102**: 4409–4413.

Ovchinnikov IV, Götherström A, Romanoval GP, Kharitonov VM, Lidén K, Goodwin W. 2000. Molecular analysis of Neanderthal DNA from the northern Caucasus. *Nature* **404**: 490–493.

Peker T, Gülekon N, Anil A, Turgut HB, Liman F. 2003. Clearing and embedding in polyester resin for demonstrating the nerve distribution pattern of skeletal muscles. *Biotech Histochem* **78**(3–4): 187–190.

Pfeiffer S, Lazenby R, Chiang J. 1995. Brief communication: cortical remodelling data are affected by sampling location. *Am J Phys Anthropol* **96**: 89–92.

Pfeiffer S, Crowder C, Harrington L, Brown M. 2006. Secondary osteon and Haversian canal dimensions as behavioural indicators. *Am J Phys Anthropol* **131**: 460–468.

Redman JE, Stewart MI, Gernaey AM. 2002. Ancient tuberculosis and lipid chemistry – odd bedfellows. *Eur J Archaeol* **5**: 112–120.

Resnick D, Niwayama G. 1995. *Diagnosis of Bone and Joint Disorders* (3rd edition). WB Saunders: London.

Sheppard CJR, Shotton DM. 1997. Image formation in the confocal laser scanning microscope. In *Confocal Laser Scanning Microscopy*. Springer: New York; 15–31.

Stout SD. 1982. The effects of long-term immobilization on the histomorphology of human cortical bone. *Calcif Tissue Int* **34**: 337–342.

Stout SD. 1983. The application of histomorphometric analysis to ancient skeletal remains. Anthropos (Greece) 10: 60–71.

Stout SD. 1989. Histomorphometric analysis of human skeletal remains. In *Reconstruction of Life from the Skeleton*, İşcan MY, Kennedy KAR (eds). Alan R Liss: New York; 41–52.

Stout SD, Lueck R. 1995. Bone remodeling rates and skeletal maturation in three archaeological populations. *Am J Phys Anthropol* **98**: 161–171.

Stout SD, Stanley SC. 1991. Percent osteonal bone versus osteon counts: the variable of choice for estimating age at death. *Am J Phys Anthropol* **86**: 515–519.

Tafforeau P, Boistel R, Boller E, Bravin A, Brunet M, Chaimanee Y, Cloetens P, Feist M, Hoszowska J, Jaeger J-J, Kay RF, Lazzari V, Marivaux L, Nel A, Nemoz C, Thibault X, Vignaud P, Zabler S. 2006. Applications of X-ray synchrotron microtomography for non-destructive 3D studies of palaeontological specimens. *Appl Phys A* **83**: 195–202.

Taylor GM, Crossey M, Saldanha J, Waldron T. 1996. DNA from *Mycobacterium tuberculosis* identified in mediaeval human skeletal remains using polymerase chain reaction. *J Archaeol Sci* **23**: 789–798.

Turner-Walker G, Nielsen-Marsh CM, Syversen U, Kars H, Collins MJ. 2002. Sub-micron spongiform porosity is the major ultra-structural alteration occurring in archaeological bone. *Int J Osteoarchaeol* **12**: 407–414.

Whittaker D. 2000. Ageing from the dentition. In *Human Osteology in Archaeology and Forensic Science*, Cox M, Mays S (eds). Greenwich Medical Media: London; 83–99.

Wittwer-Backofen U, Buba H. 2002. Age estimation by tooth cementum annulation: perspectives of a new validation study. In *Palaeodemography: Age Distributions from Skeletal Samples*, Hoppa RD, Vaupel JW (eds). Cambridge University Press: Cambridge; 107–128.

Wittwer-Backofen U, Gampe J, Vaupel JW. 2004. Tooth cementum annulation for age estimation: results from a large known-age validation study. *Am J Phys Anthropol* **123**: 119–129.

Yin G-C, Tang M-T, Song Y-F, Chen F-R, Liang KS, Duewer FW, YunW-B, Ko C-H, Shieh H-PD. 2006. Energy-tunable transmission X-ray microscope for differential contrast imaging with near 60 nm resolution tomography. *Appl Phys Lett* **88**: 241115.

8

Molecular Palaeopathology of Human Infectious Disease

Helen D. Donoghue

Centre for Infectious Diseases and International Health, Department of Infection, University College London, London, UK

INTRODUCTION

The emergence of human infectious diseases has been linked to changes in human ecology and to interactions between populations (Cockburn, 1971; Cohen and Armelagos, 1984). Hunter-gatherers were colonized with a commensal microflora that had co-evolved with humans since ancestral times. In addition, they were exposed to diseases of other animals, which infect humans only incidentally–a scenario supported by contemporary fieldwork (Owen *et al.*, 2005). The historical change to settled farming communities coincided with the appearance of diseases associated with larger, denser populations, a sedentary lifestyle, widespread domestication of animals, social stratification, and possibly a less varied diet (Armelagos *et al.*, 2005). Examples include parasitic diseases and tuberculosis, identified by traditional palaeopathology. A second major transition has been linked to the initial contacts between the civilizations in Europe and Asia, from approximately 3000–2000 BP, possibly leading to the spread of bubonic plague and leprosy to Europe. The European expansion around the globe during the last five centuries caused trans-oceanic spread of disease, most notably that of influenza, measles and smallpox to the Americas, and possibly venereal syphilis back to the Old World (McMichael, 2004).

Advances in Human Palaeopathology Edited by Ron Pinhasi and Simon Mays

Can molecular-based palaeopathology answer any of the archaeological and historical questions associated with these scenarios? Initially, the aim was simply to determine straightforward matters, such as the verification of morphological diagnosis of infectious disease. Once achieved, specific historical queries were investigated; for example, whether tuberculosis existed in pre-Columbian America. In the comparatively brief existence of palaeomicrobiology, the field has developed to such an extent that we are now exploring the co-evolution of humans and their microbial pathogens in relation to ancient migrations and genomics.

SURVIVAL MECHANISMS OF MICROBIAL HUMAN PATHOGENS

It is useful to consider human infectious diseases from the perspective of the microorganism. Microbial pathogens are believed to have originated from free-living environmental species that became adapted to a lifestyle in which they live in or on a host. Only a small proportion of microorganisms cause disease, and the interaction between different animal hosts, their microbial flora and environmental factors determines the outcome.

Hit and Run

A rapid spread of infection and a high mortality are characteristics of diseases caused by microorganisms that have just crossed a species barrier. Such infections are acquired from animals, often via insect vectors, and are known as zoonoses. Examples include bubonic plague, anthrax, and rickettsial diseases such as typhus fever. Rickettsiae are small bacteria that are obligate intracellular parasites, normally transmitted to humans by arthropods, such as ticks, mites and lice. Here, the pathogen relies upon an alternative reservoir of infection for survival, and as humans are an incidental host, a high mortality has no long-term impact. Normally the microbes spread via the bloodstream, so preferred sampling sites for molecular studies in skeletal remains are the cancellous tissue of long bones for bone marrow residues, and dental pulp regions. Vector remains can be examined for evidence of disease, as can remains of animal hosts.

Occasional Chronic Carriers

Infections where there is tolerance in a proportion of human hosts can result in high mortality or morbidity during early stages of the disease, with long-term chronic or latent infection in the survivors. The microorganisms may be found at various sites in the body and at different population levels, depending on the stage of disease. Chronic carriers may develop skeletal palaeopathological markers of infection. An example is brucellosis, where the bony pathology is easily confused with that caused by tuberculosis. Typhoid fever can result in chronic carriage of the causative organism, *Salmonella typhi*, in the gall bladder. Parasitic diseases, such as Chagas' disease, malaria and leishmaniasis, are further examples and are known to have occurred in ancient times.

Long-Term Infections

In these infections, a majority of hosts have a long or lifetime infection, but disease may be latent or have phases of activity which then subside. Pathogen and host can coexist, which provides a reservoir of infection for the pathogen and may have caused selection pressure on the survival of its human host. In tuberculosis, vulnerable groups, such as infants, may develop active disease with a high mortality. Adults may become susceptible from a variety of causes, such as malnutrition, lowered host resistance due to old age or stresses caused by warfare. Leprosy is a specific human infection that is associated with long-term close contact between individuals. Schistosomiasis or bilharzia is a debilitating disease where a life-long infection results in a constant opportunity for the parasite to be transmitted to its fresh-water snail vector. Syphilis may be a relatively recent manifestation of a disease that had its origins in endemic skin infections. In about 30% of cases, early infection progresses to complete cure without treatment, normally within 3–5 years. In another 30% of cases the infection remains latent and is characterized by high levels of antibodies. The remaining cases progress to tertiary syphilis, which is non-infectious but demonstrates host hypersensitivity responses, which are reflected in characteristic bony pathology.

Opportunistic Infections

Environmental or commensal organisms cause opportunistic infections and are associated with a lowered host resistance, including the lowering of the natural body defences by stress, wounds and trauma. A puncture or dirty wound may permit *Clostridium tetani*, a normal inhabitant of the soil and of the intestinal tract of many animals, including horses, to produce a toxin that results in tetanus. A description in an Egyptian surgical papyrus appears to be consistent with a diagnosis of tetanus, which is indirect evidence of its occurrence in ancient times (Miller, 1997). Infections, tooth decay and lack of dental hygiene may result in abscesses, thereby releasing organisms into the blood and causing septicaemia, which is fatal in the absence of treatment (Li *et al.*, 2000). Parish mortality tables from 17th-century London (Figure 8.1a and b) illustrate the impact of a major epidemic of bubonic plague on deaths from other infections. The normal commensal microflora of the intestinal tract may cause diarrhoea or bladder infections. The skin or throat microflora may cause childbed (puerperal) fever.

The biomolecular identification of this group of infections in palaeopathology is extremely problematic, for how is it possible to distinguish active infection from normal commensal carriage or environmental contamination? If microbes are found at an unusual body site, then this may simply be a result of the breakdown of the natural body defences at the point of death that leads to the commensal microflora being transmitted via the bloodstream throughout the body. This is an important stage in the natural decay process, although it may also be a sign of disseminated infection during life (Zink *et al.*, 2000).

The Diseases and Casualties this Week.

Abortive	2	French-pox	2
Aged	38	Found dead (an Infant) at St. Andrew Holborn	1
Apoplexie	1	Gowt	1
Bedridden	1	Griping in the Guts	11
Burnt by a fall into the fire at St. Giles in the Fields	1	Hanged her self at St. Saviours Southwark	1
Childbed	2	Head-mould-shot	1
Chrisoms	12	Infants	4
Consumption	95	Kild by the fall of a Scaffold at St. Bartholomew the Great	1
Convulsion	28	Lethargy	1
Cough	1	Rickets	6
Dropsie	36	Rising of the Lights	4
Drowned in a Tub of water at St. Martins in the Fields	1	Scurvy	2
Executed	2	Spotted Feaver	3
Feaver	29	Stilborn	10
Flox and Small pox	15	Stone	1
Flux	2	Stopping of the Stomach	8
		Surfeit	9
		Teeth	20
		Thrush	4
		Tissick	15
		Winde	3
		Wormes	2

Christned { Males 121, Females 112, In all 233 } Buried { Males 199, Females 197, In all 396 } Plague 0

Increased in the Burials this Week 3

Parishes clear of the Plague 130 Parishes Infected 0

The Assize of Bread set forth by Order of the Lord Maior and Court of Aldermen,
A penny Wheaten Loaf to contain Ten Ounces and a half, and three half-penny White Loaves the like weight.

Figure 8.1 An example of the impact of a major epidemic on deaths from other infections. Diseases and casualties of two different weeks during 1665, in 130 London parishes from 'London's Dreadful Visitation or, A collection of all the Bills of Mortality for the present year beginning the 27th of December 1664 and ending the 19th of December following'. (a) The week of 28 February–7 March was free from plague. The greatest numbers of deaths (95) were ascribed to 'Consumption', the term used for pulmonary tuberculosis. Note also the 20 deaths due to 'Teeth'.

The Diſeaſes and Caſualties this Week.

ABortive	6	Kingſevil	10
Aged	54	Lethargy	1
Apoplexie	1	Murthered at Stepney	1
Bedridden	1	Palſie	2
Cancer	2	Plague	3880
Childbed	23	Pluriſie	1
Chriſomes	15	Quinſie	6
Collick	1	Rickets	23
Conſumption	174	Riſing of the Lights	19
Convulſion	88	Rupture	2
Dropſie	40	Sciatica	1
Drowned 2, one at St. Kath-Tower, and one at Lambeth	2	Scowring	13
Feaver	353	Scurvy	1
Fiſtula	1	Sore legge	1
Flox and Small-pox	10	Spotted Feaver and Purples	190
Flux	2	Starved at Nurſe	1
Found dead in the Street at St. Bartholomew the Leſs	1	Stilborn	8
Frighted	1	Stone	2
Gangrene	1	Stopping of the ſtomach	16
Gowt	1	Strangury	1
Grief	1	Suddenly	1
Griping in the Guts	74	Surfeit	87
Jaundies	3	Teeth	113
Impoſthume	18	Thruſh	3
Infants	21	Tiſſick	6
Kild by a fall down ſtairs at St. Thomas Apoſtle	1	Ulcer	2
		Vomiting	7
		Winde	8
		Wormes	18

Chriſtned: Males 83, Females 83, In all 166.
Buried: Males 2656, Females 2663, In all 5319. Plague—3880

Increaſed in the Burials this Week — 1289

Pariſhes clear of the Plague — 34. Pariſhes Infected — 96

The Aſsize of Bread ſet forth by Order of the Lord Maior and Court of Aldermen,
A penny Wheaten Loaf to contain Nine Ounces and a half, and three half-penny White Loaves the like weight.

Figure 8.1 (Continued) (b) The week of 8–15 August 1665 was at the height of the bubonic plague outbreak. In addition to the 3880 deaths from plague, there were 174 deaths from 'Consumption', 10 deaths from 'Kings evil', the name used for scrofula or skin tuberculosis, and 113 deaths from 'Teeth'. Note that a penny wheaten loaf is 10% smaller, compared with the price in March. Reproduced by permission of the Wellcome Library, London

NON-DNA BIOMOLECULAR METHODS IN PALAEOPATHOLOGY

Direct Observation and Molecular Techniques

Ruffer (1910), cited in Adamson (1976), observed the characteristic eggs of the parasitic trematode worm *Schistosoma haematobium*, which led to the new discipline of palaeoparasitology (Araújo and Ferreira, 2000; Gonçalves *et al.*, 2003). Morphological changes alone can reveal data on prevalence, the time-scale of association with humans, and insights into the lifestyle and contacts between populations. For example, microscopy demonstrated a 65% prevalence of *S. haematobium* among mummies dated to AD 350–550 from the Wadi Halfa area on the Sudan–Egyptian border (Miller *et al.*, 1992). Direct observation of eggs, worms and ectoparasites aids the selection of tissue for molecular studies and indicates the likely species present. The use of thin sections and polarized light can indicate specific and non-specific infections (Schultz, 2001).

Immunological Detection of Carbohydrates and Proteins

Antigens from pathogenic microbes can be detected in mummified tissues or faecal remains (reviewed by Lowenstein (2004)), a task made easier by commercial clinical diagnostic kits. Early examples included the detection of *Salmonella* antigens in a Peruvian mummy (Sawicki *et al.*, 1976), examination of ancient Egyptian mummies for the presence of *Schistosoma* antigen (Deelder *et al.*, 1990) and for the malaria parasite *Plasmodium falciparum* (Miller *et al.*, 1994). Cerutti *et al.* (1999), using an immuno-enzymatic assay for the parasite histidine-rich protein-2 antigen, estimated that around 40% of the mummies of the Gebelen group of mummies in ancient Egypt were infected with *P. falciparum*. Similarly, a commercial enzyme-linked immunosorbent assay (ELISA) kit, based on monoclonal antibody specific for *Entamoeba histolytica* adhesin, successfully identified the presence of *E. histolytica* protein in coprolite samples from about 5300 BP (Gonçalves *et al.*, 2004). The use of thin sections and immunocytochemistry based on indirect fluorescence staining has successfully detected *S. haematobium* in mummy tissues (Rutherford, 1999, 2005).

Bone can protect proteins from degradation and it is possible to isolate extracellular matrix proteins, including collagen and immunoglobulin G molecules, from archaeological material (Schmidt-Schultz and Schultz, 2004). Molecules are separated by one-dimensional polyacrylamide gel electrophoresis and identified by Western blot analysis using antibodies linked to an enzyme- or fluorescent-detection system, or silver stained. Two-dimensional gel separation increases sensitivity. Kolman *et al.* (1999) examined 200-year-old tissue for the causative organism of syphilis, *Treponema pallidum*. High-performance liquid chromatography (HPLC) was used to isolate immunoglobulin from femoral bone tissue, which was identified using ELISA against treponemal antigen, although this did not distinguish between subspecies.

Physico-Chemical Methods for Lipids and other Organic Biomolecules

The survival of proteins in mineralized tissue can be surprisingly good (Collins *et al.*, 2002; Schmidt-Schultz and Schultz, 2004). Lipids have received less attention, but the causative organisms of tuberculosis and leprosy, *Mycobacterium tuberculosis* and *Mycobacterium leprae*, have characteristic long-chain fatty acids and other cell-wall components (Figure 8.2). Mycolic acids are hydrophobic long-chain fatty acids which are firmly attached to a cell-wall arabinogalactan polysaccharide; this is in turn linked to a structural peptidoglycan polymer. Three types of mycolic acid are found in *M. tuberculosis*: alpha-, keto- and methoxy-mycolic acids (Villeneuve *et al.*, 2005). The mycolic acid organelle interacts with a range of very complex 'free' lipids to complete a formidable permeability barrier to protect the pathogen.

HPLC has detected cell-wall mycolic acids specific for the *M. tuberculosis* complex from archaeological specimens (Gernaey *et al.*, 1998, 2001), and confirmed a finding of *M. tuberculosis* complex DNA (Donoghue *et al.*, 1998). More recently, mass spectroscopy has been used to detect such molecules (Mark *et al.*, 2006). Unfortunately, although this is potentially a method of great promise, the equipment is expensive and not widely available, and neither is the expertise.

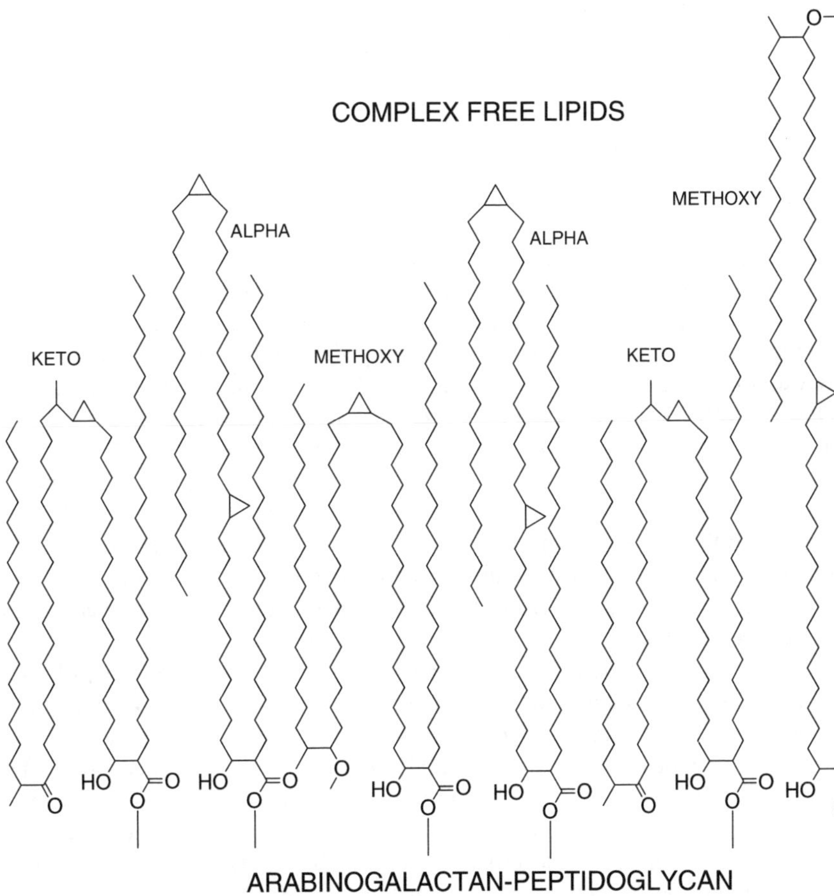

Figure 8.2 Schematic diagram of the *M. tuberculosis* cell wall mycolic acids that can serve as biomarkers of infection

ANCIENT DNA AND PALAEOMICROBIOLOGY OF HUMAN PATHOGENS

The earliest work on ancient DNA (aDNA) relied on cloning–a technique whereby a DNA fragment is introduced into a plasmid–a circular self-replicating piece of DNA, which is subsequently reproduced within a rapidly growing bacterium such as *Escherichia coli* (Higuchi *et al.*, 1984; Pääbo, 1985). The development of the polymerase chain reaction (PCR; Saiki *et al.*, 1985, 1988) transformed this field of research. PCR relies upon the exponential amplification of DNA sequences defined by oligonucleotide primers. This greatly enhances the sensitivity of detection and also enables verification by testing the reproducibility of findings. Early work resulted in an explosion of reports of the detection of DNA sequences from extinct animals and plants–some going back to former geological eras, early human migrations, and population analyses of mitochondrial DNA. Gross morphological changes and bone lesions were the basis of the first papers in which molecular techniques were used to detect DNA of an infecting pathogen (Spigelman and Lemma, 1993; Salo *et al.*, 1994). Useful summaries were published from around the 10th anniversary of the emergence of aDNA and palaeomicrobiology as distinct fields of study (e.g. Marota and Rollo, 2002; Zink *et al.*, 2002; Greenblatt and Spigelman, 2003; Pääbo *et al.*, 2004; Cipollaro *et al.*, 2005; Drancourt and Raoult, 2005).

Degradation of Ancient DNA

DNA is readily broken down by enzymatic and chemical processes, so it is important to appreciate the conditions that encourage its persistence. Enzymatic degradation begins after death, caused initially by cellular autolysis, the host commensal microflora and, thereafter, by environmental microbes. Therefore, any conditions that inhibit microbial activity favour DNA survival. Environmental conditions are more important than the actual age of the sample (Pääbo, 1989). Good preservation occurs with a continuous low temperature, a dry environment and absence of sunlight (Smith *et al.*, 2003). However, DNA also persists in material submerged by water, where the cold and lack of oxygen inhibit microbial degradation. Hot and dry environments may also lead to well-preserved DNA, if there were no fluctuating conditions or exposure to damaging radiation (Zink and Nerlich, 2003). Other important factors are the alkalinity or acidity of the surroundings, salt concentration, and the presence of natural or man-made chemical inhibitors of microbial activity, such as resins, bitumen and embalming materials.

Enzymatic activity causes strand breaks and a reduction in the overall quantity of residual DNA. Oxidative lesions damage bases and deoxyribose residues, leading to fragmented DNA and nucleotide modification. Hydrolytic damage results in loss of amino groups, which cause coding artefacts (Hofreiter *et al.*, 2001a, Stiller *et al.*, 2006). Reactions between DNA and other biomolecules can result in cross-links, such as Maillard sugar condensation products, which prevent DNA strand separation in PCR reactions. All these processes result in fragmented DNA, which may be subject to miscoding and be non-amplifiable by PCR (Pääbo *et al.*, 2004). 'Jumping PCR', where small fragments are linked together into a chimaeric PCR product (Wang and Wang, 1997), and spurious PCR products, can be detected by restriction enzyme digestion of the amplicons or, preferably, by sequencing.

Authentication of Ancient DNA

As DNA is such an unstable molecule, aDNA sequences will always be outnumbered by modern sequences. Stringent precautions must be taken to reduce extraneous contamination to a minimum, from the initial removal of samples from the archaeological site (Spigelman and Greenblatt, 1998), throughout all subsequent examinations, and eventual transport to the laboratory. These practical difficulties led to a critical reassessment of much of the early work, and criteria were drawn up to be used to verify the validity of findings (O'Rourke *et al.*, 2000; Hofreiter *et al.*, 2001b; Pääbo *et al.*, 2004).

If mishandled, even relatively impervious tissues such as teeth can become contaminated with modern DNA (Gilbert *et al.*, 2003, 2006). Many researchers routinely remove the outer surface, treat bones with bleach (Kemp and Smith, 2005) or expose bones to ultraviolet light in an attempt to remove modern contaminants, although the radiation may result in unacceptable damage to ancient sequences (Götherström *et al.*, 1995). Bouwman *et al.* (2006) have demonstrated how the knowledge of DNA degradation over time, coupled with cloning, enabled them to distinguish between modern contaminating sequences–restricted to the outer 1–2 mm of the bone surface and genuine mitochondrial DNA sequences from the specimen. This provides a useful approach in the face of persistent arguments about authenticity of data.

Strategies for Studying Ancient DNA

Selection of Specimens and Sampling Site

Some authors used the degree of racemization of amino acids in a specimen to indicate the extent of DNA preservation (Poinar and Stankiewicz, 1999). If there was evidence of amino acid degradation, then aDNA findings were regarded sceptically. Later work suggested that both hydroxyapatite and collagen yield may be better indicators of biomolecular stability (Götherström *et al.*, 2002).

The site formerly occupied by dental pulp in sound adult teeth is an excellent source of aDNA: its stability is increased by adsorption to hydroxyapatite in calcified tissue (Götherström *et al.*, 2002) and any microorganisms present in the blood will potentially be present. The presence of surface lesions may be an indication of pathogenic microbial aDNA, but this is not an infallible guide (see below). Unfortunately, the use of bleach or ultraviolet light to remove surface contamination may have the unwanted effect of removing pathogenic microbial aDNA, such as *M. tuberculosis* on an inner rib surface. Other examples of inadvertent damage to specimens are the removal of fluids surrounding pathology museum specimens (Barnes *et al.*, 2000) and the de-fleshing and cleaning methods used by museums for skeletal remains to be put on display, such as those examined by Barnes and Thomas (2006).

Precautions To Be Taken

Work on aDNA must be physically separated from other activities, especially any work with modern DNA, and pre- and post-PCR work carried out in separate laboratories. Plentiful negative controls should be included throughout DNA extraction and amplification. An inverse correlation should apply between length of target sequence and amplification efficiency, with claims of long amplicons scrutinized critically–normally, the maximum fragment length of mammalian aDNA is <100 bp (Pääbo, 1989; Stiller *et al.*, 2006). Results should be repeated in a second extract, and verified in an independent laboratory. Some authors regard quantitation of the number of amplifiable DNA molecules and cloning of PCR products as essential determinants of authenticity (Willerslev and Cooper, 2005, 2006). Others suggest a more pragmatic approach, on a 'case-by-case' basis (Gilbert *et al.*, 2005a).

DNA Extraction

Procedures must be optimized for fragmented DNA sequences. After release from the specimen, DNA is separated from other substances that co-purify and interfere with PCR, such as humic acids (O'Rourke *et al.*, 2000). Many protocols are used to disaggregate the sample: drilling bone; grinding in a pestle and mortar; successive freezing in liquid nitrogen and thawing; demineralization with Proteinase K and EDTA; and incubation in a lysis buffer based on a guanidium salt. DNA is normally captured onto silica (Boom *et al.*, 1990; Höss and Pääbo, 1993), or precipitated by isopropanol–shown to remove inhibitors (Hänni *et al.*, 1995). Repeated silica extraction is also effective (Kemp *et al.*, 2006). Finally, DNA is eluted or rehydrated into solution. DNA extracts are not stable, so are often aliquoted into 'no stick' plastic tubes, before storing at $-20\,°C$, or preferably $-80\,°C$, to avoid unnecessary freezing and thawing.

Stabilizers, such as bovine serum albumin, can bind inhibitors and/or stabilize DNA. Covalent cross-links, such as Maillard condensation products, must be cleaved to enable DNA strand separation. *N*-Phenacylthiazolium bromide has been used successfully (Poinar *et al.*, 1998), although results are inconsistent. Enzymatic repair of damaged DNA before amplification has been attempted (Di Bernardo *et al.*, 2002; Iñiguez *et al.*, 2003). Isothermal amplification can be used to increase the amount of target DNA before specific amplification of an identified target sequence (Groathouse *et al.*, 2006). Commercial spin columns are often used to capture DNA, but protocols may need to be customized, and the proportion of ethanol in washing buffers increased, to allow for the isolation of DNA fragments of <200 bp. DNA concentration from a large volume gives improved chances of recovery, and this is facilitated by the use of magnetic beads.

Polymerase Chain Reaction Amplification

Stabilizers such as bovine serum albumin can also improve PCR amplification, probably via multiple roles: masking non-specific binding sites, stabilizing DNA fragments and binding or otherwise inactivating co-purified PCR inhibitors. Betaine is another commonly used facilitator (Abu Al-Soud and Rådström, 2000). The reaction buffer can be stringent (KCl based, pH 8.3), or permissive (Tris–HCl–$(NH_4)_2SO_4$, pH 8.8). All parameters of the PCR reaction should be optimized: the higher the level of divalent cations in the mix, the more permissive is the reaction. Different enzymes vary in specificity, and excess enzyme can overcome PCR inhibitors (Sutlovic *et al.*, 2005). New enzymes are being developed specifically for the amplification of ancient or damaged DNA (McDonald *et al.*, 2006). Hot-start PCR increases specificity, as does a careful choice of primers and annealing temperature.

The choice of PCR primers is crucial. Where there is no knowledge of what is to be found in a specimen, one strategy is to use universal primers, followed by sequencing of amplified DNA, possibly after cloning (Cano *et al.*, 2000). This can be problematic, as artefacts such as chimaeric molecules can be difficult to identify. Where the whole genome has been sequenced, a more satisfactory strategy is to design highly specific primers to exclude any

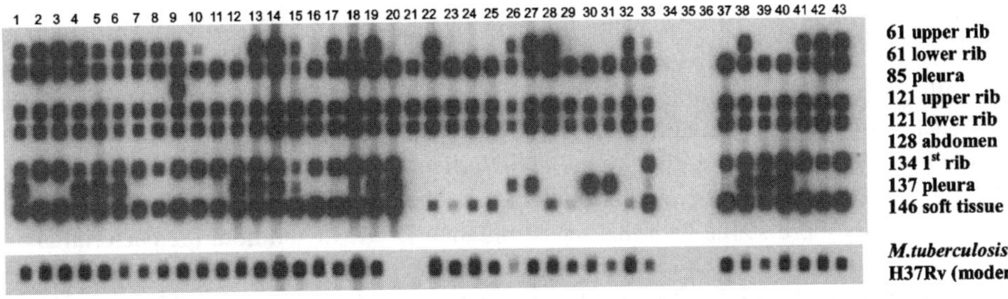

Figure 8.3 Spoligotyped samples from 18th-century AD naturally mummified bodies found in the church crypt in Vác, Hungary. Modern strains of *M. tuberculosis* lack spacers 34–36 and are distinguished by other deleted spacers. *M. bovis* strains and 'ancestral' lineages of *M. tuberculosis* have spacers 34–36. However, *M. bovis* strains lack spacers 39–43. Some of the Vác samples are very well preserved and show none of the additional deletions found in modern-day *M. tuberculosis* strains. The negative or incomplete patterns from other samples are believed to be due to poor preservation, not deletions

but the organism or sequence being sought. Generally, a target sequence of between 90 and 130 bp gives best results (Fletcher *et al.*, 2003a). Some regions of the genome are more stable than are others, which may be related to the percentage of guanine and cytosine bases (%GC), or to positional features, such as proximity to mutation or deletion hotspots. Target regions present in multiple copies within the genome will increase the likelihood of successful amplification. Real-time PCR is a useful tool to optimize reactions and to compare the efficiencies of different PCRs (Taylor *et al.*, 2006). It is based on the incorporation of a fluorescent dye into the amplified DNA, which is measured throughout the reaction. If preservation is sufficiently good, then real-time PCR can enable quantitation of the amplified aDNA (Fletcher *et al.*, 2003b).

PCR combined with hybridization to a probe is tedious to do but gives greater sensitivity. An example is the 'spacer-oligonucleotide' or spoligotyping of *M. tuberculosis* complex strains, in which PCR primers are directed at multi-copy repeat units, and the amplicons hybridized to probes for the 43 unique spacer oligonucleotides that separate these repeated units, which are on a membrane (Kamerbeek *et al.*, 1997). This method is particularly suitable for aDNA, as results can be obtained with very short fragments of DNA. Strains and the individual species of the *M. tuberculosis* complex can be distinguished due to a different pattern of deleted spacer oligonucleotides (Figure 8.3). Future developments are likely to include a similar system for other deleted regions of the genome–'deligotyping', and possibly the use of microarrays, where a DNA extract can be tested against the entire genome of a particular microbe.

Bacterial versus Human DNA

Early investigators of tuberculosis in archaeological material soon reported *M. tuberculosis* DNA in individuals with no palaeopathological signs of disease (Faerman *et al.*, 1997; Zink *et al.*, 2003). Specialists who studied mammalian aDNA interpreted such data as contamination and laboratory artefacts (e.g. Willerslev and Cooper, 2005). Also, there was initial disbelief at reports of *M. tuberculosis* DNA detected from ancient Egyptian samples, when the decay rate of DNA from the host or associated materials (papyri) from such an environment was believed to be very rapid, and the amino acid preservation in the specimens apparently poor (Marota *et al.*, 2002; Gilbert *et al.*, 2005b). In contrast, microbiologists found such results consistent with what is known of the organism and the natural history of tuberculosis (Donoghue *et al.*, 2004; Zink and Nerlich, 2005).

The DNA of *M. tuberculosis* and related species such as *M. leprae* is of high %GC, and so is intrinsically more stable than that of mammalian DNA. In addition, the bacterial DNA is contained within a lipid-rich resistant and persistent cell wall (Spigelman and Donoghue, 2003; Donoghue *et al.*, 2004). The reported distribution of *M. tuberculosis* DNA is consistent with our knowledge of the natural history of the disease, where the development of skeletal lesions is found only in 5–6% of untreated cases. Such considerations have led to a revised list of recommendations for this group of pathogens, where the persistence of human DNA and assessment of the state of preservation are less relevant for verification of authenticity. In such circumstances, where there is clear evidence of excellent preservation of pathogenic

Table 8.1 Criteria used to determine authenticity of pathogenic microbial aDNA

Criteria	Comments and examples
Sampling sites should be based on the natural history of the infectious disease	For example, inner rib surfaces, lung tissue for tuberculosis; vascularized tissues for septicaemic diseases, etc.
Choice of sampling site should also be based on likely robustness and persistence of microbial aDNA	For example, dental pulp region for less robust pathogens, low %GC DNA
Strict physical separation of pre- and post-PCR activities	To prevent cross-contamination
Prevention of modern DNA in aDNA laboratory	Modern DNA is non-fragmented and will inevitably be a major PCR amplicon
Plentiful negative controls of DNA extraction and PCR	This detects glove-tip cross-contamination
Confirmation of results by replication within laboratory	Results are often inconsistent due to uneven distribution of pathogen
Independent confirmation of results by external laboratory	Discrepant results may be genuine, so many replicates advised
Results should be consistent with natural history of the infectious disease	For example, *M. leprae* DNA may be absent in extremities as organism is localized in nerve nodes; tertiary syphilis lesions often sterile
An inverse relationship between fragment size and amount of PCR amplicons should be observed	GC-rich microbial templates can yield remarkably large PCR amplicons, e.g. 300–500 bp
aDNA sequence data should make phylogenetic sense	Direct sequencing adequate for GC-rich microbial aDNA
Cloning or multiple sequencing useful	The less robust the microbial aDNA the more useful this may be

microbial DNA, cloning is unnecessary, although it has been used to verify direct sequences (Taylor *et al.*, 2006).

The interpretation of negative results also differs in the study of microbial pathogens. Unlike host DNA, the distribution of the microbe within the host will vary according to the stage and intensity of infection. Therefore, it is essential that both positive and negative findings be subjected to independent verification, as different samples from the specimen and minor differences in laboratory procedures can result in success or failure. Removal of the outer surface of a bone may remove DNA from the infecting microbes, such as the surfaces of ribs adjacent to the lungs and pleura. The relationship of microbes to lesions is not consistent. A bony lesion may be the best place to start an investigation (Spigelman and Lemma, 1993; Spigelman *et al.*, 2003), but the centres of mummified lesions may be sterile with most organisms found around and beyond the leading edge of the lesion. In some cases lesions are the result of a host immune response and contain no microbial DNA–an important example is tertiary syphilis (Bouwman and Brown, 2005). Therefore, an additional criterion should be used: that samples should be taken from sites appropriate to what is known of the natural history of the infection (Donoghue and Spigelman, 2006) (Table 8.1).

PALAEOMICROBIOLOGY OF SPECIFIC INFECTIONS

Infections Spread by Vectors

Bubonic Plague

There are historical descriptions of three pandemics of plague: the Justinian Plague (AD 541–767), the Black Death (from AD 1346 and in subsequent epidemics), and the third pandemic that arose in the mid-19th-century AD in China (Achtman *et al.*, 1999). However, reports of *Yersinia pestis* DNA from the dental pulp region of teeth recovered from French mass graves, believed to contain 16th- and 18th-century plague victims (Drancourt *et al.*, 1998; Raoult *et al.*, 2000), were received with scepticism. *Y. pestis* is a thin-walled Gram-negative bacterium, and its DNA is unlikely to persist except in a very protected site. There was no independent verification, and attempts to repeat the work (Gilbert *et al.*, 2004a,b; Pusch *et al.*, 2004) were unsuccessful, although different samples and methodology were used (Drancourt and Raoult, 2004). Microbiologists and epidemiologists continue to suggest that there are sufficient anomalies in the historical reports and epidemiological parameters to indicate one or more other agents, such as viral haemorrhagic fevers (Duncan and Scott, 2006), may have been responsible for outbreaks traditionally ascribed to bubonic plague, and this remains a possibility, especially in northern Europe (Drancourt *et al.*, 2004; Cohn and Weaver, 2006).

However, an independent group (Wiechmann and Grupe, 2005) have verified the role of *Y. pestis* in medieval Bavarian samples by detecting its DNA. Genotyping indicates that the Orientalis biovar was involved in all of the three pandemics (Drancourt *et al.*, 2004, 2007; Raoult *et al.*, 2005; Vergnaud, 2005). The mechanisms that determine whether the disease is spread to a limited extent by fleas or by primary septicaemic spread are now understood (Sebbane *et al.*, 2006), which should answer the epidemiological concerns. The hypothesis that favourable climatic conditions in central Asia may have triggered the medieval Black Death and the 19th-century plague pandemic has received recent experimental support (Stenseth *et al.*, 2006).

Trench Fever and Epidemic Typhus Fever

Humans are the only known reservoir of *Bartonella quintana*, a small intracellular Gram-negative bacterium that is spread by the body louse. *B. quintana* causes a symptomless infection in healthy individuals, but under conditions of stress and low host resistance it can cause trench fever, which was widespread in the battlefields of north-west Europe during the First World War. The bacteria are found in the blood, and their DNA has successfully been detected and confirmed by sequencing in the dental pulp region of a 4000-year-old tooth (Drancourt *et al.*, 2005). Both *B. quintana*, and *Rickettsia prowazekii*, the causative organism of typhus fever, were detected in teeth from soldiers from the remains of Napoleon's Grand Army in retreat from Moscow; *B. quintana* DNA was also detected in fragments of body lice. In a sample of 35 human remains, 10 contained detectable DNA from one or other of these organisms, suggesting that disease played a major role in the retreat of this army (Raoult *et al.*, 2006).

Chagas' Disease and Leishmaniasis

American trypanosomiasis, or Chagas' disease, is caused by a flagellated protozoan, *Trypanosoma cruzi*, which can infect more than 100 mammalian species. *T. cruzi* has a complicated life cycle and is spread by several species of bug when they take a blood meal, usually at night. Today the disease is widespread in Central and South America. In Chagas' disease, there is an acute phase with a mortality of around 10%. In survivors, the disease becomes chronic or asymptomatic. The parasite can be found in the blood, but at relatively low levels, and is then localized in the tissues, primarily the heart, liver, spleen and bone marrow.

An excellent target region for DNA studies of all flagellated protozoans is the kinetoplast, a DNA-containing body at the base of the flagellum, which is present in thousands of copies within a single parasite. Using nested PCR, and labelled probes detected by hybridization, Aufderheide *et al.* (2004) described results obtained from 283 naturally mummified human remains. Over 41% of tissue extracts were positive for *T. cruzi*, over a time-span from 7050 BC to AD 1500. Data were analysed by human population, age at death, sex, tissue type, and comparison with today.

Leishmaniasis is caused by various *Leishmania* species, and is transmitted to humans via the bites of female sandflies. The disease, which is often acquired from infected animals, varies in its clinical manifestations from a mild cutaneous form, a more damaging mucocutaneous infection, or a fatal visceral disease known as Kala-azar. It is increasing in incidence and distribution in the present day, and is believed to have originated in East Africa. Genus-specific *Leishmania* kinetoplast aDNA has been detected by PCR in early Christian Nubian samples (Spigelman *et al.*, 2005) and in ancient Egypt (Zink *et al.*, 2006). The sandfly vector of the disease is associated with Acacia–Balanites woodland, found in Sudan, so it is more plausible that leishmaniasis in ancient Egypt resulted from cultural contact between Egypt and Upper Nubia rather than an endemic infection.

Malaria

Malaria is a protozoan disease caused in humans by four species of the genus *Plasmodium* (*P. falciparum*, *P. vivax*, *P. ovale* and *P. malariae*), transmitted by the bite of an infected female mosquito of the genus *Anopheles*. Malaria was prevalent in Europe prior to the 20th century and remains endemic in parts of Asia, Africa, Central and South America, and

Oceania (Carter and Mendis, 2002). It still kills more than one million people (most of them young children living in Africa) each year. It is characterized by extreme exhaustion associated with paroxysms of high fever, sweating, shaking chills, and anaemia. Malaria in animals is caused by other species of plasmodia. Malaria has been a major scourge throughout recorded history, and the typical bouts of fever are described in the 4th–5th centuries BC in Sicily (Sallares *et al.*, 2004). However, there are few reports of the direct detection of the parasite in archaeological material.

Following the detection of *P. falciparum* antigen (Miller *et al.*, 1994), Taylor *et al.* (1997) examined two samples from the early 20th century, two from medieval Spain and one from the ancient Egyptian Granville mummy. They used the ParaSight F test for *P. falciparum* histidine-rich protein-2 antigen and analysis for 18s rRNA sequences for the four human pathogenic species. One of the 20th-century samples gave a positive result for *P. falciparum* DNA. *P. falciparum* DNA was found in one of 40 skeletons from an infant cemetery dating to *c.* AD 450 at Lugnano in Treverina, Umbria, Italy (Sallares and Gomzi, 2001; Sallares *et al.*, 2004). Preliminary data have also been presented from ancient Egypt (Zink *et al.*, 2001). Cerutti *et al.* (2005) found initial DNA evidence of *P. falciparum* in pre-dynastic (3200 BC) Egyptian mummies that showed associated pathology such as criba orbitaria, but confirmation is required.

Schistosomiasis (Bilharzia)

It is believed that schistosomiasis originated in central Africa and spread to ancient Egypt down the Nile (Adamson, 1976). It has been extensively documented in ancient Egypt, as egg morphology enables identification of the species. It is transmitted to humans via freshwater snails. After infection, the parasites are spread via the bloodstream. They form permanent adult mating pairs, which lead to life-long shedding of eggs into the environment and are associated with a host granulomatous response. *S. haematobium* is usually located in the kidney or bladder, whilst *Schistosoma mansoni* is found in the intestinal tract and liver. Other species occur in different geographical regions, and fertile inter-species hybrids can result from mixed infections.

Bulinus spp., the snail host of *S. haematobium*, are bottom feeders, which are unaffected by the water flow rate. However, *Biomphalia* spp., the *S. mansoni* snail host, are surface feeders and cannot establish themselves except in waters with little or no current. Therefore, the presence of different *Schistosoma* species can indicate the local environment in ancient times. Recent observations based on species-specific PCR using primers based on a tandem repeat sequence and mitochondrial DNA have detected both *S. haematobium* and *S. mansoni* in ancient Egyptian mummies dating from the Middle Kingdom (approximately 3900 BP), indicating a possible co-infection (Donoghue *et al.*, 2007a).

Brucellosis (Undulant Fever)

Brucellosis is a zoonotic infection normally acquired from infected goats and cattle, by animal husbandry, or consumption of meat and dairy products. During the acute phase of infection, the thin-walled bacilli are spread via the bloodstream, so bone marrow and dental pulp sites should be examined. In later stages, granulomatous nodules, which can develop into abscesses, may occur in liver, spleen, lymphatic tissue and bone marrow. Infections often become chronic, with occasional recurrent bouts of fever. Although there is morphological evidence of brucellosis in archaeological remains, there are no published reports of the

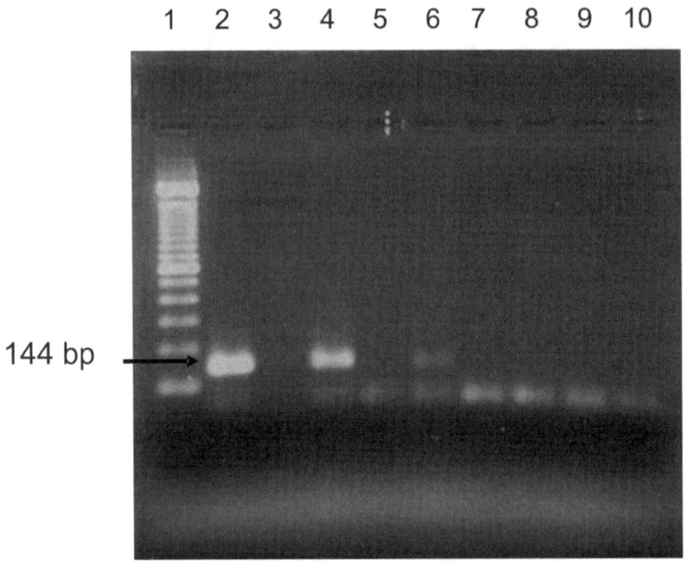

Figure 8.4 Gel electrophoresis demonstrating a 144-bp genus-specific *Brucella* IS*711* PCR product from an adult female with vertebral lytic lesions, found in a Siberian Iron Age site dated 360–170 BC. This was confirmed by sequencing (G.M. Taylor, personal communication). Lane 1: 100 bp ladder; lane 2: *Brucella suis* (positive control amplified in a separate PCR in another location); lanes 4 and 6: Siberian Iron Age sample XXX1.34 (positive); lanes 3, 5, and 8–10: other negative samples; lane 7: negative PCR control

detection of aDNA from *Brucella* species. Mays *et al.* (2001) examined nine specimens with lesions, using PCRs based on specific multicopy insertion elements to distinguish between brucellosis and tuberculosis. Results were negative for *Brucella* and positive for the *M. tuberculosis* complex–the preferred morphological diagnosis. However, preliminary data have been obtained from an adult female with vertebral lytic lesions, found in a Siberian Iron Age site dated to 360–170 BC, inhabited by semi-nomadic pastoralists (G.M. Taylor, personal communication). Using an IS*711* target site, a 144-bp PCR product was obtained in two separate extractions (Figure 8.4) and sequenced.

Long-Lasting Human Infections Spread Person to Person

Leprosy (Hansen's Disease)

Leprosy is readily identified in human skeletal remains as it results in characteristic palaeopathology. It is caused by *M. leprae*, which is harboured by healthy carriers in the nose, but which causes disease in peripheral nerves, skin and bones. Its clinical effects depend upon the host immune response and range from a slowly developing form with few bacilli, termed tuberculoid leprosy, to a more serious disease, lepromatous leprosy, where there are extremely high numbers of bacilli and much tissue destruction. The organism cannot be cultured in cell-free media, so it was an early candidate for the development of molecular techniques for clinical diagnosis.

M. leprae DNA was first detected in a metatarsal bone from AD 600 Palestine by Rafi *et al.* (1994), who used a large PCR target region of 530 bp based on the 36 kDa antigen gene. Two skulls from southern Germany with possible leprosy palaeopathology, dated to between AD 1400 and 1800 were examined by Haas *et al.* (2000, 2002). They were found to be positive for *M. leprae* DNA, using primers for repetitive regions RLEP1 and RLEP3 that gave PCR products of 372 bp and 320 bp respectively. A skull from a 13th–14th-century AD case of lepromatous leprosy in Orkney, Scotland, was positive for *M. leprae* DNA using PCR primers based on a 153-bp locus in RLEP1 (Taylor *et al.*, 2000) and confirmed by sequencing. However, the DNA preservation was clearly not good, as primers of a 175-bp target region in the same locus were negative.

By this time, the importance of using primers designed for short DNA fragments in aDNA was better recognized, and Donoghue *et al.* (2001) described two nested primer pairs for *M. leprae*: one pair was based on a 129-bp locus in the multicopy RLEP3; the other pair was for a 136-bp locus in the single copy 18 kDa antigen gene. Both sets of these primers detected *M. leprae* DNA in nasal-region samples from medieval Poland and 10th–11th-century AD Hungary. Subsequently, these primers confirmed the presence of *M. leprae* DNA in a foot bone (Spigelman and Donoghue, 2001) from an earlier case that had been the subject of unresolved debate (Hershkovitz *et al.*, 1992, 1993; Manchester, 1993). Further work, on samples from 10th–11th-century AD Hungary, yielded positive results in casum nasale samples with all primer pairs, including the primers used in the first reported finding of *M. leprae* aDNA. The sequence obtained from the 530-bp, 36-kDa antigen gene of *M. leprae* is unusually long for aDNA, but negative controls were satisfactory. Such data indicate the robustness of mycobacteria and their probable high numbers at this particular skeletal site at the time of death (Donoghue *et al.*, 2002).

Owing to concerns about the possibility of contamination occurring from the use of PCR positive controls when examining archaeological samples, Montiel *et al.* (2003) examined four metacarpal specimens from 12th-century AD Seville. DNA was amplified from two samples using nested PCR based on a 97-bp target region that contained a restriction enzyme site. The amplicons were subjected to restriction analysis, cloned and sequenced, and *M. leprae* DNA confirmed (Montiel *et al.*, 2003). The laboratory had never been used to examine modern *M. leprae*. The age and geographical range of samples positive for *M. leprae* aDNA have since been expanded to include 1st-century AD Israel, 4th-century AD Egypt, 10th–13th-century AD Sweden (Donoghue *et al.*, 2005b) and 12th-century AD Bohemia (Likovsky *et al.*, 2006).

Leprosy has occasionally been described in infants, normally in areas of high prevalence and in the absence of effective treatment (Brubaker *et al.*, 1985). However, the bony changes are often subtle, especially in young children, and Lewis (2002) suggested that the paucity of evidence might relate to the difficulty in diagnosis of the less severe forms of the disease. A case was reported recently dating from 8th–9th-century AD Byzantine Turkey (Donoghue *et al.*, 2005a). The specimen was of an infant cranium, with non-specific pathology indicating new bone formation (Figure 8.5). However, the sample was positive for *M. leprae* DNA (Figure 8.6), and leprosy in the community was additionally confirmed by specific morphology and *M. leprae* DNA in three adults from the same burial site. This case could only have been detected by molecular methods.

An exciting recent development is molecular typing techniques that enable different strains of *M. leprae* to be distinguished. In the first publication of *M. leprae* molecular fingerprinting from archaeological material, Taylor *et al.* (2006) used 'variable-number tandem repeat'

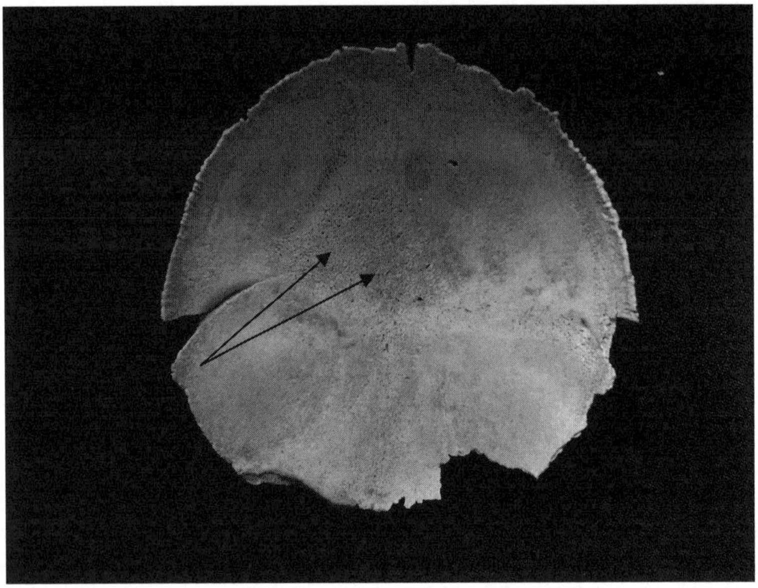

Figure 8.5 Occipital bone from 5-month-old infant, from Kovuklukaya, Turkey, dated to the Byzantine period (770–970 AD). There is new bone formation (arrowed), a non-specific sign of chronic inflammation

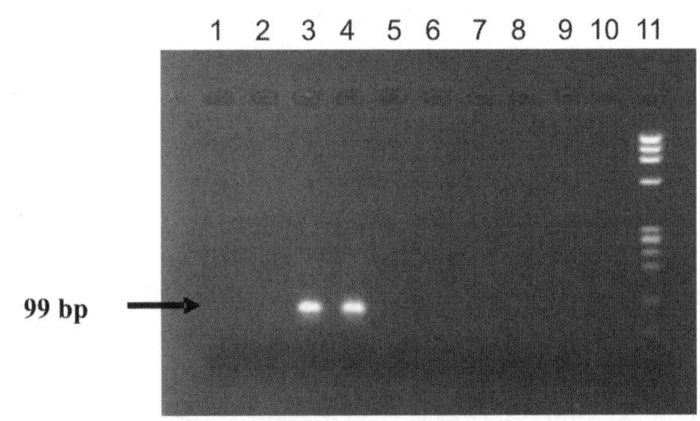

Figure 8.6 Gel electrophoresis demonstrating a 99-bp RLEP sequence specific for *M. leprae*, from the specimen shown in Figure 8.5 (lane 3), and from nasal scrapings of an adult male with typical palaeopathology of lepromatous leprosy from the same archaeological site (lane 4). Lanes 1, 2 and 5–8: other samples (negative); lane 9: negative DNA extraction control; lane 10: negative PCR control; lane 11: PhiX174*Hae*III molecular markers

(VNTR) methods, which detected strains with different numbers of microsatellite AGT and TTC repeat sequences, and a 21-bp repeat minisatellite sequence. This was a two-centre study, which also compared results obtained by direct sequencing of PCR products and results obtained by cloning. The authors concluded that the few clones with discrepancies of VNTR numbers probably reflected ‘replication slippage’ or ‘slipped strand mispairing’, which can result in an increase or decrease in the copy number of the repeat element during cell division, rather than a mixed population of genotypes. Broader genotypes have been described, which are based on single nucleotide polymorphisms (SNPs), and it has been suggested that these may have been derived from a single clone and spread by prehistoric migrations (Monot *et al.*, 2005). The data can support more than one evolutionary scenario (Pinhasi *et al.*, 2006), and the relative stability of *M. leprae* DNA may one day enable the origin of this disease to be determined by direct analysis.

Tuberculosis

The earliest work on the DNA of pathogenic microorganisms was on *M. tuberculosis*, and this has been the most studied. Why should this be so? Tuberculosis is associated with typical skeletal palaeopathology, so is believed to be an ancient disease that was widespread. *M. tuberculosis* is still a major pathogen and approximately one-third of the world population is currently infected, with over 2 million deaths per year (Fernando and Britton, 2006). As the organism is extremely slow growing, the current intense research activity has led to the development of many well-characterized molecular diagnostic methods and typing schemes. The complete genomes of more than one strain of *M. tuberculosis*, *Mycobacterium bovis* and other closely related species have been determined, and there are international databases of gene sequences from thousands of isolates that have led to the identification of families and lineages of modern strains. This is extremely useful for the palaeomicrobiologist who wishes to investigate this disease.

The earlier molecular work has been reviewed (Spigelman and Donoghue, 2003; Donoghue *et al.*, 2004), but questions and controversies remain. There is continuing interest in the relationship of disease to the presence of bony lesions (Mays *et al.*, 2002; Zink *et al.*, 2005a; Santos and Roberts, 2006). Genotypic analysis has continued (Taylor *et al.*, 2005), and it is clear that 'modern strains' of *M. tuberculosis* were present in ancient Egypt, alongside more ancestral strains (Zink *et al.*, 2003, 2005b). *M. bovis* infects many host species, so both wild and domesticated animals can act as reservoirs of infection. *M. bovis* appeared to be absent from archaeological material until reported by Taylor *et al.* (2007) in semi-nomadic pastoralists from the Siberian Iron Age. However, its scarcity is reflected by the low levels of *M. bovis* infection reported in the absence of effective control measures today, where *M. bovis* causes only around 6 % of deaths due to tuberculosis (Grange, 1995). This suggests that a long-term close association with an infected herd, such as the Siberian pastoralists probably experienced, is the most likely scenario for the detection of this organism. So far, mainly European and Near-Eastern archaeological samples have been subjected to further molecular analysis, and it would be of great interest to determine the nature of the indigenous tuberculosis found in archaeological material from Africa, the Indian subcontinent, the Far East and pre-Colombian America.

INTERACTIONS BETWEEN INFECTIONS AND HOST FACTORS

Early human populations would have been exposed to many parasites, bacteria and viruses, so it is likely that co-infections or multiple infections were common, yet we are only starting to realize the implications of this. Intestinal parasites, such as worms, have a profound effect upon the host immune response, and can reverse intestinal inflammation (Elston, 2006). Mycobacteria, many species of which are ubiquitous in soil and water, are also strong immune modulators and are believed to play an important role in determining whether a protective cell-mediated immune response develops (Rook and Rosa Brunet, 2005).

The recognition of the importance of interactions should lead to research on their occurrence in the past, to improve our understanding of the interrelationships between microorganisms, and between them and their host. One example is the interaction between leprosy and tuberculosis. Co-infections had been reported in the 20th century, primarily from historical records and archival material that predates adequate chemotherapy. However, when a collection of archaeological specimens positive for lepromatous leprosy were re-examined for evidence of tuberculosis, a significant number were found to be co-infected, suggesting a possible reason for the historical decline of leprosy in western Europe (Donoghue *et al.*, 2005b).

The relationship between infectious diseases and the host natural resistance to infection was mentioned in relation to endogenous infections. Nutrition plays an important role, and Spigelman *et al.* (2005) have reported preliminary data on the relationship between nutritional stress markers, coprolite analysis and molecular detection of tuberculosis in early Christian Nubia. Another important consideration is host genetic susceptibility to infectious disease. In addition to population-based genomic studies, it is possible to identify specific host mutations linked to increased susceptibility (Larcombe *et al.*, 2005). Neoplasms also have a detrimental effect on host resistance. For example, Langerhans cell histiocytosis (LCH), recently recognized as a neoplasm, has distinct palaeopathology and has been diagnosed in skeletal remains (reviewed by Mays and Nerlich (1997)). The relationship of LCH to immune dysfunction and an increased risk of acquired infections has been recognized (Zink and Nerlich, 2001; Oxenham *et al.*, 2005) and is demonstrated in the recent report of a case of LCH in an infant who also had molecular evidence of tuberculosis infection (Spigelman *et al.*, 2006).

MICROBIAL GENOMICS AND PALAEOMICROBIOLOGY

Evolution of Human Microbial Pathogens

The sequencing of several whole genomes has led to a better understanding of the relationship between pathogenic microbes and their most closely related free-living relatives. In thin-walled Gram-negative bacteria, horizontal gene transfer occurs, allowing for gene acquisition as well as loss. The acquired genes may be recognized by their distinct %GC and have been termed ‘pathogenicity islands’. They are often on mobile elements, such as plasmids, and encode for virulence factors, such as those enabling tissue invasion, toxins, resistance to host defence mechanisms, or infection of vectors (Chain *et al.*, 2004; Groisman and Casadesús, 2005). Another strategic change occurs by losing genetic material, known as reductive evolution (Dagan *et al.*, 2006). This has occurred in many microbial pathogens, such as *S. typhi*, *Y. pestis* and *Rickettsia* spp., and is the prime mechanism found in the slow-growing, thick-walled mycobacteria. Many of the eliminated genes encode for metabolic pathways of intermediate metabolism and amino acid biosynthesis. *M. leprae* is an extreme example of this process and has lost about one-quarter of its genes (Cole *et al.*, 2001). The high degree of sequence conservation in pathogenic mycobacteria and the lack of clonal variation have been interpreted as the result of an evolutionary bottleneck (Frothingham, 1999; Gutierrez *et al.*, 2005).

Human Palaeogeography and Ancestral Sequence Inference

Our knowledge of the stability of modern molecular typing methods and their rate of change makes it possible to estimate the rate of evolutionary change under different scenarios. Thin-walled Gram-negative bacteria that permit horizontal gene transfer show a much greater rate of change than the slow-growing mycobacteria undergoing reductive evolution. However, the strong selection pressure caused by modern antimicrobial therapy has noticeably speeded up the loss of mycobacterial genetic sequences, mediated by recombination events between repetitive sequences.

The calculation of the age of a pathogenic microbe is the outcome of a genomic process termed 'ancestral sequence inference'. Comparing the present genome sequence of the plague bacillus, *Y. pestis*, with its closest relative, *Yersinia pseudotuberculosis*, the estimated age of *Y. pestis* is calculated to be up to ~22 000 years (Achtman *et al.*, 1999). Yersiniae are rapid growers and are capable of acquiring genetic material by horizontal gene transfer, so this period of time is likely to be one of the shortest. Further analysis has identified unique genomevars confined to specific geographical regions and led to the conclusion that modern plague evolved from Yunnan Province in China, due to the emergence of biovar Orientalis from biovar Antiqua (Zhou *et al.*, 2004). This is consistent with the direct aDNA findings of Drancourt *et al.* (2004, 2007).

The evolution of *Plasmodium* spp. as human parasites has received similar analysis. *P. falciparum*, *P. vivax* and the other human malarial parasites are believed to have evolved independently from other primate hosts. By examination of their complete mitochondrial genomes, Jongwutiwes *et al.* (2005) estimated the age of the most recent common ancestor of *P. falciparum* and *P. vivax* at around 200 000 to 300 000 years ago. This is close to the estimated origin of *Homo sapiens* and suggests that these *Plasmodium* species were parasites of the hominid lineage and that the population expansion of modern man coincided with a population expansion of these parasites.

Whole genome sequencing has led to a revision of our earlier ideas about the evolution of the *M. tuberculosis* complex. Because these bacteria evolve by the accumulation of deletions, it appears that the human pathogen *M. tuberculosis* has evolved from a more ancestral lineage than *M. bovis* (Brosch *et al.*, 2002). The detection of an ancestral-like *M. tuberculosis* spoligotype from a Pleistocene bison (Rothschild *et al.*, 2001) is consistent with current views but remains controversial. There are large international databases of the molecular characteristics of these organisms, based on numbers of repetitive sequences, spoligotypes and SNPs. Recent meta-analyses of these databases have led to distinct lineages of *M. tuberculosis* strains being recognized, which are associated with different geographical regions and human populations (Baker *et al.*, 2004; Hirsh *et al.*, 2004; Gagneux *et al.*, 2006), possibly contemporaneous with early hominids in Africa (Gutierrez *et al.*, 2005). Analysis of one *M. tuberculosis* genotype has led to the hypothesis that it may have originated in central Asia in humans migrating from the Middle East during a second out-of-Africa migration in the Upper Palaeolithic, 45 000 to 30 000 years ago (Mokrousov *et al.*, 2005).

FUTURE AREAS OF RESEARCH

Palaeomicrobiology is an exciting area of research. There is still much controversy over authenticity of some findings, but there are now sufficient numbers of well-planned and executed studies to give confidence in the reports of ancient DNA from human microbial pathogens. Animal microbial diseases were not included in the remit for this book; yet, even if they had been, there is very little work that has successfully detected aDNA of microbial pathogens. Animal palaeopathology has been described, and we need better understanding of the role of animal microbial pathogens in human zoonoses, or as potential reservoirs for new human infections. Another area ripe for detailed study is that of viruses. Beyond the visual detection of viral skin lesions or of integrated virus sequences in human DNA, the only well-characterized example is that of the 1918 strain of influenza virus (Taubenberger and Morens, 2006). A recent preliminary report has demonstrated hepatitis B virus (HBV) in the liver of a 'wet' Korean mummy from 350–400 years ago (Shin *et al.*, 2003; Donoghue *et al.*, 2007b). Three independent centres detected HBV by PCR for the core protein gene, the *prec/c* region and the S-gene. Two centres confirmed that the HBV DNA was of genotype C. In addition, two centres demonstrated co-infection with the *M. tuberculosis* complex. These data suggest that examination of well-preserved organs for biomarkers of similarly robust viruses should prove fruitful.

In conclusion, it is clear that we live in an era where there is an increase in infectious diseases, due to their re-emergence, or the emergence of new diseases (Weiss and McMichael, 2004). The efforts made to determine the nature of the 1918 influenza virus are a graphic example of the maxim that in order to understand the future we need to understand the past.

ACKNOWLEDGEMENTS

Palaeomicrobiology is a fascinating field of study and I am eternally grateful to those who gave me the opportunity to work in it. By its nature collaborations are developed–with palaeopathologists who supply specimens of interest, and with colleagues who assist in the independent verification of data. My grateful thanks go to every one of you, and our joint publications show the fruits of our efforts. Professor David Minnikin (University of Birmingham) kindly supplied Figure 8.2. Dr Ruth McNerney and her colleagues at the London School of Hygiene & Tropical Medicine facilitated the spoligotyping shown in Figure 8.3. I acknowledge the thoughtful discussions and assistance given by my colleague Dr G. Michael Taylor (formerly of Imperial College London) and thank both him and Dr Eileen Murphy (Queen's University, Belfast) for permission to use their unpublished data on *Brucella* in this chapter, and for Figure 8.4.

REFERENCES

Abu Al-Soud W, Rådström P. 2000. Effects of amplification facilitators on diagnostic PCR in the presence of blood, feces and meat. *J Clin Microbiol* **38**: 4463–4470.

Achtman M. Zurth K, Morelli G, Guiyoule A, Carniel E. 1999. *Yersinia pestis*, the cause of plague, is a recently emerged clone of *Yersinia pseudotuberculosis*. *Proc Natl Acad Sci U S A* **96**: 14 043–14 048.

Adamson PB. 1976. Schistosomiasis in antiquity. *Med Hist* **20**: 176–188.

Araújo A, Ferreira LF. 2000. Paleoparasitology and the antiquity of human host-parasite relationships. *Mem Inst Oswaldo Cruz* **95**(Suppl 1): 89–93.

Armelagos GJ, Brown PJ, Turner B. 2005. Evolutionary, historical and political economic perspectives on health and disease. *Soc Sci Med* **61**: 755–765.

Aufderheide AC, Salo W, Madden M, Streitz J, Buikstra J, *et al.* 2004. A 9,000-year record of Chagas' disease. *Proc Natl Acad Sci U S A* **101**: 2034–2039.

Baker L, Brown T, Maiden MC, Drobniewski F. 2004. Silent nucleotide polymorphisms and a phylogeny for *Mycobacterium tuberculosis*. *Emerg Inf Dis* **10**: 1568–1577.

Barnes I, Thomas MG. 2006. Evaluating bacterial pathogen DNA preservation in museum osteological collections. *Proc R Soc B* **273**: 635–653.

Barnes I, Holton J, Vaira D, Spigelman M, Thomas MG. 2000. An assessment of the long-term preservation of the DNA of a bacterial pathogen in ethanol-preserved archival material. *J Pathol* **192**: 554–559.

Boom R, Sol CJA, Salimans MMM, Jansen CL, Wertheim-van Dillen PME, van der Noordaa J. 1990. Rapid and simple method for purification of nucleic acids. *J Clin Microbiol* **28**: 495–503.

Bouwman AS, Brown TA. 2005. The limits of biomolecular palaeopathology: ancient DNA cannot be used to study venereal syphilis. *J Archaeol Sci* **32**: 703–713.

Bouwman AS, Chilvers ER, Brown KA, Brown TA. 2006. Brief communication: identification of the authentic ancient DNA sequence in a human bone contaminated with modern DNA. *Am J Phys Anthropol* **131**: 428–431.

Brosch R, Gordon SV, Marmiesse M, Brodin P, Buchrieser C, *et al.* 2002. A new evolutionary scenario for the *Mycobacterium tuberculosis* complex. *Proc Natl Acad Sci U S A* **99**: 3684–3689.

Brubaker ML, Meyers WM, Bourland J. 1985. Leprosy in children one year of age and under. *Int J Lepr Other Mycobact Dis* **53**: 517–523.

Cano RJ, Tiefenbrunner F, Ubaldi M, Del Cueto C, Luciani S, *et al.* 2000. Sequence analysis of bacterial DNA in the colon and stomach of the Tyrolean Iceman. *Am J Phys Anthropol* **112**: 297–309.

Carter R, Mendis KN. 2002. Evolutionary and historical aspects of the burden of malaria. *Clin Microbiol Rev* **15**: 564–594.

Cerutti N, Marin A, Massa ER, Savoia D. 1999. Immunological investigations of malaria and new perspectives in palaeopathological studies. *Boll Soc Ital Biol Sper* **75**: 17–20.

Cerutti N, Santovito A, Turina M, Crobu F, Rabino Massa E. 2005. Detection of *Plasmodium falciparum* ancient DNA in Egyptian mummies. *J Biol Res* **80**: 88–90.

Chain PS, Carniel E, Larimer FW, Lamerdin J, Stoutland PO, *et al.* 2004. Insights into the evolution of *Yersinia pestis* through whole-genome comparison with *Yersinia pseudotuberculosis*. *Proc Natl Acad Sci U S A* **101**: 13 826–13 831.

Cipollaro M, Galderisi U, Di Bernardo G. 2005. Ancient DNA as a multidisciplinary experience. *J Cell Physiol* **202**: 315–322.

Cockburn TA. 1971. Infectious diseases in ancient populations.*Curr Anthropol* **12**: 45–62.

Cohen MN, Armelagos GJ. 1984. Paleopathology at the origins of agriculture: editor's summation. In *Paleopathology at the Origins of Agriculture*, Cohen MN, Armelagos GJ (eds). Academic Press: Orlando, FL; 585–601.

Cohn Jr SK, Weaver LT. 2006. The Black Death and AIDS: CCR5-Δ32 in genetics and history. *Q J Med* **99**: 497–503.

Cole ST, Eiglmeier K, Parkhill J, James KD, Thomson NR, *et al.* 2001. Massive gene decay in the leprosy bacillus. *Nature* **409**: 1007–1011.

Collins MJ, Nielsen-Marsh CM, Hiller J, Smith CI, Roberts JP, *et al.* 2002. The survival of organic matter in bone: a review. *Archaeometry* **44**: 383–394.

Dagan T, Blekhman R, Graur D. 2006. The 'domino theory' of gene death: gradual and mass gene extinction events in three lineages of obligate symbiotic bacterial pathogens. *Mol Biol Evol* **23**: 310–316.

Deelder AM, Miller RL, de Jonge N, Krijger FW. 1990. Detection of schistosome antigen in mummies. *Lancet* **335**: 724–725.

Di Bernando G, Del Gaudio S, Cammarota M, Galderisi U, Cascino A, *et al.* 2002. Enzymatic repair of selected cross-linked homoduplex molecules enhances nuclear gene rescue from Pompeii and Herculaneum remains. *Nucleic Acids Res* **30**: e16.

Donoghue HD, Spigelman M. 2006. Comment. Pathogenic microbial ancient DNA: a problem or an opportunity? *Proc R Soc B* **273**: 641–642.

Donoghue HD, Spigelman M, Zias J, Gernaey-Child AM, Minnikin DE. 1998. *Mycobacterium tuberculosis* complex DNA in calcified pleura from remains 1400 years old. *Lett Appl Microbiol* **27**: 265–269.

Donoghue HD, Holton J, Spigelman M. 2001. PCR primers that can detect low levels of *Mycobacterium leprae* DNA. *J Med Microbiol* **50**: 177–182.

Donoghue HD, Gladkowska-Rzeczycka J, Marcsik A, Holton J, Spigelman M. 2002. *Mycobacterium leprae* in archaeological specimens. In *The Past and Present of Leprosy*, Roberts CA, Lewis ME, Manchester K (eds). British Archaeological Reports, BAR International Series 1054. Archaeopress: Oxford; 271–285.

Donoghue HD, Spigelman M, Greenblatt CL, Lev-Maor G, Bar-Gal GK, *et al.* 2004. Tuberculosis: from prehistory to Robert Koch, as revealed by ancient DNA. *Lancet Infect Dis* **4**: 584–592.

Donoghue HD, Erdal YS, Pinhasi R, Spigelman M. 2005a. A possible case of congenital leprosy in a 5-month-old child from Byzantine-Age Turkey. *In Proceedings of the 26th Annual Congress of the European Society of Mycobacteriology*, Istanbul, Turkey, 26–29 June; 177.

Donoghue HD, Marcsik A, Matheson C, Vernon K, Nuorala E, *et al.* 2005b. Co-infection of *Mycobacterium tuberculosis* and *Mycobacterium leprae* in human archaeological samples: a possible explanation for the historical decline of leprosy. *Proc R Soc B* **272**: 389–394.

Donoghue HD, Matheson CD, David AR, Spigelman M. 2007a. *Schistosoma haematobium, Schistosoma mansoni* and family relationship in two Egyptian mummies. In *VIth World Congress on Mummy Studies*, Teguise, Lanzerote, Canary Islands, Spain, 20–24 February (presentation 71).

Donoghue HD, Spigelman M, Grant P, Klein A, Shouval D, *et al.* 2007b. Microbiological investigation of a Korean medieval child mummy found in Yangju. In *VIth World Congress on Mummy Studies*, Teguise, Lanzerote, Canary Islands, Spain, 20–24 February (presentation 126).

Drancourt M, Raoult D. 2004. Molecular detection of *Yersinia pestis* in dental pulp. *Microbiology* **150**: 263–264.

Drancourt M, Raoult D. 2005. Palaeomicrobiology: current issues and perspectives. *Nat Rev Microbiol* **3**: 23–35.

Drancourt M, Aboudharam G, Signoli M, Dutour O, Raoult D. 1998. Detection of 400-year-old *Yersinia pestis* DNA in human dental pulp: an approach to the diagnosis of ancient septicemia. *Proc Natl Acad Sci U S A* **95**: 12 637–12 640.

Drancourt M, Roux V, Dang LV, Tran-Hung L, Castex D, *et al.* 2004. Genotyping, Orientalis-like *Yersinia pestis*, and plague pandemics. *Emerg Infect Dis* **10**: 1585–1592.

Drancourt M, Tran-Hung L, Courtin J, Lumley H, Raoult D. 2005. *Bartonella quintana* in a 4000-year-old human tooth. *J Infect Dis* **191**: 607–611.

Drancourt M, Signoli M, Dang LV, Bizot B, Roux V, *et al.* 2007. *Yersinia pestis* Orientalis in remains of ancient plague victims. *Emerg Infect Dis* **13**: 332–333.

Duncan CJ, Scott S. 2006. What caused the Black Death? *Postgrad Med J* **81**: 315–320.

Elston DM. 2006. The hygiene hypothesis and atopy: bring back the parasites? *J Am Acad Dermatol* **54**: 172–179.

Faerman M, Jankauskas R, Gorski A, Bercovier H, Greenblatt CL. 1997. Prevalence of human tuberculosis in a medieval population of Lithuania studied by ancient DNA analysis. *Ancient Biomol* **1**: 205–214.

Fernando SL, Britton WJ. 2006. Genetic susceptibility to mycobacterial disease in humans. *Immunol Cell Biol* **84**: 125–137.

Fletcher HA, Donoghue HD, Holton J, Pap I, Spigelman M. 2003a. Widespread occurrence of *Mycobacterium tuberculosis* DNA from 18th–19th century Hungarians. *Am J Phys Anthropol* **120**: 144–152.

Fletcher HA, Donoghue HD, Taylor M, van der Zanden, Spigelman M. 2003b. Molecular analysis of *Mycobacterium tuberculosis* DNA from a family of 18th century Hungarians. *Microbiology* **149**: 143–151.

Frothingham R. 1999. Evolutionary bottlenecks in the agents of tuberculosis, leprosy and paratuberculosis. *Med Hypotheses* **52**: 95–99.

Gagneux S, DeRiemer K, Van T, Kato-Maeda M, de Jong BC, *et al.* 2006. Variable host–pathogen compatibility in *Mycobacterium tuberculosis. Proc Natl Acad Sci U S A* **103**: 2869–2873.

Gernaey AM, Minnikin DE, Copley MS, Power JJ, Ahmed AMS, *et al.* 1998. Detecting ancient tuberculosis. *Internet Archaeol* **5**: http://intarch.ac.uk/journal/issue5/gernaey_index.html.

Gernaey AM, Minnikin DE, Copley MS, Dixon RA, Middleton JC, *et al.* 2001. Mycolic acids and ancient DNA confirm an osteological diagnosis of tuberculosis. *Tuberculosis (Edinb)* **81**: 259–265.

Gilbert MTP, Willerslev E, Hansen AJ, Barnes I, Rudbeck L, *et al.* 2003. Distribution patterns of postmortem damage in human mitochondrial DNA. *Am J Hum Genet* **72**: 32–47. Erratum: *Am J Hum Genet* **72**: 779.

Gilbert MTP, Cuccui J, White W, Lynnerup N, Titball RW, *et al.* 2004a. Absence of *Yersinia pestis*-specific DNA in human teeth from five European excavations of putative plague victims. *Microbiology* **150**: 341–354.

Gilbert MTP, Cuccui J, White W, Lynnerup N, Titball RW, *et al.* 2004b. Response to Drancourt and Raoult. *Microbiology* **150**: 264–265.

Gilbert MTP, Bandelt H-J, Hofreiter M, Barnes I. 2005a. Assessing ancient DNA studies. *Trends Ecol Evol* **20**: 541–544.

Gilbert MTP, Barnes I, Collins MJ, Smith C, Eklund J, *et al.* 2005b. Long-term survival of ancient DNA in Egypt: response to Zink and Nerlich (2003). *Am J Phys Anthropol* **128**: 110–114.

Gilbert MTP, Hansen AJ, Willerslev E, Turner-Walker G, Collins M. 2006. Insights into the processes behind the contamination of degraded human teeth and bone samples with exogenous sources of DNA. *Int J Osteoarchaeol* **16**: 156–164.

Gonçalves MLC, Araújo A, Ferreira LF. 2003. Human intestinal parasites in the past: new findings and a review. *Mem Inst Oswaldo Cruz* **95**(Suppl 1): 103–118.

Gonçalves MLC, da Silva VL, de Andrade CM, Reinhard K, da Rocha GC, *et al.* 2004. Amoebiasis distribution in the past: first steps using an immunoassay technique. *Trans R Soc Trop Med Hyg* **98**: 88–91.

Götherström A, Fischer C, Lindén K, Lidén K. 1995. X-raying ancient bone: a destructive method in connection with DNA analysis. *Laborativ Arkeologi* **8**: 26–28.

Götherström A, Collins MJ, Angerbjörn, Lidén K. 2002. Bone preservation and DNA amplification. *Archaeometry* **44**: 395–404.

Grange JM. 1995. Human aspects of *Mycobacterium bovis* infection. In *Mycobacterium bovis Infection in Animals and Humans*, Thoen CO, Steele JH (eds). Iowa State University Press: Ames: IA; 29–46.

Greenblatt C, Spigelman M (eds). 2003. *Emerging Pathogens: Archaeology, Ecology and Evolution of Infectious Disease*. Oxford University Press: Oxford.

Groathouse NA, Brown SE, Knudson DL, Brennan PJ, Slayden RA. 2006. Isothermal amplification and molecular typing of the obligate intracellular pathogen *Mycobacterium leprae* from tissues of unknown origin. *J Clin Microbiol* **44**: 1502–1508.

Groisman EA, Casadesús J. 2005. The origin and evolution of human pathogens. *Mol Microbiol* **56**: 1–7.

Gutierrez MC, Brisse S, Brosch R, Fabre M, Omais B, *et al.* 2005. Ancient origin and gene mosaicism of the progenitor of *Mycobacterium tuberculosis*. *PLoS Pathog* **1**: 55–61.

Haas CJ, Zink A, Pálfi G, Szeimies U, Nerlich AG. 2000. Detection of leprosy in ancient human skeletal remains by molecular detection of *Mycobacterium leprae*. *Am J Clin Pathol* **114**: 428–436.

Haas CJ, Zink A, Szeimies U, Nerlich AG. 2002. Molecular evidence of *Mycobacterium leprae* in historic bone samples from south Germany. In *The Past and Present of Leprosy*, Roberts CA, Lewis ME, Manchester K (eds). British Archaeological Reports, BAR International Series 1054. Archaeopress: Oxford; 287–292.

Hänni C, Brousseau T, Laudet V, Stehelin D. 1995. Isopropanol precipitation removes PCR inhibitors from ancient bone extracts. *Nucleic Acids Res* **23**: 881–882.

Hershkovitz I, Spiers M, Katznelson A, Arensberg B. 1992. Unusual pathological conditions in the lower extremities of a skeleton from ancient Israel. *Am J Phys Anthropol* **88**: 23–26.

Hershkovitz I, Spiers M, Arensberg B. 1993. Leprosy or Madura foot? The ambiguous nature of infectious disease in paleopathology: reply to Dr Manchester. *Am J Phys Anthropol* **91**: 251–253.

Higuchi R, Bowman B, Freiberger M, Ryder OA, Wilson AC. 1984. DNA sequences from the quagga, an extinct member of the horse family. *Nature* **312**: 282–284.

Hirsh AE, Tsolaki AG, DeRiemer K, Feldman MW, Small PM. 2004. Stable association between strains of *Mycobacterium tuberculosis* and their human host populations. *Proc Natl Acad Sci U S A* **101**: 4871–4876.

Hofreiter M, Jaenicke V, Serre D, Haeseler AvA, Pääbo S. 2001a. DNA sequences from multiple amplifications reveal artifacts induced by cytosine deamination in ancient DNA. *Nucleic Acids Res* **29**: 4793–4799.

Hofreiter M, Serre D, Poinar HN, Kuch M, Pääbo S. 2001b. Ancient DNA. *Nat Rev Genet* **2**: 353–359.

Höss M, Pääbo S. 1993. DNA extraction from Pleistocene bones by a silica-based purification method. *Nucleic Acids Res* **21**: 3913–3914.

Iñiguez AM, Araújo A, Ferreira LF, Vicente ACP. 2003. Analysis of ancient DNA from coprolites: a perspective with random amplified polymorphic DNA-polymerase chain reaction approach. *Mem Inst Oswaldo Cruz* **98**(Suppl 1): 63–65.

Jongwutiwes S, Putaporntip C, Iwasaki T, Ferreira MU, Kanbara H, *et al.* 2005. Mitochondrial genome sequences support ancient population expansion in *Plasmodium vivax*. *Mol Biol Evol* **22**: 1733–1739.

Kamerbeek J, Schouls L, Kolk A, van Agterveld M, van Soolingen D, *et al.* 1997. Simultaneous detection and strain differentiation of *Mycobacterium tuberculosis* for diagnosis and epidemiology. *J Clin Microbiol* **35**: 907–914.

Kemp BM, Smith DG. 2005. Use of bleach to eliminate contaminating DNA from the surface of bones and teeth. *Forensic Sci Int* **154**: 53–61.

Kemp BM, Monroe C, Smith DG. 2006. Repeat silica extraction: a simple technique for the removal of PCR inhibitors from DNA extracts. *J Archaeol Sci* **33**: 1680–1689.

Kolman CJ, Centurion-Lara A, Lukehart SA, Owsley DW, Tuross N. 1999. Identification of *Treponema pallidum* subspecies *pallidum* in a 200-year-old skeletal specimen. *J Infect Dis* **180**: 2060–2063.

Larcombe LA, Nickerson P, Hoppa RD, Matheson C. 2005. Detection of a single nucleotide polymorphism in the IL-6 promoter region of ancient nuclear DNA. *Infect Genet Evol* **5**: 117–122.

Lewis ME. 2002. Infant and childhood leprosy: present and past. In *The Past and Present of Leprosy*, Roberts CA, Lewis ME, Manchester K (eds). British Archaeological Reports, BAR International Series 1054. Archaeopress: Oxford; 163–170.

Li X, Kolltveit KM, Tronstad L, Olsen I. 2000. Sytemic diseases caused by oral infection. *Clin Microbiol Rev* **13**: 547–558.

Likovsky J, Urbanová M, Hájek M, Cerny V, Cech P. 2006. Two cases of leprosy from Zatec (Bohemia) dated to the turn of the 12th century and confirmed by DNA analysis for *Mycobacterium leprae*. *J Archaeol Sci* **33**: 1276–1283.

Lowenstein EJ. 2004. Paleodermatoses: lessons learned from mummies. *J Am Acad Dermatol* **50**: 919–936.

Manchester K. 1993. Unusual pathological condition in the lower extremities of a skeleton from ancient Israel. *Am J Phys Anthropol* **91**: 249–250.

Mark L, Patonai Z, Molnár E, Marcsik A. 2006. High-throughput mass spectroscopic determination of ancient pathological biomarkers. In *16th Paleopathology Association European Meeting*, Santorini, Greece, 28 August–1 September.

Marota I, Rollo F. 2002. Molecular paleontology. *Cell Mol Life Sci* **59**: 97–111.

Marota I, Basile C, Ubaldi M, Rollo F. 2002. DNA decay rate in papyri and human remains from Egyptian archaeological sites. *Am J Phys Anthropol* **117**: 310–318.

Mays S, Nerlich AG. 1997. A possible case of Langerhans' cell histiocytosis in a mediaeval child from an English cemetery. *J Paleopathol* **9**: 73–81.

Mays S, Taylor GM, Legge AJ, Young DB, Turner-Walker G. 2001. Paleopathological and biomolecular study of tuberculosis in a medieval skeletal collection from England. *Am J Phys Anthropol* **114**: 298–311.

Mays S, Fysh E, Taylor GM. 2002. Investigation of the link between visceral surface rib lesions and tuberculosis in a medieval skeletal series from England using ancient DNA. *Am J Phys Anthropol* **119**: 27–36.

McDonald JP, Hall A, Gasparutto D, Cadet J, Ballantyne J, *et al.* 2006. Novel thermostable Y-family polymerases: applications for the PCR amplification of damaged or ancient DNAs. *Nucleic Acids Res* **34**: 1102–1111.

McMichael AJ. 2004. Environmental and social influences on emerging infectious diseases: past, present and future. *Phil Trans R Soc Lond B* **359**: 1049–1058.

Miller RL. 1997. Tetanus after cranial trauma in ancient Egypt. *J Neurol Neurosurg Psychiatry* **63**: 758.

Miller RL, Armelagos GJ, Ikram S, De Jonge N, Krijger FW, *et al.* 1992. Palaeoepidemiology of *Schistosoma* infection in mummies. *Br Med J* **304**: 555–556.

Miller RL, Ikram S, Armelagos GJ, Walker R, Harer WB, *et al.* 1994. Diagnosis of *Plasmodium falciparum* infections in mummies using the rapid manual *Para*Sight™-F test. *Trans R Soc Trop Med Hyg* **88**: 31–32.

Mokrousov I, Ly HM, Otten T, Lan NN, Vyshnevskyi B, *et al.* 2005. Origin and primary dispersal of the *Mycobacterium tuberculosis* Beijing genotype: clues from human phylogeography. *Genome Res* **15**: 1357–1364.

Monot M, Honoré N, Garnier T, Araoz R, Coppée J-Y, *et al.* 2005. On the origin of leprosy. *Science* **308**: 1040–1042.

Montiel R, Garcia C, Canadas MP, Isidro A, Guijo JM, *et al.* 2003. DNA sequences of *Mycobacterium leprae* recovered from ancient bones. *FEMS Microbiol Lett* **226**: 413–414.

O'Rourke DH, Hayes MG, Carlyle SW. 2000. Ancient DNA studies in physical anthropology. *Ann Rev Anthropol* **29**: 217–242.

Owen IL, Gomez Morales MA, Pezzotti P, Pozio E. 2005. *Trichinella* infection in a hunting population of Papua New Guinea suggests an ancient relationship between *Trichinella* and human beings. *Trans R Soc Trop Med Hyg* **99**: 618–624.

Oxenham MF, Thuy NK, Cuong NL. 2005. Skeletal evidence for the emergence of infectious disease in Bronze and Iron Age northern Vietnam. *Am J Phys Anthropol* **126**: 359–376.

Pääbo S. 1985. Molecular cloning of ancient Egyptian mummy DNA. *Nature* **314**: 644–645.

Pääbo S. 1989. Ancient DNA: extraction, characterization, molecular cloning and enzymatic amplification. *Proc Natl Acad Sci U S A* **86**: 1939–1943.

Pääbo S, Poinar H, Serre D, Jaenicke-Despres V, Hebler J, *et al.* 2004. Genetic analyses from ancient DNA. *Ann Rev Genet* **38**: 645–679.

Pinhasi R, Foley R, Donoghue HD. 2006. Reconsidering the antiquity of leprosy. *Science* **312**: 846.

Poinar HN, Stankiewicz BA. 1999. Protein preservation and DNA retrieval from ancient tissues. *Proc Natl Acad Sci U S A* **96**: 8426–8431.

Poinar HN, Hofreiter M, Spaulding WG, Martin PS, Stankiewicz BA, *et al.* 1998. Molecular coproscopy: dung and diet of the extinct ground sloth *Nothrotheriops shastensis*. *Science* **281**: 402–406.

Pusch CM, Rahalison L, Blin N, Nicholson GJ, Czarnetzki A. 2004. Yersinial F1 antigen and the cause of Black Death. *Lancet Infect Dis* **4**: 484–485.

Rafi A, Spigelman M, Stanford J, Lemma E, Donoghue H, *et al.* 1994. DNA of *Mycobacterium leprae* detected by PCR in ancient bone. *Int J Osteoarchaeol* **4**: 287–290.

Raoult D, Aboudharam G, Crubézy E, Larrouy G, Ludes B, *et al.* 2000. Molecular identification by 'suicide PCR' of *Yersinia pestis* as the agent of medieval Black Death. *Proc Natl Acad Sci U S A* **97**: 12 800–12 803.

Raoult D, Drancourt M, Fournier M, Ogata H. 2005. Discussion: *Yersinia pestis* genotyping. *Emerg Infect Dis* **11**: 1318–1319.

Raoult D, Dutour O, Houhamdi L, Jankauskas R, Fournier PE, *et al.* 2006. Evidence for louse-transmitted diseases in soldiers of Napoleon's Grand Army in Vilnius. *J Infect Dis* **193**: 112–120.

Rook GA, Rosa Brunet L. 2005. Microbes, immunoregulation, and the gut. *Gut* **54**: 317–320.

Rothschild B, Martin L, Lev G, Bercovier H, Greenblatt C, *et al.* 2001. *Mycobacterium-tuberculosis*-complex DNA from an extinct bison dated 17,000 years before the present. *Clin Infect Dis* **333**: 305–311.

Ruffer MA. 1910. Note on the presence of *Bilharzia haematobia* in Egyptian mummies of the XXth dynasty (1250–1000 BC). *Br Med J* **1**: 16.

Rutherford P. 1999. Immunocytochemistry and the diagnosis of schistosomiasis: ancient and modern. *Parasitol Today* **15**: 390–391.

Rutherford P. 2005. Schistosomiasis in modern and ancient tissues. *J Biol Res* **80**: 80–83.

Saiki RK, Scharf S, Faloona F, Mullis KB, Horn GT, *et al.* 1985. Enzymatic amplification of beta-globulin genomic sequences and restriction site analysis for diagnosis of sickle cell anemia. *Science* **230**: 1350–1354.

Saiki RK, Geland DH, Stoffel S, Scharf SJ, Higuchi R, *et al.* 1988. Primer-directed enzymatic amplification of DNA with a thermostable DNA polymerase. *Science* **239**: 487–491.

Sallares R, Gomzi S. 2001. Biomolecular archaeology of malaria. *Ancient Biomol* **3**: 195–213.

Sallares R, Bouwman A, Anderung C. 2004. The spread of malaria to southern Europe in antiquity: new approaches to old problems. *Med Hist* **48**: 311–328.

Salo WL, Aufderheide AC, Buikstra J, Holcomb TA. 1994. Identification of *Mycobacterium tuberculosis* DNA in a pre-Columbian Peruvian mummy. *Proc Natl Acad Sci U S A* **91**: 2091–2094.

Santos AL, Roberts CA. 2006. Anatomy of a serial killer: differential diagnosis of tuberculosis based on rib lesions of adult individuals from the Coimbra Identified Skeletal Collection, Portugal. *Am J Phys Anthropol* **130**: 38–49.

Sawicki VA, Allison MJ, Dalton HP. 1976. Presence of *Salmonella* antigens in feces from a Peruvian mummy. *Bull N Y Acad Med* **52**: 805–813.

Schmidt-Schultz TH, Schultz M. 2004. Bone protects proteins over thousands of years: extraction, analysis, and interpretation of extracellular matrix proteins in archaeological skeletal remains. *Am J Phys Anthropol* **123**: 30–39.

Schultz M. 2001. Paleohistopathology of bones: a new approach to the study of ancient diseases. *Yrbk Phys Anthropol* **44**: 104–147.

Sebbane F, Jarrett CO, Gardner D, Long D, Hinnebusch BJ. 2006. Role of the *Yersinia pestis* plasminogen activator in the incidence of distinct septicemic and bubonic forms of flea-borne plague. *Proc Natl Acad Sci U S A* **103**: 5526–5530.

Shin DH, Youn M, Chang BS. 2003. Histological analysis on the medieval mummy in Korea. *Forensic Sci Int* **137**: 172–182.

Smith CI, Chamberlain AT, Riley MS, Stringer C, Collins MJ. 2003. The thermal history of human fossils and the likelihood of successful DNA amplification. *J Hum Evol* **45**: 203–217.

Spigelman M, Donoghue HD. 2001. Brief communication: unusual pathological condition in the lower extremities of a skeleton from ancient Israel. *Am J Phys Anthropol* **114**: 92–93.

Spigelman M, Donoghue HD. 2003. Paleobacteriology with special reference to pathogenic mycobacteria. In *Emerging Pathogens: Archaeology, Ecology and Evolution of Infectious Disease*, Greenblatt C, Spigelman M (eds). Oxford University Press: Oxford; 175–188.

Spigelman M, Greenblatt CL. 1998. A guide to digging for pathogens. In *Digging for Pathogens: Ancient Emerging Diseases – their Evolutionary, Anthropological, and Archaeological Context*, Greenblatt CL (ed.). Rehovot: Balaban Publishers: Rehovot; 345–361.

Spigelman M, Lemma E. 1993. The use of polymerase chain reaction to detect *Mycobacterium tuberculosis* in ancient skeletons. *Int J Osteoarchaeol* **3**:137–143.

Spigelman M, Matheson C, Lev M, Greenblatt C, Donoghue HD. 2003. Confirmation of the presence of *Mycobacterium tuberculosis*-complex specific DNA in three archaeological specimens. *Int J Osteoarchaeol* **12**: 393–401.

Spigelman M, Greenblatt C, Vernon K, Zylber MI, Sheridon S, *et al.* 2005. Preliminary findings on the paleomicrobiological study of 400 naturally mummified human remains from Upper Nubia. *J Biol Res* **80**: 91–95.

Spigelman M, Pap I, Donoghue HD. 2006. A death from Langerhans cell histiocytosis and tuberculosis in 18th century Hungary – what palaeopathology can tell us today. *Leukemia* **20**: 740–742.

Stenseth NC, Samia NI, Viljugrein H, Krausrud KL, Begon M, *et al.* 2006. Plague dynamics are driven by climate variation. *Proc Natl Acad Sci U S A* **103**: 13 110–13 115.

Stiller M, Green RE, Ronan M, Simons JF, Du L, *et al.* 2006. Patterns of nucleotide misincorporations during enzymatic amplification and direct large-scale sequencing of ancient DNA. *Proc Natl Acad Sci U S A* **103**: 13 578–13 584. Correction: *Proc Natl Acad Sci U S A* **103**: 14 977.

Sutlovic D, Gojanovic MD, Andelinovic S, Gugic D, Primorac D. 2005. Taq polymerase reverses inhibition of quantitative real-time polymerase chain reaction by humic acid. *Croat Med J* **46**: 556–562.

Taubenberger JK, Morens DM. 2006. 1918 Influenza: the mother of all pandemics. *Emerg Infect Dis* **12**: 15–22.

Taylor GM, Rutland P, Molleson T. 1997. A sensitive polymerase chain reaction method for the detection of *Plasmodium* species DNA in ancient human remains. *Ancient Biomol* **1**: 193–203.

Taylor GM, Widdison S, Brown IN, Young D. 2000. A mediaeval case of lepromatous leprosy from 13–14th century Orkney, Scotland. *J Archaeol Sci* **27**: 1133–1138.

Taylor GM, Young DB, Mays SA. 2005. Genotypic analysis of the earliest known prehistoric case of tuberculosis in Britain. *J Clin Microbiol* **43**: 2236–2240.

Taylor GM, Watson CL, Bouwman AS, Lockwood DNJ, Mays SA. 2006. Variable number tandem repeat (VNTR) typing of two palaeopathological cases of lepromatous leprosy from mediaeval England. *J Archaeol Sci* **33**: 1569–1579.

Taylor GM, Murphy E, Hopkins R, Rutland P, Chitov Y. 2007. First report of *Mycobacterium bovis* DNA in human remains from the Iron Age. *Microbiology* **153**: 1243–1249.

Vergnaud G. 2005. *Yersinia pestis* genotyping. *Emerg Infect Dis* **11**: 1317–1318.

Villeneuve M, Kawai M, Kanashima H, Watanabe M, Minnikin DE, *et al.* 2005. Temperature dependence of the Langmuir monolayer packing of mycolic acids from *Mycobacterium tuberculosis*. *Biochim Biophys Acta Biomembranes* **1715**: 71–80.

Wang GC, Wang Y. 1997. Frequency of formation of chimeric molecules as a consequence of PCR coamplification of 16S rRNA genes from mixed bacterial genomes. *Appl Environ Microbiol* **63**: 4645–4650.

Weiss RA, McMichael AJ. 2004. Social and environmental risk factors in the emergence of infectious diseases. *Nature Med* **10**(12 Suppl): S70–S76.

Wiechmann I, Grupe G. 2005. Detection of *Yersinia pestis* DNA in two early medieval skeletal finds from Aschheim (Upper Bavaria, 6th century A.D.). *Am J Phys Anthropol* **126**: 48–55.

Willerslev E, Cooper A. 2005. Review paper. Ancient DNA. *Proc Biol Sci* **272**: 3–16.

Willerslev E, Cooper A. 2006. Reply. Pathogenic microbial ancient DNA: a problem or an opportunity? *Proc R Soc B* **273**: 643.

Zhou D, Han Y, Song Y, Huang P, Yang R. 2004. Comparative and evolutionary genomics of *Yersinia pestis*. *Microbes Infect* **6**: 1226–1234.

Zink A, Nerlich AG. 2001. A case of Langerhans cell histiocytosis in an infant of a late Roman cemetery. *J Palaeopathol* **13**: 67–73.

Zink A, Nerlich AG. 2003. Molecular analyses of the 'pharaos': feasibility of molecular studies in ancient Egyptian material. *Am J Phys Anthropol* **121**: 109–111.

Zink AR, Nerlich AG. 2005. Long-term survival of ancient DNA in Egypt: Reply to Gilbert *et al. Am J Phys Anthropol* **128**: 115–118.

Zink A, Reischl U, Wolf H, Nerlich A. 2000. Molecular evidence for bacteremia by gastrointestinal pathogenic bacteria in an infant mummy from ancient Egypt. *Arch Pathol Lab Med* **124**: 1614–1618.

Zink A, Haas CJ, Herberth K, Nerlich AG. 2001. PCR amplification of *Plasmodium* DNA in ancient human remains. *Ancient Biomol* **3**: 293.

Zink AR, Reischl U, Wolf H, Nerlich AG. 2002. Molecular analysis of ancient microbial infections. *FEMS Microbiol Lett* **213**: 141–147.

Zink A, Grabner W, Reischl U, Wolf H, Nerlich AG. 2003. Molecular study on human tuberculosis in three geographically distinct and time delineated populations from ancient Egypt. *Epidemiol Infect* **130**: 239–249.

Zink AR, Grabner W, Nerlich AG. 2005a. Molecular identification of human tuberculosis in recent and historic bone tissue samples: the role of molecular techniques for the study of historic tuberculosis. *Am J Phys Anthropol* **126**: 32–47.

Zink AR, Köhler S, Motamedi N, Reischl U, Wolf H, *et al.* 2005b. Preservation and identification of ancient *M. tuberculosis* complex DNA in Egyptian mummies. *J Biol Res* **80**: 84–87.

Zink AR, Spigelman M, Schraut B, Greenblatt GL, Nerlich A, *et al.* 2006. Leishmaniasis in ancient Egypt and Upper Nubia. *Emerg Inf Dis* **12**: 1616–1617.

9

Databases

William White

Museum of London, 150 London Wall, London EC2Y 5HN, UK

INTRODUCTION

Historically, as the study of human remains in archaeological or anthropological collections has advanced, it has been marred by major problems of accessibility of research materials. Chief among these was that the holdings of the said collections were not well known, nor was their existence properly publicized. Would-be students and their supervisors, as well as post-doctoral researchers, have had to expend much time and effort in tracing skeletal samples of the desired size, date, quality and other characteristics. Another problem was that data collection was not regulated or standardized, so that some inter-site comparisons that were made were of questionable validity. In the latter years of the 20th century, however, much progress was made toward providing remedies to these and associated problems. A number of themes converged to produce this welcome result, including political initiatives at government level. In Europe, much of the improvement in the overall situation, especially with regard to data collection, has been informed by, and developed in response to, salient events in North America.

HISTORICAL COLLECTIONS OF HUMAN REMAINS IN BRITAIN

The types of collections of human remains curated in UK museums are of unparalleled diversity. Thus, it is salutary to survey these holdings first.

In Britain, human remains were traditionally held in medical collections (notably in pathology museums within schools of medicine; for example, in the museums of the Royal College of Surgeons of England, the Royal College of Surgeons of Scotland and the Wellcome Trust), anthropology collections (for example, the Natural History Museum, London, and the Duckworth Collection at the University of Cambridge), charnel deposits (Hythe, Kent, and Rothwell, Northamptonshire) and in general collections in provincial museums.

Advances in Human Palaeopathology Edited by Ron Pinhasi and Simon Mays

The internal market in archaeological excavation in England and Wales led to burgeoning collections of archaeological skeletons amid the new commercial archaeological units (English Heritage, 1991). The type of inventory used for collections of human skeletal remains varied from the tried-and-trusted card index, increasingly into spreadsheets and other electronic information systems (for example, at the Natural History Museum, Duckworth Collection and the Hunterian Museum at the Royal College of Surgeons of England), even if there was no standardization as yet. The traditional paper records used in-house did not lend themselves to sharing information with researchers, except those who called in person. The increasing use of MS Excel and Access spreadsheets improved this situation, although there is as yet no agreed standard for the recording and presentation of the data. This will be discussed below. In continental Europe, the situation appeared no better. Knowledge of the existence of the collections and their inventories likewise depended largely upon word-of-mouth communications, which was an unsatisfactory state of affairs (Ron Pinhasi, personal communication).

Data collection

Throughout Europe up to the mid-20th century, the study of human skeletal remains was typically performed by anatomists, physicians, dentists and those in related disciplines. Each scholar used an idiosyncratic suite of criteria for ageing and sexing the remains and for the diagnosis of pathology. In the absence of formal guidance for data collection the results lacked standardization and their value for use in comparative studies was questionable.

Meanwhile, in the USA the provision of common standards was being driven by the growing pressure for the reburial of collections of human remains, most of which were of indigenous origin. The Native American Graves Protection and Repatriation Act (1990) required US museums and laboratories to evaluate their collections of human remains, produce inventories and consult local tribes with a view to repatriation as appropriate. The overall results could not be foreseen, but, as reburial was the likely outcome in many cases, it was necessary to promote full analysis and recording of material at risk. Under these exceptional circumstances, a rational set of minimum standards for the recording process was devised and published (Buikstra and Ubelaker, 1994).

In Britain, the state of affairs became similarly dynamic, but not entirely for the same reasons. The Institute of Field Archaeology issued a guidebook for the archaeological excavation and post-excavation treatment of human remains (Roberts and McKinley, 1993). Guidelines for writing osteological reports at the post-excavation analysis phase of an archaeological project were subsequently published by English Heritage (Mays *et al.*, 2002). Finally, the British Association for Biological Anthropology and Osteoarchaeology (BABAO) produced its own recommended standards for osteological recording (Brickley and McKinley, 2004). For further discussion of issues connected with data collection, see Chapter 4.

THE DEVELOPMENT OF DATABASES IN MUSEUMS OF PATHOLOGY, ANTHROPOLOGY AND ARCHAEOLOGY

As was touched upon above, museums of pathology and anthropology often maintain databases for their own internal uses. There is, however, a new kind of ethos prevailing in the museum world, especially in connection with archaeology, that the databases ought to be made more widely available and preferably interlinked. The Museums and Libraries Association held a conference in London in 2006 on the theme 'Towards a strategy for the curation of archaeological archives'. This was partly in recognition that very large datasets have accumulated from archaeological interventions in England and Wales. Indeed, English Heritage commissioned the Archaeology Data Service (ADS) to investigate preservation and management for exceptionally large data formats, termed 'Big Data' (http://ads.ahds.ac.uk/project/bigdata/). The ADS provides access to certain human bone databases (see Appendix A).

These initiatives have been made against a background of unusual interest by the British Government. As the result of a meeting in 2001 between the British and Australian prime ministers, an agreement was made that Britain would expedite the repatriation to Australia of the skeletal remains of the latter's indigenous peoples from public and private collections. The political process continued with the setting up of a working group on human remains in museum collections by the Cultural Property Unit of the Governmental Department of Culture, Media and Sport (DCMS). The working group was charged with examining all aspects of the care, safekeeping, legal status and requests for return of human remains curated within the publicly funded museums and galleries of England and Wales (DCMS, 2003). This report stated that there were 132 institutions that among them held the remains of about 61 000 individuals (DCMS, 2003: 11). Both of these figures are known to be vast underestimates and paid no heed to the skeletal holdings in some pathology museums, university departments and commercial archaeological units. They also, by definition, ignore the position in Scotland (Historic Scotland, 1997), Northern Ireland and the Irish Republic (O'Sullivan and Killgore, 2003). Although, the DCMS committee produced a flawed and biased report, slanted heavily toward the repatriation of human remains of overseas origin and paying only lip service to the research value of such collections, the consultation process finally produced useful guidelines (DCMS, 2005). The new document, *Guidance for the Care of Human Remains in Museums*, contained a requirement (DCMS, 2005: 22) that:

> Museums should have a policy to compile and make public an inventory of their holdings of human remains. This should include known information about the date and provenance of the remains and their exact nature and the circumstances of their acquisition.

Thus, driven originally by the vicissitudes of the repatriation process, actions initiated by both the UK and US governments have led to a situation in which museum holdings of human remains are required to be declared publicly. In turn, skeletal data collection has been reformatted to conform to minimum standards because the skeletal material may subsequently be lost to science. Those museums that maintain databases of their holdings of human remains are arguably better placed to make their inventories available externally. In continental Europe different situations apply, but here there are also some important initiatives, such as the History of Health in Europe and the SYNTHESYS database project (see below).

DATABASE TYPES

Human bones databases vary from ad hoc compilations, made solely for internal use by the institution concerned, to online sophisticated relational databases of great depth and utility. Are all databases equally useful for research in palaeopathology? Obviously, much depends upon the approach of the researcher. Sometimes it is sufficient merely to know the size and date of the skeletal sample and then to handle the material so identified and screen for one or more pathological conditions of interest. Such projects require only the minimum and rawest of data to be recorded in the database. By contrast, large-scale comparative projects may require greater detail regarding bone condition, age and sex, provisional disease process classification or detailed description with differential diagnosis. This last admits the further possibility of making direct use of the data recorded in the database or of challenging provisional diagnoses, either by re-evaluation of the stated criteria or by re-examination of the material. The available databases accordingly reflect a range of utility, and may be classified into their different types as described below.

A database of Type 1 would be the simple inventory, intended to help researchers locate skeletal collections. A Type 2 database is likewise an inventory, but one that allows researchers to locate specific bone specimens or skeletons of particular interest within the larger collections. Type 3 databases are those wherein data have been compiled with the intention that researchers use the existing data directly. This applies in particular where the human remains have been repatriated, reburied or are otherwise no longer accessible. Type 4 databases are those resulting from specific, problem-orientated research projects. In effect, this is secondary dissemination of material from specific research projects. Finally, Type 5 databases are multipurpose, combining several of the above functions. Specific examples of these different types of database are described below and in Appendix A.

DEVELOPMENT OF RELATIONAL DATABASES FOR HUMAN REMAINS

Databases ought not to be mere inventories maintained by an institution for its own internal purposes. A secondary aim, of growing importance, is the dissemination of information about skeletal collections. In turn, this serves palaeopathology by drawing attention to large skeletal samples available for analysis. Type 1 databases provide a blank canvas, in that researchers can approach the skeletal material to be screened for pathology or pathologies of interest, without preconceptions. The provision of further detail (Type 2 and Type 3 databases) allows research to be further refined or targeted.

Following the passage of 'The National Museum of the American Indian Act' (1989) in the USA, the evaluation of the vast archive of approximately 16 000 palaeoindian remains was required, and these remains began to be recorded on the 'Standardized Osteological Database' (SOD), an MS-DOS-based system devised by (and still maintained at) the Center for Advanced Spatial Technologies at the University of Arkansas. This is a Type 3 database, but there is no evidence of its recorded data being employed in published research. The fact that it uses an obsolete MS-DOS-based system means probably that it will never be used to the full. Recently, in recognition of the limitations of the above system, all recording was updated onto an Oracle relational database and SOD was superseded by the Repatriation Osteological Laboratory (ROL) database at the National Museum of Natural History (NMNH) in Washington, DC. Bone inventories, age and sex estimations, etc., have been integrated with the palaeopathological information in a hierarchical database (Ousley *et al.*, 2006). The use of a relational database allows the manipulation of large amounts of data, links data from different fields and permits the comparison of many records at once.

The Wellcome Osteological Research Database (WORD) at the Museum of London is a Type 5 database. It has links to the archaeological site information, such as geographic information systems (GISs), stratigraphy and finds. It is used below to illustrate the use of a complex relational database.

The Wellcome Osteological Research Database

The Museum of London, through the London Archaeological Archive Research Centre, curates more than 17 000 human skeletons. This vast archive has accumulated during urban archaeology in the Greater London area during the past 30 years. These collections are not the result of prospecting for human remains, but have been acquired during rescue (salvage) archaeology in advance of building development or redevelopments.

The skeletal material forms a diachronic collection, covering all periods in London, from prehistoric to post-medieval, constituting an invaluable resource for research and teaching. Indeed, this collection currently comprises the largest scientifically excavated and documented human bone assemblage known from any city in the world. This material provides a unique opportunity to follow trends through time in a major urban centre. However, a major failing, in common with other holdings of skeletal and other materials in museums, universities and archaeological units, was that the human remains from many London archaeological sites were largely unknown, unpublished and, hence, little studied.

The purpose of the WORD database is to unlock the research potential of human skeletons from the larger archaeological sites in London. The database was developed in-house, using recording forms redesigned by osteologist Brian Connell and converted into an electronic version by Peter Rauxloh, the museum's IT manager (Connell and Rauxloh, 2002). The database is an Oracle relational database. As an initial step, a diachronic assemblage of about 5000 skeletons has been analysed and recorded directly onto it. Archaeological site summaries and interpretative reports have also been integrated into the database, with the aim of making the data available via the internet for remote searching online. Moreover, remote researchers will be able to access the resultant datasets in order to abstract information on demography, skeletal measurements and prevalence of pathological conditions. Here, WORD functions as a Type 3 database. Furthermore, the database will be available to assist investigators in formulating research designs. The database will allow the identification of a specific archaeological site or sites with sample sizes and other characteristics suitable for the needs of the researcher. It will also draw attention to available comparative samples from other archaeological sites. The documentation of osteological indicators of disease will similarly permit scholars to identify particular specimens suitable for their own projects (i.e. it is functioning here as a Type 1/Type 2 database).

All palaeopathology recording in WORD is free text and comprises an accurate description of lesion location, type of bone alteration and distribution in the skeleton (Roberts and Connell, 2004). Additionally, observed pathology is classified into the following categories: congenital, infectious, joint disease, trauma, metabolic, endocrine, neoplastic or circulatory. The multiple criteria will allow rapid calculation of prevalences across an archaeological site or comparisons between sites (see Chapter 4).

The development of WORD involved a launch online in 2007. Accordingly, as details of external user experience are not available for evaluation, a description of the pre-launch in-house usage at the Museum of London will have to suffice.

Case Study: the Royal Mint Site

Excavations in 1986–1988 on the site of the Royal Mint, London, revealed substantial skeletal samples from several phases of medieval burial. The earliest was a large catastrophe cemetery associated with the Black Death (AD 1348–1350). It covered approximately 2 ha in which 750 burials were clustered in two groups some 50 m apart: the Eastern and Western cemeteries. Of these, 634 Black Death victims were available for analysis and recording. The Western cemetery also contained later graves associated with the Cistercian Abbey of St Mary Graces (AD 1353–1540), from which a further 400 burials were recorded.

The skeletal remains from the Royal Mint site have proved of great interest to academic researchers, not least because they include victims of the Black Death. Teeth from those people interred in the Black Death catastrophe cemeteries were included in a Europe-wide project to try to find genetic evidence of *Yersinia pestis*, ostensibly the causative organism for plague. These results unfortunately were negative (Thomas *et al.*, 2004). Dental research has also concerned the plight of those who died in the mid-14th-century plague epidemic, having lived through the Great Famine during AD 1315–1317 (Antoine and Hillson, 2005).

The data on 634 victims of the Black Death from this archaeological site recorded in WORD were manipulated to inform the site publication (Grainger *et al.*, 2008). Of these skeletons, 259 were burials in individual graves, whereas 375 were interred in mass-burial trenches as the epidemic began to make itself felt. Interrogation of the database allowed palaeodemographic comparisons of the individuals in the 11 rows of graves with the occupants of the trenches. Detailed queries then provided information on stature and general skeletal preservation, together with palaeopathology, including congenital conditions, degenerative changes, trauma, circulatory disease, cribra orbitalia, specific and non-specific infections, metabolic disease, neoplasms, dental caries, calculus, periodontitis, ante-mortem tooth loss and linear dental enamel hypoplasia. Data were exported into Excel spreadsheets that were then manipulated to produce the desired output, especially overall disease prevalences. These results, for the individual graves, the mass trenches, and the combined Black Death victims, appear in the site publication (Grainger *et al.*, 2008).

Future Developments

As was discussed above, the criteria for data collection have been well established, although further refinements will continue to be made (Ousley *et al.*, 2006). The same cannot be said of databases of human remains, where there is no agreed standard, which perhaps is inevitable given the differing priorities and funding arrangements among those organizations that curate human remains.

It might be expected that database designers will standardize on English as the *lingua franca* (at least until such time as outputs can be released freely in any language of academic use). It is interesting, therefore, that the one international initiative so far, the SYNTHESYS database, is being launched with English as the common language.

However, the manner in which databases will be developed is less easy to foresee. At present, the relational database would appear to be the way forward, as foreshadowed by the ROL database and WORD. These represent significant advances on the simple, all-purpose spreadsheet. The primary need is thus the building of a hierarchical system to facilitate prevalence-orientated quantitative methods in palaeopathology. This type of information system also provides an efficient means for the comparison of data among groups using multiple criteria. Users will draw upon the raw data accessed online directly for their research purposes. Some enquiries can be customized in available subroutines; others may use interrogative tools such as MSQuery. Archaeological data pertinent to individual skeletons (such as spatial information, stratigraphy and associated finds) would be accessed by 'drilling down'.

It is, of course, impossible to anticipate all the purposes for which future researchers will wish to utilize the databases. However, for non-archaeological users, an interface would provide a high level of osteological, palaeopathological and other interpretative information, allowing users to build queries using keyword lists, including interpretative groupings and keywords supplied for specialist users. Those with a particular concern with archaeology will also be able to view spatial distribution, archaeological site information and burial data via GISs. Expert users would be able to interrogate the database directly themselves.

Another use of comprehensive databases like WORD is that researchers may use them to identify those osteoarchaelogical samples best tailored to their own research goals. This would take into account archaeological site summaries of bone preservation, sample size and demography, availability of biometric data, disease pattern, pathology types, etc.

Much will depend upon commitment to the long-term maintenance of established databases. It is expected that a relational database with an archaeological basis will be maintained and regularly upgraded as a matter of course. Here, the financial position of the institution concerned in maintaining the database is of paramount importance. Again, when there is an archaeological background, there will not merely be a responsibility toward sustained maintenance; beyond that there will be a drive to add to the collection itself and hence to expand and develop the database. For databases with a very narrow focus or period specificity, however, there may be greater problems in keeping them up to date. The decision to set up a relational database also has cost implications. Normally only government-run institutes or other large establishments with adequate funding will be able to undertake investments of resources in this manner. Accordingly, many museums and comparable holders of human skeletal remains may continue to afford, at best, little more than an annotated catalogue.

CONCLUSIONS

In recent years there has, for a variety of reasons, been an increase in the availability and accessibility of databases that cover human remains collections. Although guidelines are now available for data collection itself, one of the major problems remaining is that there is, as yet, no common standard for such databases. Consequently, they are of variable quality, depth and utility. Some are little more than ad hoc catalogues of bones, others are more concerned with recording burials per se, whereas others contain information on sample size, date(s), demography and palaeopathology in the collection, and yet others are relational databases having further supporting contextual and other data. The construction of databases has multiple purposes. First, a comprehensive and up-to-date database of human remains holdings is an important element in collections care and management. Second, the maintenance of such databases is vital for the dissemination of information concerning the characteristics of the collections, whether by responses to researcher enquiries or more proactively.

This dissemination is particularly important, given that the lead time to the appearance of osteological reports in the site publication can often be very long. At a time of often global initiatives in project design, the availability of online databases is particularly pertinent in facilitating transnational comparisons.

APPENDIX A: EVALUATION OF A SELECTION OF HUMAN REMAINS DATABASES

The Global History of Health Project

Anthropologists from 16 European countries have joined this project, initiated by Rick Steckel, Jerome Rose, Clark Larsen and Philip Walker, recording human skeletons in accordance with the criteria in the *Data Collection Codebook* (Steckel and Rose, 2002). Potentially, approximately 60 000 skeletons from about 350 locations will be recorded onto three large databases (http://global.sbs.ohio-state.edu/) in order to assemble a global history of health during the past 10 000 years. This is a 'stand-alone' Type 5 database. It yields secondary information rather than primary data.

A Database of Archaeological Sites Yielding Human Remains in England

This is a database compiled by Simon Mays of the Ancient Monuments Laboratory, English Heritage. It is a dBase III+ file, listing more than 1000 English archaeological sites that have produced human remains in England. Information such as number of burials, date, location of published reports, curation of skeletons and archived associated records are included (Mays, 1991). Thus, it is a Type 1 database. The database does not include the demographic and other detail found in certain other databases. However, its content is ideal for those seeking the location of collections of human remains available for the purposes of research.

Biological Anthropology Research Centre

The Biological Anthropology Research Centre (BARC) is located at the University of Bradford, within the Department of Archaeological Sciences. It was set up with support from the Art and Humanities Research Council and contains the largest collection of human skeletal remains within a UK university department. The approximately 3000 skeletons stored in this facility include important samples such as that from the hospital of St Mary and St James, Chichester, with its high proportion of leprous material, the Black Death victims from Hereford Cathedral and the war dead from the Battle of Towton. The BARC database is a Type 2/Type 3 database and was under development and only partially accessible at the time of writing (www.barc.bradford.ac.uk/).

Synthesys

Over 20 museums of natural history and botanic gardens in Europe are involved in networking activities focused on producing a single 'virtual museum' service. An important activity within the project is the development of databases of their collections, and a working group is looking at an extension into developing databases to hold anthropology collection data: http://www.synthesys.info. This is likely to be a Type 1/Type 2 Microsoft Access system, and the museums whose human remains holdings are likely to be inventorized in the database include the Hungarian Natural History Museum, Budapest, the Natural History Museum, London, the Naturhistorisches Museum, Vienna, and the Institut Royal des Sciences Naturelles, Brussells.

Hunterian Museum, Royal College of Surgeons of England

A large sample of 5000 human remains and body parts is being recorded onto 'Surgicat', an online database: http://surgicat.rcseng.ac.uk/. This is a database of Type 2.

Living with the Dead

This is a Type 1 database of British and Irish human remains dating from the Neolithic. It is maintained by the ADS: http://ads.ahds.ac.uk/catalogue/specColl/lwtd/.

Early Anglo-Saxon Census Project

As the title suggests, this is a Type 1 database devoted to human skeletons of 6th to 9th century AD date from England. Currently it covers only the county of Kent, but is likely to be extended. It is maintained by University College London (UCL): www.ucl.ac.uk/~ucylelh/EASCP.htm.

University College London Museums and Collections Human Bones Databases

UCL's human remains holdings are dispersed throughout many different buildings. The collections comprise those in the Anatomy, Biological Anthropology, Ethnography and Histopathology departments, as well as the Galton Collection and the Grant and Petrie Museums. Each of these maintained its own database, but now these holdings are united internally in a Microsoft Access-based system, Type 1 database.

Glasgow University Archaeology Department

This is a database of UK Anglo-Saxon burials, but currently without osteological detail: www.glasgow.ac.uk/archaeology/resources. It is a Type 1 database.

The Archaeology Data Service Monastic Burials Database

This is an Oracle database designed to be used within a GIS. It comprises the plan data and a structured database on monastic burials in Britain (Gilchrist and Sloane, 2005: 231–234). There is much archaeological detail, but the osteological information goes little further than some demography and palaeopathology (leprosy, tuberculosis, DISH, etc.). It is a database of Type 1. It is maintained by the ADS: http://ads.ahds.ac.uk/catalogue/resources/html?cemeteries_ahrb_2005/index.cfm.

The Sedgeford Historical and Archaeological Research Project

This is a circumscribed database dealing with skeletal material from the Romano-British, Saxon, Medieval and post-Medieval periods excavated in the area of Sedgeford, Norfolk. It is a Microsoft Access system: http://www.sharp.org.uk/.

British and Irish On-Line Database Index to Excavated Skeletons

This is a project proposed by BABAO for a database to encompass human remains from archaeological sites in England, Wales, Scotland, Northern Ireland and the Republic of Ireland. It would subsume the English Heritage database mentioned above and go into a greater level of detail, including date and location of excavation, historical period, number of individuals (subtotals for females, males, adults of unknown sex, juveniles, neonates, etc.), metadata on archaeology, osteology and palaeopathology, accessibility (curated or reburied), publications and location of archives. When set up, the British and Irish On-line Database Index to Excavated Skeletons database would be maintained by the British Association for Biological Anthropology and Osteoarchaeology: www.babao.org.uk/.

The Standard Osteological Database

The NMNH, Smithsonian Institution, Washington, DC, USA, curates approximately 16 000 skeletons, the largest museum collection of Native American human remains in the USA. The database is maintained by the Center for Advanced Spatial Technologies at the University of Arkansas (http://www.cast.uark.edu./cast/sod/), under a grant from the National Science Foundation. It is a Type 3 database. The main disadvantage of this database is that it is MS-DOS based and does not work in the now more widely available Microsoft Windows environment. It therefore faces obsolescence.

The Smithsonian Institution's Repatriation Osteology Laboratory

As a substantial upgrade of SOD was required, advantage was taken at the Smithsonian Institution to provide a new relational database that would allow data to be organized in a hierarchical fashion and the use of multiple criteria for comparing results among groups. The integrated information system so developed includes, along with palaeopathology data, skeletal age and sex estimates, taphonomic observations, cranial and postcranial measurements, dental inventory/pathology/wear, dental morphology traits, skeletal non-metric traits, macromorphoscopics, cultural modification and general summary information. The ROL database will be accessible online: www.nmnh.si.edu/anthro. The vicissitudes caused by potential repatriation determined that the material be comprehensively photographed and radiographed where prudent, in order that the remains be recorded as completely as possible. The utilization of this vast database in practice could not be evaluated at the time of writing. Because of the risk that much of the skeletal material faces reburial, the database will have a role in preserving 'virtual skeletons'.

The Wellcome Osteological Research Database

This database has been described in detail in the main body of this chapter. A current disadvantage with the database is that not all the collection is yet accessible online. The input on the diachronic sample of 4840 individuals is complete, but a further sample of around 5500 individuals from the medieval Priory and Hospital of St Mary Spital will not be accessible until 2009; then information on the remaining approximately 7000 skeletons from smaller archaeological site samples will be rolled out a little further down the line. Furthermore, because the core of the skeletal collection is intended to be retained as a resource for research, there was not the pressing need to proceed immediately to exhaustive photography of the remains. The latter work will be pursued in due course; but, in the meantime, the online availability of photographs and radiographs is of a lower order.

REFERENCES

Antoine D, Hillson S. 2005. The Great Famine, Black Death and health in the 14th century. *Lond Archaeol Int* **8**: 26-28.

Brickley M, McKinley JI (eds). 2004. *Guidelines to the Standards for the Recording of Human Remains*. IFA Paper No. 7. British Association for Biological Anthropology and Osteoarchaeology/Institute of Field Archaeologists: Reading.

Buikstra J, Ubelaker D. 1994. *Standards for Data Collection from Human Skeletal Remains*. Arkansas Archeological Survey Research Series No. 44. Arkansas Archaeological Survey: Fayetteville, AR.

Connell, B, Rauxloh P. 2003. *A rapid method for recording human skeletal data.* Unpublished manuscript held on file at the Museum of London. Museum of London Archaeology Service: London.

DCMS. 2003. *The Report of the Working Group on Human Remains*. Department of Culture, Media and Sport: London.

DCMS. 2005. *Guidance for the Care of Human Remains in Museums*. Department of Culture, Media and Sport: London.

English Heritage. 1991. *Management of Archaeological Projects* (second edition). English Heritage: London.

Gilchrist R, Sloane B. 2005. *Requiem: The Medieval Monastic Cemetery in Britain*. Museum of London Archaeology Service: London.

Grainger I, Hawkins D, Cowal L, Mikulski R. 2008. *The Black Death Cemetery, East Smithfield, London*. Museum of London Archaeology Service: London.

Historic Scotland. 1997. *The Treatment of Human Remains in Archaeology*. Operational Policy Paper No. 5. Historic Scotland: Edinburgh.

Mays S. 1991. The Ancient Monuments Laboratory. *Paleopathol Assoc Br Sect News* **11**: 3–4.

Mays S, Brickley M, Dodwell N. 2002. *Human Bones from Archaeological Sites: Guidelines for Producing Assessment Documents and Analytical Reports*. English Heritage: Swindon.

O'Sullivan J, Killgore J. 2003. *Human Remains in Irish Archaeology*. An Chomhairle Oidhreachta/The Heritage Council: Dublin.

Ousely S, Dudar JC, Jones E, London M, Madden G, Mulhern D, Wilczak C. 2006. Standardizing the standards: computerized documentation of skeletal pathology at the Smithsonian Institution. *Workshop presented at the Annual Meeting of the Paleopathology Association*, Anchorage, AK.

Roberts CA, Connell B. 2004. Guidance on recording palaeopathology. In *Guidelines to the Standards for the Recording of Human Remains*, Brickley M, McKinley JI (eds). IFA Paper No. 7. British Association for Biological Anthropology and Osteoarchaeology/Institute of Field Archaeologists: Reading; 34–41.

Roberts CA, McKinley JI. 1993. *Excavation and Post-Excavation Treatment of Cremated and Inhumed Human Remains*. IFA Technical Paper No. 13. Institute of Field Archaeologists: Oxford.

Steckel R, Rose J. 2002. *The Backbone of History: Health and Nutrition in the Western Hemisphere*. Cambridge University Press: Cambridge.

Thomas M, Gilbert P, Cuccui J, Lynnerup N, Titball RW, White WJ, Cooper A, Prentice MB. 2004. Absence of *Yersinia pestis*-specific DNA in human teeth from five European excavations of putative plague victims. *Microbiology* **150**: 341–354.

PART 2

Diagnosis and Interpretation of Disease in Human Remains

10

Differential Diagnosis of Skeletal Lesions in Infectious Disease

Donald J. Ortner

Department of Anthropology, Smithsonian Institution, Washington, DC 20560, USA

INTRODUCTION

Disease agents, particularly infectious pathogens, have undoubtedly played a major role in the evolution of modern humans. Like the anatomical changes apparent in the evolution of the hominines, the effect of infectious pathogens on the skeletal remains provides an important source of information on this aspect of biological changes associated with the development of modern humans. Unfortunately, a major source of data on past human societies, the human skeleton, is singularly resistant to the effects of infectious pathogens. Acute infection with rapid death, a major cause of mortality in human groups, rarely affects the skeleton. Chronic infection can affect the skeleton, but this manifestation usually occurs only in a relatively small percentage of individuals with the disease. On the basis of modern clinical data, primarily before the advent of antibiotics, the relatively few chronic infectious diseases that can affect the skeleton do so somewhere in the range of 5–20 % of patients who have the disease. It is this small number of skeletal infectious disease cases that provides at least some insight regarding the role of infection in past human groups. When these data are combined with other data from skeletal samples, such as demographic analysis, an incomplete but useful picture emerges about the relative health of the population represented by the sample.

Advances in Human Palaeopathology Edited by Ron Pinhasi and Simon Mays

Table 10.1 Common infectious diseases that have the potential to affect the human skeleton

Osteomyelitis
Periostitis/periostosis
Tuberculosis
Leprosy
Treponematosis
Brucellosis
Mycosis and mycosis-like diseases
Smallpox
Echinococcosis
Leishmaniasis[a]
Septic arthritis[a]

[a]Infectious disease not discussed in this chapter.

Infection is one of several general disorders that affect the skeleton. It is not always possible to distinguish between infection and one or more of the other disorders. When a diagnosis of infection can be made in an archaeological human burial, distinguishing between the various infectious disorders that can affect the skeleton is sometimes a challenging exercise and may not be possible. Table 10.1 provides a list of infectious disorders that will be reviewed in some detail in this chapter. There are two additional infectious disorders included in Table 10.1 that also affect the skeleton but which are rare and will not be discussed here.

The human skeleton is affected by infectious diseases primarily when the disorder does not result in rapid death. People with an optimal immune reaction to infectious pathogens, as well as those who succumb quickly to infectious organisms, are unlikely to exhibit skeletal evidence of infectious disease. Distinguishing between these two responses in a skeletal sample is not possible with the present state of knowledge regarding human skeletal palaeopathology. Reconstructing the prevalence of infectious diseases on the basis of skeletal evidence in archaeological human skeletal samples will always involve making inferences in which many of the important variables that affect disease prevalence are inaccessible. Nevertheless, data on the prevalence of skeletal infection, along with other diseases that can affect the skeleton, do provide important insight regarding the role of disease in past human populations and deserve careful attention by the bioarchaeologist and biological anthropologist.

The options in differential diagnosis of skeletal lesions potentially attributable to infectious disease include trauma, metabolic diseases, reticuloendothelial diseases, haematopoietic disorders and tumours. Distinguishing between these diagnostic options will not always be possible on the basis of skeletal evidence. However, careful attention to the type and distribution of lesions will permit differential diagnosis to at least a general category of disease in most cases. Diagnosis of a specific infectious disease is more challenging.

Skeletal manifestations of disease include abnormal bone formation, abnormal bone destruction, abnormal size and abnormal shape. Although all of these abnormalities can occur in the skeletal response to infection, the two most common pathological processes that occur in bone are abnormal bone formation and abnormal bone destruction. The focus in this chapter will be on the expression of these two abnormal conditions in infectious disease.

The limitations of space for this chapter will prevent a comprehensive review of all infectious diseases that can affect the skeleton. Because of this I will limit my discussion to those infectious diseases that I have been able to study in modern pathology collections and/or in archaeological skeletal samples. Despite this limitation, most of the more common infectious diseases that produce skeletal abnormalities will be discussed and the reader should have a good base of information for interpreting evidence of skeletal abnormalities caused by infectious diseases. For a more comprehensive discussion of these infectious diseases, and others that can affect the skeleton, one should consult a major reference work on skeletal radiology or skeletal palaeopathology (e.g. Aufderheide and Rodriguez-Martin, 1998; Resnick, 2002; Ortner, 2003).

Both abnormal bone formation and abnormal bone destruction can and often do occur in the same case of infection and may occur in the same lesion. In the latter situation, the observer should distinguish between a lytic focus surrounded by bone formation and bone formation where there is a focus in which abnormal bone failed to form perhaps due to a soft-tissue mass, such as a granuloma, that inhibited bone formation in part of a lesion. Both of these abnormalities can be seen in cases of treponematosis. An important step in differential diagnosis of lesions that combine both formation and destruction is to determine, if possible, the primary nature of the lesion. In lesions of the skull in treponematosis, for example, there is a central lytic focus that is the initial lesion. Subsequently, this lesion will form a crater-like depression surrounded by reactive bone formation. These lesions are virtually pathognomonic for treponematosis.

There are different types of bone formation and destruction that are largely dependent on local neurovascular conditions stimulated by the immune response to a pathogen. These conditions affect the speed with which either of these basic lesions forms. Rapid bone formation produces woven bone that typically is the initial stage in many abnormal bone-forming lesions caused by infection. Normally, woven bone is remodelled into compact bone in later stages of the disease process or during remodelling when the pathogenic stimulus has been eliminated or neutralized. Slow bone formation creates compact bone and is associated with chronic infection, as seen in long-standing cases of osteomyelitis. Both types of abnormal bone can occur in the same burial and in the same lesion. The latter is often expressed as a central focus of woven bone with compact bone on the periphery of the lesion, but the converse also occurs.

Other neurovascular and inflammatory conditions result in bone destruction. Here again, the speed of this destruction is apparent in the anatomy at the margin of the lesion. With very rapid destruction the margin is poorly defined and typically it is difficult to determine the boundary between normal bone and bone being destroyed. Rapid destruction has poorly defined margins and little, if any, evidence of compact reactive bone formation (sclerosis). Slow destruction is usually accompanied by substantial sclerosis at the margin of the lesion. Keep in mind that a destructive lesion initially may be rapid with subsequent reduction in speed of destruction and evidence of sclerosis.

It is variation in the type and distribution pattern of lesions within the skeleton that provides the evidence on which the researcher depends for differential diagnosis (Table 10.2). A fundamental principle in skeletal palaeopathology is to describe carefully the lesions one encounters with an emphasis on both the type and distribution before attempting a diagnosis. With this information, those reading a report or paper can evaluate diagnostic options for themselves. Even when a specific diagnosis is plausible, it is good practice to include reasonable diagnostic alternatives in one's description of a case.

Table 10.2 Summary of characteristic skeletal abnormalities associated with infectious diseases discussed in this chapter

	Occurrence[a]							
	Tuberculosis	Leprosy	Treponematosis	Periostitis	Osteomyelitis	Brucellosis	Smallpox	Mycosis
Abnormal bone formation	++	+++	+++	+++	+++	+	+++	++
Abnormal bone destruction	+++	+++	++	+	+++	+++	+	++
Sclerosis on lytic margins	++	++	++	+	+++	+++	+	++
Central lytic/peripheral forming	−	+	+++	+	++	+	−	++
Bilateral	+	++	+++	+	+	++	+++	−
Symmetrical	?	++	++	+	++	+	++	−
Rhinomaxillary remodelling	+	+++	+	−	−	−	−	−
Axial involvement	+++	++	++	+	+++	+++	−	+++
Appendicular involvement	+	+++	+++	+++	+++	++	+++	+++
Clavicular involvement	−	−	+	−	−	−	−	+
Elbow predilection	−	−	−	−	−	−	+++	−

[a] +++: common manifestation; ++: occurs occasionally; +: occurs but is uncommon; −: does not occur or is rare; ?: insufficient evidence.

In chronic skeletal infectious disease the overwhelming majority of lesions are either bone forming or bone destroying. Occasionally, chronic infection will affect bone growth and abnormalities in size and shape will occur, but this is uncommon. Also, in compound fracture, infection is a common complication that, in combination with poor alignment of the fracture, may result in abnormal shape associated with an infectious process.

By far the most common cause of skeletal infectious disease is pathogenic bacteria, including staphylococcus, streptococcus, mycobacteria, treponemal organisms, and brucella. Infection by fungi can affect the skeleton. Viral diseases rarely affect the skeleton. One exception to this is smallpox, but skeletal involvement is uncommon. Multicelled parasites, such as the larval stage of the flatworm that causes echinococcosis, can affect the skeleton as well. This review will highlight the basic characteristic features that one must consider in the diagnosis of infectious disease.

In more than 40 years of conducting research on skeletal disease, I have encountered many examples of skeletal disorders where I have been fairly certain of the specific diagnosis. However, I often have not been able to do more than establish a reasonable probability that one of the general categories of skeletal disease is present in a burial. Nevertheless, even this general information provides useful insight into the health of past human populations.

INFECTIOUS DISEASES OF THE HUMAN SKELETON

Osteomyelitis

Osteomyelitis simply means bone infection in which the primary skeletal focus is in the marrow. This is in contrast with osteitis, which has a focus within compact bone, and periostitis, which primarily affects the outer layer of bone and its associated soft tissue, the periosteum. It is fairly common for all three areas of bone to be involved in an infectious disorder, and determining the primary skeletal site in an archaeological burial may not be possible.

Most of the infectious diseases that affect bone do so via the circulatory system and marrow and may be called osteomyelitis. Particularly in the older medical literature, a modifier is added before the term, as in smallpox osteomyelitis or treponemal osteomyelitis. However, there is another less specific usage in which osteomyelitis, lacking a modifier, is caused most commonly by staphylococcus, although streptococcus and other organisms may be implicated as well. In the following discussion of osteomyelitis the emphasis will be on the disorder caused by staphylococcus or streptococcus pathogens.

One of the most certain diagnoses in skeletal palaeopathology is seen in classic cases of osteomyelitis. In these cases, three very characteristic features are present: one or more cloacae; sequestrated bone; and involucrum (Figure 10.1). A cloaca is a hole through the cortex that provides drainage from the marrow for pus and necrotic tissue. Sequestrated bone has been isolated from its blood supply by the pathology, which results in tissue death. Osteoclasts in adjacent living tissue attempt to destroy the sequestrated bone and, if the necrotic area is small enough, may do so leaving little, if any, evidence. Sequestrated bone varies in size ranging from one or more small areas within a bone to the entire diaphysis. Small sequestra may be removed from the body through cloacae in the pus that flows from the marrow to the outer surface of the body. Larger pieces of sequestrated bone cannot be destroyed completely by osteoclasts and cannot pass through cloacae. They can be identified by the margins of the sequestrum, which will usually have a scalloped edge indicative of osteoclastic activity.

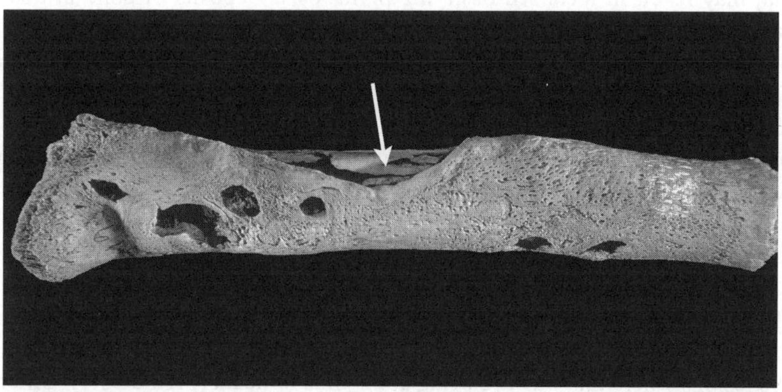

Figure 10.1 Osteomyelitis of the right tibia with sequestrum (arrow), involucrum surrounding the sequestrated bone and cloacae. Hunterian Museum, Royal College of Surgeons of England catalogue no. HM RCS P618

Involucrum is the product of a periosteal reaction to the death of bone, usually part or all of a long-bone diaphysis. Although the cortical bone of the diaphysis is dead, the periosteum may be viable and forms new bone to ensure continued biomechanical function of the limb. Involucrum tends to be formed quickly and typically has a very irregular surface. Not all cases of osteomyelitis manifest all three features; but, when they do occur, one has a high certainty that the disorder is osteomyelitis.

Osteomyelitis can occur with any combination of the three common features described above. However, osteomyelitis may exhibit none of these features, with only reactive bone formation in the marrow and on the outer cortex. Generally there will be reactive, often poorly organized bone formed within the marrow, but the periosteum also tends to be activated and may form either smooth or lumpy compact bone. This variation in the expression of osteomyelitis is the result of several factors, including age of onset, the size and nature of the initial infection and the immune response of the individual.

Periostitis (Periostosis)

The germinative inner layer of the periosteum retains the potential to form bone throughout life. It plays a crucial role in fracture healing, but it also has the potential to react to a neurovascular stimulus caused by many pathological conditions in addition to infection and trauma. Abnormal periosteal bone formation may be expressed in several different ways, from one or more layers of woven or compact bone to spicules of bone that are perpendicular to the surface of bone (Resnick, 2002: 4884).

Bone-forming lesions occurring on the outer, periosteal surface are among the most common abnormalities encountered by the skeletal palaeopathologist. Although local or systemic infection is a common cause, other disorders, including trauma and tumour, are significant options in differential diagnosis. When there is a reasonable certainty that the cause of the lesion is an infectious agent, the term periostitis is appropriate. If one is in doubt about the cause of a lesion, then a better term is periostosis, which simply means a bone-forming abnormality of the outer bone surface. When bone is close to the skin surface, as occurs with the medial anterior tibia and the posterior ulna, a blow, accidental or intentional, may result in localized ossifying periostosis. Deposits of bone in this situation do not have the chronic stimulus seen in many infectious diseases affecting the periosteum and tend to be well-organized compact bone and limited in size. They often remodel and become completely incorporated and continuous with the original cortex, leaving only an abnormal lump of compact bone.

Like osteomyelitis, a periosteal bone-forming lesion is an abnormal condition often seen in general infectious disease categories. It is one of the skeletal manifestations of most specific infectious diseases. Occasionally, it does occur as a localized disorder that is not linked with a systemic infectious disease.

There is a continuum of periosteal lesions ranging from rapidly formed woven bone (Figure 10.2) to dense compact bone lesions (Figure 10.3). In chronic conditions, woven bone formed during an acute phase tends to be remodelled into compact bone, and one encounters lesions where this was occurring at the time of death (Figure 10.4). Unremodelled woven bone is always indicative of active abnormal bone formation at the time of death, since woven bone is rapidly converted to compact bone in chronic or healing stages of a disorder.

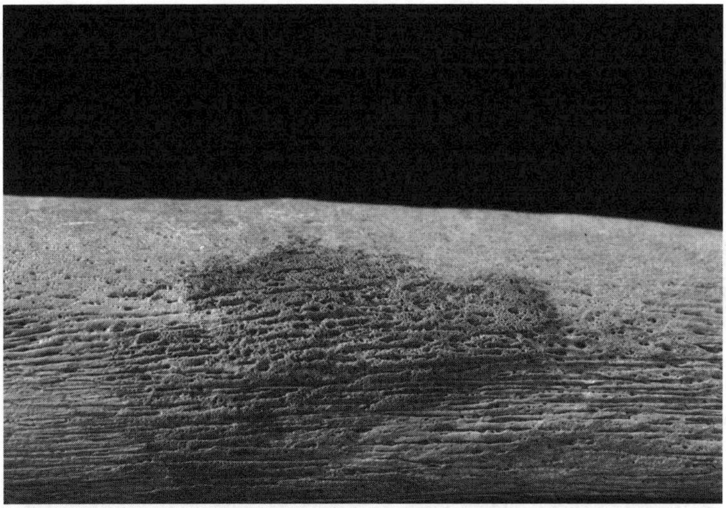

Figure 10.2 Reactive periosteal woven bone on the medial diaphysis of the left tibia in an adult from the medieval Blackfriars site, Gloucester, England

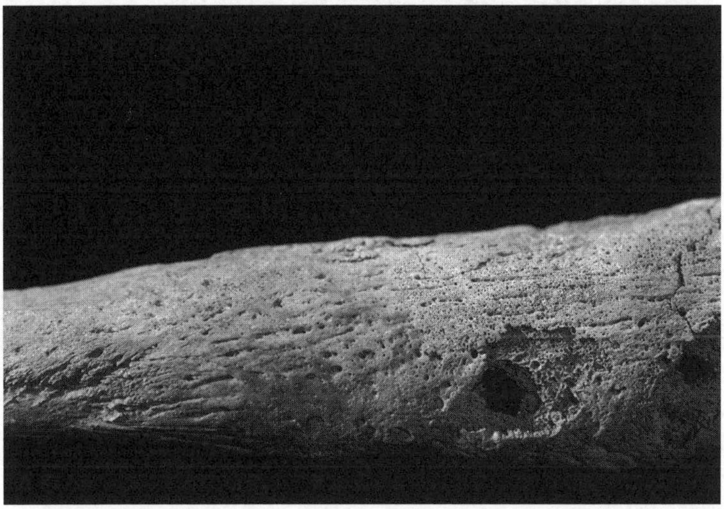

Figure 10.3 A central focus of infection on the left distal diaphysis of the radius. This medieval case of probable treponematosis is an adult from the Magistrate's Court site, Hull, England. Note the woven bone surrounding the cloaca and the woven bone remodelled into compact bone on the periphery of the lesion. Woven bone is indicative of active infection at the time of death

As in other skeletal diseases, acute conditions may stimulate the formation of woven bone that can later remodel into compact bone. Chronic conditions tend to stimulate compact bone formation. However, chronic infectious diseases often have one or more acute phases, particularly during the initial stage of the disease. Attachment to underlying bone varies from fine spicules connecting a fairly solid layer of woven or compact bone to the compact bone of the original periosteal surface, to layers of compact bone that are added directly to

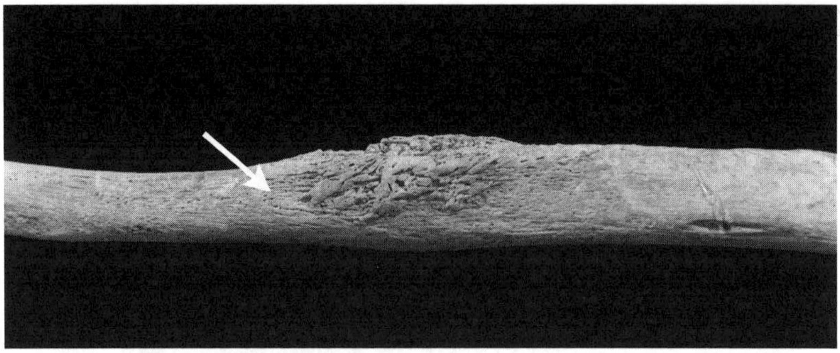

Figure 10.4 A reactive periosteal bone lesion on the medial diaphysis of the right ulna from the same individual as Figure 10.2. The central portion of the lesion is woven bone being remodelled into compact bone. There is a zone of woven bone on the periphery of the lesion (arrow)

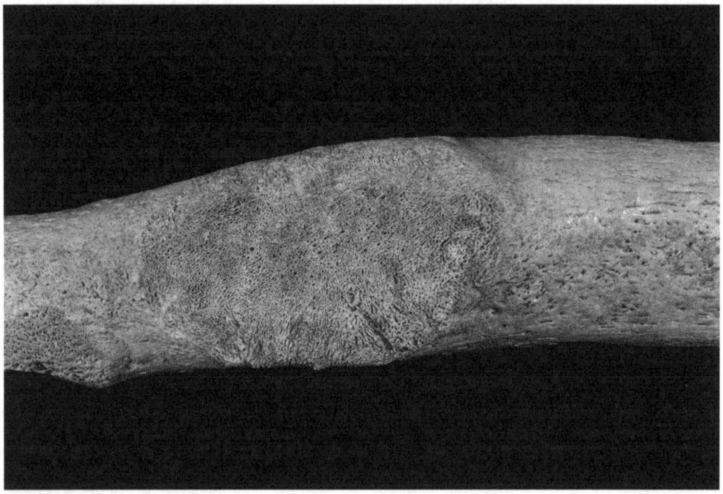

Figure 10.5 Localized periostitis associated in life with an overlying skin ulcer on the right tibia of an adult from the Juhle Site in Maryland, USA (NMNH catalogue no. 384299) dated AD 1400–1480. Note the well-defined margin between reactive, porous bone and the adjacent normal cortex

the normal cortex. Rapidly formed woven bone is poorly organized and always has a porous appearance due to the loose organization of the mineralized osteoid fibres. This porous surface also tends to contain more vascular spaces than compact bone, which reflects the vascular supply needed for rapid bone formation. These vascular channels, combined with the loose organization of the bone fibres, create bone that has poor biomechanical function. Conditions that stimulate woven bone formation may continue for a long time in chronic conditions, but the tendency is for woven bone to remodel into compact bone, which is a much stronger tissue.

In many, if not most, skeletal infectious diseases, abnormal periosteal bone formation will be a manifestation of these disorders. In most cases the characteristics of the periosteal lesions overlap between infectious diseases so that differential diagnosis on the basis of periosteal lesions is often not possible. However, there are exceptions to this, one of which is the periosteal reaction to an overlying skin ulcer (Figure 10.5). The most common site for this lesion is the anterior medial surface of the tibial diaphysis. Because the ulcer is usually limited to an oval or circular area with its shortest axis often between 5 and 10 cm in diameter, the underlying area of reactive bone formation is similarly limited, usually with a fairly distinct boundary with the surrounding bone. Skin ulceration is typically a very chronic condition stimulating bone formation over a period of months to years, and the area of reactive bone formation may rise 2 mm or more above the original cortical surface. A circumscribed lesion of this type is a clear indication of an overlying skin ulcer. The most common causative microorganisms are staphylococcal bacteria, which are found on the skin surface of virtually all individuals. They readily contaminate underlying tissue when the skin is broken. This type of ulcer is a common complication of diabetes.

Although the most common skeletal reaction to ulceration is limited to the site of the ulcer itself, there may be secondary periostitis that can affect adjacent bone, including the fibula. However, it is almost always possible to identify the primary site of infection because of the clearly demarcated bony margin associated with the boundary of the ulcer. Within the primary site of abnormal bone formation the bone surface typically is characterized by fine porosity reflecting hypervascularity related to the chronic nature of the lesion.

Tuberculosis

Tuberculosis can affect virtually any part of the skeleton; but, unlike treponematosis discussed below, it predilects the axial skeleton and particularly the spine, although the major joints are a fairly common site as well. Because of its skeletal involvement, in some cases it has been the subject of considerable research by palaeopathologists (e.g. Pálfi *et al.*, 1999; Roberts and Buikstra, 2003). Skeletal lesions in tuberculosis tend to be destructive more than formative. However, the generalization that reactive bone formation is uncommon simply does not reflect reality. Indeed, reactive bone formation can be extensive in some cases, and the margins of destructive lesions in tuberculosis typically will exhibit at least some sclerosis.

Vertebral tuberculosis has the primary focus in vertebral bodies and only rarely affects the vertebral arch. The destructive process commonly results in kyphosis, creating the classic Pott's deformity (Figure 10.6). Like leprosy and treponematosis, tuberculosis can cause bone destruction of the rhinomaxillary region (lupus vulgaris). Skull vault lesions can occur, and these tend to originate on the inner table of the vault. Because of this the size of the destructive lesion usually is larger in the inner table (Figure 10.7) than the outer, and this provides an important diagnostic feature in differentiating tuberculosis from infectious conditions of the skull, in which the primary focus is the outer table.

In the pre-antibiotic past, the peak age of onset for skeletal tuberculosis was in childhood (Ortner, 2003), and in some of these cases skeletal involvement known as dactylitis (or spina ventosa) resulted in greatly enlarged metacarpals and phlanges (Figure 10.8). However, this disorder also occurs in other infectious diseases, including congenital treponematosis.

Pus from a destructive focus in a vertebral body may drain between the psoas muscle and the vertebral column. Reactive bone formation may develop on vertebrae or the innominates, where the connective tissue sack encapsulating the abscess is in close contact with bone. The association with the psoas muscle makes it relatively easy to reconstruct the path of the abscess.

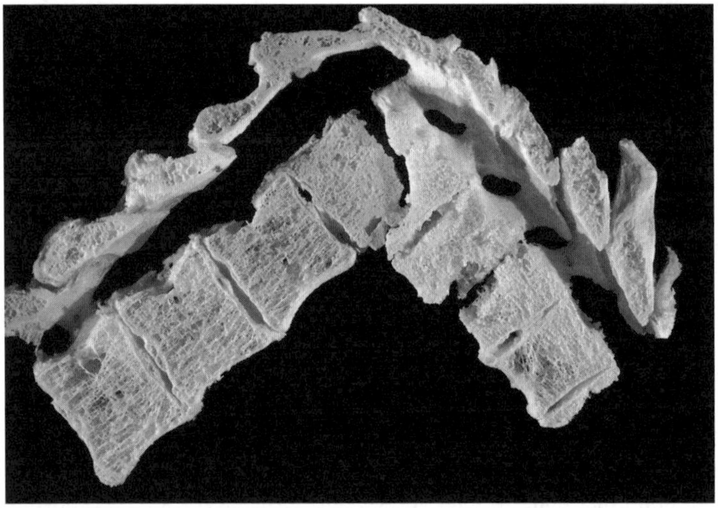

Figure 10.6 Severe vertebral body destruction from tuberculosis resulting in kyphosis. The infectious focus was in the lower thoracic vertebrae of a modern male, 65 years old. This case is from the Galler Collection (no. S246/51) currently curated at the Natural History Museum in Basel, Switzerland

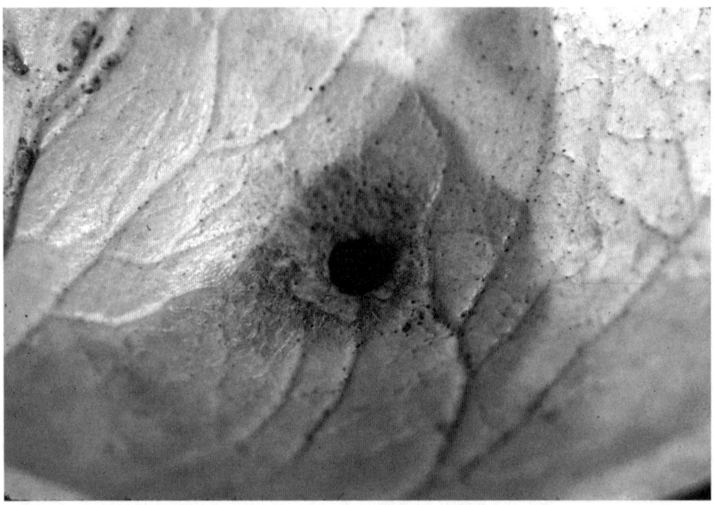

Figure 10.7 Endocranial view of a destructive lesion of the right parietal affecting both tables and the diploë, caused by tuberculosis in a modern female, age 55 years. Note that the largest diameter of the lesion is in the inner table, indicating the primary location of the tuberculous granuloma. National Museum of Anthropology, Prague, Czech Republic, catalogue no. 2480

Leprosy

Leprosy is caused by bacteria (*Mycobacterium leprae*) in the same genus as the bacterium that causes tuberculosis (*Mycobacterium tuberculosis*). Although both disorders tend to be chronic if the skeleton is involved, the characteristic skeletal lesions in leprosy are different from

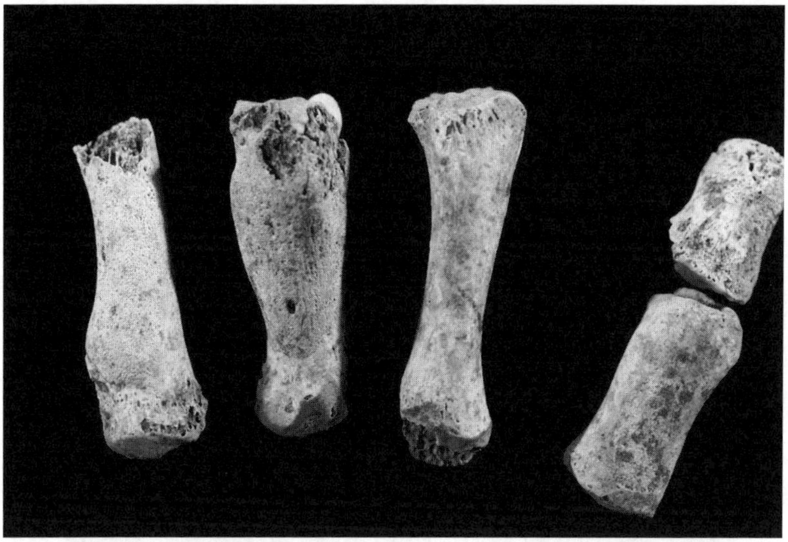

Figure 10.8 Dactylitis of the left metacarpals in a child 3–5 years of age probably caused by systemic infection (e.g. tuberculosis or treponematosis). Note particularly the substantial enlargement of the diaphyses. The case is from the medieval site of St Gregory's Priory, Canterbury, England, and is curated by the Canterbury Archaeological Trust

the typical type and distribution of lesions in other infectious diseases and often involve the face (rhinomaxillary remodelling) and the distal appendicular skeleton (Møller-Christensen, 1953, 1961, 1978; Andersen and Manchester, 1992). Rhinomaxillary abnormalities occur in tuberculosis (lupus vulgaris) but are uncommon. Abnormal rhinomaxillary remodelling is also seen in treponematosis, discussed later in this chapter, as well as other pathological conditions, such as carcinoma. However, carcinoma of the skeleton is a more aggressive condition and typically is associated with rapid bone destruction. The lesions usually do not have the remodelled compact bone margins seen in leprosy. Skull vault lesions usually do not occur in leprosy (Møller-Christensen, 1965: 604).

Møller-Christensen (1953) applied the term facies leprosa to rhinomaxillary abnormalities in leprosy and provided a careful description based on his study of skeletons excavated from a cemetery associated with a medieval leprosy hospital in Denmark. In the face, the margins of the pyriform aperture tend to be rounded and usually, although not always, enlarged, and the anterior nasal spine is destroyed (Figure 10.9) in some cases. The anterior maxillary alveolar process associated with the premaxilla may also undergo a destructive remodelling that reduces the alveolar sockets and may result in the loss of the maxillary incisors. Differential diagnosis, when possible, is enhanced by careful attention to the type and distribution of lesions in other parts of the skeleton.

The orbital roof in leprous skeletons may exhibit cribra orbitalia (Figure 10.10). This is one of several diseases associated with this symptom. Its presence in leprosy highlights the importance of not attributing cribra orbitalia to a specific disease, such as anaemia, unless additional evidence supports this diagnosis. Orbital lesions in leprosy probably reflect a

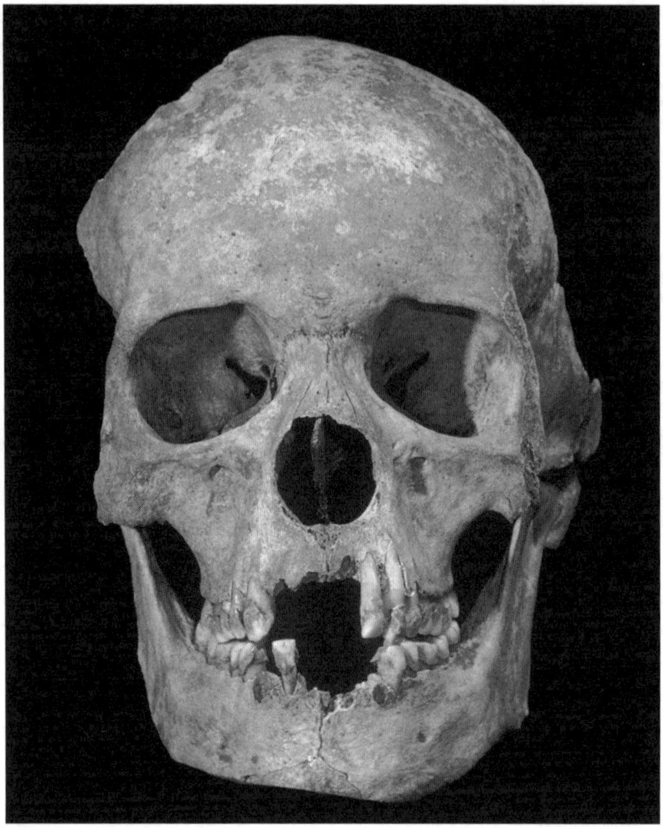

Figure 10.9 Destructive remodelling of the bone forming the pyriform aperture in a case of leprosy in an adult from the medieval site of St James and St Mary Magdalene, Chichester, England. Note the rounding of the margins of the normally sharp edges of the pyriform aperture

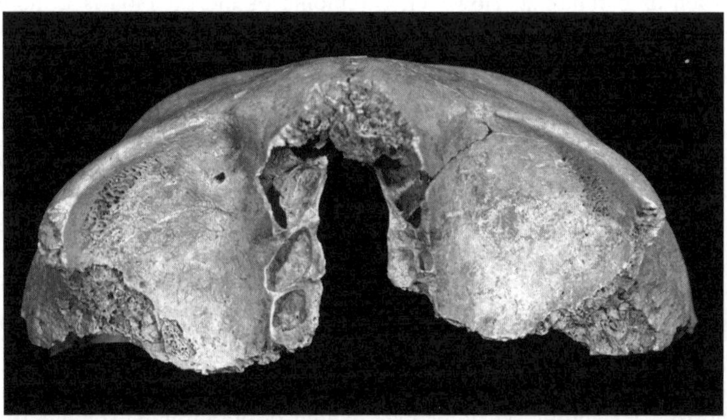

Figure 10.10 Porosity of the orbital roof (cribra orbitalia) in a burial with leprosy. Porosity is probably due to a vascular response to chronic infection of the eye. Burial of an adult from the medieval site of St James and St Mary Magdalene, Chichester

vascular response to chronic infection of the eye that is known to occur in some cases of leprosy and leads to blindness.

Particularly in the hands, neurologic destruction associated with leprosy may result in severe flexion contractures of the fingers, resulting in pressure erosion of metaphyseal cortex where the joint margin of the distal component of a joint has pressed on the adjacent metaphysis (Andersen and Manchester, 1987). The tubular bones of the hands and feet also undergo destructive remodelling, typically expressed as concentric atrophy (Figure 10.11). This remodelling begins in the distal ends of bone and in some cases may proceed to the point where the diaphysis and even the proximal metaphysis are completely destroyed. In the feet, another manifestation of destructive remodelling of the metatarsal bones can result in a blade-like diaphysis of the bone (Figure 10.12). This abnormality is virtually pathognomonic

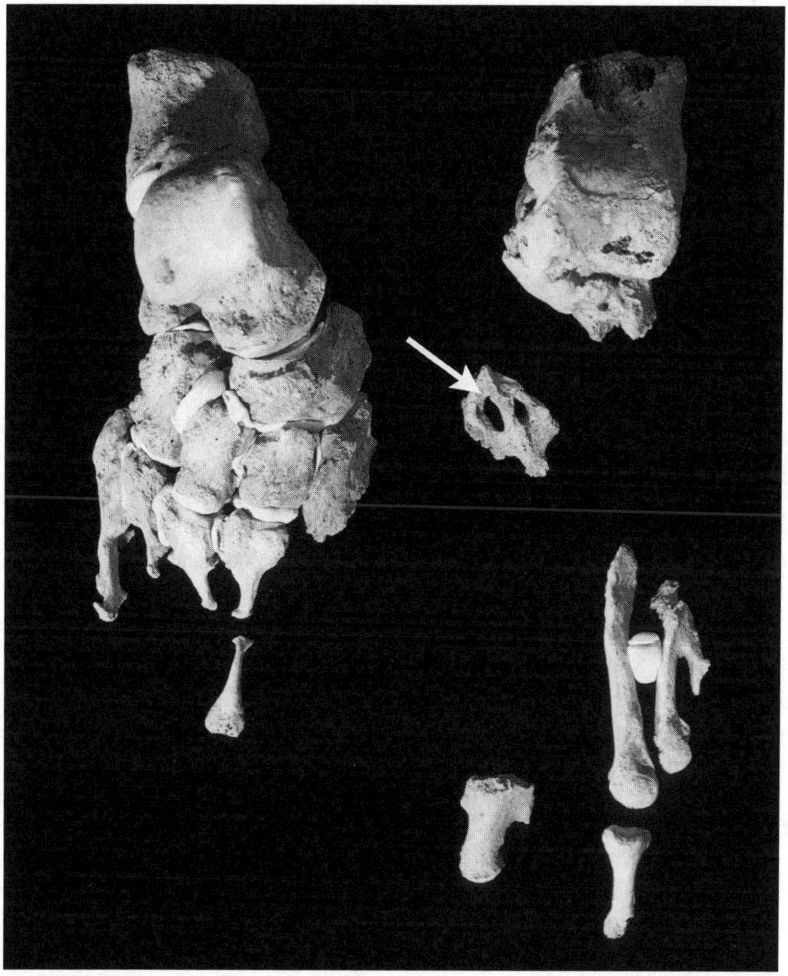

Figure 10.11 Concentric destructive remodelling of the metacarpals and phalanges of the feet in the same case of leprosy as in Figure 10.9. Note the evidence of osteomyelitis in the deformed tarsal bone of the left foot (arrow)

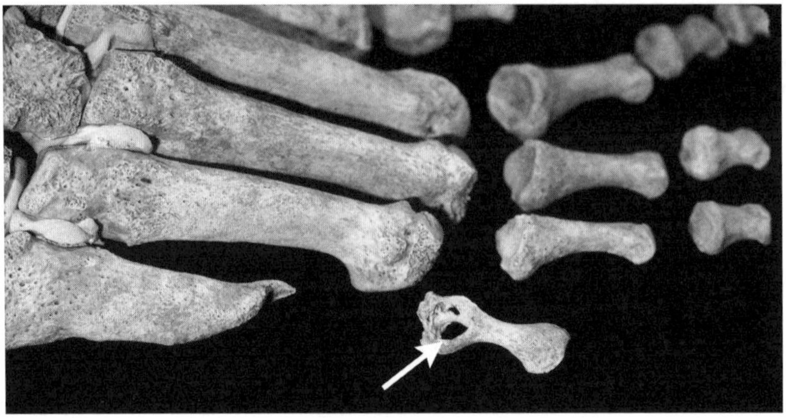

Figure 10.12 Blade-like destructive remodelling of the right fifth metatarsal in a case of leprosy in an adult from the medieval site of St James and St Mary Magdalene, Chichester. Note the evidence of osteomyelitis in the fifth right proximal phalanx (arrow)

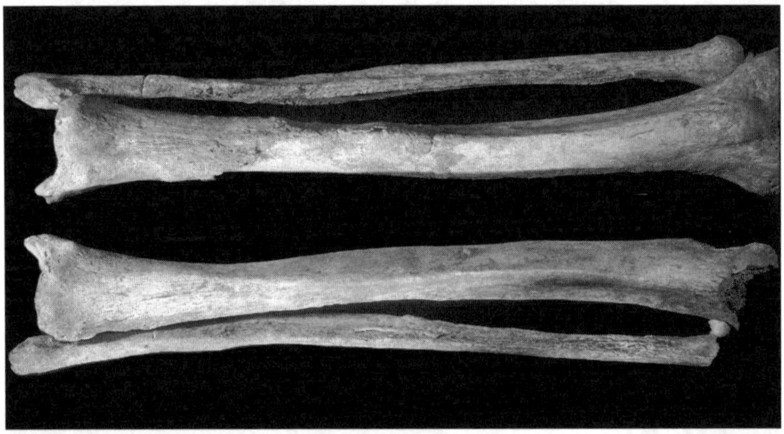

Figure 10.13 Reactive woven-bone formation on the distal diaphysis and metaphysis of the right and left tibia and fibula in the same case of leprosy as in Figure 10.12

for leprosy, although other disorders, including diabetes, psoriasis and frostbite, should be considered if there are no other bone disorders apparent in other areas of the skeleton.

The destruction of sensory nerves that occurs in some patients with leprosy is accompanied by circulatory dysfunction. Because of this abnormality, injury to the feet may provide a portal of entry for secondary infection with a poor immune response related to an inadequate vascular response. Most commonly, the resulting chronic infection is caused by staphylococcal organisms and not *M. leprae*. Reactive woven bone formation may be stimulated by infection, and this woven bone may undergo remodelling into compact bone. Because the primary focus of the secondary infection typically is the foot, bone involvement is greatest in the distal tibia and fibula, but it can extend well beyond the midshaft (Figure 10.13). This distal location is not typical of lower extremity involvement in other specific infectious diseases, such as treponematosis.

Treponematosis

Three syndromes of treponematosis affect the skeleton: syphilis, bejel and yaws. No subject in human palaeopathology has generated the publications and controversy stimulated by the history of syphilis (Dutour *et al.*, 1994; Powell and Cook, 2005). Syphilis is a sexually transmitted disease with an age of onset that tends to be in late adolescence or early adulthood. Yaws and bejel normally are not sexually transmitted and in most cases are acquired in childhood through transmission between an affected child and another child via open wounds or sores. All three syndromes can result in a congenital variant, but this is far more common in syphilis than the other two syndromes. The key to transmission across the placental barrier is age of onset and the stage of the disease in the mother at the time of pregnancy. In most cases the organism only crosses the placental barrier during the early acute phase of treponematosis. Since the age of onset for yaws and bejel is ordinarily in childhood, in most cases the disease has become chronic when pregnancy first occurs and the developing infant is unaffected. Since syphilis is a sexually transmitted disease, the age of onset is usually later and women are much more likely to be in the acute phase of the disease the first time they become pregnant.

The skull vault, forearm and lower leg are the predilected sites of involvement in all three syndromes. Rhinomaxillary involvement does occur and may be similar to leprosy although destructive remodelling of the alveolar process is not a common manifestation of facial participation. In leprosy, the skull vault normally is not involved and major long-bone lesions tend to be limited to the distal tibia and fibula rather than the midshaft, the typical location in treponematosis.

In thc skull vault, the lesions and their developmental stages have been carefully described by Hackett (1976). The classic appearance of skull vault lesions is caries sicca, a crater-like lesion with a central destructive focus and reactive, compact bone formation on the margins of the lesion (Figure 10.14). The lesion begins as a cluster of fine holes through the outer

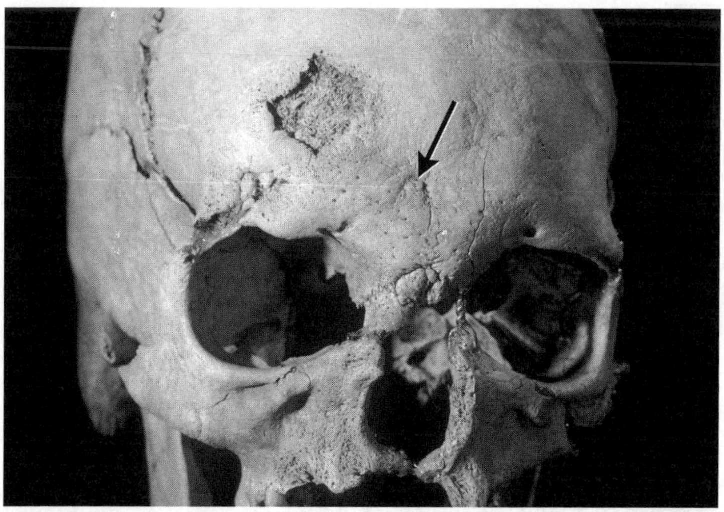

Figure 10.14 Multiple lytic lesions of the frontal bone with destructive rhinomaxillary remodelling in the same burial illustrated in Figure 10.2. The large destructive lesion of the right frontal is probably a coalescent lesion. A classic healed caries sicca lesion is apparent just superior to the nasion (arrow)

table (Figure 10.15) characteristic of the early destructive phase of the lesion. However, most of the lesions one encounters have a very chronic appearance and most of the surface of the lesion is compact bone. Within this lesion there often are fine, thread-like lines of compact bone that radiate from the centre of the lesion. The combination of a crater-like lesion with the radiating lines is pathognomonic for treponematosis.

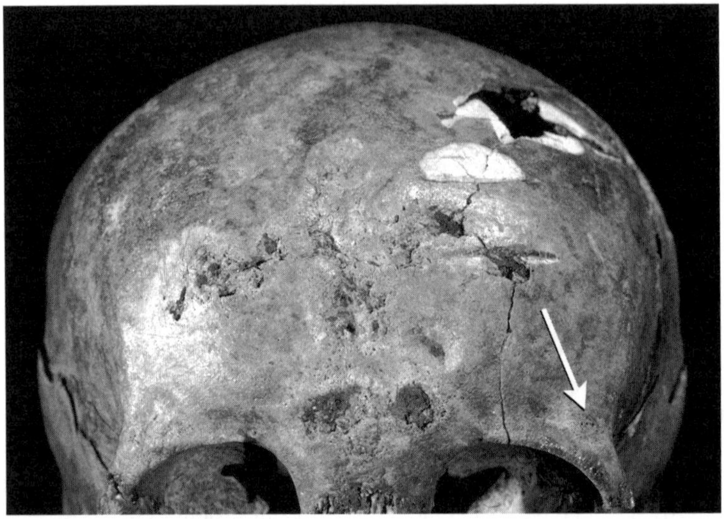

Figure 10.15 Multiple lytic lesions of the frontal bone. This case of probable treponematosis is the same case illustrated in Figure 10.3. An example of early-stage caries sicca is apparent on the left antero-lateral portion of the frontal bone (arrow) with a circular cluster of fine holes measuring about 12 mm in diameter

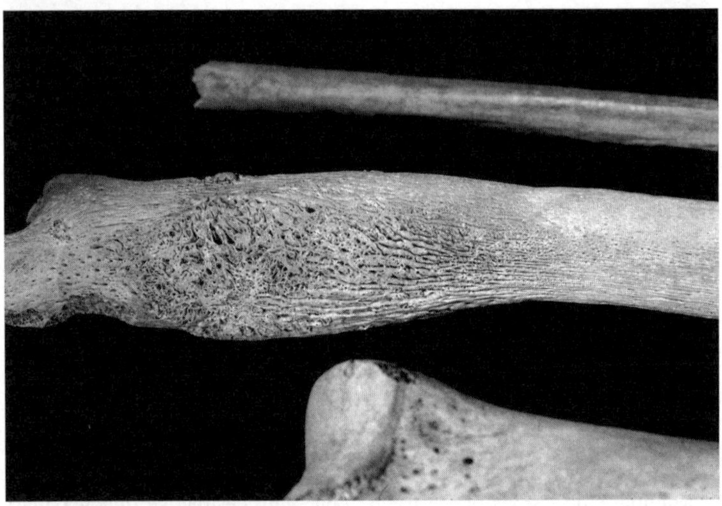

Figure 10.16 A large woven-bone lesion of the distal right tibia that is being remodelled into compact bone, from the same burial illustrated in Figure 10.2

Reactive bone formation on the diaphyseal long-bone surface characterizes the abnormalities apparent in the appendicular skeleton (Figure 10.16) but can also be found in the axial skeleton, including the ribs and the clavicles. Bone abnormality in the long bones, clavicle and ribs may also involve destructive lesions or a destructive focus surrounded by bone formation. This needs to be distinguished from a failure to form bone in the central focus of a bone-forming lesion (Figure 10.17), which also occurs in treponematosis and is usually found in the long bones. Clavicular involvement is uncommon in other systemic skeletal infectious diseases.

Bone-forming lesions may be composed of woven bone, but far more typically they consist of compact bone reflecting the chronic nature of the disorder. Lesions of the tibia can stimulate considerable reactive bone formation that usually consists of compact bone. The anterior

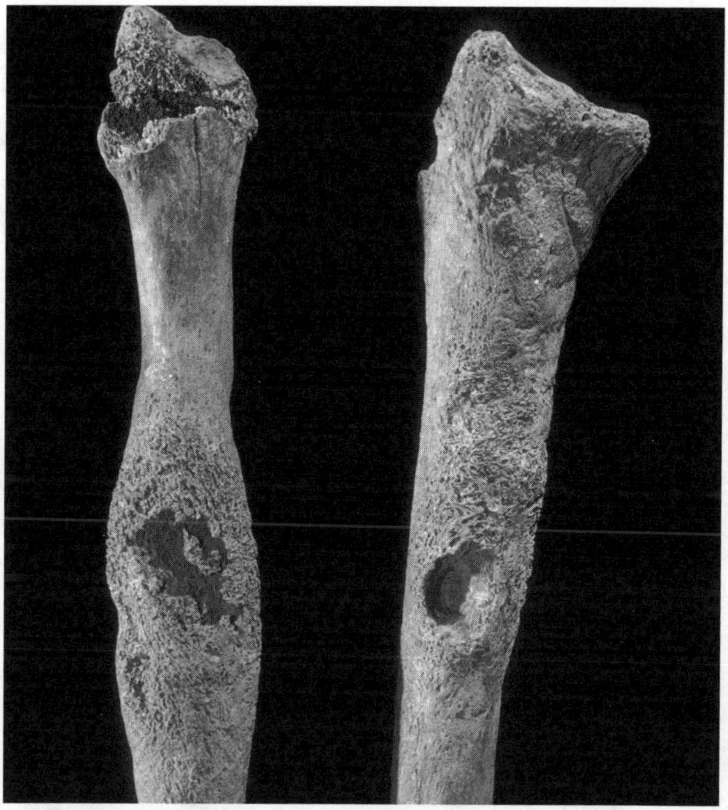

Figure 10.17 Bilateral periosteal reactive bone formation in the diaphyses of the radii from a probable case of chronic infection from the medieval site at the Magistrate's Court, Hull, England. Note that the lesion of the right diaphysis does not communicate with the marrow space. This is probably a central focus of infection, perhaps a granuloma, that has inhibited bone formation but is surrounded by reactive woven bone. A similar lesion occurs on the left diaphysis, although the central depression may have penetrated through the cortex to the marrow space. The lesion on the right radius, and possibly on the left as well, illustrates the caution needed to distinguish between a true cloaca, where a hole penetrates through the lesion to the marrow space, and a hole created by the failure to form reactive bone in a central focus while reactive bone is being formed around the focus

location of the reactive bone gives rise to the appearance of anterior bowing. However, no bowing (in the sense of abnormal bending) occurs in adult manifestations of acquired syphilis. This can easily be confirmed through careful observation of the interosseous line on the lateral diaphysis of the tibia. The line normally is straight and remains so in adult acquired syphilis despite the extensive abnormal bone formation on the anterior surface. The basic pattern of skull vault, forearm and tibia/fibula involvement is virtually pathognomic of treponematosis. However, one can have skull lesions without postcranial involvement and vice versa.

Skeletal participation in congenital treponematosis tends to have the same pattern of involvement seen in adult manifestations. However, I have not seen caries sicca lesions of the skull vault in childhood manifestations of congenital syphilis, although reactive bone-forming lesions do occur (Figure 10.18). If the child survives to adulthood, then the skeletal

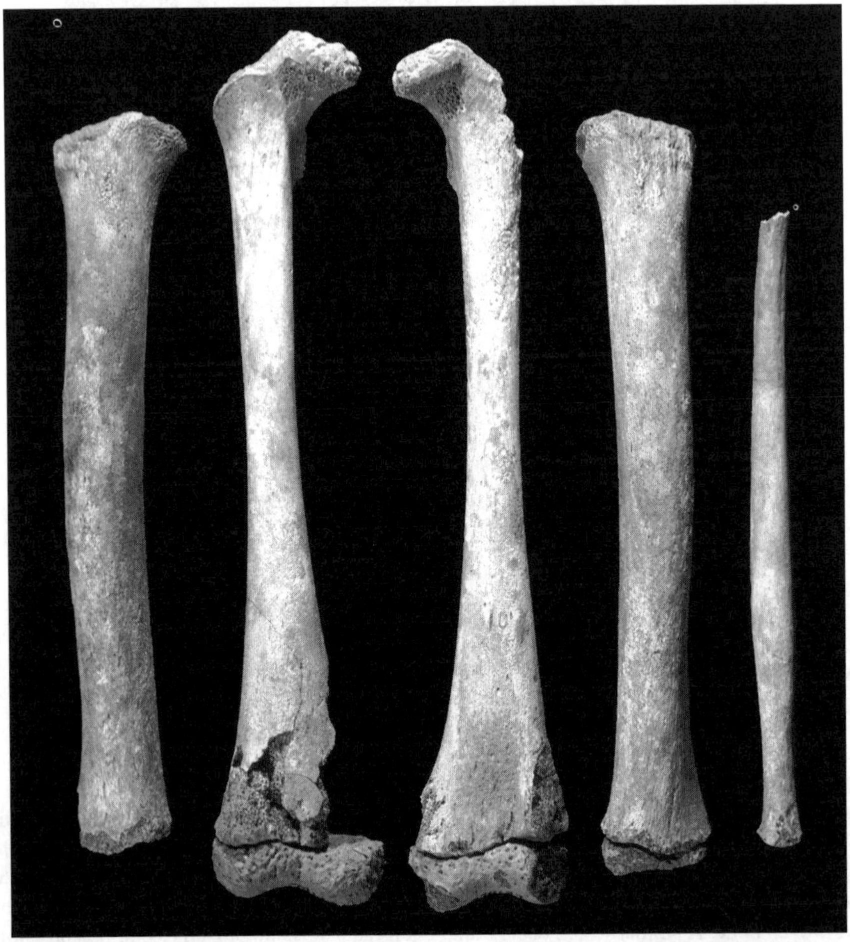

Figure 10.18 Bilateral periostitis with reactive bone formation of both tibias in a child with probable congenital treponematosis about 6 years of age from the Jones Site, Virginia, USA (NMNH catalogue no. 379177) dated to AD 1400–1500. Note that the midshaft diameter of the tibias is much larger than the midshaft diameter of the femora

manifestations of lesions formed after childhood are indistinguishable from those seen in acquired syphilis. In some cases of childhood congenital syphilis there is true bowing of the tibia (Resnick, 2002: 2558). It is not clear why this occurs. It is, of course, possible for a child to have congenital syphilis and rickets, which would explain the bending. Another option is that the stimulus resulting in the anterior periosteal reactive bone formation also stimulates differential growth of the anterior tibia relative to the posterior component, leading to abnormal curvature.

Some efforts have been made to distinguish between the skeletal manifestations of the three syndromes of treponematosis that affect the skeleton (e.g. Steinbock 1976). There may be slight differences that might be apparent in adequate skeletal samples of the three syndromes, but this research remains to be done. However, it is clear that there is considerable overlap and, on the basis of current evidence, it is, in my judgement, premature to distinguish between the acquired syndromes on the basis of skeletal evidence. Even congenital expression of treponematosis cannot be assumed to be caused by syphilis, although the probabilities certainly favour this diagnosis. Furthermore, distinguishing between an adult manifestation of congenital syphilis and syphilis acquired in late adolescence or early adulthood is likely to be impossible.

Brucellosis

Brucellosis is another bacterial infection of considerable potential interest to biological anthropologists and palaeopathologists, in part because so many mammals utilized for food are vectors for the disease. These vectors include goats, pigs, sheep, cattle, bison, reindeer and many others. It is highly likely that brucellosis has been a significant pathogen for humans for many millennia. Brucellosis is a disease that typically has no more than moderate morbidity and low mortality. It is a very chronic disease and does affect the skeleton. The sacro-iliac joint (El-Desouki, 1991) is a common site, as is the vertebral column (Resnick, 2002: 2522). The radiological feature most characteristic of brucellosis is an osteophyte-like projection from the anterior-superior margin of a vertebral body (Figure 10.19), known in the medical literature as a 'parrot's beak'. This abnormality is the result of a focus of infection in the vertebral body that may result in partial destruction of the end-plate and disc cartilage. In addition to the parrot's beak osteophyte there is reactive sclerosis of the adjacent spongy bone in the vertebral body.

The problem is to distinguish between osteophyte formation associated with degenerative disc disease and osteophytes caused by brucellosis. Osteophytes associated with degenerative disc disease tend to involve both the superior and inferior portions of the vertebral body, and limitation to the superior margin would be unusual. In ongoing unpublished research, colleagues and I are evaluating the presence of trabecular sclerosis in cases of vertebral degenerative disc disease with significant osteophyte formation. Trabecular sclerosis can be seen in extremely severe cases of osteophyte formation or in association with other spinal disorders, but it does not occur in the less severe cases of vertebral degenerative disc disease that represent the large majority of cases.

Brucellosis is also associated with large lytic lesions that have marginal sclerosis. The primary site for these lesions is the spine and sacro-iliac joint, but they can occur in other areas of the skeleton. Unlike tuberculosis, vertebral lesions in brucellosis often involve multiple vertebrae that are not contiguous, and the vertebral arch is also a common site of involvement.

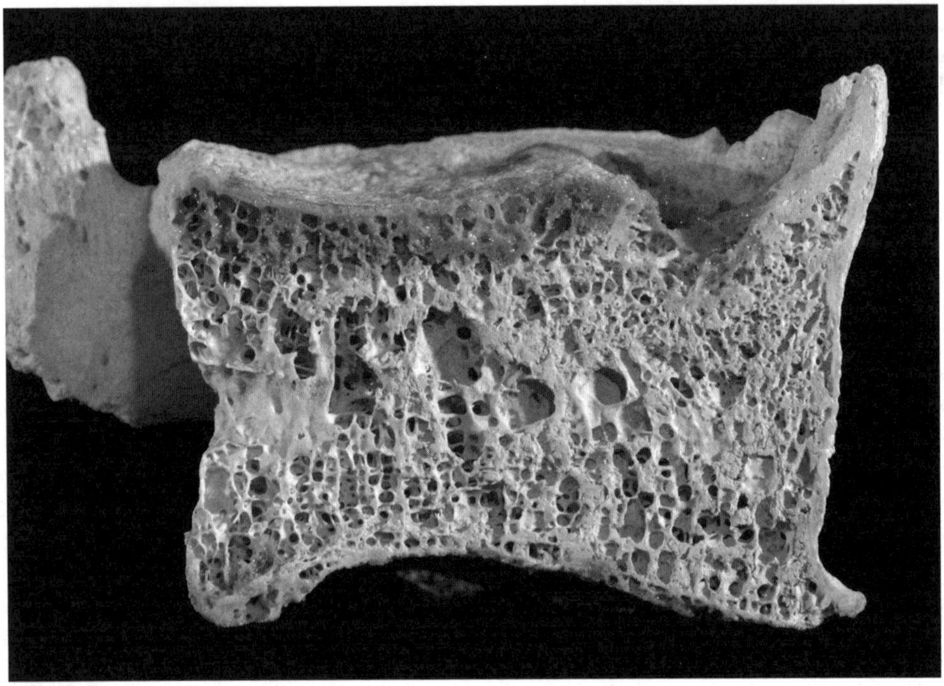

Figure 10.19 Sagittal section through a lumbar vertebral body from an adult from Bâb edh-Dhrâ (*c.* 3300 BC), Jordan. Note the osteophyte projecting from the antero-superior margin of the body and the zone of greatly increased trabecular density adjacent to the osteophyte that is suggestive of an inflammatory stimulus. Chronic infection seems likely and brucellosis is a possible cause

Echinococcosis

Echinococcosis is caused by infection by the larval stage of a flatworm. It affects bone in about 2 % of cases (Ivanissevich, 1934: 17) and the spine and innominate are the most common skeletal sites. A common vector is the dog, particularly when dogs are associated with herding sheep. The larvae are conveyed to tissues and organs within the body by the vascular system. The soft organs, particularly the liver, are most frequently affected; but the larvae can affect marrow in bone, and the spine is the most common site of involvement in the skeleton (Ivanissevich, 1934; Resnick, 2002: 2602). Skeletal involvement rarely affects more than one bone or area of the skeleton (Resnick, 2002: 2602).

Foci of larvae in the spine and marrow of other bones result in a destructive lesion with a sclerotic margin. Distinguishing these lesions from other chronic destructive disorders of the spine may not be possible on the basis of skeletal evidence. The presence of dog bones at the site would be evidence supporting the possibility of echinococcosis but would not be conclusive.

Mycosis

Mycotic (fungal) infection of bone is uncommon. There are several types of fungus that are known to cause disease in humans (Resnick, 2002: 2563). The best-known mycotic pathogens that can affect bone include blastomycosis, coccidioidomycosis and histoplasmosis. We are routinely exposed to fungi through our respiratory and gastrointestinal systems. These organisms also enter the body through wounds to the skin. Most of the time the human immune system eliminates mycotic pathogens easily and no morbidity results from exposure. However, particularly in situations where the immune system is compromised by any of several factors, such as chronic infection or malnutrition, disease may result. Morbidity can also occur in an otherwise healthy person if there is unusually elevated exposure to mycotic organisms.

Most often, mycotic pathogens are disseminated within the body through the vascular system and skeletal involvement can occur from this source. However, bone involvement usually occurs secondary to an adjacent soft tissue focus (Resnick, 2002: 2563). Unlike most other infectious pathogens, mycotic organisms do not have a preferred area within the body, and they can and do affect virtually any area of the skeleton. Because they may affect any part of the body, mycotic infectious lesions tend to have a random distribution that is bilateral only because of chance factors and is rarely symmetrical. The lesions can be bone forming or bone destroying, although the latter is more common. There is also considerable variation in the morphology and size of the lesions.

Actinomycosis is not a mycotic disease, despite the suffix; it is caused by a higher bacterium. However, the effect on bone resembles that in mycotic infection (Figure 10.20) and this diagnostic option does need to be included in differential diagnosis of multifocal, randomly distributed lesions. Because the organism is normally found in the mouth, involvement of the jaws, particularly the mandible, is common. However, distinguishing this disorder from true mycotic infection on the basis of skeletal evidence is

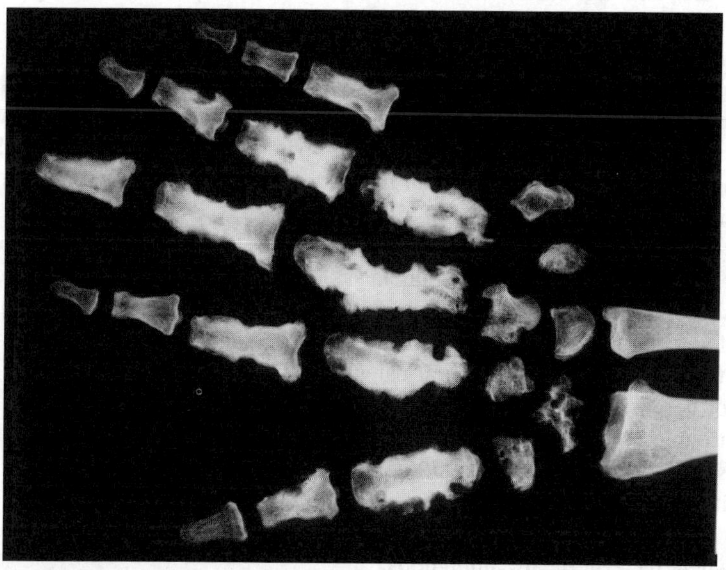

Figure 10.20 A modern case of actinomycosis in an adult male from Iran. Note the scalloped cortex from multiple infectious foci and the substantial sclerosis apparent in the medullary cavity in affected bones. (Case courtesy of Dr Bruce D. Ragsdale)

likely to be impossible in most cases. A tropical variant of actinomycosis is maduromycosis. In this infection, the primary focus is the foot through a portal created by injury to the skin.

Smallpox

Smallpox is one of a very few viral infectious pathogens that can affect the human skeleton. It only does so when the age of onset occurs during the developmental period and it can only affect the skeleton if the child survives the disease. The primary reason for including smallpox in this chapter is that it provides an excellent example of the importance of the type and pattern of bone lesions in differential diagnosis. Diagnosis in typical cases has a high probability of being correct.

In juvenile smallpox of the skeleton the bones of the elbow (Figure 10.21) are the predilected site. Lesions tend to be bone forming, but this may be accompanied by considerable destructive remodelling. The crucial diagnostic factor is that bone involvement is limited to the bones of the elbow in typical cases and tends to be bilateral. This distribution is very unusual and, when encountered in an archaeological human burial, a diagnosis of smallpox is likely.

CONCLUSIONS

The first step in any attempt to identify a disease present in an archaeological skeleton is to describe the abnormalities carefully with an emphasis on the types of lesions and their distribution. Because archaeological burials are often incomplete or comingled with other burials, it may be impossible to obtain a complete picture of the type and distribution of abnormalities that were present during life in a specific individual. However, a major effort is needed to try to determine, at the very least, whether the abnormality is multifocal and, if it is, whether the lesions are predominantly axial or appendicular. Equally important is whether or not the abnormalities are unilateral or bilateral and, if the latter, whether they are symmetrical. That is, are the lesions about the same size and located at about the same place on the contralateral bone?

Infection is one of several general disease categories all of which have at least limited potential to cause abnormalities in the human skeleton. Within this category, several specific infectious diseases can affect the human skeleton and, thus, provide information on the presence of infectious pathogens in human bioarchaeological populations. Distinguishing between infection and the other general categories of skeletal disease is often challenging and may not be possible in some cases. Diagnosis of a specific infectious disease is often even more troublesome, and the palaeopathologist should be very careful to avoid unjustified diagnostic specificity. The more general the disease category used, the higher the probability is of the diagnosis being correct (Waldron, 1994: 28–41; Miller *et al.*, 1996). This argues for caution in making very specific diagnoses rather than a general category of disease in situations where the criteria are ambiguous.

It is always a useful strategy to identify the significant diagnostic options in any case description. To be useful this needs to be done with some care. It makes little sense simply to list all or most of the diagnostic possibilities that exist. There needs to be a reasonable

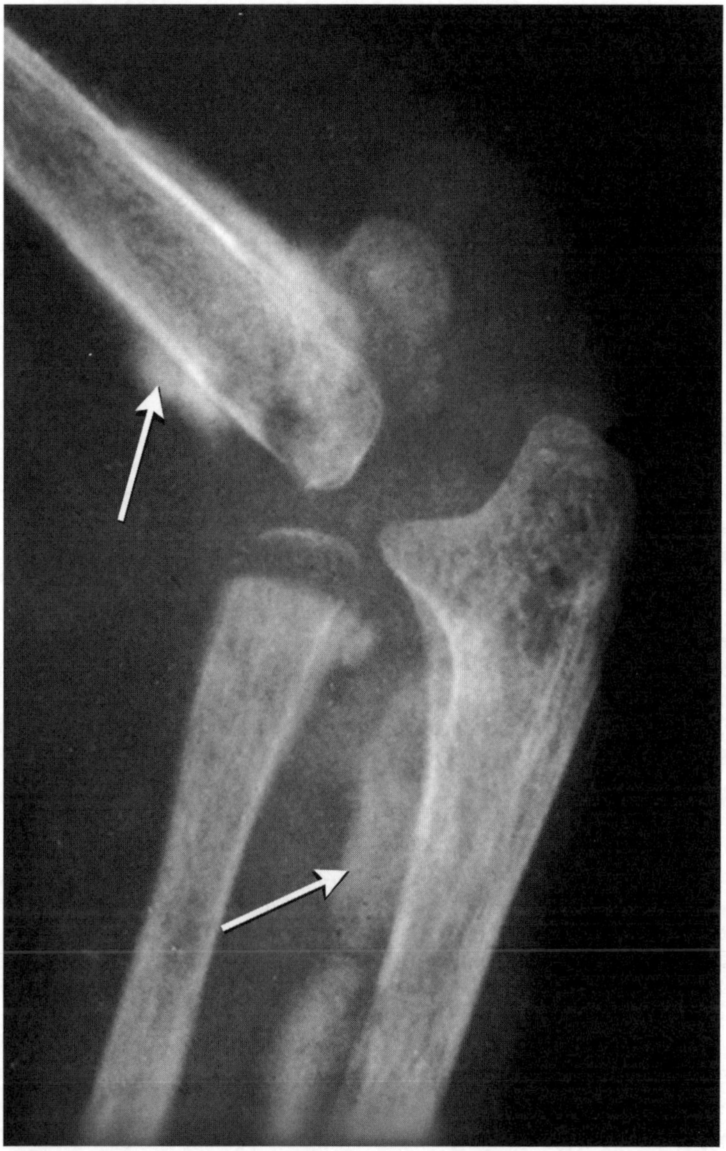

Figure 10.21 Radiograph of an elbow showing smallpox osteomyelitis with reactive bone formation (arrows). (Subadult, modern case, AFIP, courtesy of Dr Mark Kransdorf)

likelihood that a diagnostic option could, in fact, cause the lesion or lesions that are present. If there is overwhelming evidence that a specific diagnosis is likely, then adding unlikely diagnostic options to a report contributes confusion rather than clarity. It is also important to recognize that an individual may have more than one disease affecting the skeleton. This is not a common occurrence, but it does happen and the failure to evaluate this possibility can lead to misdiagnosis.

Like any research endeavour, the quality of the data and the rigour in collecting them are major factors affecting the legitimacy of the conclusions reached in any study of skeletal palaeopathology. The first decision in the study of an archaeological burial is whether or not there are any abnormalities present. This requires a sound knowledge of what a normal skeleton looks like for both sexes and at all stages of development. Once a decision has been made that an abnormal condition exists, careful description of the type and distribution of lesions is basic. On the basis of this information it is often possible to establish at least the general category of disorder that is the most probable diagnostic option. In many cases, with experience and ongoing study, it may be possible to establish a specific diagnosis. The accumulation of data on the prevalence of all evidence of disease in a skeletal sample provides an important index regarding the health of the people represented by the sample.

REFERENCES

Andersen JG, Manchester K. 1987. Grooving of the proximal phalanx in leprosy: a palaeopathological and radiological study. *J Archaeol Sci* **14**: 77–82.

Andersen JG, Manchester K. 1992. The rhinomaxillary syndrome in leprosy: a clinical, radiological and palaeopathological study. *Int J Osteoarchaeol* **2**: 121–129.

Aufderheide A, Rodríguez-Martín C. 1998. *The Cambridge Encyclopedia of Human Paleopathology.* Cambridge University Press: Cambridge.

Dutour O, Pálfi G, Berato J, Brun J-P (eds). 1994. *L'Origine de la Syphilis en Europe Avant ou Aprè 1493?* Centre Archéologique du Var, Éditions Errance: Tulon.

El-Desouki M. 1991. Skeletal brucellosis: assessment with bone scintigraphy. *Radiology* **181**: 415–418.

Hackett C. 1976. *Diagnostic Criteria of Syphilis, Yaws and Treponarid (Treponematosis) and of Some Other Diseases in Dry Bones.* Springer: Berlin.

Ivanissevich O. 1934. *Hidatidosis Ósea.* Imprenta Amorrortu: Buenos Aires.

Miller E, Ragsdale B, Ortner D. 1996. Accuracy in dry bone diagnosis: a comment on palaeopathological methodology. *Int J Osteoarchaeol* **6**: 221–229.

Møller-Christensen V. 1953. *Ten Lepers from Naestved in Denmark.* Danish Science Press: Copenhagen.

Møller-Christensen V. 1961. *Bone Changes in Leprosy.* Munksgaard: Copenhagen.

Møller-Christensen V. 1965. New knowledge of leprosy through paleopathology. *Int J Lepr* **33**: 603–610.

Møller-Christensen V. 1978. *Leprosy Changes of the Skull.* Odense University Press: Odense.

Ortner DJ. 2003. *Identification of Pathological Conditions in Human Skeletal Remains*, second edition. Academic Press: Amsterdam.

Pálfi G, Dutour O, Deák J, Hutás I (eds). 1999. *Tuberculosis Past and Present.* Golden Books/Tuberculosis Foundation: Szeged.

Powell M, Cook D (eds). 2005. *The Myth of Syphilis: the Natural History of Treponematosis in North America.* University Press of Florida: Gainesville, FL.

Resnick D. 2002. *Diagnosis of Bone and Joint Disorders*, fourth edition. WB Saunders: Philadelphia, PA.

Roberts C, Buikstra J. 2003. *The Bioarchaeology of Tuberculosis.* University Press of Florida: Gainsville.

Steinbock R. 1976. *Paleopathological Diagnosis and Interpretation.* Charles C. Thomas: Springfield, IL.

Waldron T. 1994. *Counting the Dead: The Epidemiology of Skeletal Populations.* Wiley: New York.

11

Metabolic Bone Disease

Simon Mays

English Heritage Centre for Archaeology, Fort Cumberland, Eastney, Portsmouth PO4 9LD, UK

INTRODUCTION

Metabolic bone disease may be defined as disease causing disruption of normal bone formation, remodelling or mineralization, or a combination of these (Albright and Riefenstein, 1948). In this contribution, four metabolic conditions are discussed: vitamin D deficiency disease (rickets, osteomalacia), vitamin C deficiency disease (scurvy), osteoporosis and Paget's disease of bone (PDB). The first two conditions were selected because they are sensitive indicators of past lifeways and diet and, hence, form important tools in biocultural studies of past populations. Vitamin D deficiency disease is intimately related to aspects of culture and lifestyle which affect the exposure of skin to natural light. Scurvy is an indicator of dietary imbalance or poor food preparation techniques. Osteoporosis and PDB were chosen because they demonstrate the potential of palaeopathology to contribute to our understanding of currently important diseases. Osteoporosis is the most important bone disease affecting modern populations; it has been estimated that 40% of American white females over 50 will suffer an osteoporotic fracture during the remainder of their lives (Gass and Dawson-Hughes, 2006). In most Western populations, PDB is the second leading skeletal disorder after osteoporosis, affecting up to 5% of those over 55 years of age (Kanis, 1998: 1–2). Despite their high prevalence, both osteoporosis and PDB are poorly understood, and palaeopathology has the potential to increase our knowledge of them.

For each condition, the pathophysiology of the disease is described, issues in palaeopathological diagnosis are considered, and themes in its biocultural study in past populations are discussed. Finally, problems and potential for the future palaeopathological diagnosis and biocultural study of these conditions are considered.

VITAMIN D DEFICIENCY: RICKETS AND OSTEOMALACIA

Vitamin D deficiency may arise from deficient acquisition of vitamin D, congenital or acquired malabsorption syndromes or chronic renal disease (Mankin, 1974a). Malabsorbtion

Advances in Human Palaeopathology Edited by Ron Pinhasi and Simon Mays

syndromes are rare, and neither these nor renal failure are likely to have been survived for long enough for them to have been a significant cause of skeletal rickets and osteomalacia in past populations. Therefore, rickets and osteomalacia in archaeological remains are essentially indicators of deficient acquisition of vitamin D. Vitamin D is naturally present in only minor quanities in most foods (although oily fish and egg yolk do contain significant amounts). In man, most vitamin D is synthesized in the body, a process initiated by the action of ultraviolet rays in sunlight upon a chemical precursor in the skin (Henry and Norman, 1992). Therefore, the occurrence of rickets and osteomalacia in the past in essence reflects inadequate exposure of the skin to natural light.

Vitamin D plays a major role in calcium homeostasis (Resnick and Niwayama, 1988: 2086–2119). In the absence of adequate vitamin D, there is reduced absorption of calcium from the gut, resulting in a decrease in the bodily pool of calcium. In order to maintain normal plasma calcium levels, parathyroid hormone is secreted, promoting release of calcium from the skeleton via increased osteoclastic bone resorption. This prompts an osteoblastic response, but there is insufficient calcium and phosphate to mineralize the osteoid laid down by the osteoblasts, so that abnormal amounts of unmineralized osteoid accumulate (Mankin, 1974b).

The skeletal effects of vitamin D deficiency are broadly those of deficient mineralization. They are most marked in the growing years, when they are termed rickets. Rickets rarely appears before 4 months of age, as stores of vitamin D are present from birth. Exceptions may occur if the mother was vitamin D deficient, in which case the infant may be born with rickets (Maiyegun *et al.*, 2002). Rickets is a disease of the rapidly growing infant and young child, and so rarely appears after about the age of 4 years (Ortner, 2003: 393). The skeletal effects of vitamin D deficiency in the mature skeleton are termed osteomalacia, and are generally less marked than in rickets.

Diagnosis in Palaeopathology

Bone changes due to rickets can be divided into those that are a direct result of metabolic disturbance and those that are due to mechanical deformation of weakened, poorly mineralized bone. These features have been described in detail elsewhere (Ortner and Mays, 1998; Mays *et al.*, 2006a), so only a brief outline will be given here. Changes that are a direct result of metabolic abnormality consist of inadequate mineralization of newly deposited bone during growth. Deficient mineralization of bone deposited upon the end of the diaphysis during endochondral bone growth leads to porosis/roughening of the bone underlying the growth plate (Figure 11.1). Defective mineralization of bone deposited in appositional growth leads to porosity of cortex. *In vivo*, the pores and other defects in the subchondral and cortical surfaces contain unmineralized osteoid. As the individual recovers from vitamin D deficiency, these are filled in with bone and obliterated. The identification of the porotic changes discussed above enables individuals who died with active rickets to be distinguished from those with healed lesions (Ortner and Mays, 1998; Mays *et al.*, 2006a). Radiographically, in active rickets, diffuse osteopaenia, coarsening and thinning of trabecular structure, and loss of cortico-medullary distinction are seen (Mays *et al.*, 2006a); these alterations are gradually removed by remodelling as the individual recovers from vitamin D deficiency. Histologically, signs of secondary hyperparathyroidism are present (Adams, 1997; Mays *et al.*, 2007).

Mechanical forces acting upon a softened skeleton lead to multiple bone deformities, including spreading and concavity of metaphyses, and diaphysial bending of long-bones.

Figure 11.1 Sequence to show increasing severity of porosity and roughening of the diaphysial bone underlying the epiphysial growth plates in active rickets (archaeological specimens). The specimens in (b)–(e) illustrate increasing severity of pathological changes. (a) Proximal end of a tibia of a 9-month infant, showing normal morphology. (b) Distal radius, 18–24-month child, showing slight roughening, giving a 'velvety' texture. (The exposed trabecular bone in the anterior part of the surface (toward the bottom of the photograph) is a post-depositional artefact.) (c) Distal end of a femur from a 3-month infant, showing more marked roughening. (d) Distal end of a radius from a 4–6-month infant showing roughening and some pitting. (e) Distal end of a radius from a 6–12-month infant showing extreme roughness and porosis

Diaphysial bending deformity often takes the form of accentuation of normal curvatures. In some cases, following recovery from rickets, bone deformity is progressively removed by remodelling, but in some instances it may remain into adult life. The frequency with which residual deformity remains depends upon the severity of childhood lesions; but, according to Hess (1930), 10–25% of cases of childhood rickets may retain noticeable deformity in later life.

In osteomalacia there is bone deformity, and pseudofractures (or Looser's zones) may be present. Bending deformities may occur in long-bone diaphyses and resemble those in rickets. More frequently, deformities occur in the trunk skeleton, particularly in areas rich in trabecular bone, which, owing to its higher remodelling rate, is preferentially affected (Ortner, 2003: 398–401). Typical deformities include angular sacral kyphosis, buckling of scapular body and pubic rami, and compression of the vertebral bodies leading to spinal scoliosis or kyphosis (Brickley *et al.*, 2005).

Looser's zones are seams of unmineralized osteoid. They may develop from stress fractures which fail to heal and they may progress to complete fractures (Resnick and Niwayama, 1988: 2098–2099). They are visible grossly as minor fissures in the bone, and radiographically as linear radiolucencies. Characteristic locations for Looser's zones are in the pubic rami, in the scapula at the lateral border and base of the scapular spine (Figure 11.2), and the neck and subtrochanteric area of the femur (Brickley *et al.*, 2005).

Histologically, the deficient mineralization of osteomalacic bone is visible as increased resorption spaces, defectively mineralized cement lines and areas of poorly mineralized bone (Brickley *et al.*, 2007) (Figure 11.3). As in rickets, hyperparathyroidism is present (Adams, 1997).

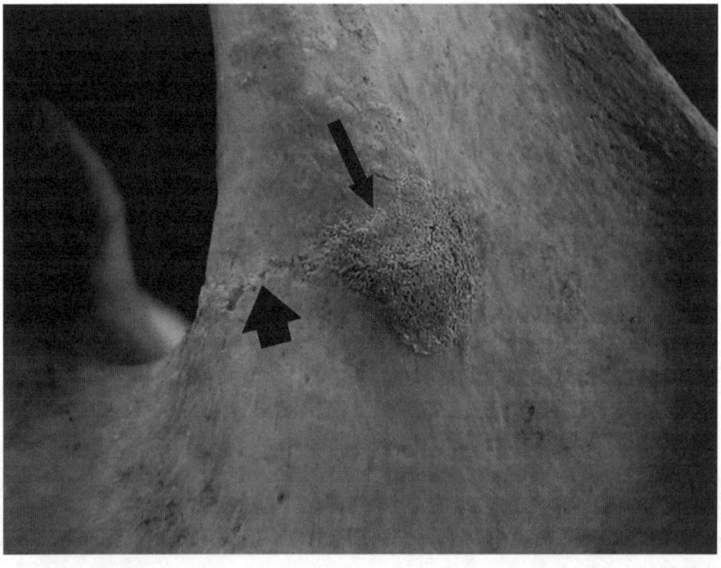

Figure 11.2 Infero-posterior aspect of the base of the scapular spine from an archaeological case of osteomalacia. There is a fissure in the bone (short arrow) with reactive marginal bone formation toward one end (long arrow). The fissure is an osteomalacic pseudofracture; the marginal bone proliferation is attempted repair

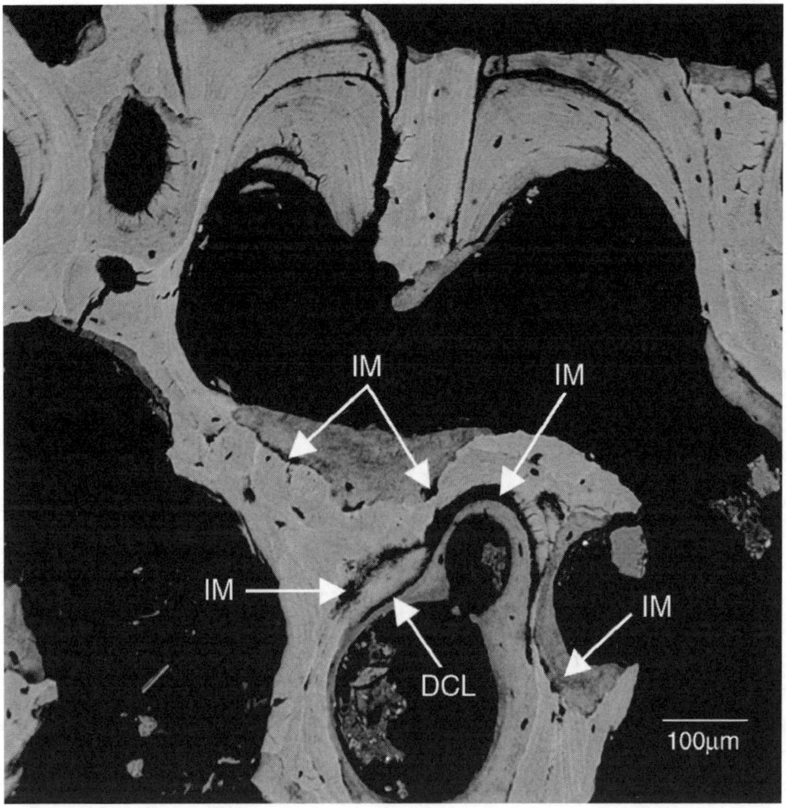

Figure 11.3 Scanning electron micrograph from a rib from an archaeological case of osteomalacia. Multiple areas of incomplete mineralization (IM) and defective cement lines (DCL) are visible

Vitamin D Deficiency Disease in Palaeopopulations

Documentary Sources

Historical sources provide a rich body of evidence concerning vitamin D deficiency disease, particularly rickets. The first convincing description of rickets dates from 1st-century AD Rome (Jackson, 1988: 38), but the first good clinical treatises come from 17th century England (by Whistler in 1645 and Glisson in 1650) (Clarke, 1962), and it was then that it first seems to have posed a regular health problem. In 17th- and 18th-century England, rickets seems mainly to have affected the children of the richer social classes, perhaps because they tended to coddle their children indoors whilst the poor were still bound to the land and so retained the outdoor lifestyle of earlier generations (Gibbs, 1994). The prevalence of the disease increased with industrialization, and the social pattern changed so that it became a disease of the urban poor (Owen, 1889). By the early 20th century, the frequency of rickets in infants and young children reached 90% in industrial cities in northern Europe (Steinbock, 1993). Changes in general living conditions during the Industrial Revolution acted severely to limit the exposure to sunlight of urban dwellers, particularly the poor, and this was the factor chiefly responsible for the rise of rickets at that time (Loomis, 1970). Sunlight failed to penetrate the narrow alleys which ran between tenements, and air pollution by industrial processes attenuated solar ultraviolet so that even when it did penetrate to ground level it was of little potency. In the 1920s, the advent of prophylaxis and treatment using vitamin-D-rich cod-liver oil enabled the disease to be combated. This, coupled with a decline in industrial air pollution, led to a rapid decline in rickets, so that by the mid-20th century it was rare (Clements, 1989).

Historical sources on osteomalacia are fewer than for rickets. It appears to have been recognized by Arab physicians in the 7th century AD (Sherman, 1950). By the 18th century, the pelvic deformities which occur in severe cases were recognized as an obstetric hazard (Hernigou, 1995). It began to be suspected in the late 18th century that osteomalacia was the adult counterpart of rickets, and this was confirmed in the late 19th century. In the 1920s, when cod-liver oil therapy was introduced for rickets, it was generally recognized by physicians to be beneficial in cases of osteomalacia as well (Sherman, 1950).

Palaeopathology

Palaeopathological diagnoses of rickets have most often been made on adult or older juvenile remains, the presence of residual bending deformities being taken as indications of healed childhood disease. Diagnosing residual rickets on this basis is potentially problematic. A variety of conditions other than rickets may cause increased bone curvature (Stuart-Macadam *et al.*, 1998). In addition, it is often difficult to determine unambiguously whether bending is due to disease or whether it simply reflects normal morphological variation. Relatively few convincing cases have been identified (Mays, 2003; Ortner, 2003: 401), and most of these come from the later historic period. The palaeopathological evidence for osteomalacia is even more sparse (Ortner, 2003: 402–404).

The majority of published palaeopathological studies of vitamin D deficiency disease come from England. The earliest English case appears to date from the 10th century AD, but the disease seems rare prior to the 17th century. For example, meta-analysis of published data relating to over 5000 burials from large medieval (10th–16th century AD) burial grounds in England revealed a prevalence of rickets of 0.3% (Mays, 2003). Although the existence of some medieval cases indicates that, contrary to the beliefs of 17th-century physicians, rickets was not a completely new disease in England when they first recognized it, the palaeopathological data are broadly consistent with the view obtained from written sources that the disease had not posed a great health threat prior to this.

Although relatively few osteological reports on skeletal material from British post-medieval burial grounds have been fully published, there is evidence for a rise in rickets during this period. In London, Broadgate (AD 1569–1720), St Benet Sherehog (late 16th–mid 19th century AD) and Christ Church Spitalfields (AD 1729–1852) each show prevalences of about 4% (White, personal communication; Molleson and Cox, 1993), and at St Martin's Churchyard, Birmingham (late 18th–late 19th century), the figure was 8% (Brickley *et al.*, 2006). These prevalences are 10–20 times that in the medieval period. The Broadgate churchyard served a poor parish in London (Pinhasi *et al.*, 2006), whereas the broadly contemporaneous St Benet Sherehog served predominantly those on middle incomes. Although documentary sources suggest that, in the 17th century, rickets mainly affected the children of the wealthier classes, the elevated prevalence at Broadgate suggests that it also posed a significant health threat for some of the urban poor as well. The frequency of rickets at Broadgate and at St Benet Sherehog may reflect the rise of urban pollution; written sources comment on reduced visibility in London in the 17th century due to atmospheric pollution from industrial processes (Brimblecombe, 1982). Industrial airborne pollution was also likely a factor in rickets in the Christ Church Spitalfields people. Another factor may have been that the fairly wealthy families who used the Spitalfields crypt for burial tended to keep their children indoors in inclement weather. In addition, many of those interred there worked in the silk industry; the work was home-based and workers and their families spent long hours indoors (Molleson and Cox, 1993).

At the time St Martin's Churchyard was in use, Birmingham was an expanding industrial centre, so overcrowding and airborne pollution were likely implicated in the high level of rickets here. In addition to the rickets cases, signs of osteomalacia were found in seven adult skeletons from St Martin's Churchyard (Brickley *et al.*, 2007). At this site, richer members of society were interred in brick-lined vaults, whereas the poorer classes made do with simple earth-cut graves. All cases of osteomalacia came from the earth-cut graves, suggesting that it was a disease of the poorer classes. Adult members of the richer classes may have had less crowded living conditions, with more outdoor open spaces and gardens, and they may also more often have travelled to less polluted environs.

In addition to the English cases, sporadic instances of rickets have been reported from Asia (e.g. Littleton, 1998), North America (e.g. Angel *et al.*, 1987), Africa (e.g. Pfeiffer and Crowder, 2004), and continental Europe (e.g. Bennike, 1985: 213–214; Blondiaux *et al.*, 2002). Most of these publications are studies of single or some few cases, but Littleton (1998) presents a study of 10 cases of rickets in infants and children from early Bahrain (1000 BC–AD 250). She interprets the occurrence of rickets in that group as indicating that cultural avoidance of sunlight may have a long history in that region.

Some workers have attempted a more nuanced understanding of rickets in past populations by looking at aspects of the manifestations of the disease in addition to prevalence rates. A further study of vitamin D deficiency disease at St Martin's Churchyard, Birmingham (Mays *et al.*, 2006a), involved comparison of rickets in subadults in that group and in a medieval English churchyard from the sparsely populated rural parish of Wharram Percy. There was a higher prevalence (13%) among children from Birmingham than in those from Wharram Percy (2%). That the prevalence of rickets was greater at Birmingham was as expected given the urban/industrial nature of the locale. That rickets should appear at all at Wharram Percy was noteworthy, given the outdoor lifestyle of that community. All cases at Wharram Percy were active infantile disease. It was suggested that these cases were of children who were otherwise sickly and, therefore, were kept indoors in dark, smoky houses and so developed rickets. This is consistent with the observation that all died in infancy, despite the fact that rickets is not a lethal disease, and with the observation that no individuals from this site who survived their early years showed healed rickets or osteomalacia. At Birmingham, there were healed as well as active rickets cases. The presence of healed disease, coupled with the overall high prevalence, suggests that, in contrast to Wharram Percy, rickets at Birmingham was a common ailment which affected children whose health was not otherwise seriously compromised. This is also supported by the observation that the changes due to the direct metabolic effects of rickets were more marked in the Wharram Percy cases than at Birmingham, indicating that whereas rickets at Birmingham was a regular part of growing up in a low-sunlight environment marginal for adequate vitamin D synthesis, the Wharram Percy cases resulted from more complete exclusion from natural light for a few individuals.

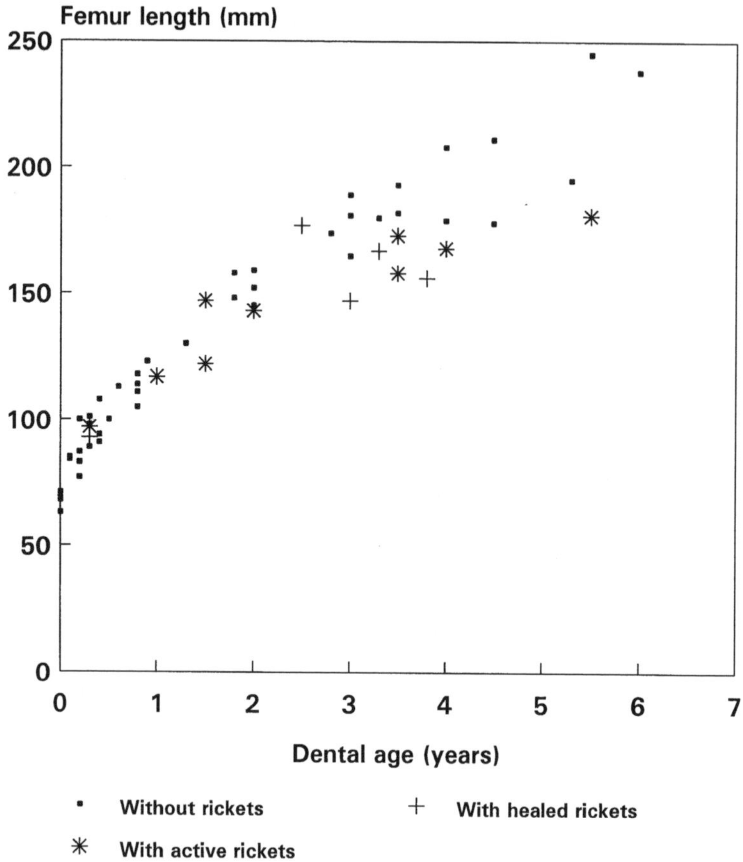

Figure 11.4 Femur length versus dental age for juveniles from St Martin's churchyard, Birmingham

Rickets may retard skeletal growth (Mankin, 1974b), and some workers have investigated whether this was so in palaeopopulations. Among rickets cases at St Martin's churchyard, Birmingham, there was no sign of a growth deficit in those under about 2 years, but in those over this age those with rickets tended to be short for their ages (Figure 11.4). In industrial urban centres rickets tended to be a seasonally recurring disease, with peaks at winter seasons (Schmorl, 1909). Perhaps a reason why only older children with rickets showed growth deficit was because growth stunting was a cumulative effect of multiple disease episodes with insufficient recovery in between: older children with rickets are more likely to have had multiple episodes of vitamin D deficiency as they would have lived through more vitamin D deficient winter seasons (Mays *et al.*, under review). By contrast, among rickets cases from Broadgate and Christ Church Spitalfields, no evidence for growth retardation was found (Pinhasi *et al.*, 2006; Chapter 16). This may reflect some difference in living environment between post-medieval London and Birmingham, but it is noteworthy that all the London cases of rickets were under about 3 years old. Perhaps they had experienced too few episodes of vitamin D deficiency for growth deficit to be apparent.

VITAMIN C DEFICIENCY: SCURVY

Humans are unable to synthesize vitamin C, so it has to be acquired from the diet. Prime sources of vitamin C are fresh fruit and vegetables, although it is found to a smaller extent in other foods such as fish and dairy produce. Boiling or prolonged storage of foods reduces their vitamin C content. Deficiency of vitamin C, therefore, reflects inadequate diet or faulty food preparation.

Vitamin C is involved in the synthesis of collagen. Collagen is the main structural protein of connective tissues, including bone. When vitamin C is deficient in growing individuals, newly formed bone may be osteopaenic. Deficiency of vitamin C also causes weakness of blood vessel walls, which leads to haemorrhage. Haemorrhage may, if it occurs adjacent to bone, provoke an osteological response, and this is the most important way in which scurvy may be recognized in skeletal remains. Prolonged deficiency of vitamin C is necessary to produce disease. Even if there is a total absence of vitamin C in the diet, the first symptoms (generally lethargy) do not generally appear until 1–3 months, and haemorrhages only after about 6 months (Stuart-Macadam, 1989; Beck, 1997; Aufderheide and Rodríguez-Martin, 1998: 310; Ortner, 2003: 383–384).

Diagnosis in Palaeopathology

In adults, skeletal scorbutic lesions tend to be fairly minor and non-specific and, consequently, scurvy is rather difficult to diagnose. The bleeding and swelling of the gums characteristic of the disease may cause loss of teeth and may potentially lead to inflammatory changes in alveolar bone (Ortner, 2003: 387). However, a complex array of different conditions may result in ante-mortem tooth loss (Mays, 1998: 155) or periodontal disease (Hillson, 1986: 310–312), most of which have nothing to do with scurvy, so scurvy cannot be inferred from the presence of these dental lesions in skeletal remains. Ossification of localized areas of haemorrhage may occur, particularly on sub-periosteal surfaces of long-bones (Aufderheide and Rodríguez-Martin, 1998: 313), and the presence of multiple ossified haematomas is a more reliable means of identifying scurvy in adult remains.

Scorbutic skeletal lesions are generally more prominent in infants and young children than they are in older individuals. In the growing bone, radiolucent zones form in the metaphyses, and in severe cases there may be metaphysial fracture (Ortner, 2003: 384). However, the most frequent and characteristic changes in scurvy in the growing skeleton are due to haemorrhage, and Ortner and co-workers in a series of studies (Ortner and Eriksen, 1997; Ortner *et al.*, 1999, 2001; Ortner, 2003) describe a constellation of haemorrhagic changes in the cranial and post-cranial skeleton which, they argue, are likely indicative of infantile scurvy. Haemorrhage adjacent to bone may stimulate the periosteum to produce new bone, and it may also provoke a localized inflammatory response, the vascular component of which may result in the proliferation of capillaries in the affected area. This latter may result in locally increased bone porosity to provide pathways for these blood vessels through bone. The distribution of these lesions in the skeleton is discussed in detail by Ortner and co-workers (Ortner and Eriksen, 1997; Ortner *et al.*, 1999, 2001; Ortner, 2003), but, in summary, typical locations are the external surface of the skull vault, the orbital walls (Figure 11.5), the greater wing of the sphenoid, the posterior surface of the maxilla and

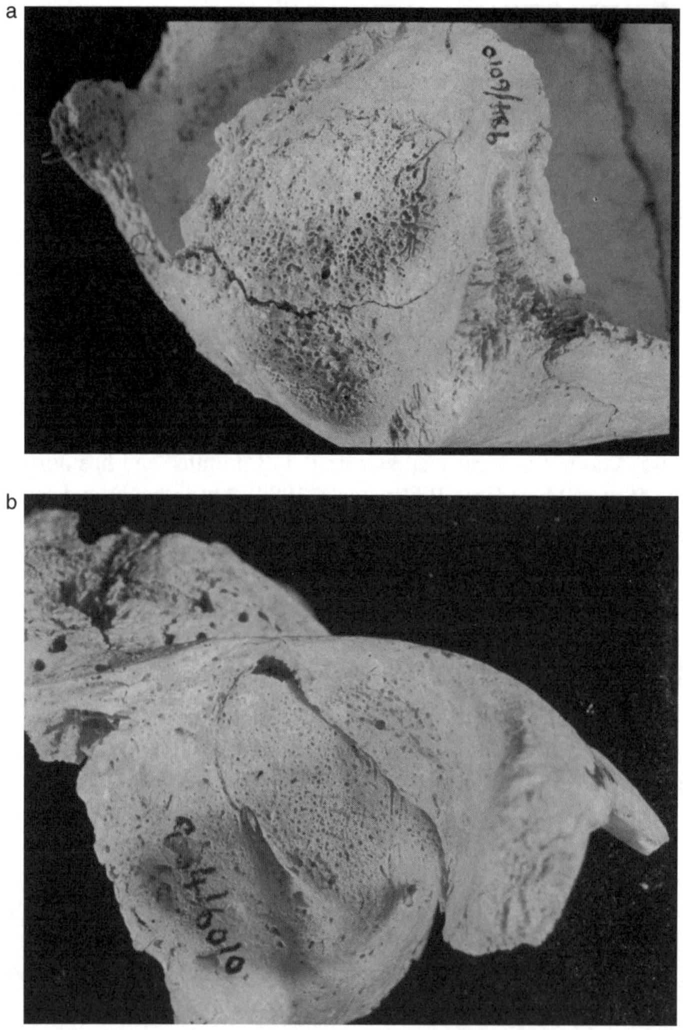

Figure 11.5 Scurvy, in a 2-year-old child (archaeological case). (a) Roof of right orbit, showing thick deposits of porous bone. (b) Inferior wall of right orbit showing abnormal porosity

the inferior surface of the hard palate. In the post-cranial skeleton, the infra- and supra-spinous fossae of the scapulae and the metaphyses of long bones are favoured sites for lesions.

The diagnostic criteria of Ortner and co-workers for infantile scurvy are based not on documented cases of the disease, but on undocumented archaeological material. Melikian and Waldron (2003) argue that because Ortner and co-workers' cases were undocumented, the relationship of the changes they observed to scurvy must be considered dubious, and further point out that Ortner and co-workers' cases do not, in general, resemble documented cases of scurvy in bones curated in pathology museums. Melikian and Waldron (2003) suggest that palaeopathological diagnostic criteria be derived from only known cases. Whilst one has some sympathy with this view, specimens of documented cases of infantile scurvy upon which dry bone changes can be assessed are few, and pathology museum collections generally show a bias toward specimens with more severe or spectacular manifestations of disease. It is important to recognize that, in all likelihood, a majority of cases encountered palaeopathologically are likely to be rather less severe than the specimens selected for retention in pathology museums. There is, thus, a need to progress beyond diagnosis based upon comparison with 'classic' cases. Ortner and co-workers' publications appear to be an attempt to do this, taking into account clinical evidence for disease and anatomical relationships between bone and soft tissue, using specimens that, whilst undocumented, bear constellations of changes that cannot credibly be explained by conditions other than vitamin C deficiency.

Scurvy in Palaeopopulations

Documentary Sources

Various early sources – for example, the Ebers papyrus (Egypt, 1500 BC) and the Hippocratic texts (Greece, 5th century BC) (Beck, 1997) – mention conditions which may be scurvy. The disease appears to have occurred sporadically in soldiery from early times: for example, in the armies of Imperial Rome (1st century AD) (Still, 1935) and in European armies during the Crusades (12th–13th century AD) (Sauberlich, 1997). It was certainly a problem in some 18th–19th century European armies and in the American Civil War (Sauberlich, 1997; Beck, 1997). However, during most of the historic period, scurvy appears to have been mainly a maritime problem, it being apparently uncommon on land other than during famine or war (Pimentel, 2003). It was first noted as a problem for sailors with the advent of European voyages of exploration from the 15th century AD onwards (Carpenter, 1986: 1–28). Although the value of fresh fruit in combating the disease was known from at least the 16th century, this failed to become common knowledge, and scurvy was a major problem on long nautical voyages (Carpenter, 1986: 1–2.). In the mid 18th century, James Lind, a British naval surgeon, fed scorbutic sailors on different diets and demonstrated that citrus fruits were the most effective treatment (Rajakumar, 2001). Nevertheless, it was not until nearly 50 years later that the British navy instituted effective measures toward combating the disease by making lemon juice a regular part of naval rations (Carpenter, 1986: 95).

Although sporadic cases of infantile scurvy were reported in the 16th and 17th centuries (Still, 1935; Carpenter, 1986: 160; Lomax, 1986), it does not then appear to have been a widespread problem. Prolonged breastfeeding, and weaning of children onto vegetable-based foods, may have acted to protect infants from scurvy. Infantile scurvy first began to be noted as a regular problem from about the 1870s, when wealthier social classes began to feed their infants on bread and milk sterilized by heating (Carpenter, 1986: 158–172). In 1914, Hess demonstrated that scurvy in infants fed on sterilized milk could be cured by including raw milk, orange juice or potatoes in their diet, and during the early 20th century the frequency of infantile scurvy fell (Carpenter, 1986: 172). By 1932, vitamin C had been isolated and characterized (Beck, 1997).

Palaeopathology

Given that written sources suggest that, for most of the historical period, scurvy was primarily a maritime problem and, to a lesser extent, a disease of land-based armies, it is natural to look to early maritime and military human remains for evidence for the condition. Turning first to maritime remains, During (1997) reported no evidence of scurvy among 21 skeletons from the *Vasa*, which sank off Stockholm in 1628, and Stirland (2000) found no clear cases among remains of 179 individuals from the *Mary Rose*, which sank off Portsmouth in 1545. The lack of skeletal cases of scurvy in this material may reflect the fact that the *Vasa* and the *Mary Rose* were war ships in waters close to their home ports, rather than vessels used for extended sea voyages. By contrast, evidence of scurvy has been found in a group of 17th–18th-century Dutch whalers buried on the subarctic island of Spitsbergen. Thirty-nine out of 50 burials showed dark staining on the bones and/or tooth roots. Immuno-enzymatic staining and microscopic analysis suggested that the dark colouration was the remains of haematomas, which had survived burial presumably thanks to the cold, subarctic environment (Maat, 1982, 2004; Maat and Uytterschaut, 1984). The results are consistent with written sources which note that scurvy was a problem on early Dutch whaling expeditions to the arctic (Maat, 1982).

Turning to human remains associated with land-based armies, there appears to be no evidence for scurvy among skeletons (numbering some 1500 individuals in total) excavated from mass graves associated with medieval battlefields (Ingelmark, 1939; Cunha and Silva, 1997; Coughlin and Holst, 2000). One might expect that vitamin C deficiency would become a more regular problem for the military during the post-medieval period, when the greater size of armies, and the prosecution of more extended military campaigns, meant that maintenance of adequate provisioning became more problematic. Documentary sources suggest that scurvy may have been a problem in Napoleon's armies (Beck, 1997), but studies of skeletal remains, totalling about 400 individuals, from battlefields from the 18th–19th-century Napoleonic period in Europe (Etxeberria, 1999; Horácková and Vargová, 1999; Rollo, 1999; Lunardini *et al.*, 2002; Meyer, 2003) have failed to identify any cases. In part this may reflect the difficulties in reliably identifying scurvy in adult skeletons, but it may also be that scurvy arose during specific reversals which caused dislocation of military provisioning, such as the retreat from Moscow (Beck, 1997), rather than being a more general problem for troops in the Napoleonic Wars.

There is, in general, little convincing palaeopathological evidence for infantile scurvy in early European populations, although sporadic cases have been identified from the Neolithic period onwards (e.g. Roberts, 1987; Carli-Thiele, 1996; Mays, 2007a). To some extent this may be a problem of underdiagnosis, but it may also be that it was genuinely rare until the late post-medieval period. This would seem to be in accord with documentary sources and with some systematic studies of historic period infant and juvenile remains. Connell and Mays (1997) examined 183 subadult skeletons preserving cranial bones from the medieval peasant community at Wharram Percy and, applying Ortner and Eriksen's (1997) criteria for diagnosing scurvy, found no evidence for it. Melikian and Waldron (2003) likewise found no evidence for the disease among 123 subadult skeletons from 3rd–16th-century AD Britain. Recently, Brickley and Ives (2006) described six 19th century cases of infantile scurvy in the skeletal remains from St Martin's Churchyard, Birmingham. All were from simple, earth-cut graves, identifying them as from the lower social classes. The authors suggest that the mid-19th-century potato famine, caused by blight to the crop, may have been a factor, as potatoes were one of the few affordable sources of vitamin C for the 19th century urban poor.

In contrast to the paucity of evidence for scurvy in Europe, studies in the Americas indicate that it was a regular health problem for some New World populations. To some

Table 11.1 Prevalences of scurvy in subadults from Native American historic and protohistoric archaeological sites[a]

	Frequency of scurvy		
	N_s	N_t	Prevalence (%)
Midatlantic USA	10	169	6
Plains USA	0	54	0
Southeast USA	6	16	38
Southwest USA	7	318	2
Peru	38	363	10

[a] N_s: individuals showing scurvy; N_t: total number of individuals examined. Data from Ortner (2003: tables 15.1 and 15.2).

extent, the patterning in the New World data (Table 11.1) can be understood in terms of what is known of Native American diet. For example, maize, which is low in vitamin C, was an important dietary staple. However, the presence of scurvy in the US south-east was somewhat unexpected, as fresh sources of vitamin C would be available year-round (Ortner *et al.*, 2001).

OSTEOPOROSIS

Osteoporosis is a disease of the elderly characterized by decreased bone mass and microarchitectural deterioration of bone tissue with resultant decreased bone strength and increased risk of fracture (Christadoulou and Cooper, 2003). Decreased bone mass rather than microarchitectural deterioration appears the more important determinant of fracture risk (Moyad, 2003). In osteoporosis, bone is lost from the endosteal envelope, leading to thinning of the trabecular structure of cancellous bone and thinning of cortical bone from the endosteal surface. Although both trabecular and cortical bone are lost, trabecular bone is lost more rapidly, probably due to its faster turnover (Riggs and Melton, 1986). Trabecular bone strength is more compromised than cortical bone strength, so osteoporotic fracture tends to occur at sites rich in trabecular bone, specifically the femur neck, distal radial metaphysis (Colles' fracture) and vertebral bodies.

Osteoporosis is associated with advancing age in adults of both sexes, but is more severe in females. In women, the principal cause of osteoporosis is the hormonal changes which accompany menopause. In men, the causes are less well documented, but it appears that, as in women, age-related decline in sex hormones plays an important role (Anderson *et al.*, 1998). In addition, in both sexes, various extraneous lifestyle factors, such as cigarette smoking and lack of exercise, are held to exacerbate the disease (Ross, 1996), but genetic factors are also important (Prentice, 2001; Ginsberg *et al.*, 2001; Rapuri *et al.*, 2004). Today, the severity and frequency of osteoporosis varies widely among different world populations. Those of European origin are most affected (Villa, 1994), but differences exist between different European groups – for example, populations from northern Europe are more at risk than those from the south (Kanis *et al.*, 2002).

Diagnosis in Palaeopathology

Palaeopathological studies of osteoporosis have focused on bone mass ('bone quantity') and, to a lesser extent, on aspects of bone microstructure ('bone quality'), and on fracture patterns.

Bone Quantity

In palaeopathological studies, measurement of bone quantity in skeletal populations is dominated by measurements of cortical bone thickness taken from radiographs ('radiogrammetry') and by various measures of bone density, chiefly dual X-ray absorptiometry (DXA). Radiogrammetry and DXA are discussed in full in Chapter 5; but, to summarize, both are used in clinical studies of osteoporosis, facilitating comparisons of ancient and modern data. Measures of cortical thickness using radiogrammetry of dry bones potentially produce results that are directly comparable to measures taken on living subjects. This facilitates comparison both of peak cortical bone thickness (attained in early adulthood) and of patterns of loss with age. DXA produces a measure of bone mineral density (BMD), but the question of comparability between results on dry bones and on living subjects is somewhat more complex than for radiogrammetry. Because skeletal remains lack marrow and soft tissue, absolute levels of BMD measured on dry bones are not directly comparable to those in living subjects. Therefore, peak BMD cannot be compared between skeletal and living populations. However, patterns of age-related loss of BMD can, in cases where diagenetic effects on archaeological bone density can be excluded, be compared between skeletal and living populations.

Bone Quality

Bone quality comprises bone microarchitecture, mineralization and mechanical properties (Grynpas, 2003). The aspect of bone quality that can most readily be investigated in palaeopathological specimens is trabecular microarchitecture. In palaeopathological studies, bones are generally thick-sectioned (Figure 11.6) and images of the exposed trabeculae captured by photography or using scanning electron microscopy, or the section is radiographed (Kneissel *et al.*, 1994; Brickley and Agarwal, 2003). Precise comparisons between the results of archaeological work and modern autopsy studies are difficult due to differences in methodologies; however, it is possible to compare broad age-related patterns (Brickley and Howell, 1999; Agarwal *et al.*, 2004). Micro-computed tomography provides an alternative, non-invasive method of capturing three-dimensional images of trabecular structure (Macho *et al.*, 2005) and may also facilitate comparisons between results on archaeological skeletal remains and living subjects.

Fractures

It is of interest, from the point of view of assessing the health impact of osteoporosis on earlier populations, to investigate whether age-related decline in bone status led to increased risk of fracture, as it does today. If fractures at skeletal locations typical of those in osteoporosis (hip, wrist, spine) can be demonstrated to be associated with low bone mass at a population level, then a general link with osteoporosis can reasonably be inferred. However, this is not straightforward. The skeleton bears a cumulative record of the fractures suffered during life, so an older age cohort will tend to show more healed fractures of all types than a younger one, simply because they have had more years at risk of incurring fracture. Fracture prevalence

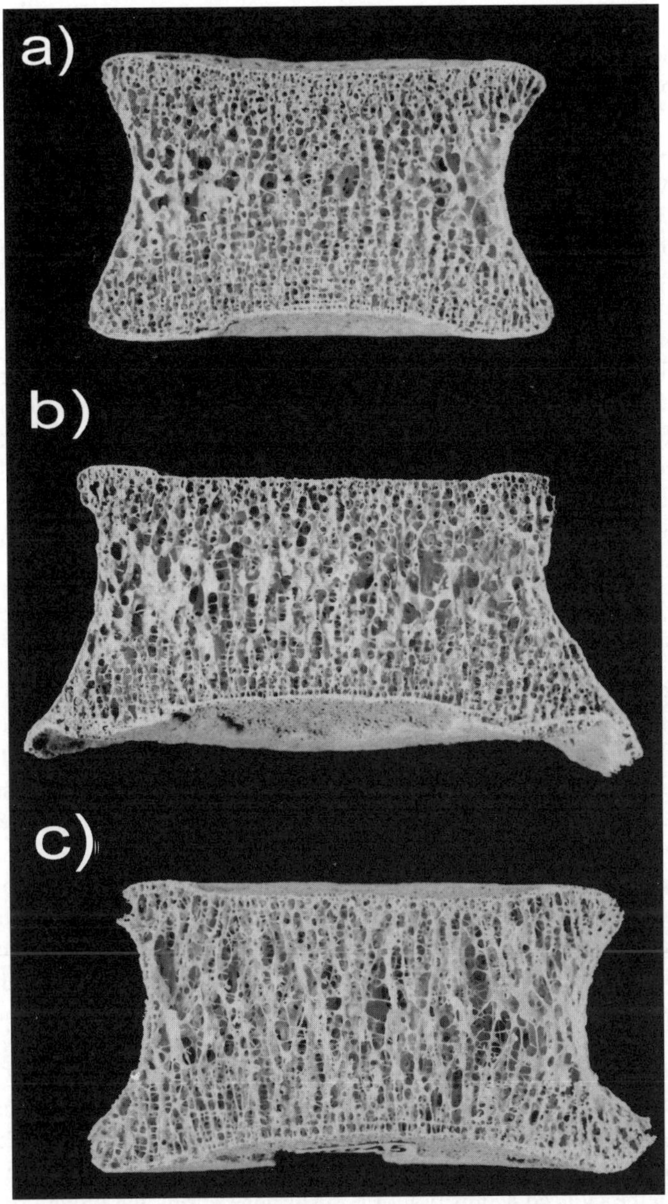

Figure 11.6 Vertical section of L4 vertebral bodies from female adults (a) aged 18–30 years, (b) 30–50 years and (c) >50 years (archaeological specimens) to demonstrate some of the age changes in trabecular bone structure. (a) A three-zonal arrangement of cancellous bone, characteristic of a young adult. Superior and inferior zones of dense cancellous bone surround a central band of more open trabecular structure. (b) Thinning of trabecular structure, especially by loss and thinning of horizontal trabeculae, is evident. (c) Further rarification of cancellous bone is apparent. Some very long, slender trabeculae are visible, and the loss of many horizontally orientated trabecular elements means that the course of some vertically orientated struts can be followed through most of the thickness of the vertebral body

and low bone mass might be expected to be associated in a skeletal population, simply as an epiphenomenon of the association between both these variables and individual age. In order to demonstrate that typical osteoporotic fractures are collectively distinct from other fracture types and are truly associated with low bone mass, one would need to demonstrate that they show a stronger relationship with low bone mass than do other fractures. This can be done by comparing bone mineral status in those with osteoporotic fractures and with other fracture types (e.g. Mays *et al.*, 2006b) or, if individual age at death is known exactly (e.g. from coffin plates), by regression analysis.

Comparison of prevalences of osteoporotic fractures between skeletal and living populations is problematic. Modern studies of osteoporotic fracture usually report incidence, focus on age at occurrence, and are carried out on selected populations. In skeletal remains, only fracture prevalences, not incidences, can be calculated, and age at occurrence of fracture in individual cases cannot usually be inferred. A few studies exist on modern subjects which report, for different age cohorts unselected for fracture risk, prevalences of different fracture types sustained at any time during life (e.g. van der Voort *et al.*, 2001). These types of study would seem to be the best choices for comparison of fracture prevalence with skeletal material; however, problems still remain, not least due to the incomplete and fragmentary nature of excavated material.

Osteoporosis in Palaeopopulations

The importance of osteoporosis today has helped stimulate interest in its occurrence in past populations, but historical sources are of little value in this respect. Although fragility fractures in old age appear to have been recognized since about the mid 19th century (Drezner, 1997), no link could be established with age-related bone loss in patients until the discovery of X-rays in 1895. To investigate osteoporosis in the past we are generally reliant on palaeopathology.

The most frequent approach in population-based palaeopathological work on osteoporosis has been to compare bone status in ancient populations with that found today. Comparisons of peak cortical bone (assessed using measures of cortical thickness from radiogrammetry) in early adult life between ancient and modern data have been restricted mainly to European material. These have generally shown lower amounts of cortical bone in the archaeological series (Table 11.2). Although, other than *in extremis*, there is no relationship in adults between skeletal maintenance and diet either in terms of protein or caloric intake (Garn, 1970), a consistent link has been found between poor childhood nutrition and deficient appositional bone growth (Adams and Berridge, 1969; Garn *et al.*, 1969; Barr *et al.*, 1972; Himes *et al.*, 1975). It seems reasonable to suppose that the likely poorer nutrition of earlier populations played a part in the reduced acquisition of cortical bone during the growth period, although other, as yet unidentified, factors may also be involved.

Some studies which compare age-related bone loss in adults in ancient and modern populations are summarized in Table 11.3. Most studies with adequate sample size have reported an age-related decline in bone substance, showing that this phenomenon has a long history in mankind. The greatest amount of work has been done on UK populations. Most of the UK archaeological studies seem to show that age-related bone loss was similar to or greater than that seen today, and it is worth evaluating this finding with respect to lifestyle factors commonly thought to exacerbate age-related bone loss in osteoporosis. These include excess caffeine intake, cortico-steroid use, cigarette smoking, sedentary lifestyle, and

Table 11.2 Some investigations of peak cortical bone which compare measures of cortical bone thickness in archaeological and modern populations[a]

Location	Date	Site in skeleton investigated	Results	Reference
England: Ancaster	3rd–4th century AD	2nd metacarpal	F lower peak cortical bone than moderns; M not investigated	Mays (2006)
England: Wharram Percy	10th–16th century AD	2nd metacarpal, femur	M and F lower peak cortical bone than moderns	Mays (1996, 2007b), Mays *et al.* (1998)
England: London Spitalfields	18th–19th century AD	2nd metacarpal	M and F lower peak cortical bone than moderns	Mays (2000, 2001)
Poland: unspecified locations	Medieval	2nd metacarpal	M and F lower peak cortical bone than moderns	Rewekant (1994)
Poland: Cedynia and Słboszewko	12th–17th century AD	2nd metacarpal	M and F lower peak cortical bone than moderns	Rewekant (2001)
Sweden: Stockholm	14th–15th century AD	2nd metacarpal	M and F lower peak cortical bone than moderns	Ekenman *et al.* (1995)
		femur	M and F higher peak cortical bone than moderns	

[a]Results, peak cortical bone (measured by radiogrammetry) found in young adults in relation to that seen in modern Europeans; M, males, F, females.

deficiencies of calcium and vitamin D (Ross, 1996). For most of the populations studied, the first three substances listed above were not available. In the absence of mechanized labour-saving devices, levels of physical activity would be expected generally to have been greater in earlier than in current UK populations. There is no reason to believe that the populations sampled were calcium deficient. For most of the study populations, there is no indication of widespread vitamin D deficiency. In general, differences in risk factors outlined above would have led one to anticipate less age-related bone loss in ancient than in modern British populations; yet this is not the case. This may indicate that lifestyle factors may only have a minor effect on bone loss in osteoporosis.

Studies of females from medieval Germany (Hammerl *et al.*, 1990), Denmark (Poulsen *et al.*, 2001), England (Mays *et al.*, 1998) and Norway (Mays *et al.*, 2006b) suggest that age-related decline in BMD in women may have begun earlier than it does today, some loss being evident by the 30–50 year age group. This implies significant pre-menopausal loss of BMD. The above writers suggest that this may reflect the effects of high parity and relatively late weaning of offspring, which appear to have been customary in medieval times (Shahar, 1990; Benedictow, 1992; Richards *et al.*, 2002). In modern populations, minor reductions in BMD may occur during pregnancy and lactation but are quickly recovered (Kalkwarf and Specker, 1995; Kohlmeier and Marcus, 1995; Laskey and Prentice, 1999). In the past,

Table 11.3 Some investigations of age-related loss of bone substance that make comparison between archaeological and modern populations

Location	Date	Method of estimating age at death[a]	Method for monitoring bone loss	Site in skeleton investigated	Results[b]	Reference
Europe						
Austria: Franzhausen	2000BC	Complex method	DXA	L4, femur neck	M and F similar loss to moderns	Kneissel *et al.* (1994)
Denmark: Nordby	AD 1000–1250	Multiple bone methods	DXA	Femur neck	M no demonstrable loss; F less loss than moderns	Poulsen *et al.* (2001)
Germany: Bockenheim	5th–7th century AD	–	DXA	Femur neck	M no demonstrable loss; F loss similar to or greater than moderns	Hammerl *et al.* (1990)
Poland: unspecified locations	Medieval	–	Cortical thickness	2nd metacarpal	M and F less loss than moderns	Rewekant (1994)
Sweden: Stockholm	14th–15th century AD	Multiple bone methods	DXA, cortical thickness	Radius, femur, 2nd metacarpal	M and F no demonstrable loss	Ekenman *et al.* (1995)
UK: Ancaster	3rd–4th century AD	Dental wear	Cortical thickness	2nd metacarpal	F greater loss than moderns; M not investigated	Mays (2006)

UK: Wharram Percy	10th–16th century AD	Dental wear	DXA, cortical thickness	2nd metacarpal, radius, femur	M and F loss generally similar to or greater than moderns	Mays (1996, 2007b), Mays *et al.* (1998), McEwan *et al.* (2004)
UK: London Farringdon St.	18th–19th century AD	Multiple bone methods	Density, physical measures	L4	M no demonstrable loss; F similar loss to moderns	Brickley and Waldron (1998)
UK: London Spitalfields	18th–19th century AD	Documented age	DXA, cortical thickness	Femur neck	F less loss than moderns	Lees *et al.* (1993), Mays (2000, 2001)
				2nd metacarpal	M and F similar loss to moderns	
Africa						
Sudan: Wadi Halfa	350 BC–AD 1400	–	Cortical thickness	Femur	M no demonstrable loss; F earlier loss than moderns	Armelagos *et al.* (1972)

Table 11.3 (Continued)

Location	Date	Method of estimating age at death[a]	Method for monitoring bone loss	Site in skeleton investigated	Results[b]	Reference
North America						
Alaska: St Lawrence Island	Precontact Eskimo	Histology	Photon-absorptiometry	Radius	M and F similar loss to moderns	Laughlin *et al.* (1979)
Alaska: Kodiak Island	500 BC–AD 1700 Eskimo	Pubic symphysis	Cortical thickness	Femur	M similar loss to moderns; F no demonstrable loss	Gunness-Hey (1986)
USA: Indian Knoll, KY	2500–200 BC Native American	Multiple bone methods, dental wear	Photon absorptiometry	Radius	M and F similar loss to moderns	Perzigian (1973)
USA: Pete Klunk, IL	50 BC–AD 250 Native American	Multiple bone methods, dental wear	Photon absorptiometry	Radius	M and F greater loss than moderns	Perzigian (1973)
USA: various locations	10th–19th century AD Native American	Pubic symphysis	Cortical thickness	Humerus, femur	M and F generally similar loss to moderns	Eriksen (1976)
USA: Campbell site, MO	AD 1540–1700 Native American	–	Cortical thickness	Femur	M and F similar loss to moderns	Van Gerven *et al.* (1969)

[a]Complex method, technique of Nemeskéri *et al.* (1960); documented age, age at death known exactly from coffin plates; –, no information given.
[b]Bone loss found in relation to that seen in modern reference populations; M, males, F, females.

recovery of bone mass following pregnancy and lactation may often have been slowed by poor maternal nutrition (Turner-Walker *et al.*, 2001). This loss of BMD pre-menopausally in addition to post-menopausal loss may be a cause of the somewhat greater bone loss by old age tentatively identified in females from some early populations compared with modern females (e.g. Hammerl *et al.*, 1990; Mays *et al.*, 1998; Mays, 2007b).

Some workers have investigated age-related deterioration of aspects of bone quality in palaeopopulations. Among 6th–10th-century AD Nubians, age-related patterns of microstructural deterioration in the fourth lumbar vertebral body commenced earlier than in modern populations (Kneissel *et al.*, 1997). Studies of fourth lumbar vertebrae in Bronze Age Austrians (Kneissel *et al.*, 1994) and 18th–19th century Londoners (Brickley and Howell, 1999) demonstrated age patterns of deterioration of microstructure similar to those today. At medieval Wharram Percy, age-related deterioration of bone quality in fourth lumbar vertebrae had occurred by about middle age, with little further change thereafter (Agarwal *et al.*, 2004).

A further question which may be addressed using palaeopathological evidence is whether current geographic patterning seen in osteoporosis has arisen recently, perhaps due to differences in lifestyle between populations, or whether the same differences existed in ancient times when lifeways were very different, which might imply inherent population differences in susceptibility to the disease. Modern Eskimos show lower bone mass and greater age-related bone loss than Caucasians (Lazenby, 1997). Nutritional factors, particularly the high protein content of Eskimo diet, have often been suggested as a cause (Lazenby, 1997). Looking at Eskimo skeletal remains dating from pre-European contact, Laughlin *et al.* (1979) found that age-related bone loss was greater than in Caucasians, demonstrating the antiquity of this difference. These workers also found that skeletons from Aleut archaeological sites had higher bone mineral profiles than did Eskimos. This is in contrast with the present-day pattern, whereby Aleuts resemble Eskimos in bone mass and its age-related loss (Mazess *et al.*, 1985). It may be that Eskimos and Aleuts have converged fairly recently in their experience of osteoporosis.

Modern populations from Norway and England differ in osteoporosis, the former showing lower levels of bone mass and higher fragility fracture rates. Comparison of hip BMD in medieval English and Norwegian skeletal populations (Mays *et al.*, 2006b) showed similar peak BMD and patterns of age-related loss in the two groups, although the medieval Norwegians had a higher fragility fracture prevalence. This may suggest that the population difference in BMD has arisen recently.

A few studies have investigated the health impact of osteoporosis in the past by investigating whether bone loss was associated with increased fracture risk. In women from 3rd–4th-century and 10th–16th-century British populations at Ancaster and Wharram Percy respectively (Mays, 1996, 2006), and in 12th–16th-century AD Trondheim, Norway (Mays *et al.*, 2006b), fragility-type fractures were associated with low metacarpal cortical index and/or low femur neck BMD. By contrast, in women from 18th–19th-century AD London Spitalfields, fragility fractures were absent, despite the observation that radiogrammetric study of metacarpal bone showed that peak cortical bone was lower than moderns and age-related loss was similar, so that all age classes were deficient in cortical bone compared with their modern counterparts (Mays, 2000). Interestingly, BMD study of the femur neck (Lees *et al.*, 1993) at Spitalfields showed conservation of trabecular bone mass compared with modern elderly, so perhaps this had a protective effect against osteoporotic fracture.

It has been noted that hip fractures are generally rare in palaeopopulations (e.g. Agarwal and Grynpas, 1996). This is so even in groups where reduction of hip BMD by old age matches or exceeds that seen today (e.g. Mays *et al.*, 1998). Reasons for the dearth of hip fractures are poorly understood, and it is likely that multiple factors are involved. One factor may simply be that too few individuals in the past reached the advanced old age (>75 years) at which hip fractures typically occur for the frequency of this type of injury to be appreciable. Studies on the Wharram Percy bones have suggested two additional factors that may be relevant. Study of bone quality suggested that bone microarchitecture may be conserved into old age even though BMD was not and that this may have had a protective effect (Agarwal *et al.*, 2004). The relative length of the femur neck was less at Wharram Percy than in a modern reference group (Chumley *et al.*, 2004); reduced femur neck length tends to have a protective effect with regard to hip fracture (Cummings *et al.*, 1994).

PAGET'S DISEASE OF BONE

PDB is a disorder characterized by excessive and abnormal bone remodelling and formation of bone which is structurally abnormal. It is a focal rather than generalized skeletal disorder. It is generally confined to middle-aged or elderly individuals and has a steeply increasing prevalence with age in the over 50s (Davie *et al.*, 1999; van Staa *et al.*, 2002). The primary cell affected in PDB is the osteoclast. In Pagetic lesions, osteoclasts are increased both in number and size and show increased numbers of nuclei (Roodman and Windle, 2005). Because of the elevated osteoclastic activity, Pagetic lesions are characterized initially by increased bone resorption. This provokes an osteoblastic response which results in rapid formation of chaotically organized new bone. After a variable period, osteoclastic activity slows so that the balance shifts in favour of bone formation and bone becomes sclerotic. Eventually, osteoclastic activity also returns to normal and the lesion becomes quiescent (Hamdy, 1981: 22–36; Resnick and Niwayama, 1988: 2130–2131).

The causes of PDB are unclear, but investigators have focused on both genetic and environmental factors. Support for a genetic component is compelling. Some 15–40% of individuals with PDB have a first-degree relative with the disease, and, reflecting this, PDB is 7–10 times as common in first-degree relatives of PDB patients than in controls (Daroszewska and Ralston, 2005). However, a solely genetic basis seems unlikely. It is unclear how a genetic mutation affecting all cells could produce focal disease, and it would seem difficult to account for the rapid changes in PDB frequency which have been observed recently in some countries. Among non-genetic factors, the most persuasive evidence seems to be for a viral agent, particularly the measles virus. Various workers (e.g. Mills *et al.*, 1984; Reddy *et al.*, 1995; Friedrichs *et al.*, 2002) have found evidence for measles virus nucleocapsid (protein coat and nucleic acids) inclusions in cells in Pagetic lesions. Introduction of measles virus nucleocapsid protein into osteoclasts in mice has been found to induce Pagetic-like histological and histomorphometric bone changes (Kurihara *et al.*, 2006). If a viral aetiology is implicated, then PDB must be a late manifestation of a persistent viral infection, since measles is generally contracted in the early years and PDB is a condition of the elderly. However, a viral role in PDB remains unproven, as other workers (e.g. Birch *et al.*, 1994; Helfrich *et al.*, 2000; Ooi *et al.*, 2000) have failed to find viral inclusions in cells in Pagetic lesions. The general consensus currently appears to favour action of environmental factor(s) upon a genetically susceptible host, but basic factors such as the focal nature of the disease and its late age at onset remain to be adequately explained.

PDB is generally more common in males, sex ratios between 1:1 and 1.6:1 being reported from large series (Schmorl, 1932; Pygot, 1957; Barker *et al.*, 1980; Cooper *et al.*, 1999; Davie *et al.*, 1999). There are pronounced geographical variations in prevalence. The disease is commonest in Britain, with a prevalence of about 5% in the over 55s (Detheridge *et al.*, 1982). Elsewhere in Europe, the prevalence is much lower (approximately 0.5–2%; Detheridge *et al.*, 1982). Outside Europe, countries intensively settled by the British, such as New Zealand and Australia, also show high prevalences (approximately 3–4.5%; Gardner *et al.*, 1978; Reasbeck *et al.*, 1983); a founder effect may account for this (Lucas *et al.*, 2005). The disease seems rare in Asia (Sridhar, 1994; Ishikawa *et al.*, 1996; Wang *et al.*, 2005), much of Africa (Bohrer, 1970), in Arab populations in the Middle East (Fouda, 1998) and among Native Americans (Lawrence, 1970; Mautalen *et al.*, 1994). Within Britain, there are marked regional differences in prevalence. There is an area of high prevalence (approximately 6–8.5%) in Lancashire (Barker *et al.*, 1980).

Direct data for PDB which allow temporal trends in prevalence to be investigated are only available for the last 30 years, but these show evidence for a decline in PDB in some countries which previously showed a high prevalence. From the mid 1970s to the mid 1990s, prevalence of PDB in Britain fell by 60%, with a reduction of prominence of the Lancashire focus (Cooper *et al.*, 1999). Recent work in New Zealand (Doyle *et al.*, 2002) reports a 50% reduction between 1983 and 2001. The downward trend does not seem to apply to other countries (e.g. Rapado *et al.*, 1999; Gennari *et al.*, 2005). Reasons for the decline are unclear; for example, vaccination against childhood measles was introduced too late to account for the change (if exposure to this agent is indeed important) (Doyle *et al.*, 2002).

Diagnosis in Palaeopathology

The appearance of Pagetic bone varies according to the phase of the disease. Initial lesions are lytic and osteoporotic. Later, bones are increased in width and in cortical thickness, and show patchy rarification and sclerosis; and later still, bones are sclerotic and heavy. Macroscopically, bone surfaces appear finely pitted, and in mid- and late-phase disease are thickened (Ortner, 2003: 435–443). About 10–35% of cases are monostotic and 65–90% polyostotic (Smith *et al.*, 2002). Any bone in the skeleton may be affected, but there is a predilection for the pelvic bones, lumbo-sacral spine and long-bones, particularly of the lower limb (Davie *et al.*, 1999).

Clinically, PDB is generally diagnosed radiographically (Walsh, 2004). Radiographic lesions are sharply circumscribed and, in long-bones, generally begin in subchondral bone and advance along the shaft, with a V-shaped boundary between Pagetic and normal bone (Figure 11.7). The patchy rarification and sclerosis of mid-phase disease produces a mottled appearance on X-ray and, together with increased width and cortical thickness, there is trabecular coarsening and loss of normal cortico-medullary distinction. In late-phase disease, the V-shaped demarcation between normal and abnormal bone in long-bones is lost and sclerosis predominates (Mirra *et al.*, 1995a,b; Smith *et al.*, 2002). The disorganized structure of Pagetic bone is also evident microscopically. There are irregular cement lines separating fragments of bone, resulting in chaotically organized microstructure, the so-called mosaic pattern (Figure 11.8; Chapter 7).

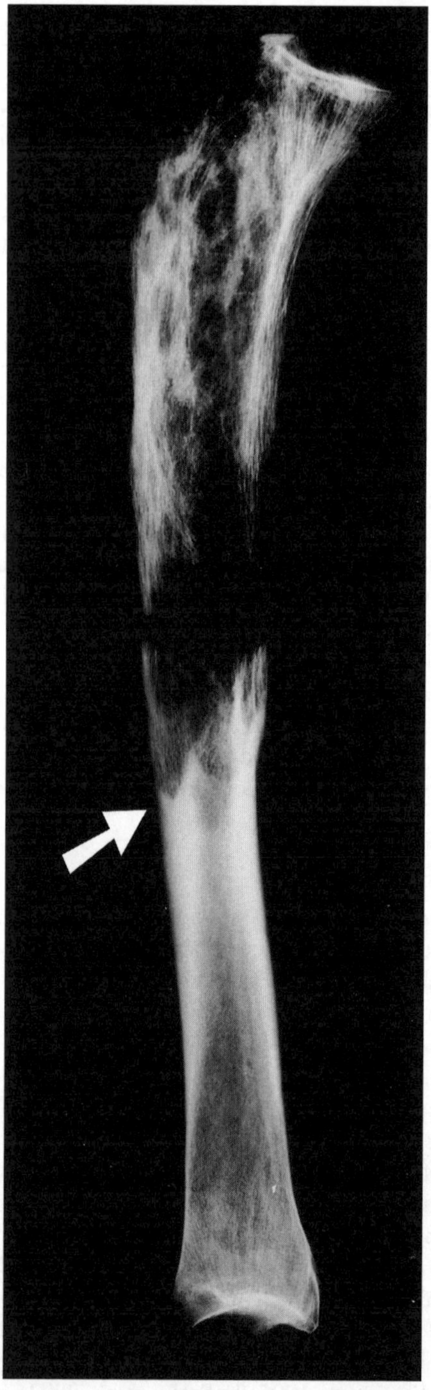

Figure 11.7 Lateral radiograph, left tibia, from an archaeological case of PDB. The diseased bone (proximal part of the tibia) has a mottled appearance and shows loss of cortico-medullary distinction. The demarcation between normal and abnormal bone (arrowed) is sharp, and the advancing front of diseased tissue is V-shaped

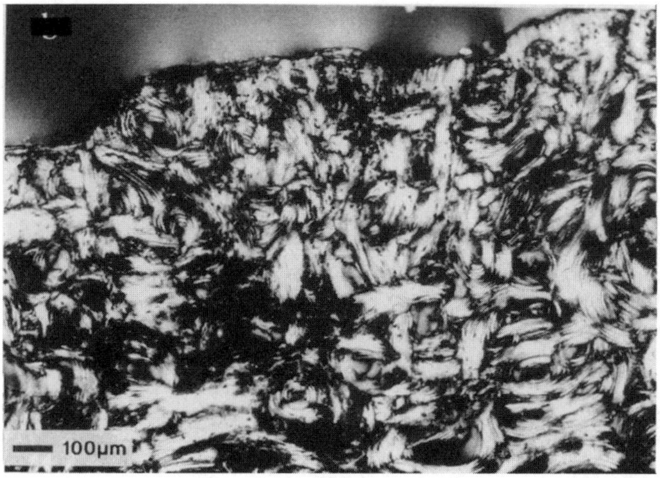

Figure 11.8 Histological view under polarized light of bone from the same skeleton as in Figure 11.7. The section shows a randomly orientated, fragmented lamellar pattern typical of PDB

Common complications of PDB include bone bowing, fracture and osteoarthritis (Hamdy, 1981: 47–56). The first two are frequent because, although in middle- and late-phase disease bone is increased in quantity, its disorganized nature means that it is structurally weak. Reasons for increased risk of osteoarthritis in Pagetic patients are less clear, but altered mechanical forces due to the weakened, distorted bone and release of substances by Pagetic bone cells promoting erosion of cartilage are likely causes (Hamdy, 1981: 55; Ankrom and Shapiro, 1998). Osteosarcoma, a rare but lethal complication, is found in less than 1% of cases (Davie *et al.*, 1999; van Staa *et al.*, 2002).

Paget's Disease of Bone in Palaeopopulations

Since PDB was only described by Sir James Paget in 1877, historical sources are of little value in investigating the disease in past populations. A sprinkling of palaeopathological cases, mostly of prehistoric date, have been reported, mainly in Europe and North America, since the late 19th century. Much of the older literature relies for diagnosis on gross appearance of specimens. This is problematic, as the pitted, thickened bones in PDB may resemble those in a variety of other conditions, particularly infectious disease. Clear radiographic views and/or microscopic demonstration of mosaic bone are needed to sustain a diagnosis. In addition, many early reports of PDB refer to isolated bones or bone fragments which, as well as presenting diagnostic challenges, are often from secondary contexts and, hence, difficult to date. Cook (1980) argues that few cases published up to 1980 stand up to rigorous scrutiny, and she indicates that there is little evidence for PDB in prehistoric times.

A review of internationally distributed periodicals was undertaken to locate palaeopathological cases of PDB. In addition, the international bibliography of palaeopathology (Tyson, 1997) was also consulted. This exercise produced 23 cases which, in the current writer's opinion, show convincing radiographic and/or microscopic alterations of PDB (Table 11.4). There are no prehistoric cases, and none from the New World. Of the total of 23 cases, 21

Table 11.4 Some palaeopathological cases of Paget's disease of bone

Location	Date	No. of cases, sex[a]	Complications?	Diagnostic method: Radiography	Diagnostic method: Microscopy	Reference
Jarrow, UK	10th century AD	1, M	Yes (f)	Yes	No	Wells and Woodhouse (1975)
Winchester, UK	10th–11th century AD	2, U	–	Yes	Yes	Price (1975)
Norwich, UK	10th–15th century AD	1, M	No	Yes	Yes	Stirland (1991); Bell and Jones (1991)
Wells, UK	10th century AD	1, M	No	Yes	Yes	Aaron *et al.* (1992)
Ipswich, UK	13th–16th century AD	1, M	Yes (f, oa)	Yes	Yes	Mays and Turner-Walker (1999)
Liseux, France	4th century AD	1, F	No	Yes	Yes	Roches *et al.* (2002)
St Pierre sur Dives, France	11th century AD	1, M	No	Yes	Yes	Roches *et al.* (2002)
Barton-upon-Humber, UK	10th–19th century AD	11, M; 3, F; 1, U	–	Yes	Yes	Rogers *et al.* (2002); Waldron, (2004)

[a]M: male; F: female; U: sex not reported.
[b]f: fracture, oa: osteoarthritis, –: no information.

come from Britain. In addition to those cited in Table 11.4, I am aware of about approximately 70 further British cases showing convincing (to the present writer) signs of PDB not published in international periodicals (chiefly, these come from published or unpublished osteological reports). Their dates range from the 4th century AD (e.g. Molleson, 1993: 196) to 19th century AD (e.g. Waldron, 1993: 81–82). These were not included in Table 11.4 so as not to bias results (it was impractical for the purposes of this article to attempt to review locally published and unpublished literature in countries besides the UK).

Currently available palaeopathological evidence would appear to be consistent with modern data in suggesting a concentration of cases in populations of British origin. It has been suggested, on the basis of geographic differences seen today, that PDB originated in Britain and was spread around the world by migration and admixture of British populations who were genetically susceptible to the disease (Cundy *et al.*, 1999). Current palaeopathological data do not contradict this model.

Most palaeopathological publications on PDB are case studies, but a few population studies have been conducted. Rogers and co-workers studied skeletons of 10th–19th century date from Barton-on-Humber, England (Rogers *et al.*, 2002; Waldron, 2004). The entry criteria for the study were that the skeleton should be at least 40% complete and that the individual should be over 35 years at death. Of 661 individuals meeting these criteria, 15 showed indications of PDB, which is a prevalence of 2.3%.

At Wharram Percy, less than 50 km north of Barton-on-Humber and of similar date, using the same entry criteria and diagnostic methods as Rogers *et al.* (2002), there were found to be no cases of PDB among 149 skeletons (Mays, 2007b). This pattern differs from that at Barton-on-Humber (Fischer's exact test: $p = 0.037$). That a difference should exist between two populations in such close proximity echoes the situation in Britain today whereby prevalence rates vary considerably, often between centres less than 100 km apart (Barker *et al.*, 1980).

Boylston and Ogden (2005) report six 12th–16th-century AD cases of PDB from Norton Priory, Cheshire, England, among 59 individuals over 35 years at death, a prevalence rate of 10.2%. Although methodological differences mean that the figures are not directly comparable, this seems to be a high prevalence, both in the light of the data from medieval Wharram Percy and Barton-on-Humber, but also with respect to modern data for Cheshire, which show a prevalence of about 4% in the over 55s (Barker *et al.*, 1980). The spatial location of the interments within the priory buildings suggested that at least some of those with PDB were probably members of a patronal family, so that familial aggregation might explain the high prevalence (Boylston and Ogden, 2005).

Pusch and Czarnetzki (2005) examined >8500 skeletons from central Europe dating from 3600 BC to AD 1500, using the same entry criteria into the study as Rogers *et al.* (2002). Although the work has yet to be fully published, they report a prevalence of 0.03 %. No case predated AD 1400. Although comparison with Rogers *et al.*'s (2002) UK results needs to be made with caution (Waldron, 2005), the German prevalence does seem lower than at Barton-on-Humber, the largest palaeopathological study of PDB in England to date.

CONCLUSIONS

Regarding the palaeopathological identification of vitamin C and vitamin D deficiency, the development of diagnostic criteria for infantile scurvy and rickets by Ortner and co-workers has been an important recent methodological development. Most of the features these writers describe are not on their own diagnostic; identification of these conditions relies on the co-occurrence of alterations at multiple locations in the skeleton. The fragmentary and incomplete nature of most archaeological infant skeletons, therefore, renders recognition of infantile scurvy or rickets difficult in many assemblages. This difficulty is compounded by the fact that some of the changes (for example, porosity/rugosity of the bone beneath the growth plates in rickets) require first-class bone preservation for their identification.

Care is required in differential diagnosis of rickets, scurvy and other conditions in infant remains. For example, rickets, scurvy and anaemia may each cause porotic lesions in the orbital roofs and cranial vault. However, in each case the final outcome of bony porosity is a result of a different process and this results in subtly different morphology. In anaemia, the surface porosity is a result of marrow hyperplasia, visible in broken sections or radiographically (Ortner, 2003: 370f). In scurvy, lesions consist of new bone deposited upon an underlying normal cortical surface or else small pores in an otherwise normal cortical surface; there is no marrow hyperplasia. In rickets, bone surfaces may be rather spicular, and superficial pores are rather larger than in scurvy and represent voids as a result of imperfect mineralization of a growing surface rather than transmitting blood vessels. Careful evaluation of morphology of lesions and of the distribution of pathological changes in the skull and post-cranial skeleton often help to advance one diagnosis at the expense of others. However, a firm diagnosis may not always be possible. A further complication is that rickets, scurvy and anaemia may often co-occur in the same individual (Mankin, 1974b; Stuart-Macadam, 1989).

Relatively few cases of vitamin C and vitamin D deficiency disease have been documented palaeopathologically. These may genuinely have been fairly rare conditions during most of the past, but the paucity of palaeopathological cases may also reflect their underrecognition, particularly prior to Ortner and co-wokers' work on infantile scurvy and rickets. Similarly, the subtle signs of osteomalacia in adult skeletons may have been overlooked. Recent work has shown that sporadic cases of vitamin D deficiency may be expected even in communities with largely outdoor lifestyles, as it may occur when, for example, cultural norms result in deliberate avoidance of exposure of skin to sunlight (Littleton, 1998), or when the chronically sick are confined indoors (Ortner and Mays, 1998). Cases of infantile scurvy may be found even in environments where fresh fruit and vegetable foods are available year round (Ortner *et al.*, 2001), perhaps reflecting their avoidance in weaning diets or the adoption of food preparation techniques which destroy vitamin C. This serves to emphasize the role of cultural factors in the expression of vitamin C and D deficiency diseases.

In palaeopathology, most biocultural population studies rely on study of disease prevalence rates. This is no less true for vitamin C and vitamin D deficiency than for other conditions. However, recent work on rickets (Mays *et al.*, 2006a) suggests that a more nuanced insight into biocultural aspects of disease may be obtained by taking into account the severity of the changes due to the direct effect of vitamin D deficiency on growing bone. Ortner *et al.* (2001) suggest that, for infantile scurvy, reactive bone formation may be a more severe response than porotic changes alone. Although further work is required to clarify matters, this may potentially provide an avenue to assess the severity as well as the frequency of vitamin C deficiency disease in palaeopopulations.

Much biocultural work on the metabolic diseases will naturally be driven by archaeological questions, such as the effect of major transitions in human society – such as the adoption of agriculture and urbanization – on disease expression. However, for scurvy and rickets, the rich documentary evidence for these conditions may also provide a framework for problem-driven palaeopathological work. For example, documentary sources imply that infantile scurvy was rare prior to the late 19th century, and that rickets in the 17th century was common in rural areas and was a disease of the wealthier classes. Work has only just begun toward evaluating whether temporal, social and geographical patterning in these diseases implied by written sources is supported by archaeological evidence.

Turning to osteoporosis and PDB, attention needs to be paid to rigorous diagnosis. Historically, many dubious cases of PDB have entered the palaeopathological literature, and have been cited as though they were good data in medico-historical reviews (e.g. Hamdy, 1981). The radiographic and microscopic appearance of PDB is pathognomonic, so palaeopathological diagnoses should be supported by radiographic and/or microscopic examination. For osteoporosis, in order to obtain a fuller picture of the disease in a palaeopopulation, indicators of bone quantity, bone quality and osteoporotic fracture need to be used in combination.

An important methodological problem in the study of age-progressive conditions, such as osteoporosis and PDB, is the limitations in present techniques for estimating skeletal age at death, particularly for older adults. Current methodologies do not permit individuals past middle age to be more precisely aged than that they are older than about 50 years (Mays, 1998: 49–66). Within this age group the prevalence of both PDB and osteoporosis increases markedly with age. If the age structure of the >50 age class in a study group differs from that in a reference population with which it is compared then this will seriously prejudice the comparison of age-related bone changes in osteoporosis or in the prevalence of PDB. For populations from the historic periods, previous approaches have been to model the age composition of the >50 age group using historical demographic sources (e.g. Mays, 1996, 2006). Clearly this is only a partial solution, and suitable historical documentation concerning mortality patterns is frequently absent or inadequate.

In the study of osteoporosis, palaeopathologists have predominantly discussed population differences in peak bone mass and age-related bone loss in terms of lifestyle factors. In recognition of the important genetic component, writers (e.g. Perzigian, 1973; Mays, 2006) have occasionally suggested that genetic factors may account for differences in bone loss in past populations, but usually only as a last resort when credible environmental explanations have been exhausted. Recently, polymorphisms in the vitamin D receptor and collagen type I genes have been implicated as affecting rates of bone loss in osteoporosis (Liu *et al.*, 2003). Potentially, these polymorphisms could be identified in palaeopopulations using ancient DNA. This would enable genetic hypotheses to be tested. For PDB, the SQSTM1 gene may be important. This encodes a protein known as p62, which plays an important role in osteoclast function. Mutations in this gene are found in 10–50% of PDB cases (Daroszewska and Ralston, 2005). Recent work (Lucas *et al.*, 2005) suggests that a founder-effect of a SQSTM1 mutation may account for the high prevalence of PDB in Australian and New Zealand patients of British descent. Study of the SQSTM1 gene could potentially be carried out using ancient DNA, and this might help explain clusters of high prevalence, such as at medieval Norton Priory (Boylston and Ogden, 2005).

The historical record for osteoporosis and PDB is meagre. For these diseases, modern epidemiological data rather than historical accounts are a source of hypotheses which might be tested using skeletal remains. One key research aim is to investigate the severity of osteoporosis and PDB in past populations compared with today to help assess the likely role of modern lifestyle factors in these conditions. Another avenue, which has to date been less explored, is to determine the degree to which modern geographical patterns in these diseases have arisen recently or else have their origins in more ancient times. Interpopulation BMD differences today can be quite marked: for example, in European populations the mean femur neck BMD in the population with the lowest values is only about 86 % of that in the group with the highest (Lunt *et al.*, 1997). Differences of this degree are potentially detectable, not only with the very large sample sizes of modern epidemiological studies, but also with the more modest samples customarily available to paleopathologists. Similarly, interpopulation differences in PDB are quite marked and, hence, their existence in past populations would be amenable to palaeopathological analysis; however, given the characteristic low absolute prevalence of PDB, large skeletal series are needed.

ACKNOWLEDGEMENTS

Thanks are due to Megan Brickley for permission to reproduce Figures 11.2 and 11.3, originally published in Brickley *et al.* (2006), *American Journal of Physical Anthropology*, volume 132, and Figure 11.4; and to Gordon Turner-Walker for permission to reproduce Figure 11.8, which originally appeared in Mays and Turner-Walker (1999), *Journal of Palaeopathology*, volume 11. Thanks are due to Bill White for making available his osteological data on the Broadgate and St Benet Sherehog remains, in advance of their full publication. Thanks are owed to the reviewers for their helpful comments which have improved this article; needless to say, any infelicities which remain are solely the author's responsibility.

REFERENCES

Aaron JE, Rogers J, Kanis JA. 1992. Paleohistology of Paget's disease in two medieval skeletons. *Am J Phys Anthropol* **89**: 325–331.

Adams JE. 1997. Radiology of rickets and osteomalacia. In *Vitamin D*, Feldman D, Glorieux F, Pike J (eds). Academic Press: New York; 619–642.

Adams P, Berridge SR. 1969. Effects of kwashiorkor on cortical and trabecular bone. *Arch Dis Child* 44: 705–709.

Agarwal SC, Grynpas MD. 1996. Bone quantity and quality in past populations. *Anat Rec* **246**: 423–432.

Agarwal SC, Dumitriu M, Tomlinson GA, Grynpas MD. 2004. Medieval trabecular bone architecture: the influence of age, sex, and lifestyle. *Am J Phys Anthropol* **124**: 33–44.

Albright F, Riefenstein E. 1948. *The Parathyroid Glands and Metabolic Bone Disease*. Williams and Wilkins: Baltimore, MD.

Anderson FH, Francis RM, Selby PL, Cooper C. 1998. Sex hormones and osteoporosis in men. *Calcif Tiss Int* **62**: 185–188.

Angel JL, Kelley JO, Parrington M, Pinter S. 1987. Life stresses of the free black community as represented by the First African Baptist church, Philadelphia, 1823–1841. *Am J Phys Anthropol* **74**: 213–229.

Ankrom MA, Shapiro JR. 1998. Paget's disease of bone (osteitis deformans). *J Am Geriatr Soc* **46**: 1025–1033.

Armelagos GJ, Mielke JH, Owen KH, van Gerven DP, Dewey JR, Mahler PE. 1972. Bone growth and development in prehistoric populations from Sudanese Nubia. *J Hum Evol* **1**: 89–119.

Aufderheide AC, Rodríguez-Martin RM. 1998. *The Cambridge Encyclopaedia of Human Paleopathology*. Cambridge University Press: Cambridge.

Barker DJP, Chamberlain AT, Guyer PB, Gardiner MJ. 1980. Paget's disease of bone: the Lancashire focus. *Br Med J* **280**: 1105–1107.

Barr DGD, Shmerling DH, Prader A. 1972. Catch-up growth in malnutrition studied in celiac disease after institution of a gluten-free diet. *Pediatr Res* **6**: 521–527.

Beck SV. 1997. Scurvy: citrus and sailors. In *Plague, Pox and Pestilence: Disease in History*, Kiple KF (ed.). Weidenfeld and Nicholson: London; 68–73.

Bell LS, Jones SJ. 1991. Macroscopic and microscopic evaluation of archaeological pathological bone: backscattered electron imaging of putative Pagetic bone. *Int J Osteoarchaeol* **1**: 179–184.

Benedictow OJ. 1992. On the origin and spread of the notion that breast-feeding women should abstain from sexual intercourse. *Scand J Hist* **17**: 65–76.

Bennike P. 1985. *Paleopathology of Danish Skeletons*. Akademisk Forlag: Copenhagen.

Blondiaux G, Blondiaux J, Secousse F, Cotten A, Danze P-M, Flipo R-M. 2002. Rickets and child abuse: the case of a two year old girl from the 4th century in Lisieux (Normandy). *Int J Osteoarchaeol* **12**: 209–215.

Bohrer SP. 1970. Osteitis deformans in Nigerians. *Afr J Med Sci* **1**: 109–113.

Boylston A, Ogden A. 2005. A study of Paget's disease at Norton Priory, Cheshire. A medieval religious house. In *Proceedings of the Fifth Annual Conference of the British Association for Biological Anthropology and Osteoarchaeology*, Zakrzewski SR, Clegg M (eds). British Archaeological Report International Series 1383. Archaeopress: Oxford; 69–76.

Birch MA, Taylor W, Fraser WD, Ralston SH, Hart CA, Gallagher JA. 1994. Absence of paramyxovirus RNA in cultures of Pagetic bone cells and in Pagetic bone. *J Bone Miner Res* **9**: 11–16.

Brickley M, Agarwal S. 2003. Techniques for investigation of age-related bone loss and osteoporosis in archaeological bone. In *Bone Loss and Osteoporosis, an Anthropological Perspective*, Agarwal SC, Stout SD (eds). Kluwer Academic/Plenum: London; 157–172.

Brickley M, Howell PGT. 1999. Measurement of changes in trabecular bone structure with age in an archaeological population. *J Archaeol Sci* **26**: 151–157.

Brickley M, Ives R. 2006. Skeletal manifestations of infantile scurvy. *Am J Phys Anthropol* **129**: 163–172.

Brickley M, Waldron T. 1998. Relationship between bone density and osteoarthritis in a skeletal population from London. *Bone* **22**: 279–283.

Brickley M, Mays S, Ives R. 2005. Skeletal manifestations of vitamin D deficiency osteomalacia in documented historical collections. *Int J Osteoarchaeol* **15**: 389–403.

Brickley M, Berry H, Weston G, Hancocks A, Richards M. 2006. The people: physical anthropology. In *St Martin's Uncovered*, Brickley M, Buteux S (eds). Oxbow Books: Oxford; 90–151.

Brickley M, Mays S, Ives R. 2007. An investigation of skeletal indicators of vitamin D deficiency in adults: effective markers for interpreting past living conditions and pollution levels in 18th and 19th century Birmingham, England. *Am J Phys Anthropol* **132**: 67–79.

Brimblecombe P. 1982. Early urban climate and atmosphere. In *Environmental Archaeology in the Urban Context*, Hall AR, Kenward HK (eds). CBA Research Report 43. Council for British Archaeology: London; 10–25.

Carli-Thiele P. 1996. *Spuren von Mangelerkrankungen an steinzeitlichen Kinderskeleten*. Fortschritte in der Paläopathologie und Osteoarchäologie; Bd. 1. Erich Goltze: Göttingen.

Carpenter KJ. 1986. *The History of Scurvy and Vitamin C*. Cambridge University Press: Cambridge.

Christodoulou C, Cooper C. 2003. What is osteoporosis? *Postgrad Med J* **79**: 133–138.

Chumley A, Petley G, Edwards CJ, Taylor P, Mays S, Sofaer-Derevenski J, Mahon P, Cooper C, Arden NK. 2004. Changes in hip geometry from medieval to modern times. *J Bone Miner Res* **19**: S86.

Clarke E. 1962. Whistler and Glisson on rickets. *Bull Hist Med* **36**: 45–61.

Clements MR. 1989. The problem of rickets in UK Asians. *J Hum Nutr Dietet* **2**: 105–116.

Connell B, Mays S. 1997. Scurvy among the juvenile skeletal remains from Wharram Percy. Unpublished manuscript held on file at English Heritage, Portsmouth.

Cook DC. 1980. Paget's disease and treponematosis in prehistoric Midwestern Indians: the case for misdiagnosis. *Ossa* **7**: 41–62.

Cooper C, Schafheutle K, Dennison E, Kellingray S, Guyer P, Barker D. 1999. The epidemiology of Paget's disease in Britain: is the prevalence decreasing? *J Bone Miner Res* **14**: 192–197.

Coughlin J, Holst M. 2000. Health status. In *Blood Red Roses: The Archaeology of a Mass Grave from the Battle of Towton AD1461*, Fiorato V, Boylston A, Knüsel K (eds). Oxbow: Oxford; 60–76.

Cummings SR, Cauley JA, Palermo L, Ross PD, Wasnich RD, Black D, Faulkner KG. 1994. Racial differences in hip axis lengths might explain racial differences in rates of hip fracture. *Osteoporos Int* **4**: 226–229.

Cundy T, Wattie D, Busch S, Rutland M, Ibbertson HK. 1999. Paget's disease in New Zealand: is it changing. *Bone* **24**(S5): S7–S9.

Cunha E, Silva AM. 1997. War lesions from the famous Portuguese medieval battle of Aljubarrota. *Int J Osteoarchaeol* **7**: 595–599.

Daroszewska A, Ralston SH. 2005. Genetics of Paget's disease of bone. *Clin Sci* **109**: 257–263.

Davie M, Davies M, Francis R, Fraser W, Hosking D, Tansley R. 1999. Paget's disease of bone: a review of 889 patients. *Bone* **24**(S5): S11–S12.

Detheridge FM, Guyer PB, Barker DJP. 1982. European distribution of Paget's disease of bone. *Br Med J* **285**: 1005–1008.

Doyle T, Gunn J, Anderson G, Gill M, Cundy T. 2002. Paget's disease in New Zealand: evidence for declining prevalence *Bone* **31**: 616–619.

Drezner MK. 1997. Osteoporosis: a disease of the age(d)s. *J Bone Miner Res* **12**: 880.

During EM. 1997. The skeletal remains from the Swedish man-of-war *Vasa* – a survey. *Homo* **48**: 135–160.

Ekenman I, Eriksson SAV, Lindgren JU. 1995. Bone density in Mediaeval skeletons. *Calcif Tiss Int* **56**: 355–358.

Eriksen MF. 1976. Cortical bone loss with age in three Native American populations. *Am J Phys Anthropol* **45**: 443–452.

Etxeberria F. 1999. Surgery in the Spanish War of Independence (1807–1819), between Desault and Lister. *J Palaeopathol* **11**(3): 25–40.

Foudha M. 1998. Paget's disease of bone. A decade since the first reported case. *Saudi Med J* **19**: 205–208.

Friedrichs WE, Reddy SV, Buder JM, Cundy T, Cornish J, Singer FR, Roodman GD. 2002. Sequence analysis of measles virus nucleocapsid transcripts in patients in Paget's disease. *J Bone Miner Res* **17**: 145–151.

Garn SM. 1970. *The Earlier Gain and Later Loss of Cortical Bone*. Charles C Thomas: Springfield.

Garn SM, Guzman MA, Wagner B. 1969. Subperiosteal gain and endosteal loss in protein-calorie malnutrition. *Am J Phys Anthropol* **30**: 153–156.

Gardner MJ, Guyer PB, Barker DJP. 1978. Radiological prevalence of Paget's disease of bone in British migrants to Australia. *Br Med J* **1**: 1655–1657.

Gass M, Dawson-Hughes B. 2006. Preventing osteoporosis-related fractures: an overview. *Am J Med* **119**(4A): S3–S11.

Gennari L, DiStefano M, Merlotti N, Martini G, Tamone C, Zatteri R, De Lucchi R, Baldi C, Vattimo A, Capoccia S, Burroni L, Geraci S, De Paola V, Calabro A, Avanzati A, Isaia G, Nuti R. 2005. Prevalence of Paget's disease of bone in Italy. *J Bone Miner Res* **20**: 1845–1850.

Gibbs D. 1994. Rickets and the crippled child: an historical perspective. *J R Soc Med* **87**: 729–732.

Ginsberg E, Škarić-Jurić T, Kobyliansky E, Malkin I, Rudan P. 2001. Evidence of major gene control of cortical index in pedigree data from Middle Dalmatia, Croatia. *Am J Hum Biol* **13**: 398–408.

Grynpas MD. 2003. The role of bone quality on bone loss and bone fragility. In *Bone Loss and Osteoporosis, an Anthropological Perspective*, Agarwal SC, Stout SD (eds). Kluwer Academic/Plenum: London; 33–45.

Gunness-Hey M. 1986. Age-related bone loss in a prehistoric Koniag Eskimo population. *Ossa* **12**: 41–47.

Hamdy RC. 1981. *Paget's Disease of Bone*. Armour Pharmaceutical: Eastbourne.

Hammerl J, Protsch R, Happ J, Frohn J, Hor G. 1990. Osteodensitometrie des femurhalses an historischen skeletten. In *Osteologie-Interdisziplinar*, Werner R, Mattiass HH (eds). Springer: Berlin; 139–142.

Helfrich MH, Hobson RP, Grabowski PS, Zurbriggen A, Cosby SL, Dickson GR, Fraser WD, Ooi CG, Selby PL, Crisp AJ, Wallace RGH, Kahn S, Ralston SH. 2000. A negative search for paramyxoviral etiology of Paget's disease of bone: molecular, immunological, and ultrastructural studies in UK patients. *J Bone Miner Res* **15**: 2315–2329.

Henry HL, Norman AW. 1992. Metabolism of vitamin D. In *Disorders of Bone and Mineral Metabolism*, Coe FL, Favus MJ (eds). Raven Press: New York; 149–162.

Hernigou P. 1995. Historical overview of rickets, osteomalacia and vitamin D. *Rev Rhum [Engl Ed]* **62**: 261–270.

Hess AF. 1930. *Rickets. Including Osteomalacia and Tetany*. Kimpton: London.

Hillson S. 1986. *Teeth*. Cambridge University Press: Cambridge.

Himes JH, Matorell R, Habicht J-P, Yarbrough C, Malina RM, Klein RE. 1975. Patterns of cortical bone growth in moderately malnourished preschool children. *Hum Biol* **47**: 337–350.

Horácková L, Vargová L. 1999. Bone remains from the common grave pit from the Battle of Austerlitz (anthropology and palaeopathology). *J Paleopathol* **11**(3): 5–13.

Ingelmark B. 1939. The skeletons. In *Armour from the Battle of Wisby, 1361*, Thordemann B (ed.). Vitterhets Historie och Antikvitets Akademien: Stockholm; 149–205.

Ishikawa Y, Tsukama H, Miller RW. 1996. Low rates of Paget's disease of bone and osteosarcoma in elderly Japanese. *Lancet* **347**: 1559.

Jackson R. 1988. *Doctors and Diseases in the Roman Empire*. British Museum Press: London.

Kalkwarf HJ, Specker BL. 1995. Bone mineral loss during lactation and recovery after weaning. *Obstet Gynecol* **86**: 26–32.

Kanis JA. 1998. *Pathophysiology and Treatment of Paget's Disease of Bone* (2nd edition). Martin Dunitz: London.

Kanis JA, Johnell O, De Laet C, Jonsson B, Oden A, Ogelsby AK. 2002. International variations in hip fracture probabilities: implications for risk assessment. *J Bone Miner Res* **17**: 1237–1244.

Kneissel M, Boyde A, Hahn M, Teschler-Nicola M, Kalchhauser G, Plenk H. 1994. Age- and sex-dependent cancellous bone changes in a 4000y BP population. *Bone* **15**: 539–545.

Kneissel M, Roschger P, Steiner W, Schamell D, Kalchhauser G, Boyde A, Teschler-Nicola M. 1997. Cancellous bone structure in the growing and ageing lumbar spine in a historic Nubian population. *Calcif Tiss Int* **61**: 95–100.

Kohlmeier L, Marcus R. 1995. Calcium disorders of pregnancy. *Endocrinol Metabol Clin N Am* **24**: 15–39.

Kurihara N, Zhou H, Reddy SV, Palacios VG, Subler MA, Dempster DW, Windle JJ, Roodman GD. 2006. Expression of measles virus nucleocapsid protein in osteoclasts induces Paget's disease-like bone lesions in mice. *J Bone Miner Res* **21**: 446–455.

Laskey MA, Prentice A. 1999. Bone mineral changes during and after lactation. *Obstet Gynecol* **94**: 608–615.

Laughlin WS, Harper AB, Thompson DD. 1979. New approaches to the pre- and post-contact history of Arctic peoples. *Am J Phys Anthropol* **51**: 579–587.

Lawrence JS. 1970. Paget's disease in population samples. *Ann Rheum Dis* **29**: 562.

Lazenby RA. 1997. Bone loss, traditional diet, and bone adaptation in Arctic populations. *Am J Hum Biol* **9**: 329–341.

Lees B, Molleson T, Arnett TR, Stevenson JC. 1993. Differences in proximal femur bone density over two centuries. *Lancet* **341**: 673–675.

Littleton J. 1998. A Middle Eastern paradox: rickets in skeletons from Bahrain. *J Paleopathol* **10**: 13–30.

Liu Y-Z, Liu Y-J, Recker RR, Deng H-W. 2003. Molecular studies of identification of genes for osteoporosis: the 2002 update. *J Endocrinol* **177**: 147–196.

Lomax E. 1986. Difficulties in diagnosing infantile scurvy before 1878. *Med Hist* **30**: 70–80.

Loomis WF. 1970. Rickets. *Sci Am* **223**: 79–91.

Lucas GJA, Hocking LJ, Darozewska A, Cundy T, Nicholson GC, Walsh JP, Fraser WD, Meier C, Hooper MJ, Ralston SH. 2005. Ubiquitin-associated domain mutations of SQSTM1 in Paget's disease of bone: evidence for a founder effect in patients of British descent. *J Bone Miner Res* **20**: 227–231.

Lunardini A, Rollo F, Fornaciari G. 2002. Bone lesions from the ossuary of the Napoleonic Battle of Marengo, northern Italy (14th June 1800). *J Paleopathol* **14**: 69–75.

Lunt M, Felsenberg D, Adams J, Benevolenskaya L, Cannata J, Dequeker J, Dodenhof C, Falch JA, Johnell O, Khaw K-T, Masaryk P, Pols H, Poor G, Reid D, Scheidt-Nave C, Weber K, Silman AJ, Reeve J. 1997. Population-based geographic variations in DXA bone density in Europe: the EVOS study. *Osteoporos Int* **7**: 175–189.

Maat GJR. 1982. Scurvy in Dutch whalers buried at Spitsbergen. In *Proceedings of the Paleopathology Association 4th European Meeting*, Middleburg–Antwerp, Haneveld GT, Perizonius WRK, Janssens PJ (eds). Paleopathology Association: Utrecht; 82–93.

Maat GJR. 2004. Scurvy in adults and youngsters: the Dutch experience. A review of the history and pathology of a disregarded disease. *Int J Osteoarchaeol* **14**: 77–81.

Maat GJR, Uytterschaut HT. 1984. Microscopic observations on scurvy in Dutch whalers buried at Spitsbergen. In *Proceedings of the Paleopathology Association 5th European Meeting*, Siena Capecci V, Rabino-Massa E (eds). Paleopathology Association: Siena; 211–218.

Macho GA, Abel RL, Schutkowski H. 2005. Age changes in bone microstructure: do they occur uniformly? *Int J Osteoarchaeol* **15**: 421–430.

Maiyegun SO, Malek AH, Devarajan LV, Dahniya MH. 2002. Severe congenital rickets secondary to maternal hypovitaminosis D: a case report. *Ann Trop Paediatr* **22**: 191–195.

Mankin H. 1974a. Rickets, osteomalacia and renal osteodystrophy. Part II. *J Bone Joint Surg Am* **56**: 352–386.

Mankin H. 1974b. Rickets, osteomalacia and renal osteodystrophy. Part I. *J Bone Joint Surg Am* **56**: 101–128.

Mautalen C, Pumarino H, Blanco MC, González D, Ghiringhelli G, Fromm G. 1994. Paget's disease: the South American experience. *Semin Arthitis Rheum* **23**: 226–227.

Mays S. 1996. Age-dependent cortical bone loss in a medieval population. *Int J Osteoarchaeol* **6**: 144–154.

Mays S. 1998. *The Archaeology of Human Bones*. Routledge: London.

Mays S. 2000. Age-dependent cortical bone loss in women from 18th and early 19th century London. *Am J Phys Anthropol* **112**: 349–361.

Mays S. 2001. Effects of age and occupation of cortical bone in a group of 18th–19th century British men. *Am J Phys Anthropol* **116**: 34–44.

Mays S. 2003. The rise and fall of rickets in England. In *The Environmental Archaeology of Industry*, Murphy P, Wiltshire PEJ (eds). Oxbow: Oxford; 144–153.

Mays S. 2006. Age-related cortical bone loss in women from a 3rd–4th century AD population from England. *Am J Phys Anthropol* **129**: 518–528.

Mays S. 2007a. A likely case of scurvy from early Bronze Age Britain. *Int J Osteoarchaeol* in press.

Mays S. 2007b. The human remains. In *Wharram: A Study of Settlement on the Yorkshire Wolds, XI. The Churchyard*, Mays S, Harding C, Heighway C (eds). York University Archaeological Publications 13. University of York: York; 77–192, 337–397.

Mays S, Turner-Walker G. 1999. A mediaeval case of Paget's disease of bone with complications. *J Paleopathol* **11**(1): 29–40.

Mays S, Lees B, Stevenson JC. 1998. Age-dependent cortical bone loss in the femur in a medieval population. *Int J Osteoarchaeol* **8**: 97–106.

Mays S, Brickley M, Ives R. 2006a. Skeletal manifestations of rickets in infants and young children in an historic population from England. *Am J Phys Anthropol* **129**: 362–374.

Mays S, Turner-Walker G, Syversen U. 2006b. Osteoporosis in a population from mediaeval Norway. *Am J Phys Anthropol* **131**: 343–351.

Mays S, Brickley M, Ives R. 2007. Skeletal evidence for hyperparathyroidism in a 19th century child with rickets. *Int J Osteoarchaeol* **17**: 73–81.

Mays S, Brickley M, Ives R. Under review. Growth and vitamin D deficiency in a population from 19th century Birmingham. *Int J Osteoarchaeol*.

Mazess RB, Barden HS, Christiansen C, Harper AB, Laughlin WS. 1985. Bone mineral and vitamin D in Aleutian Islanders. *Am J Clin Nutr* **42**: 143–146.

McEwan J, Mays S, Blake G. 2004. Measurements of bone mineral density of the radius in a mediaeval population. *Calcif Tiss Int* **74**: 157–161.

Melikian M, Waldron T. 2003. An examination of skulls from two British sites for possible evidence of scurvy. *Int J Osteoarchaeol* **13**: 207–212.

Meyer C. 2003. Osteological evidence for the Battles of Zürich, 1799: a glimpse into the soldiery of the past. *Int J Osteoarchaeol* **13**: 252–257.

Mills BG, Singer FR, Weiner LP, Suffin SC, Stabile E, Holst P. 1984. Evidence for both respiratory syncytial virus and measles virus antigens in the osteoclasts of patients with Paget's disease of bone. *Clin Orthop* **183**: 303–311.

Mirra JM, Brien EW, Tehranzadeh J. 1995a. Paget's disease of bone: review with emphasis on radiologic features. Part I. *Skel Radiol* **24**: 163–171.

Mirra JM, Brien EW, Tehranzadeh J. 1995b. Paget's disease of bone: review with emphasis on radiologic features. Part II. *Skel Radiol* **24**: 173–184.

Molleson T. 1993. The human bones. In *Excavations at Poundbury 1966–1980. Volume II: The Cemeteries*, Farwell DE, Molleson TI (eds). Dorset Natural History and Archaeological Society: Dorchester.

Molleson T, Cox M. 1993. *The Spitalfields Project. Volume 2: The Anthropology*. CBA Research Report 86. Council for British Archaeology: York.

Moyad MA. 2003. Osteoporosis: a rapid review of risk factors and screening methods. *Urol Oncol* **21**: 375–379.

Nemeskéri J, Harsanyi L, Acsàdi G. 1960. Methoden zur diagnose der lebensalters von skelettfunden. *Anthropol Anz* **24**: 70–95.

Ooi CG, Walsh CA, Gallagher JA, Fraser WD. 2000. Absence of measles virus and canine distemper virus transcripts in long-term bone marrow cultures from patients with Paget's disease of bone. *Bone* **27**: 417–421.

Ortner DJ. 2003. *Identification of Pathological Conditions in Human Skeletal Remains* (2nd edition). Academic Press: New York.

Ortner DJ, Eriksen MF. 1997. Bone changes in the human skull probably resulting from scurvy in infancy and childhood. *Int J Osteoarchaeol* **7**: 212–220.

Ortner DJ, Mays S. 1998. Dry-bone manifestations of rickets in infancy and early childhood. *Int J Osteoarchaeol* **8**: 45–55.

Ortner DJ, Kimmerle EH, Diez M. 1999. Probable evidence of scurvy in subadults from archaeological sites in Peru. *Am J Phys Anthropol* **108**: 321–331.

Ortner DJ, Butler W, Cafarella J, Milligan L. 2001. Evidence of probable scurvy in subadults from archaeological sites in North America. *Am J Phys Anthropol* **114**: 343–351.

Owen I. 1889. Geographical distribution of rickets, acute and subacute rheumatism, chorea, cancer and urinary calculus in the British Islands. *Br Med J* **1**: 113–16.

Perzigian AJ. 1973. Osteoporotic bone loss in two prehistoric Indian populations. *Am J Phys Anthropol* **39**: 87–96.

Pfeiffer S, Crowder C. 2004. An ill child among mid-Holocene foragers of southern Africa. *Am J Phys Anthropol* **123**: 23–29.

Pimentel L. 2003. Scurvy: historical review and current diagnostic approach. *Am J Emerg Med* **21**: 328–332.

Pinhasi R, Shaw P, White B, Ogden AR. 2006. Morbidity, rickets and long-bone growth in post-medieval Britain – a cross-population analysis. *Ann Hum Biol* **33**: 372–389.

Poulsen LW, Qvesel D, Brixen K, Versterby A, Boldsen JL. 2001. Low bone mineral density in the femoral neck of medieval women: a result of multiparity. *Bone* **28**: 454–458.

Prentice A. 2001. The relative contribution of diet and genotype to bone development. *Proc Nutr Soc* **60**: 45–52.

Price JL. 1975. The radiology of Saxon and mediaeval human remains from Winchester. *Clin Radiol* **26**: 363–370.

Pusch CM, Czarnetzki A. 2005. Archaeology and the prevalence of Paget's disease. *J Bone Miner Res* **20**: 1484.

Pygott F. 1957. Paget's disease of bone: the radiological incidence. *Lancet* **1**: 1170–1171.

Rajakumar K. 2001. Infantile scurvy: a historical perspective. *Pediatrics* **108**: e76.

Rapado A, Jiménez J, Morales A, Carbonell J, Del Pino J. 1999. Patterns of diagnosis of Paget's disease in Spain. *J Bone Miner Res* **14**(S2): S96–S98.

Rapuri PB, Gallagher JC, Knezetic JA, Kinyamu HK, Ryschon KL. 2004. Association between vitamin D receptor polymorphisms and the rate of bone loss in elderly women – importance of adjusting for dietary and lifestyle factors. *J Steroid Biochem Mol Biol* **89–90**: 503–506.

Reasbeck JC, Goulding A, Campbell DR, Beale LR, Stewart RDH. 1983. Radiological prevalence of Paget's disease in Dunedin, New Zealand. *Br Med J* **286**: 1937.

Reddy SV, Singer FR, Roodman GD. 1995. Bone marrow mononuclear cells from patients with Paget's disease contain measles virus nucleocapsid messenger ribonucleic acid that has mutations in a specific region of the sequence. *J Clin Endocrinol Metabol* **80**: 2108–2111.

Resnick D, Niwayama G. 1988. *Diagnosis of Bone and Joint Disorders* (2nd edn). WB Saunders: London.

Rewekant A. 1994. Aging in prehistoric and contemporary human populations. *Variabil Evol* **4**: 57–65.

Rewekant A. 2001. Do environmental disturbances of an individual's growth and development influence the later bone involution process? A study of two medieval populations. *Int J Osteoarchaeol* **11**: 433–443.

Richards MP, Mays SA, Fuller BT. 2002. Stable carbon and nitrogen isotope values of bones and teeth reflect weaning age at the medieval Wharram Percy site, Yorkshire, UK. *Am J Phys Anthropol* **118**: 205–210.

Riggs BL, Melton LJ. 1986. Involutional osteoporosis. *N Engl J Med* **314**: 1676–1686.

Roberts CA. 1987. Case report 9: scurvy. *Paleopathol Assoc Newslett* **57**: 14–15.

Roches E, Blondiaux J, Cotton A, Chastanet P, Flipo R. 2002. Microscopic evidence for Paget's disease in two osteological samples from early northern France. *Int J Osteoarchaeol* **12**: 229–234.

Rogers J, Jeffrey DR, Watt I. 2002. Paget's disease in an archaeological population. *J Bone Miner Res* **17**: 1127–1134.

Rollo F. 1999. A war wound in a skeleton from the battlefield of Tolentino (2–3 May 1815). *J Paleopathol* **11**(3): 41–46.

Roodman GD, Windle JJ. 2005. Paget disease of bone. *J Clin Invest* **115**: 200–208.

Ross PD. 1996. Prediction of fracture risk II: other risk factors. *Am J Med Sci* **312**: 260–269.

Sauberlich H. 1997. A history of scurvy and vitamin C. In *Vitamin C in Health and Disease*, Packer L, Fuchs J (eds). Dekker: New York; 1–24.

Schmorl G. 1909. Die Pathologische Anatomie der Rachtischen Knockenerkrankung mit Besonder Berücksichtingung ihrer Histologie und Pathogenese. *Ergeb Inn Med Kinderheil* **4**: 403–454.

Schmorl G. 1932. Über ostitis deformans Paget. *Virchow Archiv Pathol Anat* **283**: 694–751.

Shahar S. 1990. *Childhood in the Middle Ages*. Routledge: London.

Sherman MS. 1950. Osteomalacia. *J Bone Joint Surg Am* **32**: 193–206.

Smith SE, Murphey MD, Motamedi K, Mulligan ME, Resnik CS, Gannon FH. 2002. Radiologic spectrum of Paget disease of bone and its complications with pathologic correlation. *RadioGraphics* **22**: 1191–1216.

Sridhar GR. 1994. Paget's disease in India: is it truly rare? *Nat Med J India* **7**: 101.

Steinbock RT. 1993. Rickets and osteomalacia. In *The Cambridge World History of Human Disease*, Kiple K (ed.). Cambridge University Press: Cambridge; 978–980.

Still GF. 1935. Infantile scurvy: its history. *Arch Dis Child* **10**: 211–218.

Stirland A. 1991. Paget's disease (osteitis deformans): a classic case? *Int J Osteoarchaeol* **1**: 173–177.

Stirland AJ. 2000. *Raising the Dead: The Skeleton Crew of King Henry VIII's Great Ship, the Mary Rose*. Wiley: Chichester.

Stuart-Macadam P. 1989. Nutritional deficiency disease: a survey of scurvy, rickets and iron deficiency anaemia. In *Reconstruction of Life from the Skeleton*, Işcan MY, Kennedy KAR (eds). Alan Liss: New York; 201–222.

Stuart-Macadam P, Glencross B, Kricun K. 1998. Traumatic bowing deformities in tubular bones. *Int J Osteoarchaeol* **8**: 252–262.

Turner-Walker G, Syversen U, Mays S. 2001. The archaeology of osteoporosis. *J Eur Archaeol* **4**: 263–268.

Tyson R (ed.). 1997. *Human Paleopathology and Related Subjects: An International Bibliography*. San Diego Museum of Man: San Diego, CA.

Villa ML. 1994. Cultural determinants of skeletal health: the need to consider both race and ethnicity in bone research. *J Bone Miner Res* **9**: 1329–1332.

Van der Voort DJM, Geusens PP, Dinant GJ. 2001. Risk factors for osteoporosis related to their outcome: fractures. *Osteoporos Int* **12**: 630–638.

Van Gerven DP, Armelagos GJ, Bartley MH. 1969. Roentgenographic and direct measurement of femoral cortical involution in a prehistoric Mississippian population. *Am J Phys Anthropol* **31**: 23–38.

Van Staa TP, Selby P, Leufkens HGM, Lyles K, Sprafka JM, Cooper C. 2002. Incidence and natural history of Paget's disease of bone in England and Wales. *J Bone Miner Res* **17**: 465–471.

Waldron T. 1993. The health of the adults. In *The Spitalfields Project. Volume 2: The Anthropology*, Molleson T, Cox M (eds). CBA Research Report 86. Council for British Archaeology: York; 66–89.

Waldron HA. 2004. Recalculation of secular trends in Paget's disease. *J Bone Miner Res* **19**: 523.

Waldron HA. 2005. Archaeology and prevalence of Paget's disease. *J Bone Miner Res* **20**: 1485.

Walsh JP. 2004. Paget's disease of bone. *MJA* **181**: 262–265.

Wang W-C, Cheng Y-SL, Chen C-H, Lin Y-J, Chen Y-K, Lin L-M. 2005. Paget's disease of bone in a Chinese patient: a case report and review of the literature. *Oral Surg Oral Med Oral Pathol Oral Radiol Endod* **99**: 727–733.

Wells C, Woodhouse N. 1975. Paget's disease in an Anglo-Saxon. *Med Hist* **19**: 396–400.

12

Tumours and Tumour-like Processes

Don Brothwell

Department of Archaeology, York University, The King's Manor, York Y01 7EP, UK

INTRODUCTION

Since a review of the palaeopathological evidence for neoplasms 40 years ago (Brothwell, 1967), many new skeletal series have been examined and the literature on tumours and tumour-like structures has expanded considerably. It is beyond my current brief to review comprehensively every case that has now been reported, but I do plan to consider and summarize where we seem to be in terms of the study of this aspect of palaeopathology. We can now begin from the position that tumours are not simply diseases of recent advanced societies, but have been a health threat to much earlier urban and tribal societies. But this is not to suggest that the epidemiological situation has remained static over the millennia. World regions and human populations have experienced varying climatic change, diseases have evolved, domestication of animals and plants has radically modified food supplies, technology has vastly changed, and life expectancy has generally increased, especially since Roman times. All of this must be kept in mind in viewing the evidence for tumours, and their probable changing impact through time on human populations.

NEOPLASMS: HAVE THE CAUSES CHANGED?

The cause of malignancy may be clear or unknown, but is usually multifactorial. Modern causes include ultraviolet light, dietary factors, tobacco consumption, alcohol, genetic predisposition, occupation, infection (especially with certain viruses), radiation and certain drugs and toxins. Neoplasms may be benign or malignant, with benign forms sometimes becoming malignant. Cancer is the accumulation of cells which are abnormal in form and size. These abnormal cells are not usually recognized and eliminated by the immune system. Malignant tumours may also result in a generalized immunosuppression. The malignancy may spread

Advances in Human Palaeopathology Edited by Ron Pinhasi and Simon Mays

by aggressive local invasion, or by metastasis. In the case of bone, 75 % of people with breast or prostate cancer will eventually display metastases. These may be osteoblastic or osteolytic, or sometimes both.

There is without doubt some familial predisposition toward formation of neoplasms, and this is especially important to note if the geographic distribution of neoplasm types is being considered. Some cancer syndromes, such as breast and ovary neoplasms, retinoblastoma, neurofibromatosis, melanoma and Wilms' tumour, exhibit clear dominant inheritance. Others may be autosomal and recessive. Gene mutations and chromosome translocations may also contribute to neoplastic development.

While clinical diagnosis today may involve screening programmes and the early identification of asymptomatic or very early invasive states, identification of tumours in palaeopathology is usually reliant on substantial changes up to the time of death. This does not always mean that the malignancy was the final cause of death, although it is sometimes the case. Physical examination of the ancient dead is thus different to the examination of the living, in that we are concerned with identifying relevant pathology, rather than symptoms prior to confirmation tests. Treatment is not the driving force in palaeopathology, but rather the assembling of sufficient evidence from a body or skeleton to support a neoplastic identification and, if possible, a more precise differential diagnosis.

Tumours and Microevolution

In his remarkable book *Evolution and Disease*, Bland-Sutton (1890) argues that 'tumours' are biological phenomena which should not be left out of debates on evolution. He discusses the various forms, which he elaborates on in his larger, later work on tumours (Bland-Sutton, 1893). Classification of tumours was then far broader and all embracing of a variety of swellings, and still rather copied the diagnostic views of the ancient writers, who had no clear sense of a malignant neoplasm. Nevertheless, in the 19th century, 'cancer' was seen as the most destructive form of tissue 'swelling' and growth. It was also recognized that such neoplasms were far less common in other mammals and that there were significant geographic and ethnic differences between human populations.

This awareness of geographic and ethnic differences became increasingly clear in the 20th century, resulting in a variety of studies. Even allowing for the fact that diagnostic and statistical competence has varied from country to country, the magnitude of the differences is such that true geographic and ethnic differences clearly exist. In the case of Burkitt's lymphoma, which causes bone changes and may have a viral aetiology, it occurs especially in tropical Africa (Burkitt, 1967). Its evolutionary significance is in the fact that it particularly affects people between the ages of 2 and 15 years. Survival into reproductive age is poor; therefore, any genetic factors would have maximum effect in the population. While this is seen as a rather special case, other tumours, for instance osteosarcoma, occur notably in pre-reproductive or reproductive years.

Nasopharyngeal carcinoma incidence contrasts markedly between populations in the Far East and other regions of the world (Shanmugaratnam, 1967). It was probably first described by Chou Uen Fung, a Chinese physician of the Sui period, living around AD 589–617. The condition also affects younger individuals and especially males. Its aetiology is complex, with both genetic and environmental factors (including the influence of the Epstein–Barr virus) being involved (Ting, 1981). Again, as in Burkitt's lymphoma, if genetic variation conferred any survival advantage in the younger individuals, then some degree of differential survival could have occurred.

In the case of regional variation in other malignant tumours, as in Britain (Howe, 1970, 1972) and continental Europe (Holland, 1988), their occurrence may be mainly in the post-reproductive age groups. Nevertheless, the differences may be indicating possible links with genetic factors, infection, diet or other environmental parameters. The notable occurrence of Hodgkin's disease in Italy might be seen to result from the continuing influence of the ancient Romans, while mortality from malignancy of the uterine cervix is higher in Britain and Denmark than elsewhere (Holland, 1988), and surely demands consideration from the point of view of Viking influences or hygiene factors.

Differences between the sexes again raise questions in terms of the influence of hormones or day-to-day habits, including diet. There are certainly interesting contrasts between males and females, with evidence of fluctuating changes through time. In Miyagi, Japan, for instance, stomach cancer was one of the commonest tumours, but with females showing about half the percentage rate seen in males. This declined in both groups from 1959–1960 to 1968–1971. In contrast, rectal cancer was similar in both sexes, both showing an increase in the same period (Doll and Armstrong, 1981). Such data may seem a long way from basic palaeopathology, but, with the growing amount of information on neoplasms in the past, we need to keep in mind the general biology of tumours for the sake of a balanced interpretation.

NEOPLASMS IN MUMMY SOFT TISSUE

The identification of tumours in varyingly preserved soft tissue should theoretically result in better diagnoses than those of skeletal material. Unfortunately, thousands of well-preserved bodies, especially from Egypt and Peru, did not survive the 19th century, or if they did they received minimal conservation. Moreover, in the case of well-preserved mummies, such as those in the British Museum, investigations of the bodies have been strictly controlled, mainly resulting in surface and radiographic examination only. Nevertheless, some invasive studies have been undertaken on better-preserved ancient tissues, and the most recent review of the pathological findings has been by Aufderheide (2003). His division and classification of the soft-tissue findings will be followed here, beginning with the tumours of the head and neck.

Today, the brain and surrounding tissue are commonly involved in primary malignancies, and both children and adults can be involved. A massive haemorrhage or foci of calcification may be distinctive features. In older adults, neoplastic growth from the meninges may give rise to commonly occurring meningiomas, which fortunately for us can cause significant changes to the cranium, even though generally benign. From both Peru and Europe, cranial evidence also suggests the occurrence of a malignant form, the meningiosarcoma, although a precise diagnosis is debatable in cranial cases. As yet, there are no reports of eye neoplasms in ancient bodies, although retinoblastoma and malignant melanoma might be expected in particular. Similarly, the ear has not produced ancient cases of tumour, although auditory tori are occasionally mistaken for benign tumours. Again, although the nose is known to display considerable modification from forms of advanced leprosy and treponematosis, tumours have not been noted in ancient mummified material. Nasal polyps, not true tumours, appear to have been described in Egyptian hieroglyphic texts, as well as by Graeco-Roman physicians, but again no ancient soft-tissue cases are known, although ancient nasopharyngeal malignancy has been described in skulls.

Preservation of organs within the trunk is surprisingly variable, even when the environment appears to be similar. Bogs, for instance, may result in well-preserved intestines, heart, lung and liver, or in very little soft tissue other than skin and hair. The differences from site to site are still not fully understood. From arid environments, the soft organs may remain, if mummification techniques have not removed them, but their examination may be problematic because of their brittleness.

It seems debatable whether some of these organs have been significantly affected by tumours in the past. Carcinoma of the lung is common today, but is clearly associated with the common post-medieval habit of tobacco smoking. Mesothelioma is a malignancy that displays a significant occurrence only in older individuals working with asbestos. If a tumour is eventually described in mummies, it is far more likely to be a metastatic deposit derived from a primary malignancy elsewhere. Again, although not yet described in ancient bodies, liver carcinoma linked to the consumption of foods containing aflatoxin is well known in Africa and Asia.

On the evidence of modern epidemiological data for various world populations, intestinal tumours should not be rare in ancient bodies, but on current evidence they are. Both in the stomach and colon, tumours can achieve a large size, which would assist in their detection. Nevertheless, only two gastrointestinal tumours have been reported, one from Italy and the other from Egypt (Fornaciari *et al.*, 1993; Zimmerman, 1995).

Hypernephroma can also achieve a large size and is relatively common today, yet it has not yet been identified in the kidneys of ancient bodies. Preservation of the kidneys, however, is very variable, and the lack of neoplastic evidence could be explained by taphonomic and diagenetic factors. In the case of genital structures, the uterus seems most likely to produce ancient evidence of tumours, and the benign smooth-muscle neoplasms (leiomyomas) are not uncommon today. Cervical carcinoma appears to be associated with the human papilloma virus, so that any eventual discovery of ancient examples of the former will provide evidence also of this virus. Two uterine tumours, probably leiomyomas, have been described in early Nubian and Swiss bodies (Strouhal and Jungwith, 1977; Kramar *et al.*, 1983). In 1825, Granville, a founder of mummy studies, reported on an ancient Egyptian with uterine and ovarian changes (Granville, 1825). Aufderheide (2003) suggests that a modern diagnosis of the changes would suggest a bilateral ovarian cystadenoma or cystadenocarcinoma, although alternative diagnoses remain possible.

While benign fibroadenomas of the breast are common and benign intraductal papillomas are far less so, it is perhaps the commonly occurring malignant carcinoma of the breast which could be especially revealing in relation to earlier populations. Today, this malignant form can occur before the menopause, and thus could well occur in earlier populations with a shorter adult life expectancy. As Aufderheide (2003) points out, the untreated tumour commonly extends to the skin surface, with possible ulceration or infection. Such surface changes would be particularly easy to pick up in ancient cases. The Edwin Smith papyrus refers to a swelling interpreted as breast cancer (Baum, 1993), and other ancient writings could include reference to such malignancies. Nevertheless, there is no good ancient body evidence of such a tumour.

Although the epidermis is often missing, the skin is still one of the best-preserved organs of ancient bodies. This may be assisted by embalming procedures, although natural preservation in cold or bog environments seems to ensure the best histological condition. However, diagenetic factors resulting in post-mortem erosion of the body surface could still look remarkably like an erosive carcinoma of the skin. But while primary skin cancer has yet to be identified, a few benign tumours have been reported. An Inca child from the high Andes displays an angiokeratoma on the leg (Horne, 1986), and there is a lipoma in another body from the New World (Aufderheide, 2003).

COULD WE DETECT OCCUPATIONAL INFLUENCES IN SPACE AND TIME?

There are clearly a number of tumour questions to be asked of archaeological human remains. What is the antiquity of forms of tumour, malignant and benign? Do there appear to be geographic influences? Are the tumours distinctive in any way from what is known clinically today? Per skeletal population and age composition of the sample, do they display unusual prevalence? One of the most difficult aspects of cancer studies to evaluate, even today, is the relationship between neoplasms and occupation. In advanced industrial society, there are of course carcinogenic environments (such as asbestos-producing plants) which have not occurred in the past. But there are various occupations extending back into prehistory that could be considered from this aetiological point of view. All occupational tumours have been well surveyed and reviewed by Alderson (1986), and it is interesting to reflect on such influences in relation to the past. They have certainly not been uniform in their influence. The significance of alcohol may not have been important until the emergence of advanced industrial societies, although native beers are known to increase the risk of aflatoxin poisoning and liver cancer. Bakers are one occupational group with an increased propensity to nasal cancer, and it seems to me that the smoky environments of poorly ventilated housing, common enough in tribal societies today, would also have been similarly detrimental. Anthracosis, present in mummies as divergent as the Andes, Canary Islands and Egypt, are an indication of this. It was present in medieval Greenland, and one frozen body had an associated nasopharyngeal cancer (Ammitzbøll *et al.*, 1991).

With the evolution of farming, human populations became increasingly associated at close range with various livestock. Some zoonoses are well known (Brothwell, 1991), but there are clearly many less certain associations. Epidemiologists in the USA have suggested that poultry density may be linked to uterine and ovarian cancer and myeloma on these farms. Other farmers could be at risk from prostate, kidney, oral, nervous system cancers and lymphomas (Williams *et al.*, 1977). Yet other studies have similarly found farmers at greater risk in relation to certain other tumours. Multiple myeloma, a condition that can give rise to well-defined skeletal changes, also displayed an increased prevalence in farmers. Also related to animal contacts is the fact that leather and tannery workers show increased risk of bladder, lung, oral and pharyngeal cancers, as well as myeloid leukaemia.

Quarrymen, masons and miners have been exploiting rock for millennia, some producing more health problems than others. Although one might suggest an association between these occupations and gastrointestinal cancers, the difficulty in this category of worker is that other factors, such as tobacco chewing, could obscure the correlations. However, stomach cancer also appears significant in men above ground quarrying slate and igneous rocks. Finally, the ancient occupation of woodworking may significantly increase the risk of adenocarcinoma and nasal cancer, presumably in part due to breathing in wood dust.

The situation is no doubt more complex than suggested here, with genetic factors, variation in wood or stone, the domestic animal species most commonly handled, and length of time working at an occupation all contributing. Nevertheless, with growing numbers of archaeological cases, any evidence of work associations is worth considering. Cultural difference, too, may provide clues as to why, for instance, multiple myeloma might occur more commonly in some groups than expected.

SKELETAL EVIDENCE OF TUMOURS

Much of the archaeological evidence for tumours is represented by skeletal lesions, but before this data can be in any way reviewed, it is important to ask how precisely can diagnosis be made from bones. In modern clinical situations, even with soft-tissue histology, differential diagnosis is no easy task, so how close can we hope to get when evaluating ancient dry bone specimens? Clearly, a number of basic facts can initially guide the evaluation. Patterns of abnormal bone growth are highly variable, and the size, age of occurrence, number and position in the skeleton may give clues to the type of neoplasm. But there is still plenty of overlap to be considered, and ultimately it may be a matter of narrowing a diagnosis down to a limited number of alternative possibilities. A variety of texts are easily available, especially of a radiological kind, such as Stoker (1986) and Resnick (1995), with Ortner (2003) reviewing the dry bone pathology in considerable detail. Because tumours present such diagnostic problems, diagnosis should ideally be after consultation with specialist colleagues, especially radiologists and pathologists. Stoker (1986) provides a useful brief classification of tumours which can be kept in mind when considering possible archaeological evidence. Some of these are far commoner today than others are, and some show specific times of initial occurrence (although such factors could have changed through time). Sites of presentation also vary characteristically and, thus, also provide clues to identity. It should be noted that malignant neoplasms are all true tumours, and can be highly destructive. In contrast, benign tumours include a number of conditions which strictly speaking are only tumour-like. Secondary tumours, metastases, can mimic primary conditions and today occur most commonly in those over 50 years old. Benign tumours, unless becoming malignant, are usually slow growing, solitary and of modest size. The benign tumour can grow to be an obstruction, but it does not infiltrate other contrasting tissues. Tumours may produce considerable extra bone from dense and rounded extensions of the cortical bone (osteoma) to massive external spicular bone linked to an eroded cortical surface (as in the osteosarcoma). Destruction of bone may give rise to well-defined and sharp rounded margins (some fibromas), to what is described as a 'moth-eaten' destructive appearance (some fibrosarcomas). In the case of archaeological material, it is important to remember that the original destruction of bone by tumours can lead the way to secondary post-depositional erosion through diagenetic processes. As a result, important diagnostic detail can be lost. It is important, therefore, to keep this in mind when attempting to describe and suggest a diagnosis of bones with evidence of tumours.

Finally, a further comment on statistical aspects of tumours. It is very difficult to know how far cancer statistics on living populations can be extrapolated back in time. But even crude incidences might provide a clue as to how often to expect particular tumours or tumour groups. For instance, the incidence of primary malignant bone tumours in the south-west of England was 1.42 per 100 000, with osteosarcoma and myeloma being most common (Price, 1970). In the case of long-bone osteosarcoma, 41 % display characteristic spicular development with a Codman triangle (an angular area of new bone produced by elevation of the periosteum by the tumour) (Vermeij, 1970). In contrast, chondrosarcomas and fibrosarcomas appear to be far more variable in their reactions on long bones. Soft-tissue cancer morbidity data for England and Wales indicated that in 1965 approximately 0.1 % of the population under 65 were registered as cancer patients (Department of Health and Social Security, 1970). Even after removing the probably smoking-related cancers, about 0.06 % had tumours of the digestive system, breast and genital organs. Without treatment, a significant proportion would probably have produced metastases into the skeleton, which may well suggest that archaeological evidence of malignancy must often be of metastases. Also, as these figures excluded the older individuals, this is more comparable to life expectancy in earlier populations.

Benign Tumours of Bone

While the majority of benign tumours have not as yet been described archaeologically, it is important to be aware of the overall classification and range of variation (Figures 12.1 and 12.2). The divisions given here are those given by Stoker (1986). The benign conditions can be usefully divided as follows:

1. Of osteoid origin
   - (a) osteoma (Figure 12.3a) and osteoid osteoma
   - (b) osteoblastoma (Figure 12.2d)
2. Of chondroid origin
   - (a) chondroma (Figure 12.4c) enchondromatosis
   - (b) osteochondroma (Figures 12.1b and 12.5)
   - (c) chondroblastoma
   - (d) chondromyxoid fibroma (Figure 12.2b)
3. Cysts of bone
   - (a) solitary bone cyst (Figure 12.1d)
   - (b) aneurysmal bone cyst
4. Giant cell tumour (Figures 12.2a, 12.4a and b)
5. Of fibrous origin
   - (a) non-ossifying fibroma (Figure 12.1a)
   - (b) desmoplastic fibroma
6. Other tumours
   - (a) haemangioma (Figure 12.2c)
   - (b) glomus tumour
   - (c) lipoma
7. Tumour-like conditions
   - (a) implantation epidermoid
   - (b) fibrous dysplasia

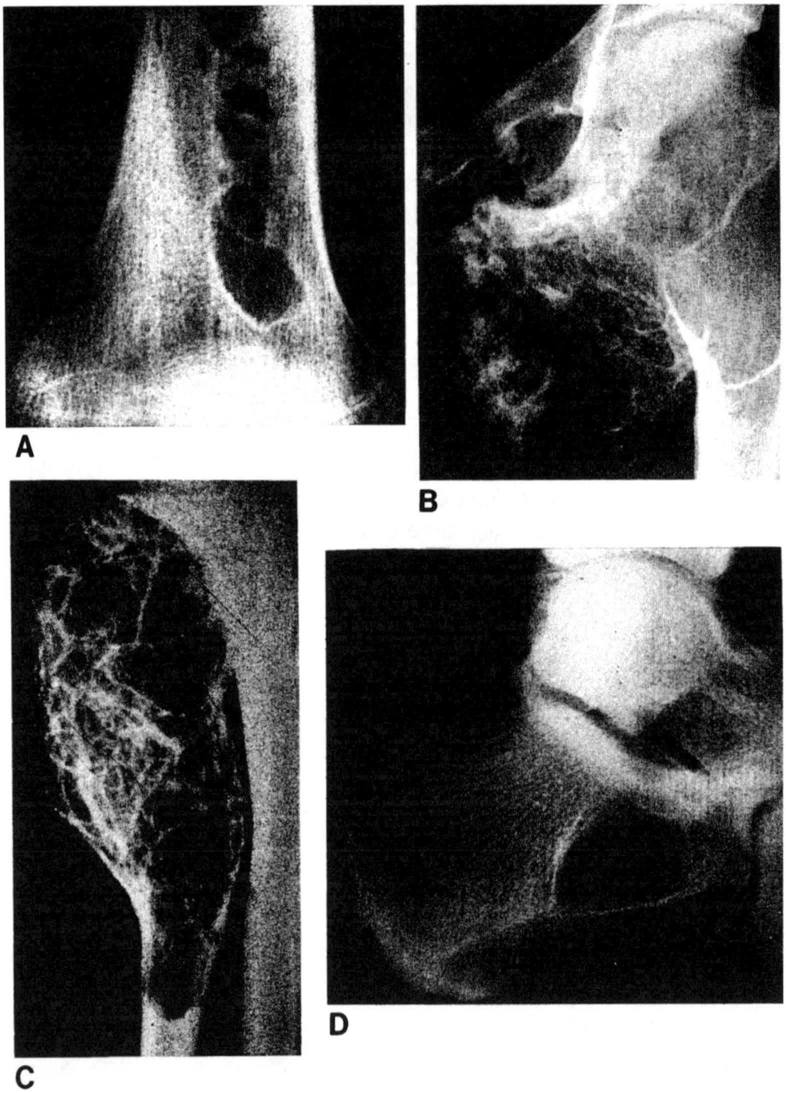

Figure 12.1 (a) Non-osteogenic fibroma, with well-defined scalloping into the dense bone. (b) Ossifying chondroma, with 'bath-sponge' new bone. (c) Fibrous dysplasia, with much trabeculation. (d) Large single cyst, well defined

The osteoma is the commonest of all archaeological tumours, and usually consists of a rounded mass of cortical bone, often on the skull surface. Most are no more than 10–15 mm in diameter. Growth is normally outwards (Figure 12.3a). If occurring in the paranasal sinuses, they can be obstructive. The osteoid osteoma is again small, but can be a painful condition. It is seen in various parts of the skeleton, but long bones of the leg are most involved. Within the periosteal, new bone may be a nidus of less dense osteoid, but in X-ray there can be much variation.

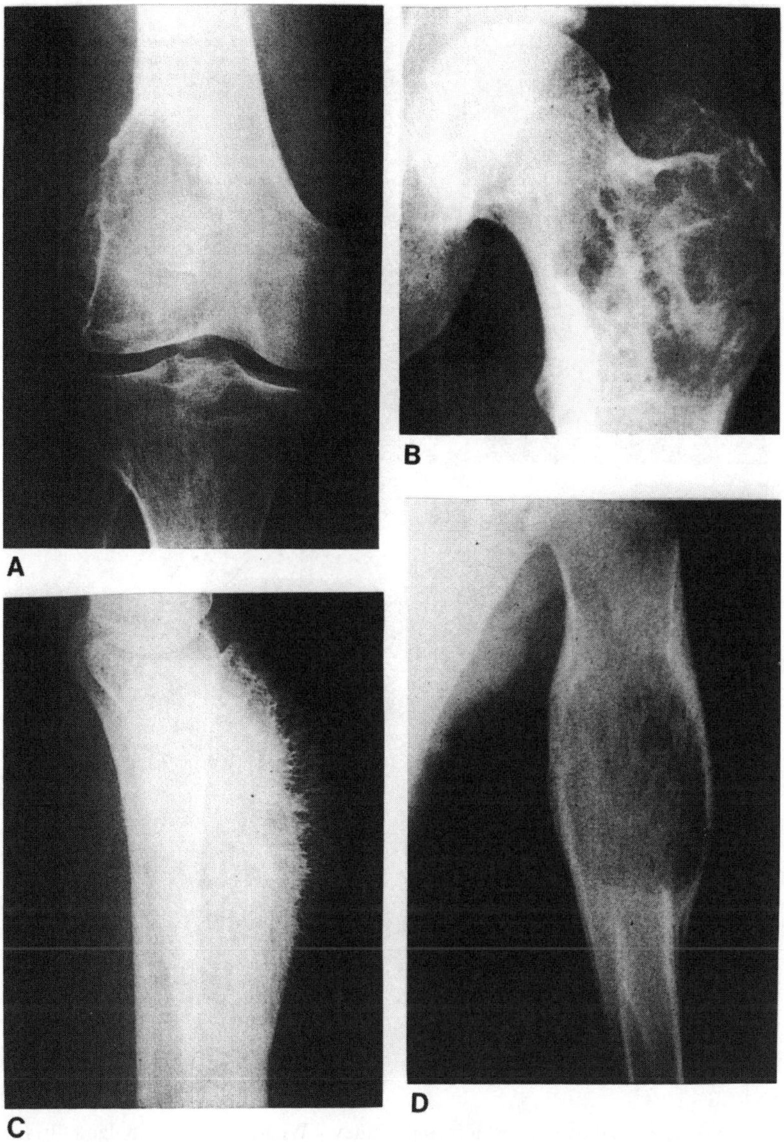

Figure 12.2 (a) Giant cell tumour, typical, near the condyle. (b) Chondromyxoid fibroma, with lobulated bone destruction. (c) Haemangioma, with spiculation which contrasts with osteosarcoma. (d) Benign osteoblastoma, with expanded thick cortical shell

The osteoblastoma (Figure 12.2d) is a rare tumour, with most cases under 30 years today. Noticeable expansion of bone may occur, with variable nodular mineralization. Tumours derived from cartilage include singular chondromas, situated within bone, or multiple tumours called an enchondromatosis. They can be found in all ages and equally in the sexes. Hands and feet are especially likely to display these tumours (Figures 12.1b and 12.4c). The osteochondroma produces a more ossified, but spongy-looking, tumour. They may occur at any

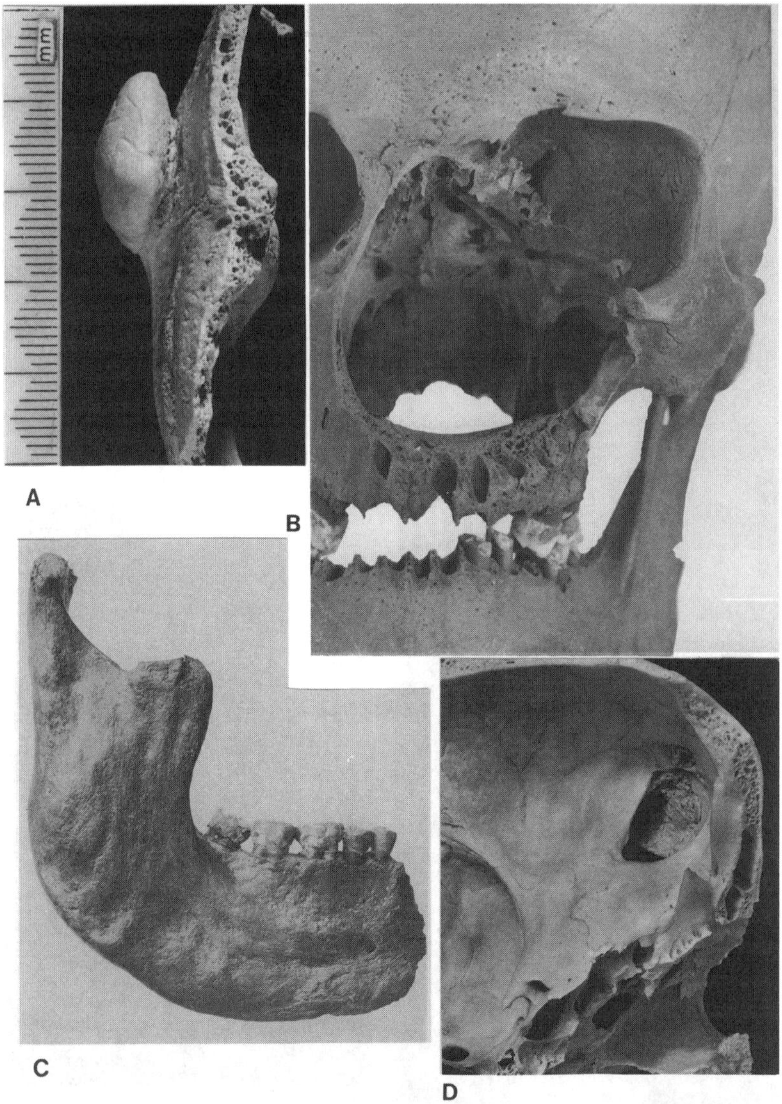

Figure 12.3 (a) Large osteoma of the skull from medieval Wharram Percy, England. (b) Large benign nasal tumour, with smooth margins. Denmark, Slagslunde. (c) Mandibular acromegaly from a pituitary tumour. Gardar, Greenland. (d) Nubian with endocranial (orbital) lesion, perhaps a dermoid cyst

age, with a slight predominance in males, and usually in larger tubular bones (Figures 12.1b and 12.5). Chondroblastomas are rare today, and are usually in the epiphyseal area of immature individuals. Twice as many males as females are affected. The tumour is usually rounded and well defined. The chondromyxoid fibroma is uncommon; it affects the lower limbs in particular, especially between 10 and 30 years. Punched-out lesions involve cortex and medulla (Figure 12.2b).

Bone cysts are usually solitary, and may not be neoplastic. The majority are found in children, with males twice as often affected as females. The proximal ends of the humerus

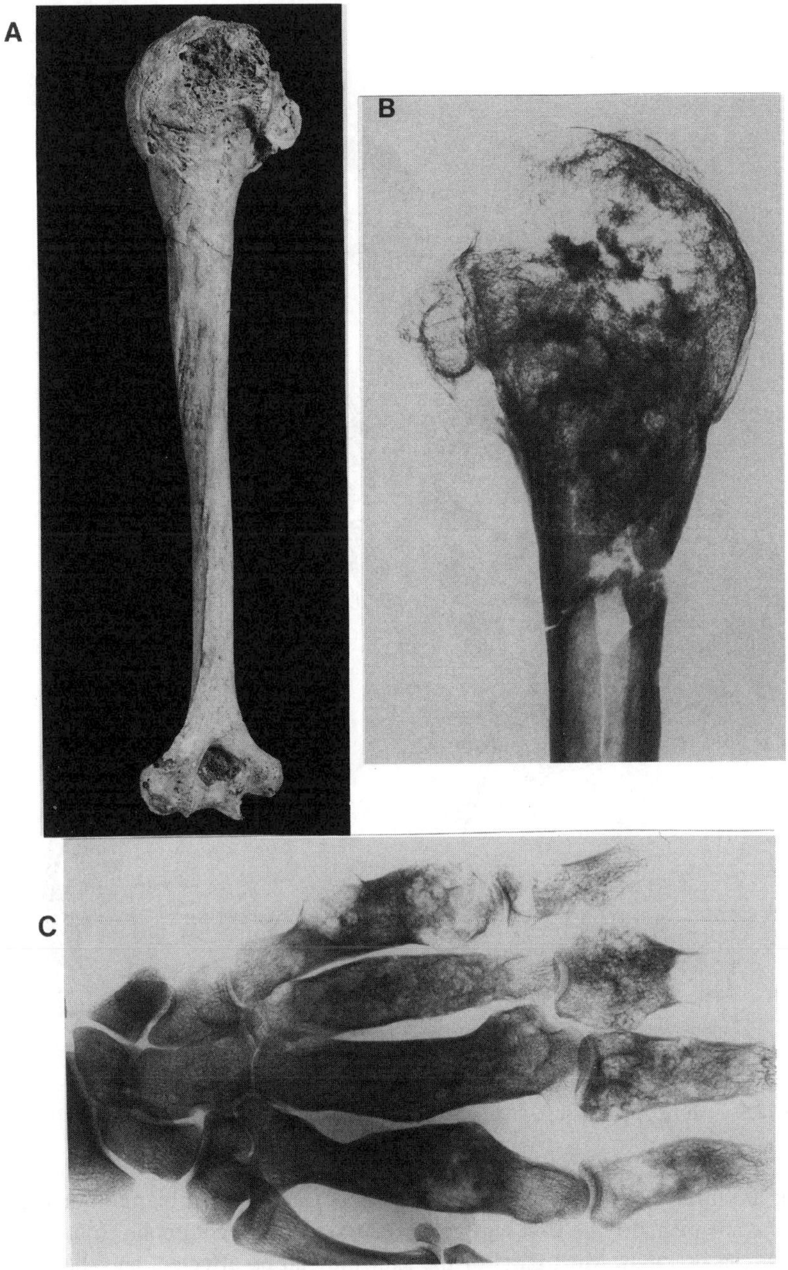

Figure 12.4 (a) Early Nubian humerus with swollen proximal end. (b) The same humerus X-rayed, and suggesting a fibrous dysplasia or possibly an osteoclastoma. (c) Recent case of enchondromatosis, with multiple cartilage tumours affecting a number of metacarpals and phalanges

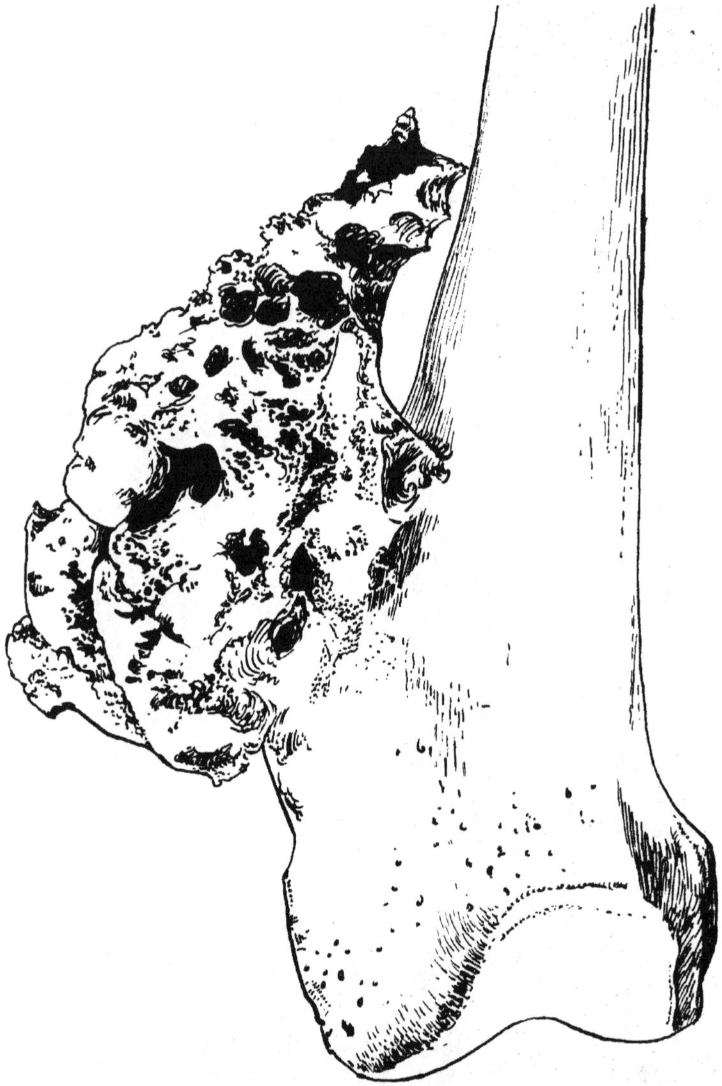

Figure 12.5 An Egyptian Vth Dynasty femur with a bone mass typical of an osteochondroma

and femur are most often involved (80 % of cases). This cyst can be multilocular. In some contrast, the aneurismal bone cyst can become massive. It is commonly found in children, and growth can be rapid. Most parts of the skeleton can be affected, and the cyst can thin the cortical bone to the point of fracture.

The giant cell tumours mainly affect young adults. The knee region is especially affected. It particularly develops in the subcortical region near the epiphyseal plate, and it is likely to cause cortical thinning (Figure 12.2a).

The non-ossifying fibroma and fibrous cortical defect are skeletally of similar appearance, but the former is larger. Sub-adults are usually involved, the majority of these fibromas being in the lower parts of the leg. The lesions are intracortical, with a lobulated 'soap bubble' appearance, and may be multiple. The desmoplastic fibroma is radiologically similar, but is very uncommon today.

The haemangioma (and equivalent lymphangioma) are considered to be of congenital origin, linked to vascular abnormality (Figure 12.2c). Cranial and vertebral lesions appear to be the commonest. In X-ray there is a coarse trabeculation, and there can be a 'soap bubble' effect, and sometimes multiple lesions. Occasionally there is spiculation. Spicules are well defined and of mature bone; this is in contrast to osteosarcoma, where there is more ragged spicular bone formation. Haemangiomatosis (massive osteolysis) is a rare and more extensive condition of a similar kind. The glomus tumour should be mentioned, as it is a soft-tissue expansion, especially of finger ends, which can cause pressure resorption of the terminal phalanx. While the lipoma is rare, it is notable for producing a poorly mineralized shell of ossified tissue around what is basically a fatty mass. The implantation epidermoid can be mentioned as it also causes changes to the distal phalanges, which may be noticeably modified in shape by cystic development. Fibrous dysplasia can be seen as a fairly rounded lytic lesion, often with denser 'rind-like' bone surrounding the changes (Figure 12.4a and b).

Malignant Tumours of Bone

Although benign tumours can occasionally mimic malignant bone tumours, there are generally considerable differences in their growth and behaviour. Benign tumours can be fast growing, but are not aggressive and invasive as, characteristically, are malignant forms. From an archaeological point of view, the malignant tumours present a greater challenge in terms of a precise diagnosis. The majority are most likely to occur during adult life and, as in benign forms, the sex ratio is variable, with different tumour types showing preference for males or females. Again, the classification adopted here is that of Stoker (1986):

1. Of osteoid origin
   (a) osteosarcoma
2. Of chondroid origin
   (a) chondrosarcoma (Figure 12.6a)
3. Of fibrous origin
   (a) fibrosarcoma (Figure 12.6c)
   (b) fibrous histiocytoma
4. Marrow tumours
   (a) Ewing's sarcoma (Figure 12.6b)
   (b) reticulum cell sarcoma of bone
5. Of notochordal origin
   (a) chordoma (Figure 12.6d)
6. Of synovial origin
   (a) synovial sarcoma

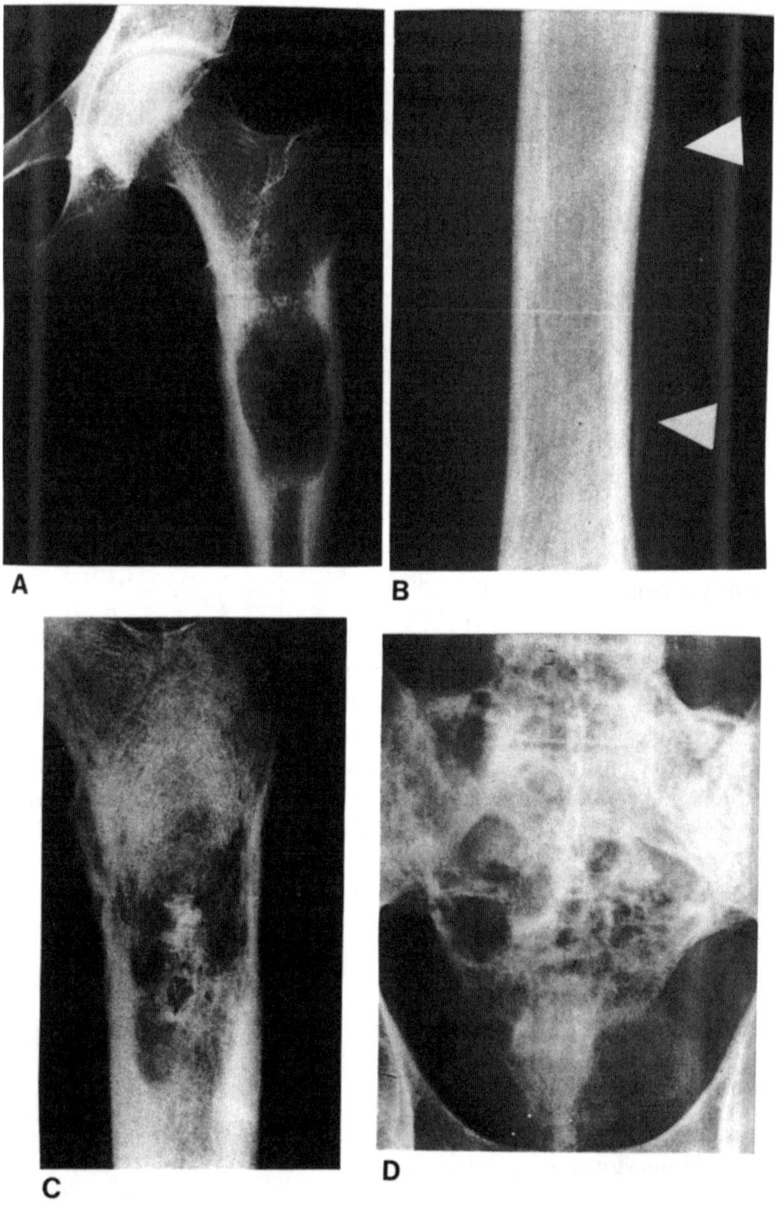

Figure 12.6 (a) Chondrosarcoma, with bone loss and some slight expansion. (b) Ewing's sarcoma, with Codman's triangles (arrowed) and early cortical destruction. (c) Fibrosarcoma, with poorly delimited bone margins to the osteolytic destruction. (d) Chordoma of sacrum, with large well-defined lytic lesion

7. Other tumours

   (a) angiosarcoma
   (b) liposarcoma
   (c) adamantinoma (ameloblastoma)

(d) myeloma (Figure 12.7a)
(e) leukaemia (Figure 12.7d)

8. Metastatic neoplasia of bone (Figure 12.7b)

The osteosarcoma is important not only because it has been recorded archaeologically, but because 80 % of cases are under 30 years of age. Male cases are twice as common as female ones. The tumour especially involves the knee, but the proximal femur and humerus are also common locations. The most characteristic change seen in the development of this tumour is the 'sunburst' spiculation, which tends to be coarse and irregular. A Codman's reactive triangle may form at the margins of the lesion when it is sufficiently large. Subdivisions of osteosarcoma have been attempted, but they are not judged important here.

As the name implies, the chondrosarcoma originates in cartilage (Figure 12.6a), but with variable bone involvement. Its peak incidence is around 50 years, with a slight preponderance of males. The proximal femur and pelvis are the commonest sites, but the knee and shoulder can be involved. Expansion of bone with 'scalloping' of the endosteal surface of the cortex is typical.

Fibrosarcoma has a tenth of the incidence of osteosarcoma, and is uncommon in children. It usually appears in younger adult years and there appears to be no sex bias. The long bones of the legs are most commonly affected, followed by the pelvis and humerus. The lesion is osteolytic and appears as destructive but expanded 'moth-eaten' bone. It may be begin in soft tissues and expand into bone. Much cortex may be destroyed, with little periosteal reaction. It can mimic lytic osteosarcoma, giant cell tumour and aneurismal bone cyst (Figure 12.6c). It is very unlikely that this tumour can be differentiated on dry bone pathology from malignant fibrous histiocytoma.

Ewing's sarcoma is an uncommon primary bone tumour (Figure 12.6b), usually affecting those under 30 years of age, and males more than females. Although normally at a single site, it may eventually involve more bones. Long bones, pelvis and ribs may all be involved. Although eventually bone destructive, it can at first be contained sub-periosteally, but it progresses to produce 'onion peel' extra bone or cortical erosion with Codman's triangles. Reticulum cell sarcoma is a rare condition with similarities to the previous tumour. It is usually an adult condition, with slightly more males affected, and with a preference for long-bone diaphyses. Bone destruction can be poorly defined in X-ray, with 'moth-eaten' cortical destruction.

The chordoma is unusual in having a predilection for the sacrum (Figure 12.6d) and skull. There is no clear differentiation by age or sex. Bone destruction can be considerable, with the margins being quite well defined. It is noteworthy that some degree of calcification can also occur. While the sacral changes might be mistaken for tubercular osteomyelitis, changes to the skull (sella turcica, nasopharynx) would be suggestive of something unusual. The other tumours listed above are likely to occur archaeologically only rarely. The synovial sarcoma is a condition of middle age, with the expansion of a large tissue mass and related destruction of nearby bone. The angiosarcoma is highly malignant, the bone destruction having a 'soap-bubble' appearance in X-ray. The liposarcoma is extremely rare and particularly affects males. The long bones of the lower limbs are especially affected. The adamantinoma of long bones is rare and especially affects the tibia shaft. The destructive lesions are well defined, with bone sclerosis being common.

In the case of metastatic neoplasms of bone (Figure 12.8a), derived from malignancies primarily affecting soft tissues in other structures of the body, these present the most difficult

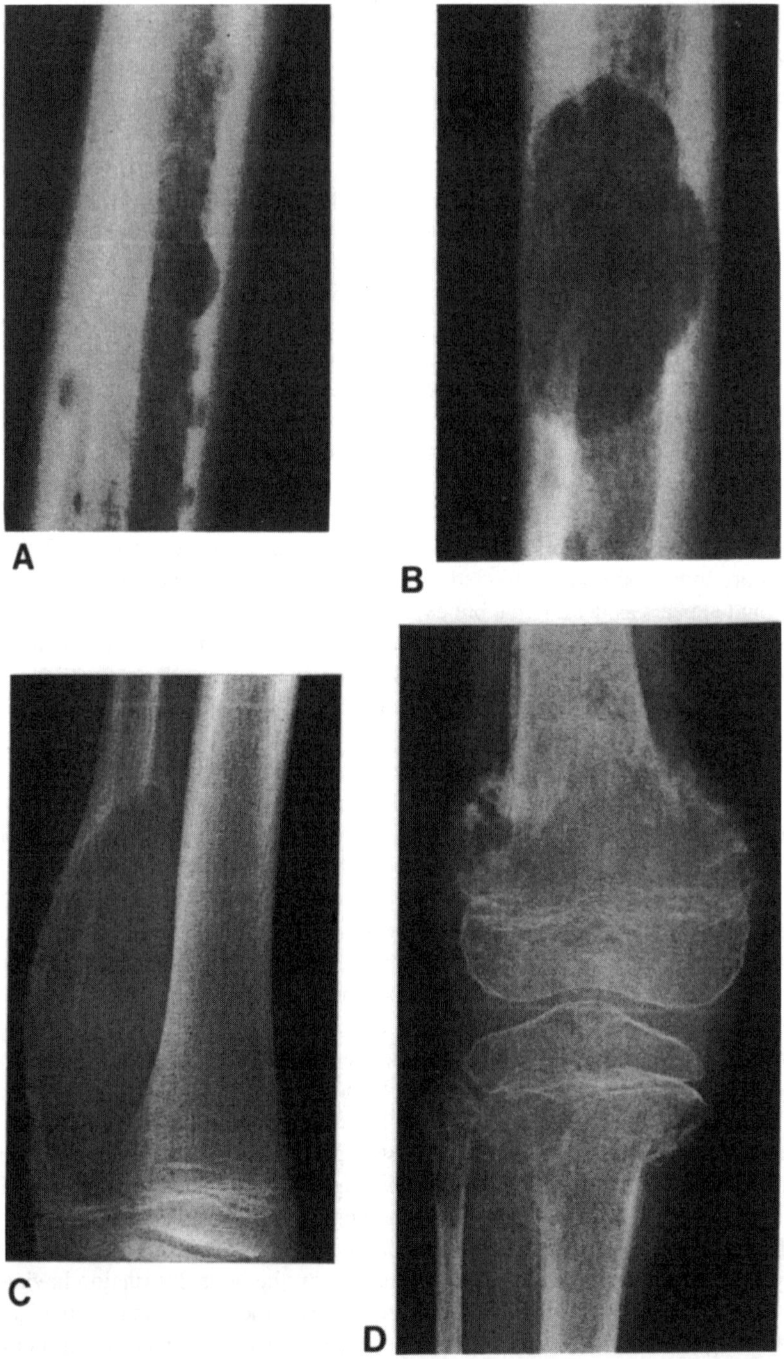

Figure 12.7 (a) Multiple myeloma, showing well-defined elliptical lytic lesions. (b) Secondary tumour in bone, with much cortical destruction. (c) Ewing-type bone malignancy, with much bone expansion and inner destruction. (d) Advanced leukaemia, with extensive bone loss, some new periosteal bone and 'moth-eaten' areas

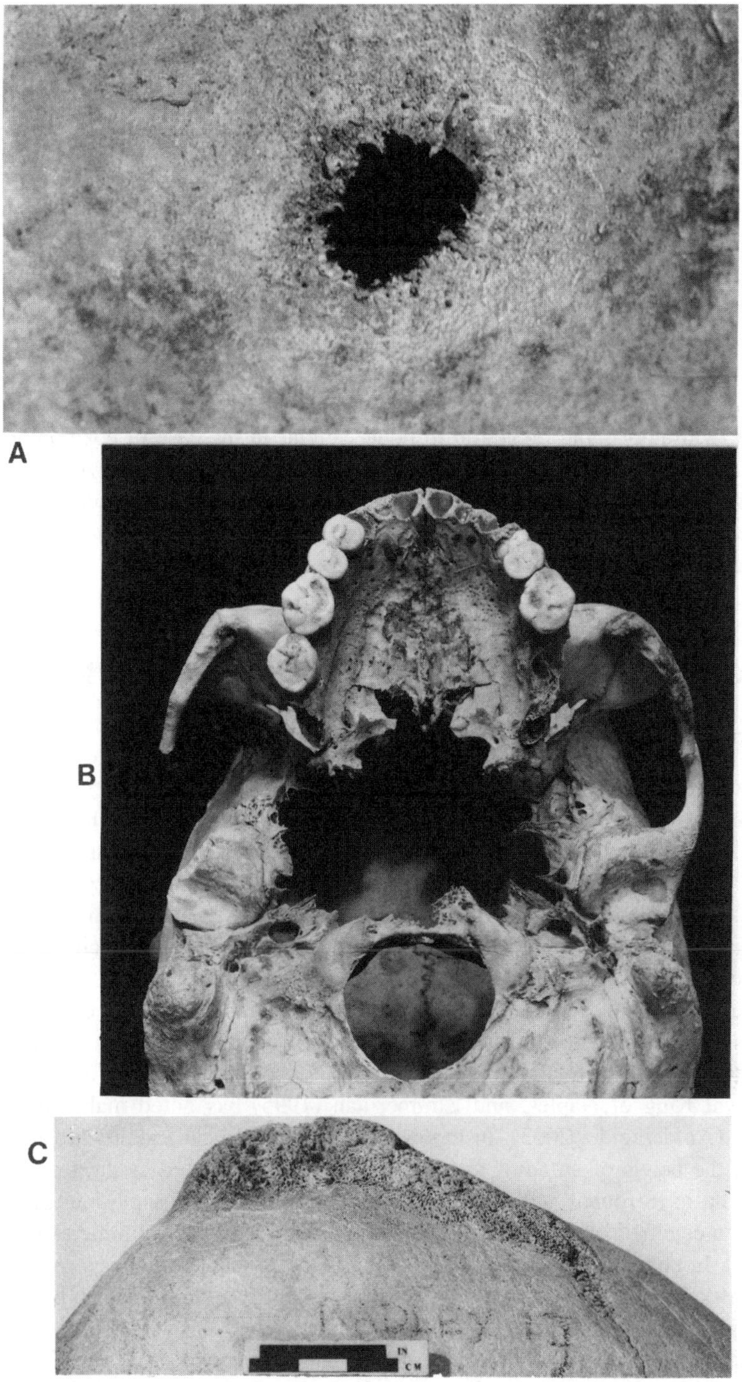

Figure 12.8 (a) Metastatic skull lesion, ragged margins. Site 68, San Joaquin Co., California (Burial 55). (b) Nubian skull (188B), with nasopharyngeal bone destruction (and marginal reaction) from a malignant tumour. (c) Roman skull from Radley, England, with external spicular mass, possibly indicative of a meningioma

aspects of a differential diagnosis. In advanced Western societies today, breast, lung and prostate cancer in particular cause the deaths of many thousands, with between 30 % and 80 % incidence of bone involvement. Approximately 9 % of bone metastases are solitary, making them at least as common as primary malignant neoplasms. This is important to remember when attempting a differential diagnosis of tumour evidence in archaeological material. However, metastatic deposits are usually multiple and have a preference for vertebrae, pelvis, skull, ribs and proximal femur and humerus. The multiple tumours are usually osteolytic (kidney, thyroid, uterus, oesophagus), but can be osteoblastic (as in prostate secondaries), and can also be mixed (breast).

In considering the differential diagnosis of primary versus secondary tumours, it is important to keep in mind the age of the individual and where the tumours are positioned in the skeleton. Metastases are not as common in children, although they do occur, perhaps especially in leukaemia and neuroblastoma. In the case of leukaemia, the secondaries can be multiple, with noticeable reduction in bone weight. In the case of eosinophilic adenoma or adenocarcinoma of the pituitary gland, there can be an excess of growth hormone, which in adults produces new bone growth, especially of the skull, ribs, vertebrae, hands and feet. Sella turcica size increases, as well as the skull sinuses. There are increases in skull thickness and the mandible is much enlarged.

THE ARCHAEOLOGICAL EVIDENCE FOR TUMOURS

Soft-Tissue Neoplasms

It is difficult from early writings to get any idea of the occurrence of neoplastic conditions, or of the early knowledge of this class of diseases. Although the Ebers Papyrus (circa 1500 BC) considers treatment for what may have included neoplastic lesions, and Hippocrates recognized tumours as distinct conditions, a more detailed and refined division of them only came later. Galen (circa 150 AD) was perhaps the first to detail a range of 'tumours', although his classification included non-neoplastic swellings (Ackerknecht, 1965).

Aufderheide (2003) has provided a valuable review of the evidence of tumours in mummified tissue. Although many hundreds of bodies, especially from arid environments, have now been studied, there are surprisingly few cases of neoplasms in soft tissue. One exception is in the gastrointestinal tract. Fornaciari *et al.* (1993) describe an adenocarcinoma in a medieval King of Naples, and Zimmerman (1995) has noted malignancy in a body from Egypt (Aufderheide, 2003). In the case of gynaecological lesions, a calcified nodular structure in the pelvis of an adult female Nubian was interpreted as the result of a benign uterine neoplasm (Strouhal and Jungwirth, 1977). Similarly, the same interpretation seems possible for a calcified mass (containing collagen) from a Swiss Neolithic skeleton (Kramar *et al.*, 1983). In both cases a uterine leiomyoma is the probable diagnosis. A further case of myoma in a skeleton is demonstrated by Brobeil *et al.* (1992) in a Spanish burial displaying a calcified nodule in the position of the uterus.

For most organs and soft tissues of the body, there is a surprising lack of good evidence of either benign or malignant neoplasms. Aufderheide (2003) reviewed all the evidence in detail, but was not able to provide clear cases of tumours of the breast, prostate, kidney, liver, lung, the eye or the brain. Regarding the skin, it is of course usual for the epidermis to be lost post-mortem, and even the dermis may be considerably changed. That tumours may still be identified is demonstrated by the detection of a small benign squamous papilloma in an Egyptian mummy hand, described by Sandison (1967).

Benign Bone Tumours

Over the years there have been a number of reviews concerned with archaeological tumours, either of a general kind (Brothwell, 1967; Ricci *et al.*, 1995; Roberts and Manchester, 1995; Aufderheide and Rodriguez-Martin, 1998; Waldron, 2001; Ortner, 2003) or more specifically concerned with geographic regions (Brothwell, 1961; Rokhlin, 1965; Strouhal, 1976, 1998, 2001; Bennike, 1985; Gladykowska-Rzeczycka, 1991; Webb, 1995; Spigelman and Bentley, 1997; Roberts and Cox, 2003). I do not intend to review here the evidence of tumours region by region, unless there are data of particular significance. Rather, it seems best to consider neoplasms in terms of their original tentative diagnoses, with further comments where they seem pertinent. Then, finally, it would be worth spanning a reasonable sample of published tumours, in relation to what we know of their biology, to see what conclusion we might draw regarding tumour studies in archaeology and palaeopathology.

Osteomas are the most commonly occurring benign tumour (Figure 12.3a), usually occurring on the skull. They can be variable in size and to some extent shape. They are found widespread in human remains from different regions and periods, with no significant differences noted yet in time and space. There are, however, questions of aetiology still for consideration. The shapes on the skull vary from a shallow rounded mound to a bulbous outgrowth with a maximum diameter well above the cortical bone surface. Could some of these mounds in fact be ossified haematomas, the result of blows to the head? Similarly, a small proportion of the cases are in the form of small multiple mounds of cortical bone, with a medieval skull from York, England, displaying 20 mounds (Stroud and Kemp, 1993). It is reasonable to ask whether any other factors, scalp infection for instance, could encourage bone elaboration of this kind. In the limited statistical information available, in a series of 3033 early British individuals, 1.05 % displayed osteomas, and the majority were males (Roberts and Cox, 2003). At Pecos Pueblo, 2.2 % were affected (Hooton, 1930). This again calls into question the possible trauma links with these bone masses, although, admittedly, I know of no case where there is a clear association with a fracture. In the case of the osteoid osteoma, fewer are described, but again they appear to occur in 1.5 % of a pooled skeletal sample. The benign osteoblastoma appears to be extremely uncommon in archaeological human remains, although it is well known in clinical situations today. Similarly, in tumours of chondroid origin, remodelled bone linked to the development of a chondroma (or enchondroma) has only been identified at one site (Anderson, personal communication). But evidence of osteochondromas has been identified at a number of sites, in the Americas, Europe and Egypt. A Vth Dynasty Egyptian example has been especially well reported (Figure 12.5), with another case from a British Neolithic site (Chamberlain *et al.*, 1992). Further cases have been identified in Britain (York, Derby), Russia (Rokhlin, 1965), Poland and Spain, but clearly they were uncommon in the past, as they are today.

Although there appear to be no certain cases of chondroblastoma in archaeological series (and indeed it is very uncommon today) it is important to remember the problematic cases of palaeopathology, to see whether a new evaluation is called for on some specimens. Thus, in the case of the humerus from West Kennet long barrow, England (Figure 12.9), it would now seem reasonable to consider a diagnosis of chondroblastoma rather than the original

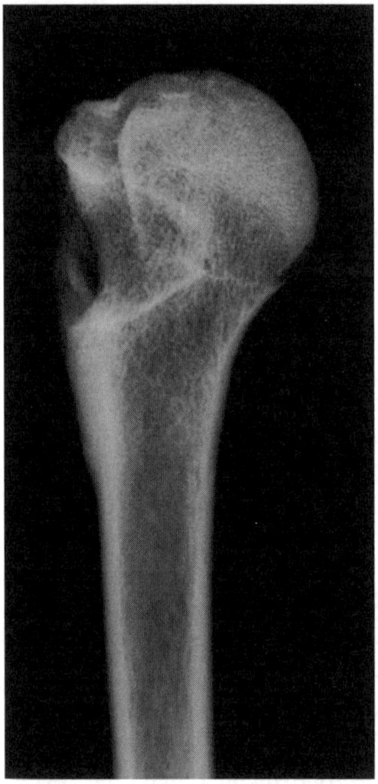

Figure 12.9 A Neolithic humerus from West Kennet, England, showing an organized hollow deep into the epiphyseal area, possibly indicating a chondroblastoma

one of trauma and infection (Brothwell, 1961), as, in this benign tumour, the rounded cavity is within the epiphyseal area, with well-defined walls. Another tumour group that deserves more critical evaluation is the bone cysts. In oral pathology, they are still poorly differentiated from apical abscesses of the jaws, and at a post-cranial level are very infrequently noted. In Czechoslovakia, for instance, two sites produced examples in a humerus and a metatarsal (in a sample of 61) (Gladykowska-Rzeczycka, 1991).

Giant cell tumour appears to have been suggested in only one long bone, a Saxon femur from Finglesham, England (Brothwell, 1967). While I still believe that this is the most likely diagnosis, and appears to be supported by the radiographic appearance (Figure 12.10), there is the question of possible post-mortem erosion of the inner cancellous bone, which could have transformed the inner appearance of the pathology. On the other hand, the proximal end of a Nubian humerus is vastly expanded, and might again be interpreted as a giant cell tumour, with marked trabeculation (Figure 12.4), but alternative diagnoses (such as fibrous dysplasia) cannot be excluded.

As a group, the fibromas are relatively uncommon today, but some tentative archaeological identifications have been made. A Czech skull displaying pressure atrophy was interpreted as having been influenced by a slow-growing fibroma or angiofibroma (Strouhal *et al.*, 1996). Uterine fibromas, usually remaining as calcified masses, have been reviewed by Verghetta and Capasso (2001). They note four previous cases from Europe and Egypt and describe two new cases. As in modern clinical studies, they are likely to be uncommon in the past,

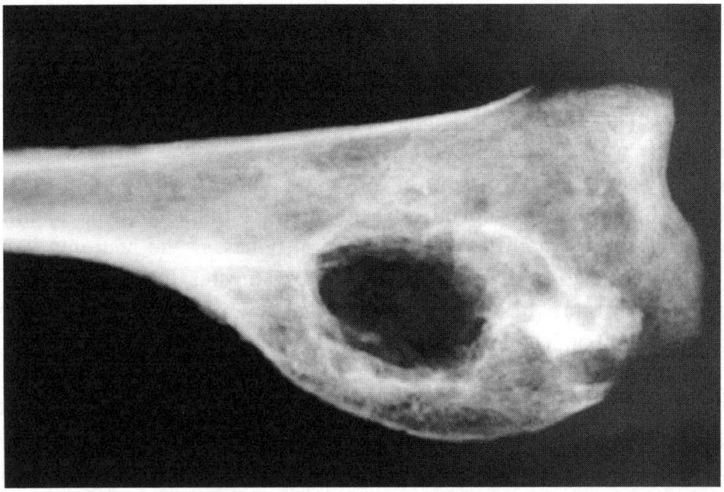

Figure 12.10 Distal femur from Saxon skeleton 44, from Finglesham, England, displaying changes possibly indicative of a giant cell tumour

and the authors note that, in about 2000 Italian burials, only 0.15 % have shown evidence of this condition. Neurofibromas at the sites of cranial nerves can also leave skeletal evidence, as indicated by a post-medieval case from London (Knüsel and Bowman, 1996).

Of the other benign tumours, little is yet known of the antiquity of the haemangioma, even though today this vascular tumour is common. An American Arikara male skull displays a small vascular lesion that has been interpreted as a haemangioma (Steele *et al.*, 1965). An Italian Iron Age case probably indicating this tumour was reported by Capasso *et al.* (1991). They also review previously described cases, which amount to a further seven. The problem in the diagnosis of this tumour is that it is not easy to distinguish from the bone changes that can occur in the development of some meningiomas (Cairns and Jupe, 1950). Indeed, there is probably a case in the differential diagnosis of some tumours for placing them in tumour groups based on eventual dry bone similarities, for we should not forget that so often our tentative diagnoses are based on the appearance of 'shadows' of the original neoplasm.

The meningioma appears to show far more variation in terms of bone response. Although it is a soft-tissue tumour developing intracranially, the bone response can be considerable. It may erode the inner table of the skull, but produce hyperostosis of bone externally (Figure 12.8c). But there is much variation in this respect, and it can be that there is little bone reaction to make a diagnosis from. Archaeological cases (Anderson, 1992) include at least seven examples from Spain (two prehistoric), at least two from Egypt, eight cases from Britain, one each from Denmark and Germany, and at least five from the Americas. In terms of meningioma occurrence in relation to burial sample size, the British cases appear to affect 0.9 % of the overall sample (Roman to medieval in date). That we must remember that such tumours probably have a Pleistocene history is exemplified by the fact that a probable case was reported in a Palaeolithic skull fragment from Grotte du Lazaret in France (de Lumley and Piveteau, 1969).

Of the other tumours of a benign nature, few have been described in archaeological material, thus suggesting a low prevalence through time. The eosinophilic granuloma is of unknown aetiology, mainly occurring in children and young adults. There can be single or multiple areas of bone rarification which can appear similar to multiple myeloma or metastases. Only four ancient cases have been described, and these are reviewed by Wada *et al.* (1987), who also provide full details of one case from Iraq.

Identification of two individuals with acoustic neuromas was made by Campillo *et al.* (1997) after a long-term search of several thousand skulls. As they point out, diagnosis is difficult because bone changes are minor, and they are unlikely to be identified without a specific search for bone modification in temporal bones. If the neuroma is not bilateral, then it may be possible to scan the internal auditory meatus for erosion and evidence of an expansion of the meatus. In view of the fact that few of us systematically examine the meatus area of the temporal bones for slow-growing tumours, we are clearly in no position to suggest how infrequent the tumours are.

Finally, as regards benign tumours that have been identified from archaeological material, the pituitary adenoma has been identified from a number of sites. About 75 % of pituitary tumours are adenomas, resulting in hypersecretion that produces gigantism or acromegaly. The first good description of an early case of the latter condition was by Keith (1931), who recognized in the vastly expanded mandible (Figure 12.3c) abnormally high temporal lines and massively thick external occipital protuberance of the skull, diagnostic features of this condition. The individual was excavated at the Norse site of Gardar in Greenland, and the man must have appeared as an impressive figure in the local community. Since then, at least five other cases have been described, three from the Americas. It is also important to note that there has been useful discussion on identifying sella turcica variation, especially from the point of view of pituitary disease (Hawkins, 1992). Sella abnormality may be associated with conditions other than acromegaly: in the case of a Bronze Age skull from Italy, the sella changes, but lack of acromegaly suggested a non-secreting adenoma (Canci *et al.*, 1992). Similarly, a medieval skull from Abingdon in England displayed an enlarged pituitary fossa that was interpreted as an expanding lesion. Hacking (1995) suggests that the tumour was either a non-secreting adenoma or even a craniopharyngioma (less likely).

Primary Bone Malignancies

While the classification and listing of tumours which are clearly malignant is considerable, discrimination at an archaeological level is, of course, far more difficult because of our inability to distinguish soft-tissue tumours of various kinds. Erosive 'shadows' of tumours that occur in particular areas (such as the nasopharynx) or which are particularly distinctive (as in leukaemia) permit perhaps a more acceptable differential diagnosis. Generally, however, beyond the few distinctive malignancies, we are left only with metastases of the skeletally undiagnostic primary soft-tissue tumours.

Osteosarcoma is a primary bone cancer that has been identified in various skeletal series. Even with this tumour, which surely is one of the most distinctive of all, there have been some debatable claims for it in the past, but the massive tumour at the knee of a young Saxon from southern England would seem to be secure (Brothwell, 1967). Another fairly certain case was reported on a proximal humerus from Switzerland, and since then further probable cases have been noted in Spain, Egypt, Czechoslovakia, Germany, England, France, Poland and Peru. An infant case of osteosarcoma (Alt *et al.*, 2002) from Germany shows that identification can be made, even when the bones are small and very immature. It is sobering to recall that the pelvic tumour from Roman Egypt, which Ruffer and Willmore (1914) believed was an osteosarcoma, does not show typical pathology and most probably is not this class of tumour.

Chondrosarcoma is a relatively common bone tumour today, but surprisingly there appears to be no evidence of it in earlier populations. This could mean that it was rare in the past, but I suspect that there are other factors involved. The pelvis is a common site of this tumour, but this is a disadvantage archaeologically as this area is often crushed and poorly preserved. The femur and humerus should also show evidence, but as yet no cases are reported in these bones. Could it be that the long-bone changes are mistaken for alternative diagnoses? The swollen shaft and inner medullary destruction could be mistaken for an infection with internal abscess formation. Without radiography, infection might be the preferred diagnosis. Or, in the tumour field, would monostotic fibrous dysplasia or a granuloma be selected in preference?

Fibrosarcoma and fibrous histiocytoma have similarly not been noted in ancient remains. There is, however, a reasonably well defined and supported case of Ewing's sarcoma from Westphalia (Löwen, 1998). Chordoma, angiosarcoma and liposarcoma remain to be described archaeologically. Two cases only of ancient ameloblastoma have been described, one being from medieval Czersk in Poland (Gladykowska-Rzeczycka, 1978). A second is from a site in Portugal, and the authors point out that these tumours, although not rare, have again probably been neglected (Dias *et al.*, 2006). So has leukaemia (Figure 12.7d), which is primarily a blood condition, but produces diffuse osteopaenia, widespread lytic lesions in bones, and enlarged vascular foramina, also failed to be recognized? In an earlier phase of the disease, could the changes be misinterpreted as osteoporosis or even nutritional hyperparathyroidism? Fortunately, leukaemia is being discussed more and more in palaeopathological literature, with Rothschild *et al.* (1997) discussing two cases in detail.

Myeloma is a tumour of neoplastic plasma cells from bone marrow, and can severely affect the skeletal tissues. In Britain, 0.001 % of the population will be diagnosed each year (about 600 cases). In advanced societies today, it is uncommon below 60 years of age, with more males affected. Skeletal changes are especially common in bones with active haemopoiesis. Usually multiple lesions occur, and it is important to distinguish these from metastatic deposits of the primary tumours (Figure 12.11). The small rounded lesions have sharp edges, whereas metastases tend to be larger and with more 'ragged' margins. Wells (1964) describes two possible medieval cases, but is hesitant to come down firmly on a myelomatosis diagnosis. Cases have occurred in the New World as well as in Europe, Egypt and beyond. Morse *et al.* (1974) describe four Amerindian cases which show variation in the size and roundness of the bone lesions. After careful consideration, they were satisfied that two of the cases were myelomatosis, but were less certain of the others. Other probable cases have been found in areas as divergent as Australia (Webb, 1995), Turkey (Pinhasi, personal communication), southern Russia (Buzhilova, 2006), and Germany (Alt and Adler, 1992). The earliest case was of Neolithic date (Strouhal, 1991). In British material, there are three cases from three sites (1873 bodies), which seems to give a crude prevalence of 0.16%, which is only slightly higher than modern clinical data suggest. The immunological confirmation of a dry-bone diagnosis of multiple myeloma has been a significant advance in the identification of tumours (Cattaneo *et al.*, 1994), and it is to be hoped that the further application of enzyme-linked immunosorbent assay or alternative techniques will help to improve on diagnosis in the future.

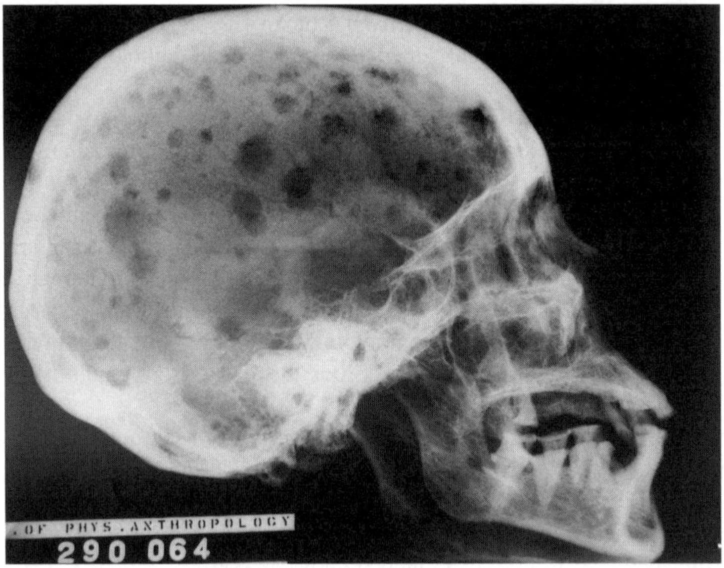

Figure 12.11 Lateral X-ray of a skull from Indian Knoll, Kentucky, showing multiple lytic lesions of varying size, most with ragged margins. Probably metastatic carcinoma rather than multiple myeloma

Metastatic Tumours in Archaeological Bone

The most difficult problem in studying tumours is in the discrimination of the secondary, metastatic tumours. In some cases, of course, it may be possible to link metastases with bone destruction that might indicate the primary malignant lesion. A good example is probably the skull of a III–Vth Dynasty Egyptian (Wells, 1963), which displays major (primary?) destruction of the left maxillo-alveolar region, part of the palate and inferior concha. This destruction is surrounded by a marginal sub-periosteal reaction, and in the skull vault generally are a series of rounded lytic perforations. Strouhal (1991) describes another probable nasopharyngeal carcinoma, and Pahl (1986) lists five other cases from Egypt. Webb (1995) also calls our attention to two Australian cases.

The position of osteolytic damage, with or without secondaries, could also at times suggest the organ or structure primarily involved. For instance, there is a Nubian skeleton with major erosion of the sacrum, which Elliot Smith and Derry (1910) suggest is due to rectal cancer (more specifically a chordoma). Linking historic records with dry-bone pathology is an innovative way of considering metastases in relation to known primary tumours, as Waldron (1997) demonstrates on a London case.

At this stage in the study of ancient neoplasms, it would be useful to have a detailed review of all the cases displaying secondary deposits. Are there any patterns to be found in terms of size, shape, number and distribution of the metastases? And what of the ages of the individuals and the sexes? Reviewing the literature suggests strongly that sample sizes are now sufficient to begin to look for patterns and potential significance. Ricci *et al.* (1994) have also argued that, in the case of childhood malignancy, there may be value in considering the age of the child in relation to the tumour. They point out that some malignancies are quite preferential in terms of age; for instance, the neuroblastoma and nephroblastoma tend to develop in children under 6 years of age. All such relationships deserve to be critically reviewed in relation to past populations, but it should be remembered that we also need to undertake further studies on the impact of cultures and the environment on past populations. Metastatic carcinoma may also occur in an individual who may already have a bone-changing disease, as in the leper from Chichester (Ortner *et al.*, 1991), and clearly it is important in such cases to separate the pathologies clearly.

From the evidence of skeletal metastases assembled by Šefčaková *et al.* (2001), Gladykowska-Rzeczycka (1991), Mays *et al.* (1996) and other studies, the total number so far noted is well in excess of 60, but detailed information on each case has yet to be provided. In a much smaller sample, where there are more data, the males and females are roughly equal. What is striking, however, is that where a fairly precise age range has been given for 15 individuals, eight were considered to be probably under 50 years, and most were judged to be under 40 years. If the age estimates (usually in a 10-year range) are correct, then metastases were commonly occurring at a younger age than we would expect today.

CONCLUSIONS

My intention has not been to provide a comprehensive and exhaustive analysis of tumour evidence from the past, but at least to review the range of tumours that have been tentatively identified in past populations. Linked to this brief survey, I have made comment on a classification of tumours and their main features from a diagnostic point of view. It can be seen that both benign and malignant forms are well in evidence, probably beginning in the Palaeolithic, and certainly by the Neolithic period. In the field of palaeopathology, diagnosis is made especially difficult by post-mortem changes and the fact that often we are only viewing the skeletal shadows of past soft-tissue tumours. Although the majority of evidence is preserved as skeletal remains, we do have a limited amount of soft-tissue evidence from mummies. We are clearly in need of more population data, so that we can get a better idea of the occurrence of varieties of tumour. Strouhal (1998, 2001) has pioneered the regional assembly of tumour data, dividing the evidence into periods spanning 6000 years, and has concluded that tumour numbers increase through time, possibly because of increases in carcinogenic agents. More information of this kind is needed, but with statistical adjustments related to the number of skeletons available for study. Unfortunately, material from the prehistoric periods tends to be less and also more fragmentary and incomplete. As Waldron (1994) has emphasized, we ultimately need to view past neoplastic data in modern epidemiological terms.

In discussing individual diagnoses for archaeological specimens, I have refrained from listing alternative diagnoses in most cases. This is not because diagnosis can be precise, but simply because lists of alternative possibilities in diminishing probability are to be avoided. In a review of this kind, nothing is to be gained by it; nevertheless, it is important to keep in mind that the preferred diagnosis follows a differential diagnosis, where various alternative diagnoses are considered. It could be that, in some cases, one of the alternative diagnoses will turn out to be correct. One of the big questions left for the future of this subject is to improve on the accuracy of diagnosis; this emphasizes the necessity for curation of collections, as only then can specimens be re-examined, and perhaps rediagnosed, by future workers.

Does this growing tumour evidence have a place in evolutionary biology? I began by mentioning the early work of Bland Sutton, and it is appropriate to end by saying that, ultimately, we should consider tumours as a part of our evolutionary biology in various ways. From studies on other mammals, it is clear that different breeding lines vary greatly in their chances of developing tumours (Nagase *et al.*, 1996). There are also significant species differences in tumour incidences. We also know that human families vary considerably in their resistance or otherwise to tumour development. The nature of these differences may well be complex, and may be more concerned with cancer suppressors rather than specific oncogene mutations (Weiss, 1999). Nevertheless, as Bodmer and Tomlinson (1996) believe, cancer is a 'somatic evolutionary process' and we need to construct potential evolutionary models not on populations of people, but of cells.

The potential value of recording ancient cases of neoplasms, especially malignant forms, is that we may be able to provide evidence that the prevalence of some tumour types has changed through time, implying that mutation rates associated with cancer may have changed significantly. At present, this is pure theory, and we have a very long way to go in palaeopathology and palaeoepidemiology before this might be achieved.

ACKNOWLEDGEMENTS

I much appreciate the assistance in various ways of the Trustees of the London Natural History Museum, the Lowie Museum of Berkeley, California, the Anthropology Laboratory of the University of Copenhagen, the Director of the Duckworth Laboratory of Cambridge University, the Smithsonian Institution of Washington, and the Photographic Studio of the University of York. The drawing of the osteochondroma is by Rosemary Powers. Figures 12.1, 12.2, 12.6 and 12.7 are derived and modified from X-rays in Simon (1965), Greenfield (1969) and Grainger and Allison (1986), and my thanks to the publishers.

REFERENCES

Ackerknecht EH. 1965. *History and Geography of the Most Important Diseases*. Hafner: New York.

Alderson M. 1986. *Occupational Cancer*. Butterworths: London.

Alt KW, Adler CP. 1992. Multiple myeloma in an early medieval skeleton. *Int J Osteoarchaeol* **2**: 205–209.

Alt KW, Adler CP, Buitrago-Tellez CH, Lohrke B. 2002. Infant osteosarcoma. *Int J Osteoarchaeol* **12**: 442–448.

Ammitzbøll T, Anderson SR, Andersson HP, Bodenhoff J, Eiken R, *et al.* 1991. The people. In *The Greenland Mummies*, Hansen JPH, Meldgaard J, Nordqvist J (eds). British Museum Publications: London; 64–101.

Anderson T. 1992. A medieval example of meningiomatous hyperostosis. *J Paleopathol* **4**: 141–154.

Aufderheide AC. 2003. *The Scientific Study of Mummies*. Cambridge University Press: Cambridge.

Aufderheide AC, Rodriguez-Martin C. 1998. *The Cambridge Encyclopedia of Human Paleopathology*. Cambridge University Press: Cambridge.

Baum M. 1993. Breast cancer 2000BC to 2000AD – time for a paradigm shift? *Acta Oncol* **32**: 3–8.

Bennike P. 1985. *Palaeopathology of Danish Skeletons: A Comparative Study of Demography, Disease and Injury*. Akademisk Forlag: Copenhagen.

Bland-Sutton J. 1890. *Evolution and Disease*. Scott: London.

Bland-Sutton J. 1893. *Tumours, Innocent and Malignant*. Cassell: London.

Bodmer WF, Tomlinson I. 1996. Population genetics of tumours. In *Variation in the Human Genome*, Chadwick D, Cardew G (eds). Wiley: Chichester; 181–193.

Brobeil SAJ, López MCB, Rodriguez LR. 1992. A contribution to medieval pathological gynaecology. *J Paleopathol* **4**: 155–161.
Brothwell D. 1961. The palaeopathology of early British man: an essay on the problems of diagnosis and analysis. *J R Anthropol Inst* **91**: 318–343.
Brothwell D. 1967. The evidence for neoplasms. In *Diseases in Antiquity*, Brothwell D, Sandison AT (eds). Charles C Thomas: Springfield; 320–345.
Brothwell D. 1991. On zoonoses and their relevance to paleopathology. In *Human Paleopathology: Current Syntheses and Future Options*, Ortner DJ, Aufderheide AC (eds). Smithsonian Press: Washington, DC; 18–22.
Burkitt D. 1967. African lymphoma. In *Racial and Geographical Factors in Tumour Incidence*, Shivas AA (ed.). Edinburgh University Press: Edinburgh; 19–23.
Buzhilova A. 2006. A possible case of multiple myeloma in a skeleton of early medieval age from northern Caucasus. In *16th European Meeting of the Paleopathology Association*, Santorini, Greece.
Cairns H, Jupe MH. 1950. Central nervous system. In *A Textbook of X-Ray Diagnosis*, Shanks SC, Kerley P (eds). Lewis: London; 3–242.
Campillo D, Carvajal A, Prim J. 1997. Acoustic neuromas in paleopathology. *J Paleopathol* **9**: 23–36.
Canci A, Repetto E, Tarli SMB. 1992. Sella pathology in a middle Bronze Age skull from southern Italy. *Int J Osteoarchaeol* **2**: 305–310.
Capasso L, Muzio MD, Tota GD, Spoletini L. 1991. A case of probable cranial hemangioma (Iron Age, central Italy). *J Paleopathol* **4**: 55–62.
Cattaneo C, Gelsthorpe K, Phillips P, Waldron T, Booth JR, Sokol RJ. 1994. Immunological diagnosis of multiple myeloma in a medieval bone. *Int J Osteoarchaeol* **4**: 1–2.
Chamberlain AT, Rogers S, Romanowski CA. 1992. Osteochondroma in a British Neolithic skeleton. *Br J Hosp Med* **47**: 51–3.
De Lumley M, Piveteau J. 1969. Les restes humains de la grotte du Lazaret. In *Mémoires de la Société Préhistorique Française*, tome 7, de Lumley H (ed.). Société Préhistorique Française: Paris; 223–232.
Department of Health and Social Security. 1970. *Digest of Health Statistics for England and Wales*. HMSO: London.
Dias GJ, Santos, AL,Prasad K. 2006. The osteological characteristics of a mandibular ameloblastoma in an identified male individual from the International Exchange Collection (Coimbra from early 20th century). In *16th European Meeting, Paleopathology Association*, Santorini, Greece.
Doll R, Armstrong B. 1981. Cancer. In *Western Diseases: Their Emergence and Prevention*, Trowell HC, Burkitt DP (eds). Arnold: London; 93–110.
Elliot Smith G, Derry DE. 1910. Anatomical report. *Archaeological Survey of Nubia*, Bulletin No. 5. Archaeological Survey of Nubia: Cairo; 11.
Fornaciari G, Castagna M, Naccaranto G, Collecci P, Tognetti A, Bevilacqua G. 1993. Adenocarcinoma in the mummy of Ferrante I of Aragon, King of Naples (AD1431–1494). *Paleopathol Newslett* **83**: 5–8.
Gladykowska-Rzeczycka J. 1978. Mandibular tumour in a male skeleton from a medieval burial ground in Czersk. *Folia Morphol* **37**: 191–196.
Gladykowska-Rzeczycka J. 1991. Tumours in antiquity in east and middle Europe. In *Human Paleopathology, Current Syntheses and Future Options*, Ortner DJ, Aufderheide AC (eds). Smithsonian Press: Washington, DC; 251–256.
Grainger RG, Allison DJ (eds). 1986. *Diagnostic Radiology*. Churchill: Edinburgh.
Granville AB. 1825. An essay on Egyptian mummies, with observations on the art of embalming among the ancient Egyptians. *Phil Trans R Soc Lond* **1**: 269–316.
Greenfield GB. 1969. *Radiology of Bone Diseases*. Pitman Medical: London.
Hacking P. 1995. A pituitary tumour in a medieval skull. *Int J Osteoarchaeol* **5**: 390–393.
Hawkins D. 1992. The diagnosis of pituitary disease from human skeletal remains. *Int J Osteoarchaeol* **2**: 51–64.

Holland WW (ed.). 1988. *European Community Atlas of Avoidable Death*. Oxford University Press: Oxford.
Hooton EA. 1930. *The Indians of Pecos Pueblo*. Yale University Press: New Haven, CT.
Horne PD. 1986. Angiokeratoma in Inca mummy. *Paleopathol Newslett* **54**: 16.
Howe GM. 1970. *A National Atlas of Disease Mortality in the United Kingdom*. Blackwell: London.
Howe GM. 1972. *Man, Environment and Disease in Britain*. David and Charles: Newton Abbott.
Keith A. 1931. *New Discoveries Relating to the Antiquity of Man*. Williams and Norgate: London.
Knüsel CJ, Bowman JE. 1996. A possible case for neurofibromatosis in an archaeological skeleton. *Int J Osteoarchaeol* **6**: 202–210.
Kramar C, Baud CA, Lagier R. 1983. Presumed calcified leiomyoma of uterus: morphological and clinical studies of a calcified mass dating from the Neolithic period. *Arch Path Lab Med* **107**: 91–103.
Löwen H. 1998. A Ewing's sarcoma from an early medieval hillside in Westphalia. *J Paleopathol* **10**: 127–132.
Mays S, Strouhal E, Vyhnánek L, Němečková A. 1996. A case of metastatic carcinoma of medieval date from Wharram Percy, England. *J Paleopathol* **8**: 33–42.
Morse D, Dailey RC, Bunn J. 1974. Prehistoric multiple myeloma. *Bull N Y Acad Med* **50**: 447–458.
Nagase H, Bryson S, Fee F, Balmain A. 1996. Multigenic control of skin tumour development in mice. In *Variation in the Human Genome*, Chadwick D, Cardew G (eds). Wiley: Chichester; 156–180.
Ortner DJ. 2003. *Identification of Pathological Conditions in Human Skeletal Remains* (2nd edition). Academic Press: London.
Ortner DJ, Manchester K, Lee F. 1991. Metastatic carcinoma in a leper skeleton from a medieval cemetery in Chichester, England. *Int J Osteoarchaeol* **1**: 91–98.
Pahl WM. 1986. Tumours of bone and soft tissue in Ancient Egypt and Nubia: a synopsis of the detected cases. *Int J Anthropol* **1**: 267–276.
Price CHG. 1970. The incidence of primary malignant bone tumours in the south-west of England. In *Symposium Ossium*, Jelliffe A, Strickland B (eds). Livingstone: Edinburgh; 331–334.
Resnick D (ed.). 1995. *Diagnosis of Bone and Joint Diseases*. WB Saunders: Edinburgh.
Ricci R, Lama R, di Tota G, Pietrangelo F, Vecchio F, Evangelista A, Capelli A, Capasso L. 1994. Skull osteolytic lesions in a XV century child: a case of childhood malignancy. *J Paleopathol* **6**: 151–159.
Ricci R, Lama R, di Tota GD, Capelli A, Capasso L. 1995. Some considerations about the incidence of neoplasms in the human history. *J Paleopathol* **7**: 5–11.
Roberts C, Cox M. 2003. *Health and Disease in Britain*. Sutton: Stroud.
Roberts C, Manchester K. 1995. *The Archaeology of Disease* (2nd edition). Cornell University Press: Ithaca, NY.
Rokhlin DG. 1965. *Diseases of Ancient Men*. Nauka: Moscow (in Russian with English summary)
Rothschild BM, Hershkovitz I, Dutour O, Latimer B, Rothschild C, Jellema LM. 1997. Recognition of leukemia in skeletal remains: report and comparison of two cases. *Am J Phys Anthropol* **102**: 481–496.
Ruffer MA, Willmore JG. 1914. Note on a tumour of the pelvis dating from Roman times (250AD). *J Path Bact* **18**: 480–484.
Sandison AT. 1967. Diseases of the skin. In *Diseases in Antiquity*, Brothwell D, Sandison AT (eds). Charles C Thomas: Springfield; 449–456.
Šefcáková A, Strouhal E, Němečková A, Thurzo M, Staššiková-Štukovská D. 2001. Case of metastatic carcinoma from end of the 8th–early 9th century Slovakia. *Am J Phys Anthropol* **116**: 216–229.
Shanmugaratnam K. 1967. Nasopharyngeal carcinoma in Asia. In *Racial and Geographic Factors in Tumour Incidence*, Shivas AA (ed.). Edinburgh University Press: Edinburgh; 169–188.
Simon G. 1965. *Principles of Bone X-Ray Diagnosis*. Butterworth: London.
Spigelman M, Bentley P. 1997. Cancer in Ancient Egypt. *J Paleopathol* **9**: 107–114.

Steele JP, Gregg JB, Clifford S, Holzhueter AM. 1965. Paleopathology in the Dakotas. *S Dak J Med* **18**: 17–29.

Stoker DJ. 1986. Bone tumours. In *Diagnostic Radiology*, Grainger RG, Allison DJ (eds). Churchill: Edinburgh; 1273–1348.

Stroud G, Kemp RL. 1993. *Cemeteries of the Church and Priory of St Andrew's, Fishergate. The Archaeology of York: Fascicule12/2.* Council for British Archaeology: York.

Strouhal E. 1976. Tumours in the remains of ancient Egyptians. *Am J Phys Anthropol* **45**: 613–620.

Strouhal E. 1991. A case of primary carcinoma from Christian Sayala (Egyptian Nubia). *J Paleopathol* **3**: 151–166.

Strouhal E. 1998. Survey and analysis of malignant tumours of past populations in England and Scotland. *J Paleopathol* **10**: 101–109.

Strouhal E. 2001. Malignant tumours in past populations in middle Europe. In *Proceedings XIIIth European Meeting of the Paleopathology Association*, Verghetta M, Capasso L (eds). Edigrafital SpA: Teramo; 265–272.

Strouhal E, Jungwirth J. 1977. Ein Verkaltes Myoma Uteri aus der Spaete Romerzeit in Egyptisch-Nubien. *Mitt Anthropol Gesellsch Wien* **107**: 215–221.

Strouhal E, Vyhnanek L, Horáčková L, Benešová L, Němečková A. 1996. Two unusual benign tumours in skulls from the ossuary at Křtiny (Czech Republic). *Int J Osteoarchaeol* **6**: 289–299.

Ting RY. 1981. Diseases of the Chinese. In *Biocultural Aspects of Disease*, Rothschild HR (ed.). Academic Press: London; 327–357.

Verghetta M, Capasso L. 2001. Paleopathology of uterine fibroma: direct and indirect sources. In *Proceedings XIIIth European Meeting of the Paleopathology Association*, Verghetta M, Capasso L (eds). Edigrafital SpA: Teramo; 145–148.

Vermeij J. 1970. The periosteal reaction in malignant bone tumours. In *Symposium Ossium*, Jelliffe A, Strickland B (eds). Livingstone: Edinburgh; 310–313.

Wada Y, Ikeda J, Suzuki T. 1987. Tumor-like lesions in a human skeleton from the Himrin basin in Iraq. *J Anthropol Soc Nippon* **95**: 107–119.

Waldron T. 1994. *Counting the Dead. The Epidemiology of Skeletal Populations.* Wiley: Chichester.

Waldron T. 1997. A nineteenth-century case of carcinoma of the prostate, with a note on the early history of the disease. *Int J Osteoarchaeol* **7**: 241–247.

Waldron T. 2001. *Shadows in the Soil: Human Bones and Archaeology.* Tempus: Stroud.

Webb S. 1995. *Paeopathology of Aboriginal Australians.* Cambridge University Press: Cambridge.

Weiss KM. 1999. *Genetic Variation and Human Disease.* Cambridge University Press: Cambridge.

Wells C. 1963. Ancient Egyptian pathology. *J Laryngol Otol* **77**: 261–265.

Wells C. 1964. Two medieval cases of malignant disease. *Br Med J* **1**: 1611–1612.

Williams RR, Stegens NL, Horm JL. 1977. *J Nat Cancer Inst* **58**: 518–524.

Zimmerman MR. 1995. Paleohistology of ancient soft tissue tumours. In *Palaeopathological Association Meeting*, Oakland, California.

13

Advances in the Palaeopathology of Teeth and Jaws

Alan Ogden

Biological Anthropology Research Centre, Department of Archaeological Sciences, University of Bradford, Bradford BD7 1DP, UK

INTRODUCTION

Teeth, being the most highly mineralized body structures, are a unique and indeed sometimes the sole surviving human remains relating to past individuals and the societies to which they belonged. Unlike bones, teeth interact directly with the environment (chewing, wear and trauma) and, after death, are highly resistant to most taphonomic processes. Their surfaces retain the characteristic wear patterns produced by different types of diet or by use as a tool or vice. The well-preserved lesions produced by diseases of teeth and their supporting structures can tell us much about health and diet, which is usually strongly related to economic and social status. Teeth, unlike bone, do not undergo remodelling in life, so they retain analysable deposits laid down incrementally in dentine and enamel during the childhood of the individual, and these can reveal detailed patterns of nutrition, disease or migration. For the above reasons, teeth always seem to take up, what seems to the novice, a disproportionate amount of space in any book on palaeopathology.

The structure of teeth and the identification and scoring of caries have recently been dealt with thoroughly in the literature (Nanci, 2003; Hillson, 2005); this chapter, therefore, will deal only with selected topics where a fresh approach would now seem to be appropriate in the light of greater clinical understanding. These comprise: gross dental enamel hypoplasia (DEH) and cuspal malformations in molars; periodontal disease, its identification, scoring and relevance to general health; and periapical voids in jaw bones and alveolar response to the death of the dental pulp.

Advances in Human Palaeopathology Edited by Ron Pinhasi and Simon Mays

GROSS DENTAL ENAMEL HYPOPLASIA AND CUSPAL MALFORMATIONS IN MOLARS

Linear or pitted defects in enamel (DEHs), have long been used as a non-specific indicator of systemic physiological stress during early life (Suckling, 1989; Goodman and Rose, 1990; Guatelli-Steinberg and Lukacs, 1999). Bouts of malnutrition, disease and fever are known to depress the activity of the ameloblasts and to result in the production of a thin and poorly calcified enamel matrix, with the formation of linearly distributed pits or grooves of defective enamel. Nearly 100 systemic conditions have also been associated with DEHs, but most are rare (Cutress and Suckling, 1982). It is important to remember, however, that minor hypoplasias and hypomineralizations also appear with surprising frequency in teeth of apparently healthy children, being present in as many as 43–64% of individuals in modern populations (Suckling and Pearce, 1984; Suckling *et al.*, 1985; Dummer *et al.*, 1986).

Enamel is not remodelled during life and every individual's enamel is a record of the first 10 to 11 years of their life when their crowns are being formed (Smith, 1991; Skinner and Goodman, 1992; Hillson and Bond, 1997). It must be remembered that the layers of enamel forming the surface of a tooth crown hide a considerable proportion of the earlier enamel deposition. For anterior teeth, the surface hides some 10–20% of crown formation time. For molars, up to one-half of crown formation time may be hidden from view beneath the cusps (Hillson, 1992). In addition, the frequency of DEH in archaeological individuals is usually underestimated, as the very high level of tooth wear in past populations removes evidence of hypoplasia from the first-formed regions of each tooth (the amelo-dentinal junction beneath the cusp tips) in those individuals who die in adolescence or adult life (Hillson, 2005), and it these very regions that we are most concerned with in this section.

This study was triggered by examination of the teeth from 45 subadults recovered from a large cemetery uncovered in 1985 in east central London during the demolition of Broad Street Railway station and the building of the Broadgate development (White, 1987; Harding, 2002). The 'New Graveyard' was founded in AD 1569 by the city as an overflow cemetery to relieve the congestion occurring in London's parish burial grounds and it continued in use until the mid 18th century (Schofield and Maloney, 1998; Harding, 2002). Only a small part of this cemetery was excavated during the development: some 388 individuals, of whom 45 were subadults, were retrieved and stored at the Museum of London (White, 1987).

Examination of the 521 teeth from the Broadgate subadults revealed a level of disturbed enamel formation that was quite exceptional, not only in prevalence, but also in severity, compared with other archaeological populations (Saunders *et al.*, 2000; Hillson, 2005). Some of this enamel hypoplasia was of the common linear or pitted character, but much of the disruption of enamel was far more extensive and profound than that usually encountered. Of the 521 teeth examined, 46 showed moderate to severe hypoplasia; and of these, 17 were deciduous second molars and 22 were permanent first molars. Gross examination and scanning electron microscopy (SEM) revealed many molars to have grossly deformed cuspal architecture (Figures 13.1–13.3). Large areas of Tomes' process pits, where the ameloblasts evidently ceased matrix production abruptly, are exposed between islands of more normal enamel (Figures 13.2 and 13.3). As soon as such teeth erupt into the mouth they will rapidly decay, as bacteria can readily lodge on the rough and incompletely mineralized enamel. With all affected molars it is striking that the cervical third of the crowns is uninvolved, often with a clear demarcation between the cuspal and the cervical enamel (Figures 13.1–13.3). The

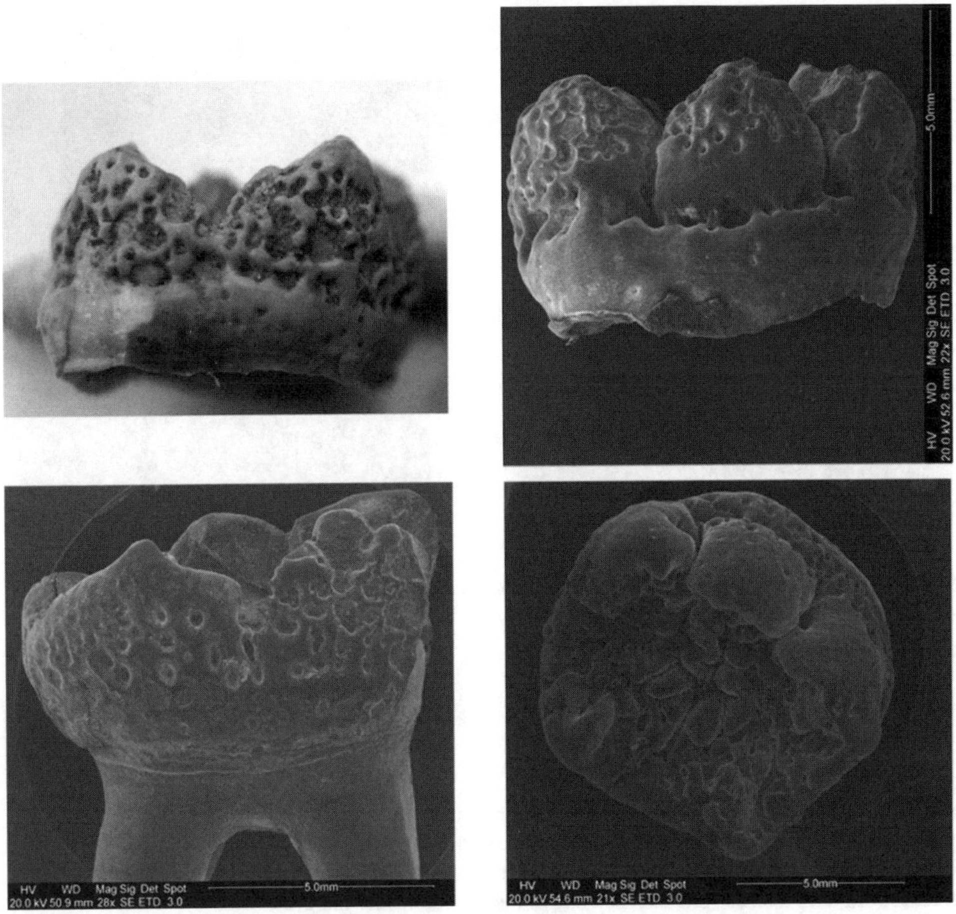

Figure 13.1 Disturbed enamel formation in upper and lower deciduous and permanent molars in different individuals from Broadgate. Note the multiple indentations in the enamel and the disturbance of the normal cusp pattern, and that the cervical third of the crowns seem relatively unaffected

smooth bulge of enamel, with perikymata grooves, apparently marks the return to normal enamel matrix formation.

The only molar illustrated in the literature with DEH resembling that in the Broadgate material is shown in Jenkins (1978: 262), and is simply labelled by him as 'gross hypoplasia' ". This type of hypoplasia, although very rare, needs to be designated as a distinct entity. With the distinctive disruption of cusp pattern along with incomplete enamel formation, this new type of hypoplasia should be described as 'cuspal enamel hypoplasia' (CEH). Figure 13.4 compares the common pitted and linear (furrowed) hypoplasia with the rare plane form and even rarer CEH.

In DEH, ameloblasts are evidently affected during the matrix secretion stage, altering or even terminating the physiochemical conditions for the commencement of the maturation process (Suga, 1989). At Broadgate, equivalent teeth in the other quadrants of the mouth are

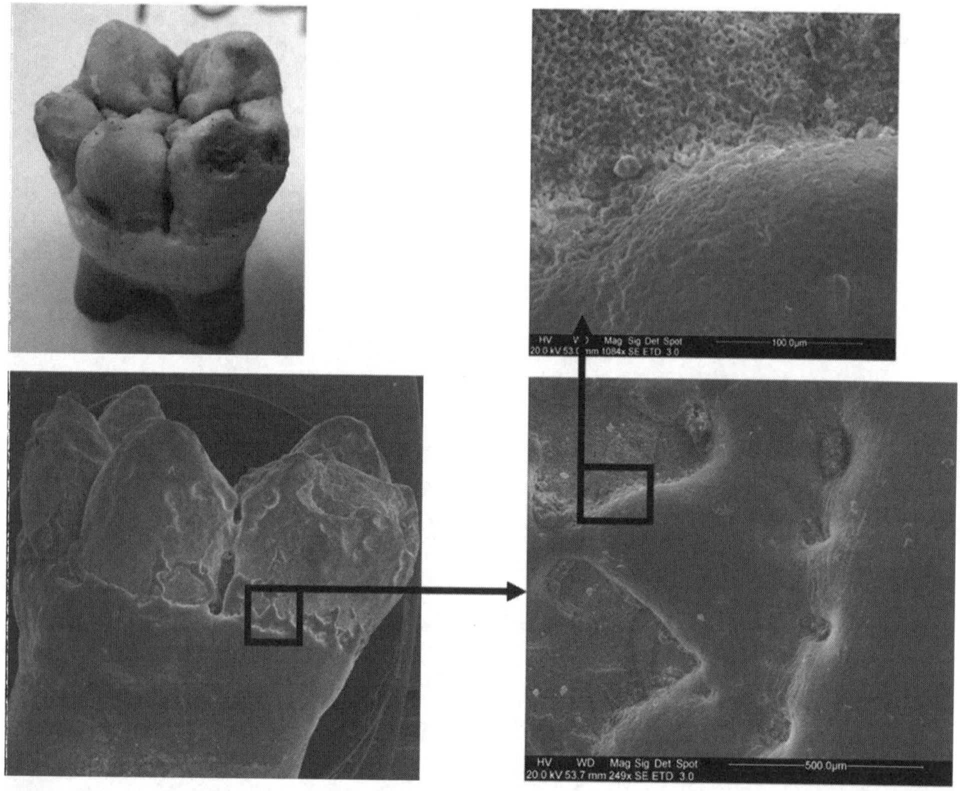

Figure 13.2 Lower right first permanent molar from Broadgate that was partially erupted, showing underdevelopment of enamel over the cuspal regions of the crown. Caries has already attacked the first erupted cusp. The scanning electron micrographs reveal the presence of incompletely mineralized enamel with deep Tomes' pits exposed above the rolled enamel margin

all affected to a varying extent; the systemic disorder must have occurred during the first 2 years of life (Moorrees *et al.*, 1963a,b; Smith, 1991), and this usually includes the period of weaning. This time is especially precarious for infants, with an increase in nutritional stress due to the sudden loss of nutrients provided by human milk, as well as a decrease in immunity due to the decreased immunoglobulin levels and the loss of immunity provided by the mother's milk. The child is also subjected to its first real extramaternal contact with the environment and new pathogens (Larsen, 1987).

The reason why first permanent molars appear most dramatically affected by CEH may be that first permanent molar crowns form in 3.8 years, i.e. half the time that permanent canines take, and so are much more vulnerable to short systemic disturbances (Ensor and Irish, 1995; Fitzgerald, 1998). As the cervical part of the enamel is usually unaffected, it could be that the insult only has potential to disturb the ameloblasts seriously during the child's earliest years. Hillson (1992) suggested that this may be due to the abrupt change from widely spaced perikymata occlusally to narrowly spaced cervically, when defects would be less extensive and obvious. Another possible explanation is that the thickness of the enamel influences the ability of the ameloblasts to resist the insult (Jälevik *et al.*, 2005).

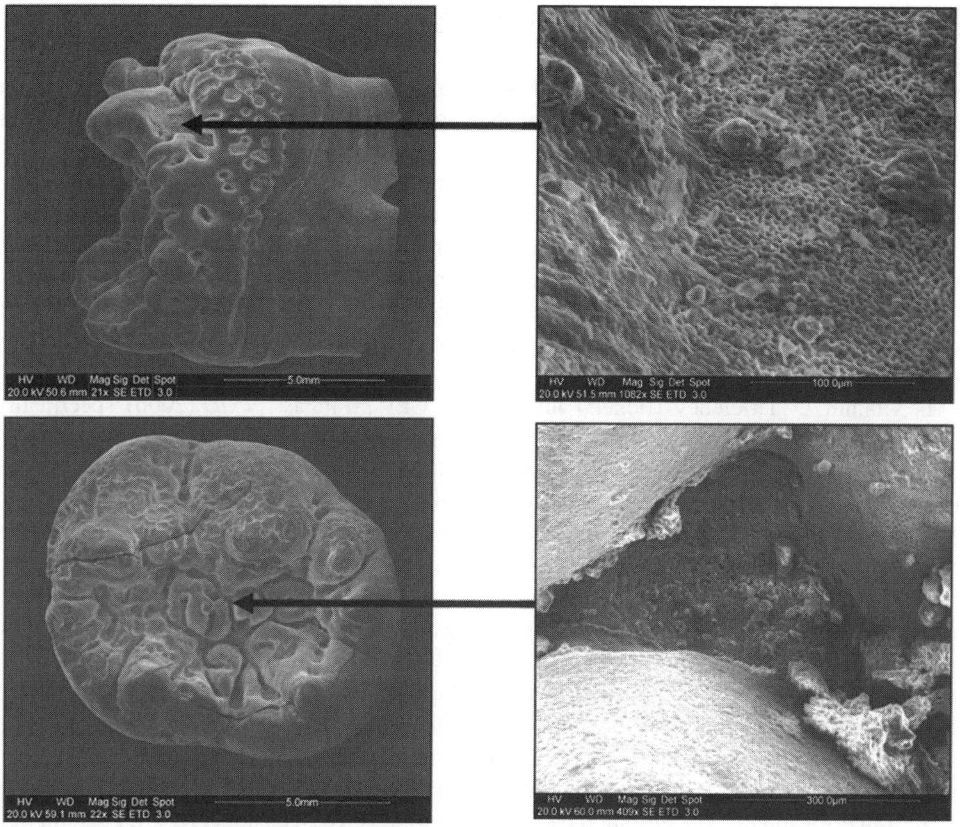

Figure 13.3 Developing upper and lower first permanent molars from different individuals from Broadgate. The abnormalities of cusp pattern and (in close-up) poorly mineralized enamel in the crevices reveal how vulnerable these teeth would be to rapid destruction by caries once erupted. It is for this reason that such teeth have seldom survived to attract attention as being hypoplastic

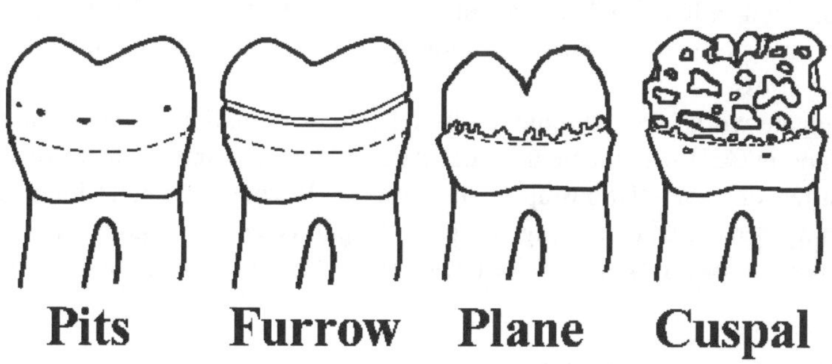

Figure 13.4 Different types of enamel hypoplasia seen on molars. Pits and furrows (linear enamel hypoplasias) are well described in the literature. Plane-form hypoplasia is much rarer and is usually bounded and overlapped by an irregular border of relatively normal enamel, but occasionally all the occlusal enamel is thin and there may be only a cervical border of relatively normal enamel. CEH is a combination of irregular plane-form defects with extensive and irregular non-linear pitting, but with a disruption of normal cusp formation and the formation of additional small cusps on the occlusal surface

Seow *et al.* (2005), studying primary dental enamel from modern pre-term children, have reported that the incisal enamel is 20% thinner than in full-term children, but none of the teeth they examined showed visible hypoplasia, although 52% showed minor defects under SEM. They did not report on molars, but their work suggests that the massive disruptions sometimes seen in molars are not related to premature birth.

A Dutch pilot study of medical records from children with first molars with developmental defects found that 48% of the cases had problems related to birth and 67% suffered from respiratory disease (van Amerongen and Kreulen, 1995). These figures were higher than normal, and the authors hypothesized that oxygen shortage might influence the mineralization of the enamel. Chronic diarrhoea or environmental toxins, dioxins in particular, taken up directly with food or via breast milk, have also been suggested as possible causes (Alaluusua *et al.*, 1996).

Is vitamin D deficiency (rickets) involved in CEH? Pinhasi *et al.* (2006) specifically examined the Broadgate subadults, but could not show a correlation between rickets and molar hypoplasia. However, it has been suggested that vitamin D deficiency may cause poor enamel mineralization without the classical skeletal signs of rickets (Kunzel, 2003). Purvis *et al.* (1973) reported on major deciduous molar hypoplasia in 50% of individuals known to have suffered from a few days neonatal tetany.

The apparently unique severity and frequency of DEH in the Broadgate population lead us to ask whether these lesions are perhaps the result of some infection, in combination with likely malnutrition. The appearance of molars with DEH at Broadgate differs from that of the 'mulberry molars' of congenital syphilis: the channels of exposed poorly mineralized enamel between the islands of more normal smooth enamel are very different, as in congenital syphilis the enamel layer appears intact and smooth, even in the crevices, and only cusp architecture is affected (Hillson *et al.*, 1998: 175). Whatever the cause, the effects appear to have lasted for months rather than days and seem to have ceased suddenly, with resumption of normal enamel production. Hillson (2005), however, considers that exposed planes 'may occupy half the crown height, but still relate to a momentary disruption of amelogenesis'.

Historical sources suggest that during the 17th and 18th centuries AD, mean age for the commencement of weaning in Britain dropped from 18 months to 7 months (Lewis, 2002). That cusp tips on deciduous and permanent molars are affected suggests that growth faltering started from birth or even *in utero*. It is unlikely, therefore, that cuspal hypoplasia is due to post-weaning stress.

The Broadgate work is an ongoing study. Histological analysis using ground sections of molars displaying CEH will enable us to clarify the formation and sequence of these defects. By examination of their relationship to the striae of Retzius, neo-natal lines and evidence of the weaning period, and by using the recently published chronology of Reid and Dean (2006), we hope to understand CEH better, in particular its aetiology and timing.

PERIODONTAL DISEASE

Periodontal disease is the result of the accumulation of bacterial plaque at the gum margins (gingivae) leading to inflammation and destruction of the periodontal tissues that anchor the tooth to the jawbone. It is prevalent in mammals and most human populations and results in significant bone destruction, with the teeth loosening, drifting and exfoliating, leading to grossly premature tooth loss in severely affected individuals (Jenkins and Mason, 1984a,b;

Baelum and Fejerskov, 1986; Sheiham, 1991; Chestnutt *et al.*, 2000; McAul *et al.*, 2000; Steele *et al.*, 2000; Dowsett *et al.*, 2001). Even in modern Britain, with our access to dentistry and oral hygiene aids, 43% of the population display active periodontal disease and 54% have at least one tooth that has lost more one-third of its support from the alveolar bone (Morris *et al.*, 2001). Much research has been carried out into the epidemiology, aetiology, prevention, and clinical management of human periodontal disease, and this has resulted in a significant increase in our understanding of the condition, but also a realization that many earlier studies are of doubtful validity.

Traditionally, periodontal disease has been divided into gingivitis and periodontitis, depending on whether destruction of the periodontal attachment has occurred.

Gingivitis

Chronic gingivitis may be defined as inflammation of the marginal gingival tissues due to the accumulation of dental plaque. In life it is characterized by redness, swelling, and bleeding from the gum margins. The condition is usually painless and is reversible if oral hygiene is improved. Many epidemiological studies have reported that chronic gingivitis is present somewhere in the mouths of more than 95% of the adult population. Some workers have even questioned whether gingivitis can be considered a disease at all, because it is a normal response to the colonization of the teeth by the commensal oral flora and it does not result in any significant damage to the host (Greene, 1986; Loë *et al.*, 1992; Marsh and Martin, 1999). The underlying periodontal ligament and alveolar bone are not involved, and so gingivitis is undetectable in skeletal material.

Periodontitis

Chronic periodontitis is plaque-induced inflammation of the periodontal tissues which, unlike gingivitis, results in destruction of the periodontal ligament and loss of crestal alveolar bone. Periodontitis is characterized by the presence of inflammation at the marginal gingivae, together with loss of attachment, which normally results in the formation of a periodontal pocket (a pathologically deepened gingival crevice, forming a cylindrical cavity around the root, extending towards the apex). Like chronic gingivitis, chronic periodontitis is usually painless. It is now accepted that gingivitis does not necessarily develop into periodontitis and that most modern populations tend to have fewer than 10% of individuals who exhibit advanced levels of destructive periodontitis (Jenkins and Kinane, 1989). The presence of dental calculus is not directly linked to periodontitis (Lieverse, 1999). Improved oral hygiene or nutrition can arrest or slow the condition; but although inflammation may be reduced, the tissue destruction seen in periodontitis is progressive and irreversible.

Periodontal disease is also now known to occur in 'bursts' in different sites, rather than as a gradual and relentless progression (Reddy *et al.*, 2000). Each potential site undergoes long periods of quiescence with short, intermittent bouts of destructive periodontal disease, and it is difficult specifically to identify active disease either in the living or in the dead. As with breakdown of the periodontal ligament, bone loss is probably not the result of the presence of any specific destructive factor, but may be more dependent on the balance between destructive bacterial factors and the level of host response (Clarke and Carey, 1985; Jenkins and Kinane, 1989).

Detecting Periodontal Disease in Skeletal Material

Prior to the 17th century in Britain, few individuals retained their dentitions intact much into the fifth decade (Kerr and Ringrose, 1998). As caries prevalence was low, this early loss of dentition has been attributed to advanced periodontal disease with the observed root exposure in skulls being interpreted as inflammatory alveolar bone loss. It has been widely assumed that a distance of greater than 2 mm between the cemento-enamel junction and the alveolar crest is indicative of periodontal disease.

Previous workers have produced many schemes for quantifying root exposure. The problem with these systems is that they ignored the normal physiological changes that are now known to produce much of the measured increase in root exposure (Costa, 1982; Clarke *et al.*, 1986; Glass, 1991; Newman, 1998). It is only relatively recently that it has been appreciated that, with severe occlusal attrition, compensatory coronal movement of the teeth occurs and that this further eruption of the teeth occurs with a stable and healthy periodontium attached. Therefore, where there is severe attrition or loss of opposing teeth, it is not possible to determine in skeletal material whether the process responsible for that root exposure is compensatory tooth eruption, alveolar bone loss, or a combination of both simply from estimates of the amount of root exposure (Figure 13.5).

Compensatory and Continuous Eruption

Gottlieb and Orban (1933) first suggested that teeth have an innate and lifelong tendency to erupt unless in contact with an antagonist, and Picton (1957) showed that the distance between the occlusal surface and the alveolar crest remained more or less unchanged throughout life. Murphy (1959) felt that loss of crown height was compensated for by eruption. Lavelle (1973) noted that softer diets led to less attrition and also less root exposure, but considered these parameters to be independent variables. Sagne and Olsson (1977) unwittingly demonstrated eruption compensating for attrition in a medieval population by charting age (calculated from attrition) against recession, producing a straight-line graph. Newman and Levers (1979) and Levers and Darling (1983) demonstrated eruption compensating for attrition in archaeological skulls.

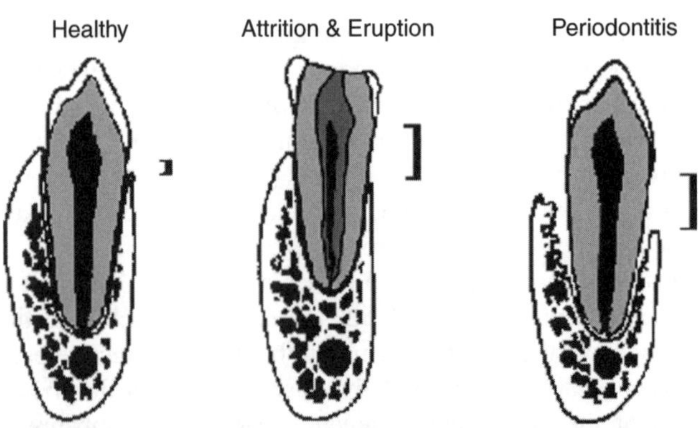

Figure 13.5 This diagram shows how the same vertical amount of root dentine can be exposed by compensatory eruption as by periodontal disease. The difference is in the quality of the alveolar margin. It is thin, rounded or knife-edged in health, but is rough and irregular in periodontal disease

Clarke and Hirsch (1991) proved, by using the inferior dental canal as a fixed point, that coronal movement of the tooth was relative to the investing tissues and resulted in the teeth moving coronally past a vertically stable periodontium. Solheim (1992) even suggested that recession of the periodontal ligament could be used as an indicator of age, because of its relationship to wear.

Attrition of the occlusal table demands continual adaptive and compensatory changes in the teeth, the facial skeleton, temporo-mandibular joint and the associated musculature (Murphy, 1959; Richards, 1985; Whittaker and Murphy, 1985; Whittaker *et al.*, 1985, 1990; Danenberg *et al.*, 1991; Newman, 1998). The formation of secondary dentine, dentinal tubule calcification, deposition of saliva- and plasma-derived calcium phosphate to maintain surface hardness, and cementum deposition to increase root length and anchorage, are all features that help to prolong the life of teeth as they wear down (Whittaker and Murphy, 1985; Maat and van de Velde, 1987; Whittaker *et al.*, 1987; Meiklejohn *et al.*, 1992; Hillson, 2000).

Attrition, if rapid or severe, brings about exposure and death of the pulp, leading to apical abscesses, and many lesions where infection of the pulp leads to destruction of the periodontium as pus tracks to the surface. It is now also appreciated that the floors of molar pulp chambers are porous, so that a pulpal infection can produce exposure of the root furcation, once thought to be pathognomonic of advanced periodontal disease. Therefore, whenever a single tooth has developed greater periodontal destruction than any of its neighbours, pulp death and pus drainage along the periodontal ligament should be suspected as the primary cause (Clarke and Hirsch, 1991; Dias and Tayles, 1997).

Whittaker *et al.*, (1990) reported that, in a group with minimal attrition, face height continued to increase throughout life. First, the lower border of the mandible continued to grow at a rate of 0.04 mm year^{-1}; second, the teeth continued to erupt some 2.8 mm over a period of 40 years. They also found that, over this period, the alveolar crest remained stable. They suggested that this was a process independent of attrition. The balance of evidence, therefore, seems to favour the concept that the human dentition undergoes both compensatory eruption in response to severe wear, and, even in the absence of attrition, there is still some continued tooth eruption with a small increase in lower face height.

Periodontal Disease at a Population Level

The evidence from several studies (Johnson *et al.*, 1988; Jenkins and Kinane, 1989; Clarke and Hirsch, 1991; Kerr, 1998a,b; Newman, 1999) suggests there is a general background level of periodontal disease in any community that is minimally affected by routine oral hygiene measures or changes in the oral environment. In the absence of other factors threatening the life span of the dentition, periodontal disease is unlikely to cause early loss of dentition, except in a minority group of any population. It is now accepted that most populations tend to have fewer than 10% of individuals who exhibit advanced levels of destructive periodontitis (Jenkins and Kinane, 1989). All individuals are likely to have their own genetically determined high- or low-risk profile which can, in turn, be influenced by other oral or non-oral factors. Population differences in periodontitis have been reported (Kerr, 1998b; Manzi *et al.*, 1999; Skrepcinski and Niendorff, 2000) and may mainly be a reflection of the proportion of individuals at the extremes of the spectrum or be due to specific items of diet (Fyfe *et al.*, 1994: Langsjoen, 1996: Hung *et al.*, 2000: Indriati and Buikstra, 2001).

Kerr (1998b) observed that, in all three skeletal populations he was studying (prehistoric, medieval and 18th century AD), a small number of individuals (6–10%) already had widespread destructive periodontitis in their late teens and early twenties, while a contrasting group in the 26–35-year age range (5–17%) appeared to be virtually immune to periodontitis. Such statistics agree with current medical concepts, which suggest that host–parasite interactions are probably unique for each individual. It has been reported that susceptibility to periodontitis varies, with about 10% of individuals particularly susceptible, a similar number resistant, while the vast majority are only moderately susceptible (Anerud *et al.*, 1983). Whittaker *et al.* (1990) and Kingsmill (1991) have confirmed that these ratios held in the 18th century AD London Spitalfields collection.

The three populations that Kerr reported in 1998 recorded an almost straight-line graph of the percentage increase in the number of sites showing periodontitis with age, suggesting that the incidence of inflammatory lesions is likely to be independent of age and is more likely to be a record of events occurring at a fairly constant level for each individual in otherwise good health, as had been suggested by Solheim (1992). The pattern of predilection of sites followed that of most reports in modern material (Anerud *et al.*, 1983). The maxilla appeared to be more susceptible than the mandible, probably because of the thin bone and the complex root-form of the upper posterior teeth. The principal exceptions to these observations were seen in the lower incisors in the prehistoric and London Spitalfields populations, which may indicate non-masticatory activity of some kind.

Figure 13.6a illustrates how, with compensatory eruption, the amount of occlusal wear approximates the length of root exposure. Figure 13.6b shows an extreme case of compensatory eruption. Figure 13.6c illustrates the range of morphology that can be exhibited by the alveolar margins.

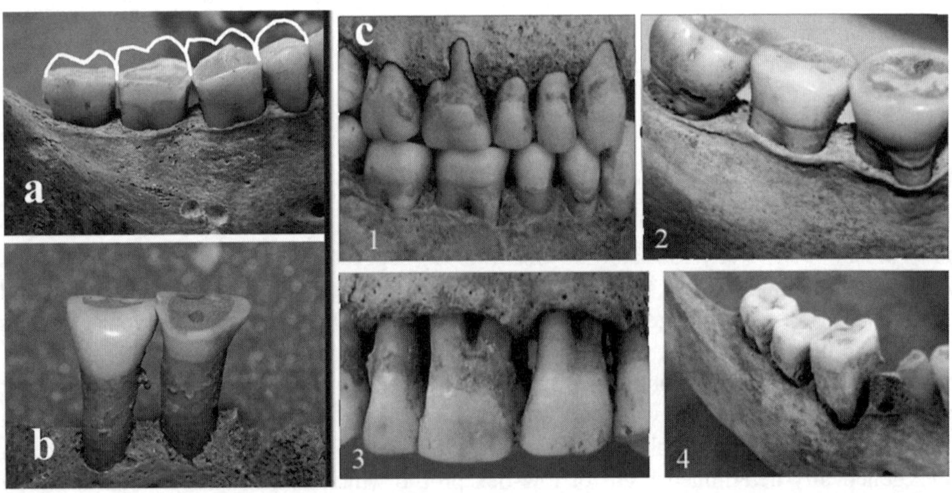

Figure 13.6 (a) Compensatory eruption in lower posterior teeth, showing that the height of crown lost by wear is approximately equal to the height of exposed root. (b) An extreme case of compensatory wear, showing massive root exposure, but a healthy alveolar margin with no evidence of periodontal disease. (c) The variable morphology that can be exhibited by alveolar margins. (This figure also illustrates the four categories of periodontal disease defined diagrammatically in Figure 13.7)

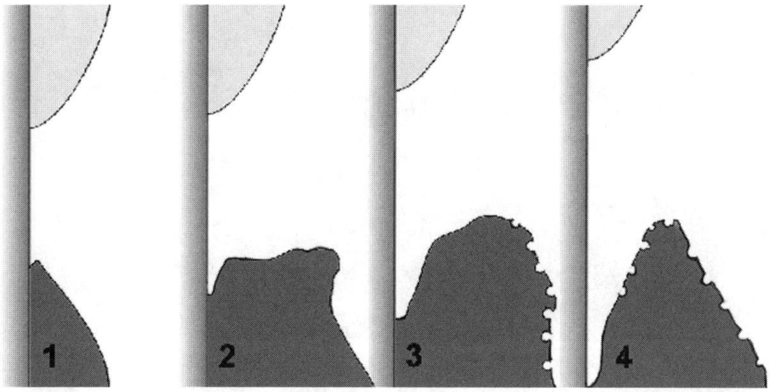

Figure 13.7 A new system for scoring periodontal disease: 0, unable to score (alveolus damaged or missing) (not illustrated); 1, alveolar margin meets tooth at a knife-edged acute angle (no disease); 2, alveolar margin is blunt and flat-topped with a slightly raised rim (mild periodontitis); 3, alveolar margin is rounded and porous, with a trough of 2–4 mm depth between tooth and alveolus (moderate periodontitis); 4, alveolar margin is ragged and porous, with an irregular trough or funnel >5 mm depth between tooth and alveolus (severe periodontitis). Note that length of root exposed is completely irrelevant, as this may simply be a function of compensatory eruption

New System for Identification and Classification of Periodontal Disease

The well-known system for recording periodontal disease in dry bones originated by Kerr (1988) is overly complex. Kerr's revised scheme (Kerr, 1998b), although simpler, has proved difficult to apply because of its concern with the interdental papillae (which are not easily observed) and its poor illustrations. In addition, this method requires that all interdental septa be viewed at $\times$ 10 magnification under a fibre-optic light source and that any lesions the operator thinks suggestive of inflammatory periodontal destruction be recorded in great detail. We have introduced a less complex screening system at Bradford, illustrated in Figure 13.7. This works well, even in the hands of the inexperienced and, most important, avoids false positives for periodontal disease, when what is being seen is simply compensatory eruption. The operator examines the buccal contour of the alveolar margins of the posterior teeth and grades them not by the amount of root exposed, but by the morphology of the alveolar margin (Figure 13.7). This screening detects jaws worthy of more detailed examination by Kerr's method, if more precise mapping of the periodontal disease is required.

PERIAPICAL VOIDS IN JAW BONES

Periapical voids seen in jaws from archaeological collections are often dramatic (Figure 13.8). They are variously described in the literature as abscesses, cysts, granulomata, fenestrations or dehiscences (Bhaskar, 1966; Lineberg *et al.*, 1972; Linn *et al.*, 1987: Nair, 2004). Dias and Tayles (1997), in a seminal paper that did not attract the attention it deserved, highlighted the need for greater precision of definition, and this section is an attempt to expand on what they suggested.

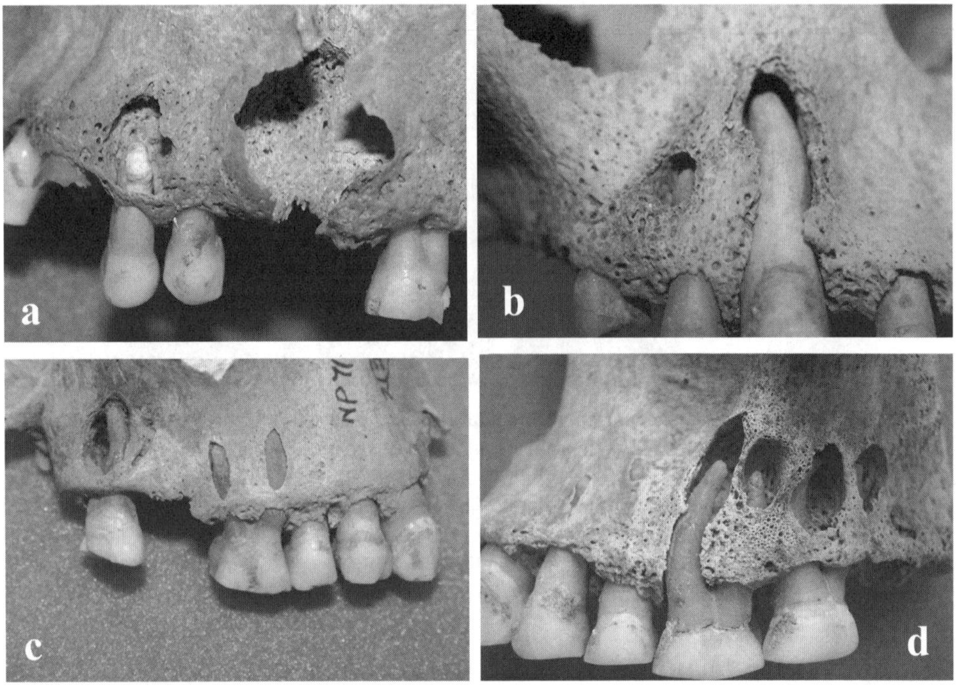

Figure 13.8 (a) Large cystic lesions with the thin alveolar wall broken away. Note the relatively few pores, indicating the presence of a lining membrane. (b) Voids left by granulomas on the apices of two teeth. Taphonomic damage has removed the thin outer bone. (c) Exposure of the roots of the first molar, but without pathology, as there are no voids around the root apices. A granuloma/cyst related to the upper third molar has become a chronic abscess–note the thickened, rounded margins. (d) Granuloma cavities above the roots of these upper molars. Their diameter (>3 mm) suggests that they were becoming cystic (Dias and Tayles, 1997), but the porosity of their walls indicates a rich blood supply

Fenestration simply means an opening, from the latin *fenestra*, a window. It says nothing about the possible cause or pathology involved. Dehiscence is often used to describe tooth roots exposed to view not because of pathology, but because the overlying paper-thin bone has been lost by taphonomic processes. A word relating to seeds, strictly speaking it means a 'gaping or a bursting open', and so is not really suited to osteology or palaeopathology. Dento-alveolar lesions in the jaws have until recently been described routinely in osteological reports as 'abscesses' with all that this implies about pain, infection and general health (Dias and Tayles, 1997; Hillson, 2005: 307–314). Extensive modern histological studies, however, indicate that such voids are relatively common, are usually painless granulomata, and fewer than a third of them may be actually infected (Table 13.1).

To understand the formation of a granuloma it is essential to realize that a dental pulp in a mature tooth has virtually no ability to repair itself when damaged, whether by trauma, caries or exposure to the oral environment. Its blood supply arrives through very narrow channels in the root, and any inflammation in the virtually sealed pulp chamber, increases the internal pressure, occludes the blood supply and leads to the death of the pulp. The necrotic pulp

Table 13.1 Histological analysis of soft-tissue lesions related to root apices

Reference	n	Granulomas (%)	Cysts (%)	Abscesses (%)
Nakade *et al.* (1989)	91	54	46	*na*
Spatafore *et al.* (1990)	1659	52	42	*na*
Nobuhara and del Rio (1993)	150	59	22	*na*
Rachmanadran Nair *et al.* (1996)	256	50	15	35
Sanchis *et al.* (1998)	125	86	14	*na*
Kuc *et al.* (2000)	788	51	47	2
Vier and Figuerdo (2002)	104	na	24	64
Ricucci *et al.* (2006a)	50	40	32	28
Ricucci *et al.* (2006b)	60	78	15	*na*

then undergoes autolysis, which will be sterile if bacteria or fungi cannot gain access. The concomitant release of breakdown products from the necrotic pulp elicits an inflammatory response (either acute or chronic) from the soft tissues surrounding the apex of the tooth. This soft-tissue lesion, a periapical granuloma, is a sphere of soft tissue surrounding the root apex and it creates a space in the surrounding bone (Cawson *et al.*, 2002: 60–61). It is this space that is seen in osteological specimens or radiographs and that we need to interpret. If there is a bony cavity related to a tooth apex, it reveals for certain that the dental pulp, or at the very least the pulp in that particular root, must have been dead for some time, even if there is no obvious sign of trauma or disease.

The essential components of a granuloma are collections of modified macrophages, usually with a surrounding zone of lymphocytes. Granulomatous inflammation seems to require the presence of indigestible or constantly replenished foreign material and/or a cell-mediated immune reaction against the injurious agent (type IV hypersensitivity reaction). In the case of dental granulomata, it would seem that breakdown products continue to leak from the apex, as polymorphs are unable to access the necrotic pulp chamber and remove its contents. Granulomata from different roots may eventually become confluent, forming larger areas. The central areas of inflammation often undergo necrosis as they become remote from their blood supply. This leads to the formation of cystic spaces within them which undergo slow enlargement over a period of years, as osmotic pressure leads to fluid build-up in the interior and as they become lined by membrane developed from the epithelial cell rests of Malassez (Cawson *et al.*, 2002: 104; Nanci, 2003: 106; Ricucci *et al.*, 2006a). Lesions from separate roots or even separate teeth are often confluent. Thus, some lesions are hybrid in nature, so that, for example, one region may be a simple granuloma while another may show signs of chronic abscess formation (Ricucci *et al.*, 2006a).

A necrotic pulp and granuloma/cyst is highly likely to become infected, either through contamination from the mouth or by blood-borne infection (Trowbridge and Stevens, 1992; Abbott, 2002, 2004). In the absence of a smooth-walled sinus tracking out through the cortical bone, which indicates the presence of a chronic abscess, all that can be proved by the presence of an apical void is that the tooth was non-vital. Whether or not it was causing pain at the time of death is impossible to tell. An acute abscess rapidly invades trabecular spaces and vascular channels, but it may not form a bone cavity because there is insufficient time for osteoclastic resorption. The pus tracks through bone, taking the path of least resistance, until it reaches an outer surface where it discharges (Dias and Tayles, 1997). These convoluted, multiple and narrow pathways are difficult to identify *in vivo*, let alone in dry bone, as they do not show on radiographs. However, the presence of a visible buccal or lingual sinus in the bone is evidence that, at some time, there was a painful abscess that had then drained and become chronic (Hillson, 2005: 307–314). Upper posterior teeth sometimes drain into the maxillary antrum and may lead to the formation of a chronic oro-antral fistula. Lower molars can also, rarely, drain through the vertical ramus of the mandible, leading to a particularly painful and intractable sub-masseteric abscess (Jones *et al.*, 2003) (Figure 13.9). Whenever a single tooth has developed greater periodontal destruction than any of its neighbours, or the equivalent tooth on the other side of the jaw, pulp death with pus drainage along the periodontal ligament should be suspected as the primary cause (Clarke and Hirsch, 1991; Dias and Tayles, 1997).

There have been reports in the literature over the years of attempts to make a differential diagnosis between cyst and granuloma based on their size and radiological features. Cysts are considered to be larger than granulomata, and if the lesion is greater than 15 mm in diameter it would certainly be considered a cyst (Mortensen *et al.*, 1970). However, Dias and Tayles (1997) considered that any void greater than 3 mm in diameter was probably cystic. A cyst was thought in the past to exhibit defined margins with a thin radio-opaque border, whereas a granuloma would show indistinct margins (Shear, 1976; Gallego Romero *et al.*, 2002). However, recent studies (Reit *et al.*, 2003; Ricucci *et al.*, 2006b) comparing radiological diagnosis with histological diagnosis show that there appears to be no correlation between the two, presumably because the radiography is so affected by the structure of the overlying bone.

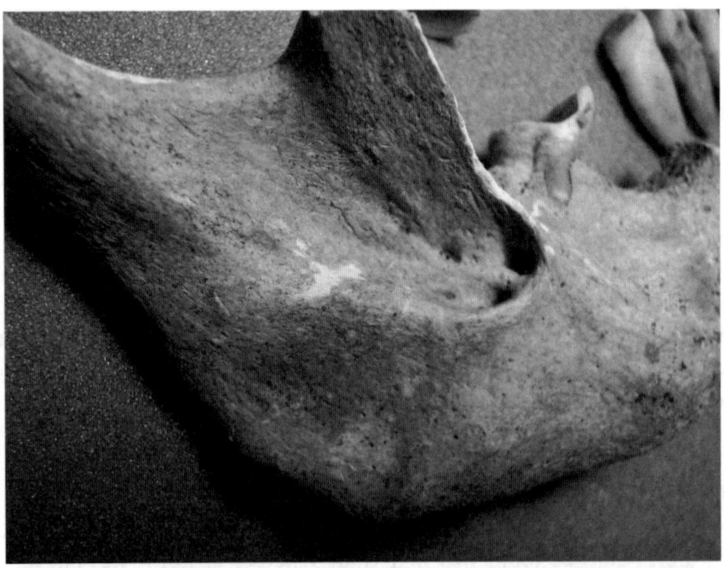

Figure 13.9 Submasseteric abscess in individual from Hereford, England. This is a view of the buccal aspect of the right mandible. A chronic infection from the lower third molar had for years been draining beneath the masseter. Note the smooth and rounded contour of the double sinus, indicating repeated bony remodelling

Large-scale histological studies of soft-tissue lesions on the apices of extracted teeth have indicated a wide range of diagnoses (Table 13.1). Note that all workers agree that the most common lesions are best described histologically as granulomata, but the percentages present varied from 40% to 86%. Cyst development had occurred in 15–47% of cases. There were, however, major disagreements about whether abscesses could be clearly identified. These pathologists have confirmed the difficulty of making a differential diagnosis even if the histology is available, let alone simply from the gross or radiographic morphology of the periapical void in the bone. For that reason alone, palaeopathologists can only make tentative judgements as to the likely diagnosis of voids in alveoli.

When examining voids in dry maxillae or mandibles, there are several characteristics to look for in order to aid diagnosis (Figure 13.10). First, the lesion must extend from, or incorporate the apices of, one or more tooth roots. If it does not clearly involve the apex of the tooth, then it may involve an accessory or lateral root canal but is more likely to be a local manifestation of a systemic disease, e.g. multiple myeloma, metastatic carcinoma, and there will be similar lesions elsewhere. Second, if the void is a granuloma it is usually 2–3 mm in diameter and is smoothly rounded in shape. Voids larger than 3 mm in diameter are more likely to be developing internal cysts (Dias and Tayles, 1997). Third, the lining of a void is smooth but porous, from the copious blood supply. The less the porosity, the greater the chance of it having developed into a cyst and having developed a lining membrane. Fourth, a void with a thin, sharp margin where it meets the cortical surface of the alveolus indicates the presence of a non-infected granuloma or cyst. If the margin is rounded or thickened and the bone appears to have been frequently remodelled, often with a halo of new bone around the orifice, then this represents a chronic abscess, which would have been seeping pus for a considerable time.

Previous osteological studies indicate that granulomata are more common in the maxilla than the mandible and explain this by the greater complexity of the upper root forms (Bhaskar, 1966; Lineberg *et al.*, 1972; Linn *et al.*, 1987; Nair, 2004). However, it may be partly because the thick cortical bone of the mandible conceals voids much more effectively than the more

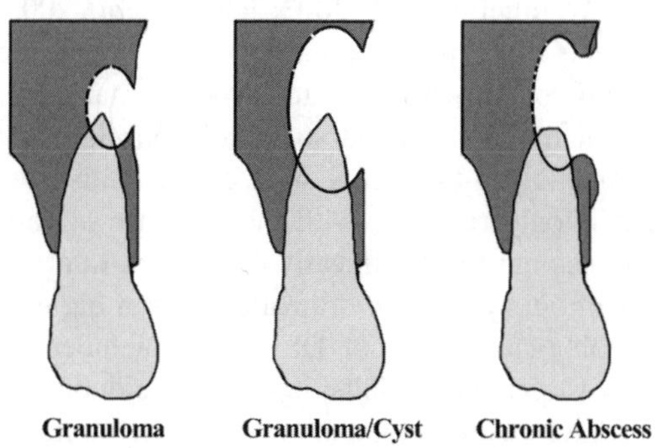

Figure 13.10 The basic characteristics of granulomas, granuloma/cysts, and chronic abscesses. The granuloma and the granuloma/cyst are exposed by taphonomic damage beneath a thin, eggshell thickness of bone with a jagged and broken edge. The chronic abscess has a rolled and thickened rim where it penetrates the cortical plate of the alveolus, and the opening is often surrounded by a 'halo' of new bone deposition. The lining is porous for the passage of blood vessels in granulomas and abscesses, but much less so in cysts, which develop a lining membrane

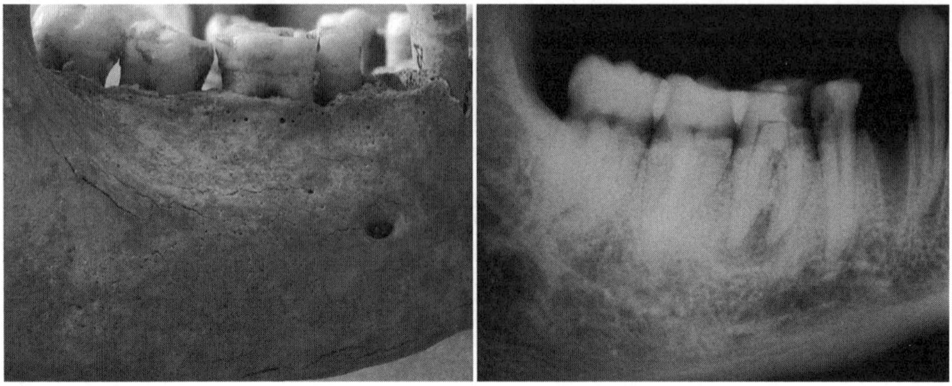

Figure 13.11 Grossly, this mandible shows no abnormality, but radiography reveals a periapical void at the first molar. This reinforces the importance of radiological examination of well-preserved mandibles

porous bone of the maxilla. Without radiographs, very many mandibular defects, and thus non-vital teeth, are missed (Figure 13.11). Radiographs, therefore, are essential if the true picture of dental health in a community is to be assessed, but it must be remembered that this will not authoritatively distinguish between granulomata and cysts (Ricucci *et al.*, 2006b).

CONCLUSIONS AND SUGGESTIONS FOR NEW RESEARCH

Cuspal Enamel Hypoplasia and Molar Incisal Hypomineralizaton

A new type of hypoplasia has recently been clinically identified, namely molar incisor hypomineralization (MIH; Weerheijm *et al.*, 2003; Jälevik *et al.*, 2005). MIH is characterized by clearly demarcated defects involving an alteration in the translucency of the enamel, with the area being white, yellow or brown in colour. When dental caries rates were extremely high, many first permanent molars decayed so rapidly during eruption that there was no possibility of clinicians diagnosing the initial state of the tooth, which had been the cause of this rampant caries. It is only recently, with the decrease in caries rates, that MIH has survived undamaged long enough to be clinically diagnosed. In archaeological material it is likely to be seen as areas of heavy wear or dramatic onset of caries.

Most published archaeological studies of DEH have included some subadults, but the majority of individuals investigated are adults, in which deciduous teeth have been shed and where caries and attrition have destroyed much potentially hypoplastic enamel on permanent teeth (Palumbeckaite *et al.*, 2002). An important aspect of this chapter, therefore, is to alert others to look actively for evidence of MIH or CEH, and to suspect them if molars are unexpectedly carious or heavily worn relative to the age of the individual for their population.

Two other areas of research have recently excited much interest in clinical dentistry: the relationship of periodontal disease to systemic conditions, and enamel erosion and its role in wear. These should also provide fertile ground for dental palaeopathological research.

Periodontal Disease and Systemic Health

Whilst the true prevalence of periodontal disease in the past is now considered to be less than was once thought, its significance as a clinical marker of health problems has dramatically increased in recent years. It was once considered a completely separate entity, in isolation from general health, except for its known connections with diabetes and pregnancy (Cutler *et al.*, 1999: Katz *et al.*, 2000: Skrepcinski and Niendorff, 2000). In recent years, with the increased emphasis on the role of host resistance, many clinical workers have found relationships between periodontal disease and a surprisingly wide range of systemic factors and diseases:

- diet (Littleton and Fröhlich, 1993; Enwonwu, 1995; Bourgeois *et al.*, 1999; Lieverse, 1999)
- vitamin D and calcium metabolism (Davideau *et al.*, 2004)
- social class and lifestyle (Donolly *et al.*, 1977; Walker and Hewlett, 1990; Sakashita *et al.*, 1997; Manzi *et al.*, 1999; Borrell *et al.*, 2006)
- psychosocial factors and stress (Montiero da Silva *et al.*, 1995, 1996; Slots 1998; Genco *et al.*, 1999)
- heart disease and strokes (Beck *et al.*, 1996; Garcia *et al.*, 1998; Seymour and Steele, 1998; Arbes *et al.*, 1999; Morrison *et al.*, 1999; Hujoel *et al.*, 2000; Geismar *et al.*, 2006; Lee *et al.*, 2006)
- arterial atheroma (Haraszythny *et al.*, 2000; Koltviet and Eriksen, 2001)
- respiratory disease (Scannapieco and Ho, 2001)
- rheumatoid arthritis (Mercado *et al.*, 2000; Marotte *et al.*, 2006)
- premature and underweight offspring (Mitchell-Lewis *et al.*, 2001; Moore, 2002)
- body mass index (Katz *et al.*, 2000)
- osteoporosis (Jeffcoat *et al.*, 2003)
- prospects for survival in the elderly (Soikkonen *et al.*, 2000).

As these studies may not be well known in palaeopathology, this list is presented in the hope that non-dental specialists may find new lines of research from investigation of the correlation between systemic conditions and periodontal disease.

Enamel Erosion

In modern European countries with virtually non-abrasive diets, acid erosion is now recognized as the major component of tooth wear (Deery *et al.*, 2000; Bartlett, 2005; Jaeggi and Lussi, 2006). Even in archaeological remains we might, therefore, expect some evidence of erosion even if the overwhelming appearance is that of abrasion from foodstuffs and milling particulates (Robb *et al.*, 1991).

Saliva provides some defence from acids by physical clearance and by chemical buffering of the acid. Acid erosion, therefore, occurs as discrete episodes when the acid softens the surface of the enamel and makes it much more vulnerable to abrasion, attrition, and abfraction (Ganss *et al.*, 2002; Moazzez *et al.*, 2004; Addy and Shellis, 2006). In palaeopathology, tooth

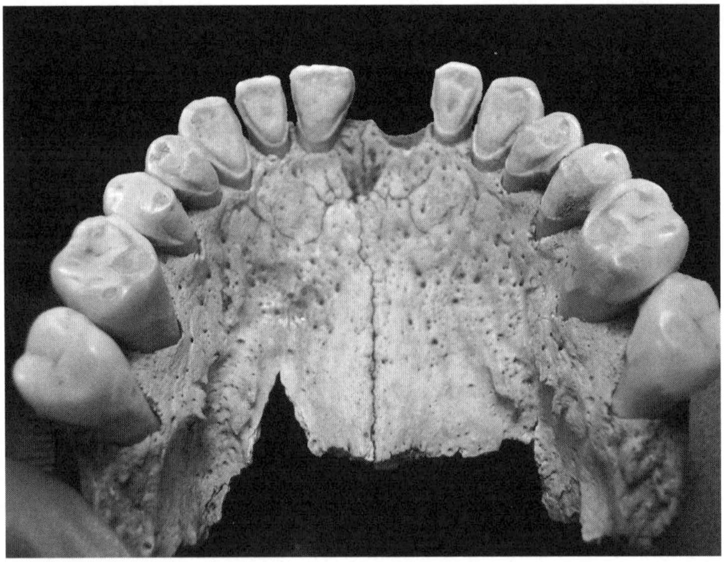

Figure 13.12 Enamel erosion in an Iron Age individual from Wetwang Slack, England, probably caused by gastric regurgitation. Note how the palatal enamel has been eroded, leaving a clear step beneath which the enamel had been protected in life by the gingival margins

wear is assumed to be closely age related, and so acid erosion may cause overestimation of individual age (Brothwell, 1989; Donachie and Walls, 1996).

Apart from fruit juices, the other main source of acid in the mouth is the stomach. Regurgitation of acid from the stomach to the oesophagus can cause irritation of the mucosal lining, presenting as the 'heartburn' that up to 65% of people will have suffered in their lifetime (Jones and Lydeard, 1989). Reflux past the upper oesophageal sphincter leads to the risk of erosion in the mouth (Jarvinen *et al.*, 1988; Bartlett *et al.*, 1994; Moazzez *et al.*, 2004). The refluxed acid commonly attacks the palatal surfaces of the upper incisors, canines and premolars (Figure 13.12). A similar appearance is seen in individuals whose regurgitation is associated with eating disorders, such as anorexia, bulimia nervosa, rumination, chronic alcoholism, or as may occur in pregnancy (Stege *et al.*, 1982; Robb *et al.*, 1995; Gilmour and Beckett, 1994; Bartlett, 2005, 2006). Such problems may have occurred throughout history and it is important that such erosive damage is detected.

REFERENCES

Abbott PV. 2002. The periapical space–a dynamic interface. *Aust Endodont J* **28**: 96–107.

Abbott PV. 2004. Classification, diagnosis and clinical manifestations of apical periodontitis. *Endodont Top* **8**: 36.

Addy M, Shellis RP. 2006. Interaction between attrition, abrasion and erosion in tooth wear. *Monogr Oral Sci* **20**: 17–21.

Alaluusua S, Lukinmaa PL, Koskimies M, Pirinen S, Holtta P, Kallio M, Holttinen T, Salmenpera L. 1996. Developmental dental defects associated with long breast feeding. *Eur J Oral Sci* **104**: 493–497.

Anerud KE, Robertson PB, Löe H, Anerud A, Boysen H, Patters MR. 1983. Periodontal disease in three young adult populations. *J Periodont Res* **18**: 655–668.

Arbes SJ, Slade GD, Beck JD. 1999. Association between extent of periodontal attachment loss and self-reported history of heart attack: an analysis of NHANES III data. *J Dent Res* **78**: 1777–1782.

Baelum V, Fejerskov O. 1986. Tooth loss as related to dental caries and periodontal breakdown in adult Tanzanians. *Comm Dent Oral Epidemiol* **14**: 353–357.

Bartlett DW. 2005. The role of erosion in tooth wear: aetiology, prevention and management. *Int Dent J* **55**: 277–284.

Bartlett DW. 2006. Intrinsic causes of erosion. *Monogr Oral Sci* **20**: 119–139.

Bartlett DW, Aggasiansah A, Owen WJ. 1994. Dental erosion:a presenting feature of gastro-oesophageal reflux disease. *Eur J Gastroenterol Hepatol* **6**: 895–900.

Beck J, Garcia R, Heiss G, Vokonas PS, Offenbacher S. 1996. Periodontal disease and cardiovascular disease. *J Periodontol* **67**: 1123–1137.

Bhaskar SN. 1966. Periapical lesions – types, incidence and clinical features. *Oral Surg Oral Med Oral Pathol* **21**: 657–668.

Borrell LN, Beck JD, Heiss G. 2006. Socioeconomic disadvantage and periodontal disease: the Dental Atherosclerosis Risk in Communities study. *Am J Public Health* **96**(2): 332–339.

Bourgeois DM, Doury J, Heskot P. 1999. Periodontal conditions in 65–74 year old adults in France. *Int Dent J* **49**: 182–186.

Brothwell DR. 1989. The relationship of tooth wear to ageing. In *Age Markers in the Human Skeleton*, Iscan MY (ed.). Charles C. Thomas: Springfield; 303–316.

Cawson RA, Odell E, Porter S. 2002. *Essentials of Oral Pathology and Oral Medicine*. Churchill Livingstone.

Chestnutt IG, Birmie VI, Taylor MM. 2000. Reasons for tooth extraction in Scotland. *J Dent* **28**: 295–297.

Clarke N, Carey S 1985. The aetiology of chronic periodontal disease: an alternative perspective. *J Am Dent Assoc* **110**: 689–691

Clarke NG, Hirsch RS. 1991. Physiological, pulpal and periodontal factors influencing alveolar bone. In *Advances in Dental Anthropology*, Kelley MA, Larsen CS (eds). Wiley-Liss: New York; 241–266.

Clarke NG, Carey SE, Srikandi W, Hirsch RS, Leppard, PI. 1986. Periodontal disease in ancient populations. *Am J Phys Anthropol* **71**: 173–183.

Costa RL. 1982. Periodontal disease in the prehistoric Ipiutak and Tigara skeletal remains from Point Hope, Alaska. *Am J Phys Anthropol* **59**: 97–110.

Cutler CW, Machen RL, Jotwani R, Jacapino AM. 1999. Heightened gingival; inflammation and attachment loss in type 2 diabetics with hyperlipidemia. *J Periodontol* **70**: 1313–1321.

Cutress TW, Suckling GW. 1982. The assessment of non-carious defects of enamel. *Int Dent J* **32**: 117–122.

Danenberg P, Hirsch R, Clarke N, Leppard P, Richards L. 1991. Continuous tooth eruption in Australian aboriginal skulls. *Am J Phys Anthropol* **85**: 305–312.

Davideau JL, Lezot F, Kato S, Bailleul-Forestier I, Berdal A. 2004. Dental alveolar bone defects related to vitamin D and calcium status. *J Steroid Biochem Mol Biol* **89–90**(1–5): 615–618.

Deery C, Wagner ML, Longbottom C. 2000. The prevalence of dental erosion in a United States and a United Kingdom sample of adolescents. *Pediatr Dent* **22**: 505–510.

Dias G, Tayles N. 1997. Abscess cavity: a misnomer. *Int J Osteoarchaeol* **7**: 548–554.

Donachie MA, Walls AWG. 1996. The tooth wear index: a flawed epidemiological tool in an ageing population group. *Comm Dent Oral Epidemiol* **24**: 152–158.

Donnelly CJ, Stiles HM, Brewer C, Thomson L, Neel J, Brunelle JA. 1977. Plaque, caries and periodontal diseases and acculturation among Yanomarno Indians, Venezuela. *Comm Dent Oral Epidemiol* **5**: 30–39.

Dowsett SA, Archila L, Segreto VA, Eckhert GJ, Kowolik MJ. 2001. Periodontal disease status of an indigenous population of Guatemala, Central America. *J Clin Periodont* **28**: 663–671.

Dummer PM, Kingdon A, Kingdon R. 1986. Distribution of developmental defects of tooth enamel by tooth type in 11–12 year-old children in South Wales. *Comm Dent Oral Epidemiol* **14**: 341–344.

Ensor BE, Irish JD. 1995. Hypoplastic area method for analyzing dental enamel hypoplasia. *Am J Phys Anthropol* **98**: 507–517.

Enwonwu C. 1995. Interface of malnutrition and periodontal diseases. *Am J Clin Nutr* **61**: 430–436.

FitzGerald CM. 1998. Do enamel microstructures have regular time dependency? Conclusions from the literature and a large-scale study. *J Hum Evol* **35**: 371–386.

Fyfe DM, Chandler NP, Wilson NHF. 1994. Alveolar bone status of some pre-seventeenth century inhabitants in Taumako, Solomon Islands. *Int J Osteoarchaeol* **3**: 29–35.

Gallego Romero D, Torres Lagares D, Garcia Calderon M, Romero Ruiz MM, Infante Cossio P, Gutierrez Perez JL. 2002. Differential diagnosis and therapeutic approach to periapical cysts in daily dental practice. *Med Oral* **7**: 59–62.

Ganss C, Klimek J, Borkowski N. 2002. Characteristics of tooth wear in relation to different nutritional patterns including contemporary and medieval subjects. *Eur J Oral Sci* **110**: 54.

Garcia RI, Krall EA, Vokonas PS. 1998. Periodontal disease and mortality from all causes in the VA Dental Longitudinal Study. *Ann Periodontol* **3**: 339–349.

Geismar K, Stoltze K, Sigurd B, Gyntelberg F, Holmstrup P. 2006. Periodontal disease and coronary heart disease. *J Periodontol* **77**: 1547–1554.

Genco RJ, Ho AW, Grossi SG, Dunford RG, Tedesco LA. 1999. Relationship of stress, distress and inadequate coping behaviours to periodontal disease. *J Periodontol* **70**: 711–723.

Gilmour AG, Beckett HA. 1994. The voluntary reflux phenomenon. *Br Dent J* **175**: 368–372.

Glass G. 1991. Continuous eruption and periodontal status in pre-industrial dentitions. *Int J Osteoarchaeol* **1**: 265–271.

Goodman AH, Rose JC. 1990. Assessment of systemic physiological perturbations from dental enamel hypoplasias and associated histological structures. *Yrbk Phys Anthropol* **33**: 59–110.

Gottlieb B, Orban B. 1933. Active and passive continuous eruptions of teeth. *J Dent Res* **13**: 214.

Greene JC. 1986. Natural history of periodontal disease in man. *J Clin Periodont*ol **13**: 441–444.

Guatelli-Steinberg D, Lukacs JR. 1999. Interpreting sex differences in enamel hypoplasia in human and non-human primates: developmental, environmental, and cultural considerations. *Yrbk Phys Anthropol* **42**: 78–126.

Haraszthy VI, Zambon JJ, Trevisan M, Zeid M, Genco RJ. 2000. Identification of periodontal pathogens in atheromatous plaques. *J Periodontol* **71**: 1554–1560.

Harding V. 2002. The Dead and the Living in Paris and London 1500–1670. Cambridge University Press: Cambridge.

Hillson S. 1992. Studies of growth in dental tissues. In *Culture, Ecology and Dental Anthropology*, Lukacs JR (ed.). *J Hum Ecol* **3** (Special Issue 2): 7–23.

Hillson S. 2000. Dental pathology. In *Biological Anthropology of the Human Skeleton*, Katzenberg MA, Saunders SR (eds). Wiley-Liss: New York; 249–285.

Hillson S. 2005. *Teeth*, 2nd edition. Cambridge University Press: Cambridge.

Hillson S, Bond S. 1997. Relationship of enamel hypoplasia to the pattern of tooth crown growth: a discussion. *Am J Phys Anthropol* **104**: 89–103.

Hillson S, Grigson C, Bond S. 1998. Dental defects of congenital syphilis. *Am J Phys Anthropol* **107**: 25–40.

Hujoel PP, Drangsholt M, Spiekerman C, DeRouen TA. 2000. Periodontal disease and coronary heart disease risk. *J Am Med Assoc* **284**: 1406–1410.

Hung S-L, Chen Y-L, Wan H-C, Liu T-Y, Chen Y-T, Ling L-J. 2000. Effects of areca nut extracts on the function of human neutrophils *in vitro*. *J Periodont Res* **35**: 186–193.

Indriati E, Buikstra JE. 2001. Coca chewing in prehistoric coastal Peru: dental evidence. *Am J Phys Anthropol* **114**: 242–257.

Jaeggi T, Lussi A. 2006 Prevalence, incidence and distribution of erosion. *Monogr Oral Sci* **20**: 44–65.

Jälevik B, Dietz W, Norén JG. 2005. Scanning electron micrograph analysis of hypomineralised enamel in permanent first molars. *Int J Paediatr Dent* **15**: 233–240.
Jarvinen V, Meurman JH, Hyvarinen H. 1988. Dental erosion and upper gastro-intestinal disorders. *Oral Surg Oral Med Oral Path* **65**: 298–303.
Jeffcoat MK, Geurs NC, Lewis CE. 2003. Osteoporosis and periodontal bone loss. *Clin Calcium* **13**: 577–581.
Jenkins GN. 1978. *The Physiology and Biochemistry of the Mouth*. Blackwell: Oxford.
Jenkins WM, Kinane DF. 1989. The 'high risk' group in periodontitis. *Br Dent J* **167**: 168–171.
Jenkins WM, Mason WN. 1984a. Periodontitis in the United Kingdom. A literature review. *Br Dent J* **156**: 43–45.
Jenkins WM, Mason WN. 1984b. Radiographic assessment of periodontitis. A study of 800 unreferred patients. *Br Dent J* **156**:170–174.
Johnson NW, Griffiths GS, Wilton JMA, Malden MFJ. 1988. Detection of high risk groups and individuals for periodontal disease. *J Clin Periodont*ol **15**: 276–282.
Jones KC, Silver J, Millar WS, Mandel L. 2003. Chronic submasseteric abscess: anatomic, radiologic, and pathologic features. *Am J Neuroradiol* **24**: 1159–1163.
Jones R, Lydeard S. 1989. Prevalence of symptoms of dyspepsia in the community. *Br Med J* **298**: 30–32.
Katz J, Chausu G, Sgan-Cohen HD. 2000. Relationship of blood glucose level to community periodontal index of treatment needs and body mass index in a permanent Israeli military population. *J Periodontol* **71**: 1521–1527.
Kerr NW. 1988. A method of assessing periodontal status in archaeologically derived material. *J Paleopathol* **2**: 67–78.
Kerr NW. 1998a. Dental pain and suffering prior to the advent of modem dentistry. *Br Dent J* **184**: 397–399.
Kerr NW. 1998b. The prevalence and natural history of periodontal disease in Britain from prehistoric to modern times. *Br Dent J* **185**: 527–535.
Kerr NW, Ringrose T. 1998. Factors affecting the lifespan of the human dentition in Britain prior to the seventeenth century. *Br Dent J* **184**: 242–246.
Kingsmill V. 1991. Chronic periodontitis in an eighteenth century population, *Br Dent J* **170**: 118–120.
Koltviet KM, Eriksen HM. 2001. Is the observed association between periodontitis and atherosclerosis causal? *Eur J Oral Sci* **109**: 2–7.
Kuc I, Peters E, Pan J. 2000. Comparison of clinical and histologic diagnoses in periapical lesions. *Oral Surg Oral Med Oral Pathol Oral Radiol Endodontont* **89**: 333–337.
Kunzel W. 2003. Die Hypomineralisation bei Molaren und Incisiven. *Zahnärztl Mitt* **93**: 1626–1629.
Langsjoen O. 1996. Dental effects of diet and coca-leaf chewing in two prehistoric cultures of northern Chile. *Am J Phys Anthropol* **101**: 475–489.
Larsen CS. 1987. Bioarchaeological interpretations of subsistence economy and behavior from human skeletal remains. *Adv Archaeol Method Theory* **10**: 339–445.
Lavelle C. 1973. Alveolar bone loss and tooth attrition in skulls from different populations. *J Periodont Res* **8**: 395–399.
Lee H-J, Garcia RI, Janket S-J, Jones JA, Mascarenhas AK, Scott TE, Nunn ME. 2006. The association between cumulative periodontal disease and stroke history in older adults. *J Periodontol* **77**: 1744–1754.
Levers RGH, Darling AI. 1983. Continuing eruption of some adult human teeth of ancient populations. *Arch Oral Biol* **28**: 401–408.
Lewis ME. 2002. The impact of industrialisation: comparative study of child health in four sites from medieval and post-medieval England (A.D. 850–1859). *Am J Phys Anthropol* **119**: 211–223.
Lieverse AR. 1999. Diet and the aetiology of dental calculus. *Int J Osteoarchaeol* **9**: 219–232.
Lineberg WB, Waldron CA, Delanne GF. 1972. A clinical, roentgenographic and histopathologic evaluation of periapical lesions. *Oral Surg Oral Med Oral Pathol* **17**: 467.

Linn J, Srikandi W, Clarke N, Smith T. 1987. Radiographic and visual assessment of alveolar pathology in first molars in dry skulls. *Am J Phys Anthropol* **72**: 515–521.

Littleton J, Fröhlich B. 1993. Fish-eaters and farmers: dental pathology in the Arabian Gulf. *Am J Phys Anthropol* **92**: 427–447.

Löe H, Anerud A, Boysen H. 1992. The natural history of periodontal disease in man: prevalence, severity and extent of gingival recession. *J Periodontol* **63**: 489–495.

Maat G, van der Velde E. 1987. The caries–attrition competition. *Int J Anthropol* **2**: 281–292.

Manzi G, Salvadei L, Vienna A, Passarello P. 1999. Discontinuity of life conditions at the transition from the Roman imperial age to the early Middle Ages: example from central Italy evaluated by pathological dento-alveolar lesions. *Am J Hum Biol* **11**: 327–341.

Marotte H, Farge P, Gaudin P, Alexandre C, Mougin B, Miossec P. 2006. The association between periodontal disease and joint destruction in rheumatoid arthritis extends the link between the HLA-DR shared epitope and severity of bone destruction. *Ann Rheum Dis* **65**: 905–909.

Marsh P, Martin M. 1999. *Oral Microbiology*. Chapman and Hall: London.

McAul LK, Jenkins WMM, Kay EJ. 2000. The reasons for extraction of permanent teeth in Scotland: a 15 year follow-up study. *Br Dent J* **190**: 658–662.

Meiklejohn C, Wyman J, Schentag C. 1992. Caries and attrition: dependent or independent variables? *Int J Anthropol* **7**: 17–22.

Mercado F, Marshall RI, Klestov AC, Bartold PM. 2000. Is there a relationship between rheumatoid arthritis and periodontal disease? *J Clin Periodont*ol **27**: 267–272.

Mitchell-Lewis D, Engebretson SP, Chen J. 2001. Periodontal infection and pre-term birth: early findings of young minority women in New York. *Eur J Oral Sci* **109**: 34–39.

Moazzez R, Aggiansah Owen WJ. 2004. Dental erosion, gastro-oesophageal reflux disease and saliva:how are they related? *J Dent* **32**: 489–494.

Montiero da Silva AM, Newman HN, Oakley DA. 1995. Psychosocial factors in inflammatory periodontal disease: a review. *J Clin Periodont*ol **22**: 516–526.

Montiero da Silva AM, Oakley DA, Newman RN, Nohl FS, Lloyd HM. 1996. Psychosocial factors and adult onset rapidly progressive periodontitis. *J Clin Periodontol* **23**: 789–794.

Moore S. 2002. The relationship between maternal periodontal disease and adverse pregnancy outcome. Unpublished PhD thesis, University of London.

Moorrees CFA, Fanning EA, Hunt EE. 1963a. Age variation of formation stages for ten permanent teeth. *J Dent Res* **42**: 1490–1502.

Moorrees CFA, Fanning EA, Hunt EE. 1963b. Formation and resorption of three deciduous teeth in children. *Am J Phys Anthropol* **21**: 205–213.

Morris AJ, Steele J, White DA. 2001. The oral cleanliness and periodontal health of UK adults in 1998. *Br Dent J* **191**: 186–193.

Morrison HI, Ellison LF, Taylor GW. 1999. Periodontal disease and risk of fatal heart and cerebrovascular disease. *J Cardiovasc Risk* **6**: 711.

Mortensen H, Winther JE, Birn H. 1970. Periapical granulomas and cysts. An investigation of 1600 cases. *Scand J Dent Res* **78**: 241–250.

Murphy T. 1959. Compensatory mechanisms in facial height adjustment to functional tooth attrition. *Aust Dent J* **4**: 312–323.

Nair PN. 2004. Pathogenesis of apical periodontitis and the causes of endodontic failures. *Crit Rev Oral Biol Med* **15**: 348–381.

Nakade O, Ohuchi T, Seki C, Kanno H, Abe H, Kaku T, Okuyama T. 1989. Survey of histopathological services in the Department of Oral Pathology, Higashi-Nippon-Gakuen University, 1979–1989. *Higashi Nippon Shigaku Zasshi* **8**: 39–46.

Nanci A. 2003. *Ten Cate's Oral Histology: Development, Structure and Function*. Mosby: St Louis, MS.

Newman HN. 1998. Factors affecting the lifespan of the human dentition in Britain prior to the seventeenth century. *Br Dent J* **184**: 229.

Newman HN. 1999. Attrition, eruption and the periodontium. *J Dent Res* **78**: 730–734.

Newman HN, Levers B. 1979. Tooth eruption and function in an early Anglo-Saxon population. *J R Soc Med* **72**: 341–350.

Nobuhara WK, del Rio CE 1993. Incidence of periradicular pathoses in endodontic treatment failures. *J Endodont* **19**: 315–318.

Palumbeckaite Z, Jankausas R, Boldsen J. 2002. Enamel hypoplasia in Danish and Lithuanian late medieval/early modern samples: a possible reflection of child morbidity and mortality patterns. *Int J Osteoarchaeol* **12**: 189–201.

Picton D. 1957. Calculus, wear and alveolar bone loss in the jaws of the sixth century Jutes. *Dent Pract* 7: 301–303.

Pinhasi R, Shaw P, White B, Ogden AR. 2006. Morbidity, rickets, and long-bone growth in post-medieval Britain – a cross-population analysis. *Ann Hum Biol* **33**: 372–389.

Purvis RJ, Barrie WJ, MacKay GS, Wilkinson EM, Cockburn F, Belton NR. 1973. Enamel hypoplasia of the teeth associated with neonatal tetany: a manifestation of maternal vitamin-D deficiency. *Lancet* **2**: 811–814.

Ramachandran Nair PN, Pajarola G, Schroeder HE. 1996. Types and incidence of human periapical lesions obtained with extracted teeth. *Oral Surg Oral Med Oral Pathol Oral Radiol Endodont* **81**: 93–102.

Reddy MS, Geurs NC, Jeffcoat RL, Proskin H, Jeffcoat MK. 2000. Periodontal disease progression. *J Periodontol* **71**: 1583–1590.

Reid DJ, Dean MC. 2006. Variation in modern human enamel formation times. *J Hum Evol* **50**: 329–346.

Reit C, Petersson K, Molven O. 2003. Diagnosis of pulpal and periapical disease. In Textbook of Endodontology, Bergenholtz G, Hørsted-Bindslev P, Reit C (eds). Blackwell Munksgaard: Oxford; 9–18.

Richards LC. 1985. Dental attrition and craniofacial morphology in two Australian aboriginal populations. *J Dent Res* **64**: 1311–1315.

Ricucci D, Pascon EA, Pitt Ford TR, Langeland K. 2006a. Epithelium and bacteria in periapical lesions. *Oral Surg Oral Med Oral Pathol Oral Radiol Endodont* **101**: 239–249.

Ricucci D, Mannocci F, Ford TR. 2006b. A study of periapical lesions correlating the presence of a radiopaque lamina with histological findings. *Oral Surg Oral Med Oral Pathol Oral Radiol Endodont* **101**: 389–394.

Robb ND, Cruwys E, Smith BGN. 1991. Regurgitation erosion as a possible cause of tooth wear in ancient British populations. *Arch Oral Biol* **36**: 595–602.

Robb ND, Smith BG, Geidrys-Leeper E. 1995. The distribution of erosion in the dentitions of patients with eating disorders. *Br Dent J* **178**: 171–175.

Sagne S, Olsson G. 1977. Studies of the periodontal status of a medieval population. *Dentomaxillofacial Radiol* **6**: 46–52.

Sakashita R, Inoue M, Inoue N, Pan Q, Zhu H. 1997. Dental disease in the Chinese Yin-Shang period with respect to relationships between citizens and slaves. *Am J Phys Anthropol* **103**: 401–408.

Sanchis JM, Penarrocha M, Bagan M, Guarinos J, Vera F. 1998. Incidence of radicular cysts in a series of 125 chronic periapical lesions. *Rev Stomatol Chirurg Maxillo-Faciale* **98**: 354–358.

Saunders SR, Hoppa RD, Macchiarelli R, Bondioli L. 2000. Investigating variability in human dental development in the past. *Anthropologie* **38**: 101–107.

Scannapieco FA, Ho AW. 2001. Potential associations between chronic respiratory disease and periodontal disease:analysis of National Health and Nutrition Examination Survey III. *J Periodontol* **72**: 50–56.

Schofield J, Maloney C (eds). 1998. *Archaeology in the City of London, 1907–1991: A Guide to Records of Excavations by the Museum of London and its Predecessors*. The Archaeological Gazetteer Series, Volume 1. Museum of London: London.

Seow WK, Young WG, Tsang AK, Daley T. 2005. A study of primary dental enamel from preterm and full-term children using light and scanning electron microscopy. *Pediatr Dent* **27**: 374–379.

Seymour RA, Steele JG. 1998. Is there a link between periodontal disease and coronary heart disease? *Br Dent J* **184**: 33–38.

Shear M. 1976. *Cysts of the Oral Regions*. John Wright: Bristol.

Sheiham A. 1991. Public health aspects of periodontal diseases in Europe. *J Clin Periodont*ol **18**: 362–369.

Skinner M, Goodman AH. 1992. Anthropological uses of developmental defects in enamel. In *Skeletal Biology of Past People: Research Methods*, Saunders SR, Katzenberg MA (eds). Wiley-Liss: New York; 153–174.

Skrepcinski FB, Niendorff WJ. 2000. Periodontal disease in American Indians and Alaska natives. *J Publ Health Dent* **60**: 261–266.

Slots J. 1998. Casual or causal relationship between periodontal infection and non-oral disease? *J Dent Res* **77**: 1764–1765.

Smith BH. 1991. Standards of human tooth formation and dental age assessment. In Advances in Dental Anthropology, Kelley MA, Larsen CS (eds). Wiley-Liss: New York; 143–168.

Soikkonen K, Wolf J, Salo T, Tilvis R. 2000. Radiographic periodontal loss as an indicator of death risk in the elderly. *J Clin Periodontol* **27**: 87–92.

Solheim T. 1992. Recession of periodontal ligament as an indicator of age. *J Forensic Odont-Stomatol* **10**: 32–42.

Spatafore CM, Griffin JA, Keyes GG, Wearden S, Skidmore, AE. 1990. Periapical biopsy report: an analysis of over a 10-year period. *J Endodont* **16**: 239–241.

Steele JG, Treasure E, Pitts NB, Morris J, Bradnock G. 2000. Total tooth loss in the United Kingdom in 1998 and implications for the future. *Br Dent J* **189**: 598–603.

Stege P, Visco-Dangler L, Rye L. 1982. Anorexia nervosa: a review including oral and dental manifestations. *J Am Dent Assoc* **104**: 648–652.

Suckling GW. 1989. Developmental defects of enamel – historical and present-day perspectives of their pathogenesis. *Adv Dent Res* **3**: 87–94.

Suckling GW, Pearce EI. 1984. Developmental defects of enamel in a group of New Zealand children: their prevalence and some associated etiological factors. *Comm Dent Oral Epidemiol* **12**: 177–184.

Suckling GW, Brown RH, Herbison GP. 1985. The prevalence of developmental defects of enamel in 696 nine-year-old New Zealand children participating in a health and development study. *Comm Dent Health* **2**: 303–313.

Suga S. 1989. Enamel hypomineralisation viewed from the pattern of progressive mineralization of human and monkey developing enamel. *Adv Dent Res* **3**: 188–198.

Trowbridge HO, Stevens BH. 1992. Microbiologic and pathologic aspects of pulpal and periapical disease. *Curr Opin Dent* **2**: 85–92.

Van Amerongen WE, Kreulen CM. 1995. Cheese molars: a pilot study of the etiology of hypocalcifications in first permanent molars. *ASDC J Dent Child* **62**: 266–269.

Vier FV, Figueirdo JA 2002. Prevalence of different periapical lesions associated with human teeth and their correlation with the presence and extension of apical external root resorption. *Int Endodont J* **35**: 710–719.

Walker P, Hewlett B. 1990. Dental health diet and social status among Central African foragers and farmers. *Am Anthropol* **92**: 383–398.

Weerheijm KL, Duggal M, Mejare M, Papagiannoulis L, Koch G, Martens LC, Hallonsten A-L. 2003. Judgement criteria for molar incisor hypomineralisation (MIH) in epidemiologic studies: a summary of the European Meeting on MIH held in Athens, 2003. *Eur J Paediatr Dent* **3**: 110–113.

White WJ. 1987. The human skeletal remains from the Broadgate site LSS85. MOLAS unpublished report HUM/REP/87/01, Museum of London Archaeological Service.

Whittaker D, Murphy W. 1985. Patterns of approximal tooth wear in Romano-British population. *J Dent Res* **64**: 698.

Whittaker D, Molleson T, Daniel A, Williams J, Rose P, Resteghini R. 1985. Quantitative assessment of tooth wear, alveolar-crest height and continuing eruption in a Romano-British population. *Arch Oral Biol* **30**: 493–501.

Whittaker D, Ryan S, Weeks K, Murphy W. 1987. Patterns of approximal wear in cheek teeth of a Romano-British population. *Am J Phys Anthropol* **73**: 389–396.

Whittaker D, Griffiths S, Robson A, Roger-Davies P, Thomas G, Molleson T. 1990. Continuing tooth eruption and alveolar crest height in an eighteenth-century population from Spitalfields, east London. *Arch Oral Biol* **35**: 81–85.

14

Trauma

Pia Bennike

Laboratory of Biological Anthropology, Institute of Forensic Medicine, University of Copenhagen Blegdamsvej 3, DK-2200 Copenhagen, Denmark

INTRODUCTION

Traditional sources of evidence for past human aggression and warfare in historic times comprise documents and archaeological evidence including human remains. When dealing with prehistoric periods, skeletal material may be the only available source of evidence. Not only the presence of trauma, but also the circumstances surrounding finds of human bodies and skeletons may help to indicate whether we are looking at a victim of violence, accident or disease. In addition, weapons or other artefacts may help us to understand the scenario, but a detailed description of these is beyond the scope of this chapter. Ritual archaeology (how corpses were treated between death and disposal) and *anthropologie de terrain* (the taphonomical approach) (Haglund and Sorg, 1997; Nilsson, 2003; Taylor, 2003) are also relevant to the consideration of trauma in the past, but are likewise largely outside the current remit.

Evidence of violence in archaeological human remains is often clearest in well-preserved mummies or bog bodies. For example, the Danish *Grauballe Man* showed clear evidence of having had his throat cut. Such an injury would not have been evident without soft tissue, as it left no mark on the bones. However, even when there is extensive soft tissue survival, the evidence for trauma is often still ambiguous and difficult to interpret (Chapter 6). In skeletal remains, injury is, of course, easier to spot on well-preserved skeletons than on fragmented bones with eroded surfaces. This may make comparative studies of bone samples with different levels of preservation problematic, as differences in frequencies of traumatic lesions between skeletal assemblages may largely reflect differences in preservation rather than in patterns of trauma. In order to circumvent this problem, some have simply excluded fragmented and/or eroded bones from comparative studies (Bennike, 1985). Another approach, for the problem of incomplete crania, is to assemble 'composite' skulls (e.g. two halves from different individuals making a whole) (Schulting and Wysocki, 2005).

Advances in Human Palaeopathology Edited by Ron Pinhasi and Simon Mays

Palaeopathological text books divide skeletal trauma into categories including fractures, crushing injuries, sharp force wounds, dislocations, deformation, scalping, mutilation, trepanation, traumatic problems arising from pregnancy (*symphysis pubica*), soft-tissue trauma, osteochondritis dissecans (fragmentation and disruption of articular cartilage consequent upon trauma), spondylolysis (fracture in the neural arch of a vertebra, usually between the superior and inferior articular processes) and tooth loss (intentional or a result of accident) (Steinbock, 1976; Knowles, 1983; Ortner and Putschar, 1985; Merbs, 1989; Rogers and Waldron, 1995; Aufderheide and Rodriguez-Martin, 1998). Steinbock (1976: 17) even included the presence of Harris lines as 'microtrauma in the growing bone induced by nutritional deficiency and disease' in his chapter on trauma, and Knowles (1983) included Schmorl's nodes. The current work takes a narrower definition of trauma, and focuses principally on fracture (including blade injuries and decapitation), as well as discussing dislocation and trepanation.

FRACTURES

A fracture may be defined as a discontinuity of, or a crack in, skeletal tissue, with or without injury to overlying soft tissue. It results from the application of mechanical forces that exceed the natural strain or elasticity of the skeletal structure (Aufderheide and Rodriguez-Martin, 1998: 20). A fracture is described as incomplete when the affected bone is not divided into separate parts, and complete when it is so. A comminuted fracture is divided into more than two parts. Fractures may be designated as simple (closed) or compound (open), indicating a much higher risk of infection. In palaeopathology, adequate description of a fracture should include the bone affected and the location and morphology of the fracture (Buikstra and Ubelaker, 1994; Boylston, 2004).

When dealing with trauma, one must begin by characterizing the bone injury. Was it inflicted *post-* (after), *peri-* (at or around) or *ante-* (before) *mortem* (death)? There are published guidelines for distinguishing between these possibilities (Merbs, 1989; Buikstra and Ubelaker, 1994; Berryman and Symes, 1998; Sauer, 1998), and applying the well-defined forensic criteria for identification may be helpful (Adams, 1972; Maples, 1986). Nevertheless, it may sometimes be impossible to distinguish between the three, even for an experienced palaeopathologist. The most useful criterion for the identification of an ante-mortem fracture is the presence of bone healing. The earliest microscopic evidence of a healing reaction is usually a slight rounding of the fracture edges, which may seem somewhat polished. It takes at least 7–10 days of ante-mortem healing before an injury can be safely identified as such, and some suggest at least 2 weeks (Maples, 1986; Aufderheide and Rodriguez-Martin, 1998).

A long-bone fracture is characterized by initial haematoma formation that becomes an organized fibrous mass (fibrous callus) delicately united with the surrounding tissue, subsequently calcifying and finally remodelling into normal bone. The acute haematoma at the fracture site coagulates in 6–8 h, after which cellular proliferation (inflammatory stage), and differentiation into fibroblasts, chondrocytes and osteoblasts, together with vasculoneogenesis, initiates repair beneath the external periosteum. The reparative stage is characterized by fibrous union, achieved by the third week, which is then calcified. The remodelling of the callus into a mature skeletal histological structure with normal anatomy is the longest stage, requiring up to an additional 6 weeks in children and 6 months in older adults. Figure 14.1

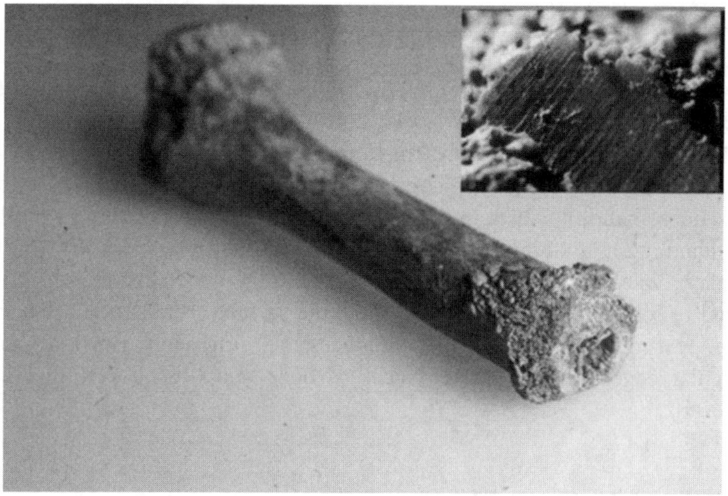

Figure 14.1 The proximal humerus is all that remains of the left arm of the skeleton of a *c.* 20-year-old woman dating to the 16th–17th century AD from Denmark. The end of the stump shows faint parallel saw marks (visible under magnification). Irregular marginal bone formation indicates that the individual survived the amputation, but only for a matter of weeks

shows new and rather unstructured callus formation at the edges of an amputated upper arm. Factors other than age that influence fracture healing rate are adequacy of vascular supply, fracture type (e.g. horizontal fractures heal slower than spiral ones), bone part (metaphysis heals faster than diaphysis), soft tissue interposition (retards healing), and type of bone tissue (cancellous bone heals more rapidly than cortical bone). Immobilization is important because movement stimulates fibrous callus formation, which takes longer to heal. Infection may delay healing, as do various underlying pathological processes that may have led to the fracture. While well-aligned healed fractures of long bones in ancient humans are not uncommon, Schultz (1967) also found that most limb fractures in wild gibbons healed without much malalignment. The neatly healed fractures in monkeys occurred without any form of treatment, so neatly healed fractures in human skeletal material do not necessarily indicate effective methods of treatment. Problems during healing may lead to complications such as osteomyelitis, pseudoarthrosis, avascular necrosis of bone, neuropathy, articular changes and bone shortening.

The slight elasticity of living bone is responsible for the difference in fracture patterns between fresh (i.e. living or recently dead) and dry bone. Bone collagen is responsible for the slight elasticity that gives living bone its tensile strength, and the gradual decay of collagen after death is responsible for the fragility of most archaeological bone. The rate at which bone elasticity is lost after death is dependent on taphonomic factors. The peri-mortem period may range from 2 weeks before death (the time without any visible bone reaction due to a healing process) until a post-mortem interval of at least 2 months (Aufderheide and Rodriguez-Martin, 1998).

In the cranium, fractures occurring in fresh bone tend to have depressed but still adhering bone at the edges of the injury due to retained bone elasticity. Some fractures may be concentric, others may be linear or stellate, the latter radiating out from the point of impact.

There may be bevelling or flaking of the inner table of the cranium at the site of impact and, on occasion, there is a *contrecoup* fracture at the side of the skull opposite the impact (Gurdjian *et al.*, 1950; Aufderheide and Rodriguez-Martin, 1998; Berryman and Symes, 1998; Novak, 2000; Schulting and Wysocki, 2005).

If the edges of a bone fracture are fresh and sharp and appear lighter in colour than the rest of the bone, this suggests post-mortem damage, probably caused during excavation or subsequent handling. It is important, therefore, that all damage to the bone during excavation, handling and storage is registered. Post-mortem breaks in dry skulls occurring long after death typically have edges at right angles to the bone surface, whilst peri-mortem fractures tend to form oblique angles with the bone surface (Buikstra and Ubelaker, 1994). Furthermore, post-mortem breaks usually produce small fragments. Dry long-bone fractures exhibit irregular edges, loss of small pieces of bone and little bevelling (Schulting and Wysocki, 2005).

Mechanism and Nature of Fracture

Different forces (e.g. bending, shearing, torsion and tension) result in different types of fracture. Force applied in the axial direction, such as in falls, results in crushing or impaction of skeletal tissue, typically of vertebrae. A depressed fracture (e.g. in the skull) is produced by a force applied to just one side of a bone, whereas a compression fracture requires forces from two sides (Aufderheide and Rodriguez-Martin, 1998: 20).

There are four main sources of injuries: interpersonal violence, accident, pathological fracture and stress fracture. The susceptibility of the bone to fracture depends upon intrinsic factors, such as elasticity, plasticity and density, which affect the ability of the bone to withstand the blow. For example, a bone of a child, which is characterized by plastic deformation, can absorb more energy before failure than a brittle bone of an elderly individual. In addition, bone can withstand a greater load if the latter is applied at a slow rate.

Interpersonal Violence

Bone injuries due to violence are usually divided into three categories related to the type of weapon that caused the injury: sharp, blunt and projectile (arrow or gunshot).

Many violent injuries involve the skull. Cranial injuries may be classified on the basis of the differing nature of the fractures (Aufderheide and Rodriguez-Martin, 1998):

- sharp-edged incisions (e.g. from metal or flint axes);
- linear fractures (e.g. from small blunt weapons);
- gross crushing injuries (e.g. from large blunt weapons);
- penetrating wounds (e.g. from projectiles or pointed weapons).

In the following discussion, both the linear fractures and gross crushing injuries are included here in the blunt trauma category.

Cut marks produced by edged weapons are the most easily identified on the skeleton. The diagnostic criteria are: linearity; a well-defined clean edge; a flat, smooth, polished cut surface; and the presence of parallel scratch marks on the bone surface when viewed by light microscopy or scanning electron microscopy (Boylston, 2000). Bladed weapons can produce either stab or cut wounds, depending on the penetration and angle of the weapon. A stab wound is characterized as being deeper than it is wide and will manifest as a puncture with a polished margin. A cut wound is the reverse (wider than it is deep) and the bone lesion has one burnished edge with parallel flaking and one roughened edge. A poleaxe blade, sword or dagger will all result in sharp force lesions (Figures 14.2 and 14.3). Sharp force injuries may show terminal fractures and, on occasion, multiple fractures that radiate from the focus of the wound, although this is less common than for blunt injuries.

When a blunt object strikes a skull with force, the bone bends inward so that there are compressive stresses acting within the outer table and tensile stresses acting within the inner table. Because bone is more susceptible to failure under tension than compression, the initial fracture usually appears on the inner table. As the force continues, the inner table fracture becomes more extensive and progresses toward the outer table. The concentric fracture around the initial impact site is formed as the fractured plates are bent inward by the blunt object. The concentric fracture associated with blunt trauma tends to be bevelled internally although this may not always be apparent. Fracture morphology varies with the magnitude of the blow, its duration, and the area of application (Berryman and Haun, 1996).

Blunt force fractures may be due to falls (Figure 14.4) or to a blow from a blunt object or weapon (Figure 14.5), but the wounds may be variable in nature. They can appear as crushing, with little evidence of the weapon used. Blunt trauma can normally be identified

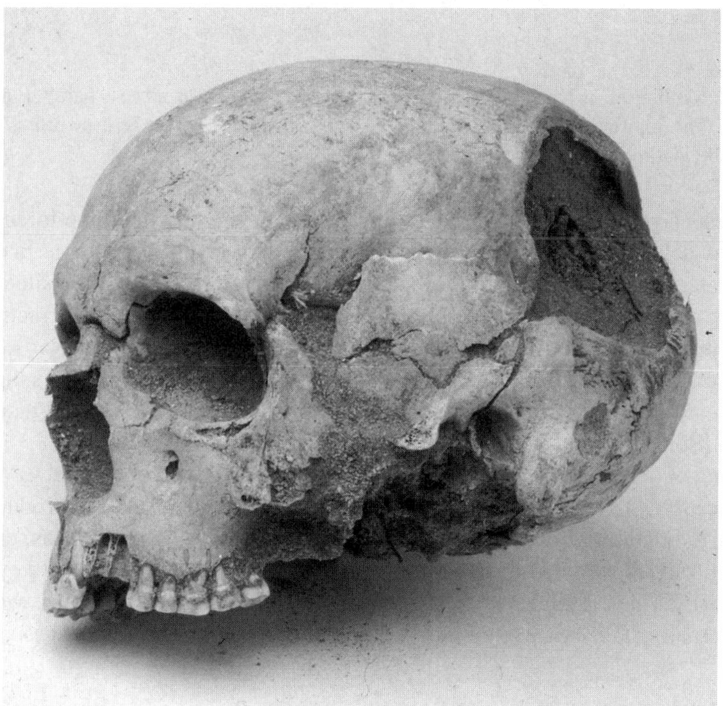

Figure 14.2 Skull from a Danish mass grave at Næstved showing a large injury from a bladed weapon

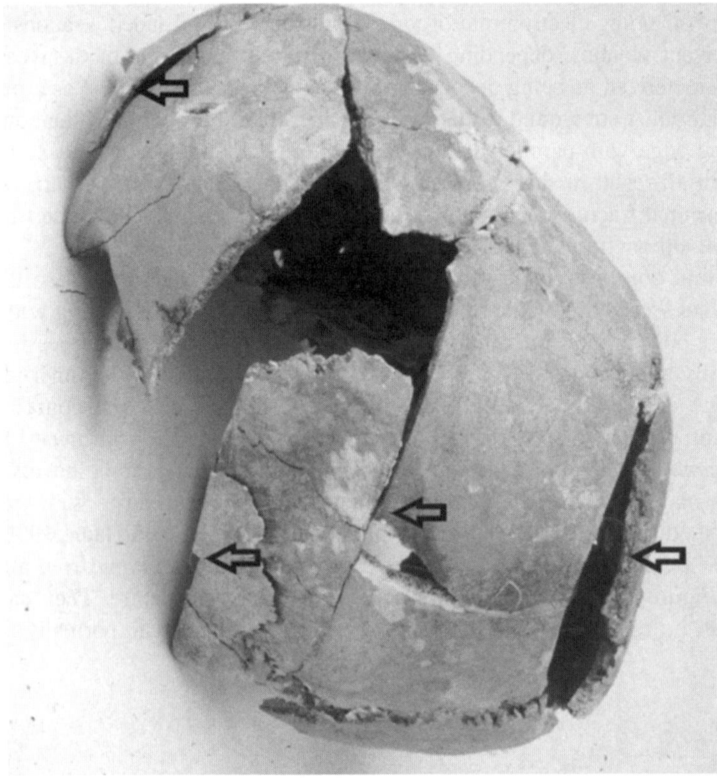

Figure 14.3 Skull from a Danish mass grave at Næstved with four almost parallel slashes on the cranial vault. The injuries were inflicted with a bladed weapon, probably a sword or an axe, and suggest that the victim was hacked to death

from the presence of concentric or radiating fractures, depending upon the force with which the blow was delivered. A relatively light blow may only result in a simple linear fracture, whereas greater force may produce multiple radiating fractures around the site of the blow. Furthermore, concentric fractures may form perpendicular to the radiating fractures. A contrecoup fracture may also occur when the force of the blow propagates towards the opposite side of the cranium from the area of the impact. The dented, cracked or splintered bone characteristically produced by blunt trauma stands in contrast to the incised nature of sharp force injury (Berryman and Symes, 1998; Boylston, 2000).

The nature of a projectile injury depends to a great extent upon the velocity of the object that caused it. Not all gunshot wounds have sufficient force to produce radiating and concentric fractures (Figure 14.6). If a bullet enters or exits the skull tangentially, then fractures will radiate away from the entrance wound and result in a typical keyhole defect (Berryman and Symes, 1998). Gunshot trauma results in a circular entrance wound that is bevelled internally. The exit wound may be irregular in shape and bevelled externally.

Accident

When bones are broken in connection with an accident like a fall (Figures 14.4 and 14.7), the risk of a break occurring in a given situation depends partly on the bone affected. Fractures

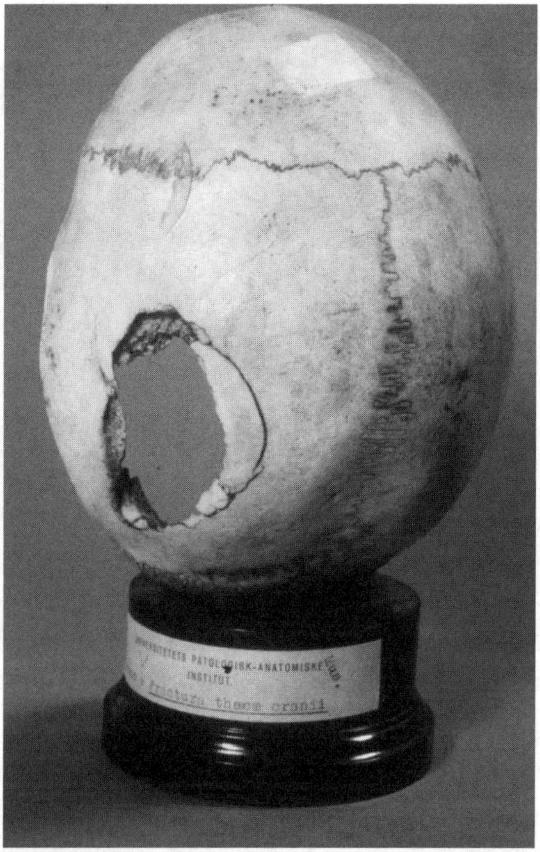

Figure 14.4 The cranium of a young sailor who fell from a ship's mast in the 19th century. He did not survive the open skull fracture. The specimen became part of a 19th century skeletal collection and is a good illustration of the fact that skull wounds are not always a result of interpersonal violence

resulting from indirect trauma comprise oblique, impacted, avulsion or burst fractures. In an impacted fracture, the bone ends are driven into each other by the force of the injury. In an avulsion fracture, a joint capsule, ligament, or tendon is pulled away from the bone, detaching a bone fragment at the site of attachment. Finally, a burst fracture is located in the spine and results from vertical compression that ruptures the intervertebral disc through the vertebral end plate, forcing disc tissue into the vertebral body. Schmorl's nodes may be considered a mild form of this injury (Lovell, 1997: 143), although some (Rogers and Waldron, 1995) dispute this.

Pathological Fractures

Spontaneous fracture may occur in bones weakened by diseases such as tumours or infection. A few diseases are accompanied by an increased risk of fracture owing to a lack of vitamins

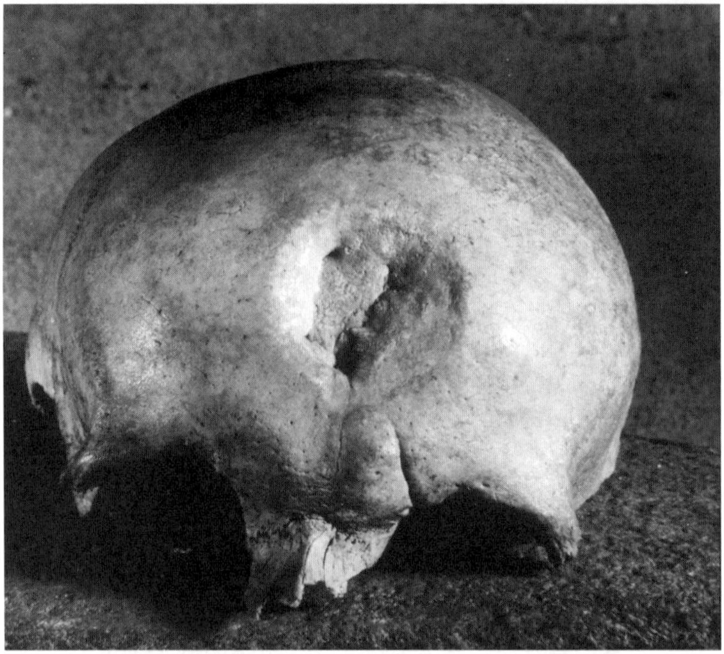

Figure 14.5 A Neolithic skull with a circular depression with smoothly rounded sloping edges. Two healed fracture lines radiate towards the orbit. At the medial fracture line, the inner table exhibits pieces of fragmented bone which have fused and extend 1–2 mm into the skull cavity. The case is a classic depressed fracture, most likely resulting from a blow from a blunt weapon

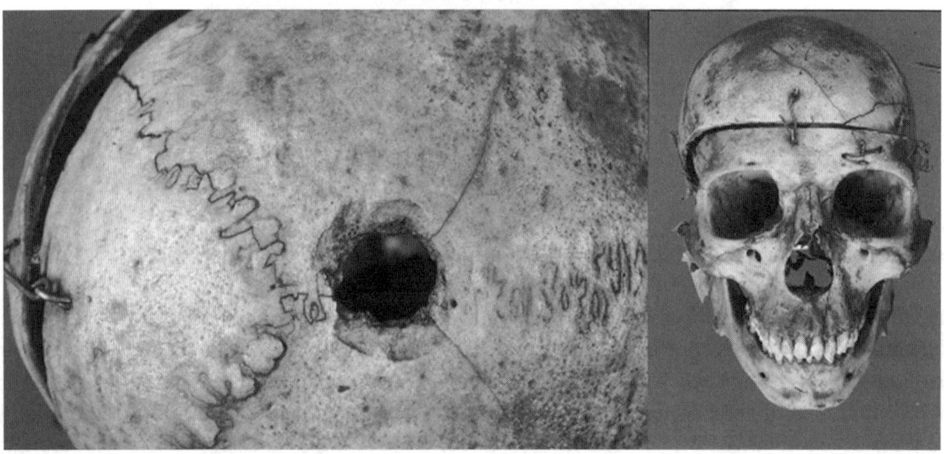

Figure 14.6 Exit wound, bevelled externally, from a gunshot. The skull is from a 19th century Danish pathological collection

or minerals or on account of a lack of ability to take up or form the substances necessary for normal bone formation. Disability resulting in non-use of an extremity may lead to osteopaenia. Many fractures in the older age groups, particularly among women, are consequent upon the loss of bone mass and deterioration of bone microstructure of osteoporosis (Chapter 11).

Figure 14.7 Specimen from a 19th century Danish pathological collection with healed compression and avulsion fractures of a vertebra due to a fall. The dense trabecular structure indicates that the vertebrae are from a young individual

Stress Fractures

Stress fractures (fatigue fractures) are caused by overuse (sustained stress or repeated micro-traumata). Stress fractures occur in individuals with normal-appearing bone, but an apparent susceptibility to fracturing under conditions of unaccustomed stress and physical activity. For example, stress fractures in metatarsal bones are often termed march fractures because of their frequent appearance in army recruits. Beginning as incomplete fractures that generally run at right angles to the long axis of the bone, stress fractures may proceed to complete fractures if the recurring stress is not eliminated (Merbs, 1989: 168).

Quantitative Study of Fractures in Skeletal Populations

Lovejoy and Heiple's (1981) quantitative approach to the analysis of palaeotrauma has inspired many colleagues to use a systematic assessment of fractures in a skeletal sample to investigate biocultural aspects of past populations (Judd, 2002). However, because determination of precise age at death in adult skeletal remains is problematic, difficulty persists in the epidemiological evaluation of ancient trauma, especially in determining which age groups were at greatest risk of injury in a given society. Buhr and Cooke (1959) observed that certain bones were prone to fracture at specific ages, some in young and active people and some in older individuals, and their analytical method has been applied in a number of palaeopathological studies (e.g. Lovejoy and Heiple, 1981). Judd and Roberts (1998) found that the frequency of fractured bones and the fracture rate among individuals from an English medieval leprosarium were much higher than those in other urban medieval cemetery samples. They relate this to possible blindness, nerve damage and bone atrophy in leprosy. In a study of English medieval rural and urban populations, the same authors (Judd and Roberts, 1999) found that males in both groups sustained a greater number of fractures, possible reflecting riskier and more diverse activities when compared with females, and that rural populations showed higher fracture rates than urban ones.

Others have studied the prevalence of multiple traumas, expressing it as the multiple injury rate for the sample (number of individuals showing multiple traumas as a proportion of the total number of individuals studied), the multiple injury rate for those with trauma (number of individuals showing multiple traumas as a proportion of the total number of those with trauma) or as the mean number of injuries (Lovell, 1990; Judd and Roberts, 1998; Judd, 2002). Some report the multiple injury rate separately for each sex (Kilgore *et al.*, 1997; Jurmain and Kilgore, 1998).

Spondylolysis

Spondylolysis is a fracture involving the separation of the neural arch from the vertebral body, more specifically a complete bilateral separation between the superior and inferior articular processes, most commonly of the fifth lumbar vertebra. Other lumbar vertebra may be affected and the separation may not be complete; sometimes the defect is merely a cleft. It is supposed to be unique to erect posture. Although the fracture has been referred to as a congenital condition, the fracture aetiology of spondylolysis is well established (Merbs, 1989).

Dislocation

Joint dislocation comprises both subluxation (partial loss of contact between joint components, with no disruption of the capsule) and luxation (complete displacement of articular surfaces and disruption of the joint capsule) (Ortner and Putschar, 1985; Merbs, 1989). Dislocation is most often seen in the joints of the shoulder and the hip. Because of its anatomy and greater mobility, in modern populations dislocation is most often at the shoulder. Shoulder dislocations probably also occurred with greater frequency than those of the hip in the past. However, archaeological examples of shoulder dislocations are not as common as might be expected (Merbs, 1989). To be detected in an archaeological specimen, the dislocation must have taken place some time before death to allow recognizable bone modification to occur. Methods for achieving reduction go back at least to the time of Hippocrates. (*c.* 430 BC) (Ortner and Putschar, 1985), and the relatively few palaeopathological cases of shoulder dislocation may reflect its regular and effective reduction in past societies.

Both congenital shoulder and hip dislocation, or later dislocations due to congenital abnormalities, may occur (Bennike *et al.*, 1987; Bennike and Bro-Rasmussen 1989). In congenital hip dislocation, the acetabulum will appear small, shallow and triangular in shape. By contrast, in hip dislocation due to trauma, the acetabulum will be fully developed but remodelled, and a new 'joint' may be recognizable on the ilium posterior to the acetabulum. Because of the lack of cartilage, the surface of the femoral head and the new 'joint' will often be marked with eburnation and degenerative changes in older individuals.

Trepanation

The word trepanation derives from the Greek *trepanon*, 'a borer'. Trepanation is an operation performed to produce a hole in the cranium. The operation was first mentioned in the works of Hippocrates, but skeletal finds that are dated to the Neolithic testify that this surgical intervention has a far older pedigree (Brothwell, 2003). Trepanations were traditionally performed in most parts of the world, often until quite recent times.

Trepanations, or burr holes, are carried out in modern surgical medicine to remove intracranial blood clots (epidural or subdural haematoma) The procedure is relatively simple, the

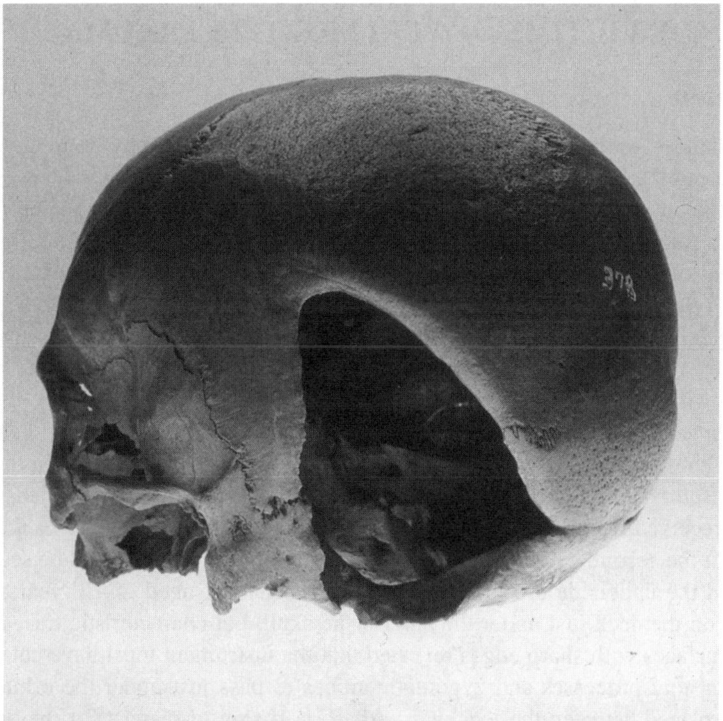

Figure 14.8 An early Neolithic male skull (Vibygårds Mose, Denmark) with a large triangular opening in the posterior part. It has been interpreted as a trepanation following an injury, such as a blow with an axe. A horizontal healed fracture line is visible on the occipital bone. The edges of the opening may have been smoothed and fragments removed, but this cannot be confirmed. All edges are healed, and it is clear that the individual survived

greatest risk being post-operative infection. Accidental damage to the brain is extremely rare. The ethnographic literature contains much information on trepanation, clarifying the motivation behind the prehistoric operation. It appears that trepanation was carried out first and foremost to remove something, for instance bone splinters or blood clots resulting from a traumatic incident, but also less palpable things such as spirits, which were believed to have taken up their abode in the brain and were the cause of headache or epilepsy.

Various methods are known from archaeological finds for carrying out a trepanation: scraping, grooving, sawing, drilling, boring or chiselling. The numerous South American examples exhibit very characteristic square shapes, whereas the many European trepanations do not appear to conform to any particular pattern. The Danish evidence for trepanning has recently been re-evaluated. This suggested that many may in fact be traces of healed blunt force injury (Figure 14.8), post-mortem damage, congenital malformations or tumours. The study showed the importance of rigorous differential diagnosis, and that careful comparison should be made with specimens with identified diseases or injuries in pathology museum and archaeological bone collections (Bennike, 2003). A recent publication (Arnott *et al.*, 2003) reviews the evidence for trepanation, to which the reader is referred for further information on this topic.

BURIALS OF VICTIMS OF PERI-MORTEM TRAUMA

Decapitation

In some instances, identification of decapitated skeletal remains may be relatively straightforward, but in others careful study may be required if errors are to be avoided. For example, at an Iron Age burial site (*c.* 300 AD) in Denmark, the richly equipped skeleton of a 13-year-old boy was found with the skull lying in a non-anatomical position, some little distance from the rest of the body. The cranium, therefore, was excavated carefully, with detailed documentation, by this author. All the cervical vertebrae were well preserved without the slightest evidence of trauma; the base of the skull and the mandible presented the same picture. One explanation may be that the individual was not decapitated but that the body was placed with the head resting on a high pillow. During decomposition, the head may have rolled away from the body. By contrast, a grave with two male Viking skeletons leaves little doubt that the two men were decapitated. Each skull was found between the legs of the individual. Both showed cut-marks on the bones concerned. One of the men, aged 30–40 years, had received a horizontal blow with a sharp instrument from the back of the neck that had cleft the second cervical vertebra. Furthermore, faint traces could be seen where the blade grazed the underside of the mandible. The other man, aged 20–30 years, exhibited a cut starting on the neck just under the base of the skull, but characteristic traces in the form of smooth surfaces with sharp edges revealed that the instrument must have cut the occipital condyles, mastoid processes and zygomatic arches to pass just under the orbits. That both skulls were placed between the legs was perhaps as a sign of disgust for the dead persons, or maybe to prevent the dead from returning (Bennike, 1985).

If the hands and/or feet are positioned very close together, the individual may have been bound. This was the case in a Viking male skeleton from Denmark lying face down and decapitated through the third cervical vertebra (Figure 14.9). Below him was a non-decapitated male skeleton.

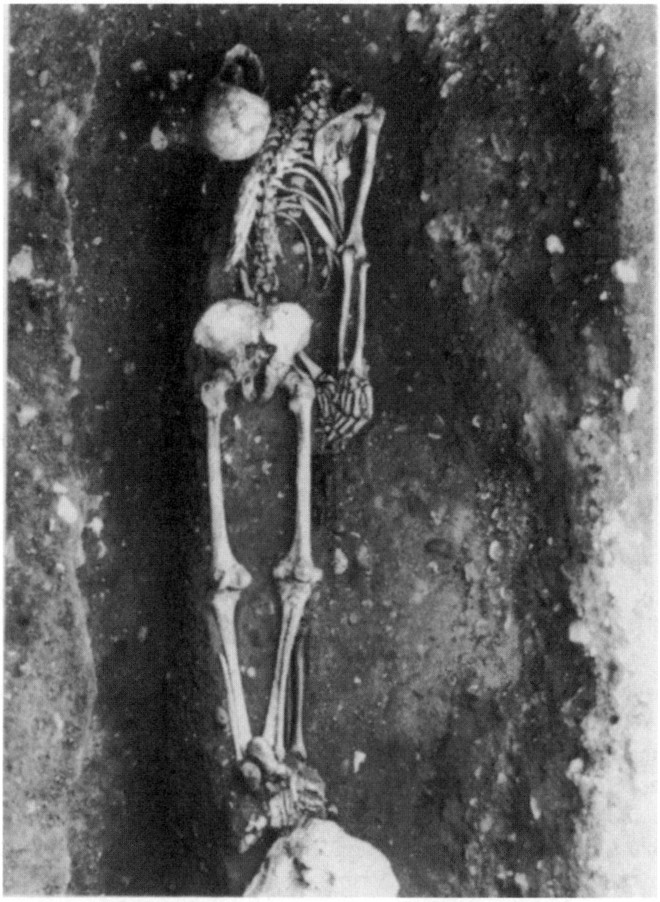

Figure 14.9 Decapitated, prone Viking male skeleton from Lejre, Denmark. The hands and feet were probably tied. Photograph: H. Andersen

Mass Graves

Mass graves are found all over the world, and can be traced to all periods in the history of mankind. Two Mesolithic skull-pits discovered at Ofnet in Germany contained the remains of around 32 individuals: 5 males, 10 females and 17 sub-adults. Because of the unusual and rather strange arrangement of the skulls, the pits were termed 'skull-nests'. Many of the skull fragments reveal patterns of blunt force trauma, and the find is thought to be related to a massacre (Frayer, 1997; Orschiedt, 2005).

Collective megalithic graves in Denmark have been dated to the Neolithic periods and represent a different kind of mass grave. In contrast to those mentioned above, collective megalithic tombs from the middle and late Neolithic were used over a long period of time. They could each contain more than 100 individuals, and the number of injuries seen on their remains is rarely unusually high. However, the reported incidence of cranial lesions was about 10% (Bennike, 1985) which is somewhat higher than the 2–5% seen as average in many North American studies (Schulting and Wysocki, 2005). The difference may perhaps be due to the good preservation of the Danish material.

Considering the number of battles and wars that have been fought in Europe, relatively few mass graves have been excavated from battlefield contexts. The following features characterize a mass grave of this type (Knüsel, 2005: 58):

1. The presence of many bodies within a grave cut (formulation modified by the author).
2. The presence of disorder in the orientation of the bodies indicating an apparent disregard for normative burial practice.
3. Bodies that are in contact with one another.
4. A commonality of cause or manner of death.

Many mass graves may still be intact, but it is also possible that victims were left on the battlefields to the mercies of wild animals, weathering and post-mortem decay, which would have destroyed or severely degraded the remains. This was the case when bones relating to a battle dated to AD 1385 were discovered in Aljubarrota, Portugal. The skeletal remains of at least 400 individuals mainly consisted of fragmented shafts of long bones (Cunha and Silva, 1997).

A recent study was carried out on 38 skeletons of victims of the battle of Towton, England (Boylston, 2000; Fiorato *et al.*, 2000; Novak, 2000; Knüsel, 2005). The battle of Towton (AD 1461) was part of the English 'Wars of the Roses', a civil war for the right of succession to the throne. The skeletal remains of the38 individuals from the battle lay in a mass grave. The 28 skulls had a total of 113 injuries (an average of 4 per individual) of which 73 were sharp, 28 were blunt and 12 were puncture wounds. In addition, 43 postcranial injuries were found.

The first mass graves to be archaeologically excavated were from Visby in Gotland, Sweden, where about 1200 individuals were buried in three mass graves that were dated to AD 1361. At that time a large part of Gotland's male population succumbed to the Danish King Valdemar Atterdag's army. The Visby skeletal collection is the largest battlefield assemblage in Europe to be subjected to anthropological study (Ingelmark, 1939). The many skeletons lay in random positions except for in one of the graves, where the uppermost 20 skeletons lay parallel to one another with their heads to the west as prescribed by medieval custom.

The skeletal remains of about 60 victims of The Battle of Good Friday (AD 1520), which took place in Uppsala, Sweden, have recently been excavated. The battle was fought between the Danish King Christian's troops and rebels loyal to Sten Sture, the Swedish national administrator (Syse, 2003; Kjellström, 2005). The skeletal remains were in three pits with different categories of bone deposits: complete skeletons, whole limbs and disarticulated bones. The remains were probably gathered some time after the battle and buried in three different pits according to their level of decomposition. The excavated area is believed to be only a small part of a much larger area with one or more mass graves which have not been excavated. The majority of the 60 excavated skeletons were males, 20–30 years of age with an average height of about 170 cm. Whereas the skulls and skull fragments (including mandibles) had 92 cut lesions can average of 1.6 cuts for all discovered skulls), few (11) lesions were found on the postcranial bones (Kjellström, 2005).

Finally, 60 male skeletons from a mass grave at Sandbjerget, Naestved, Denmark (AD 1300–1350), exhibited more than 200 injuries. It was expected that the distribution of the many unhealed lesions would follow a pattern that reflected the type of battle, the types

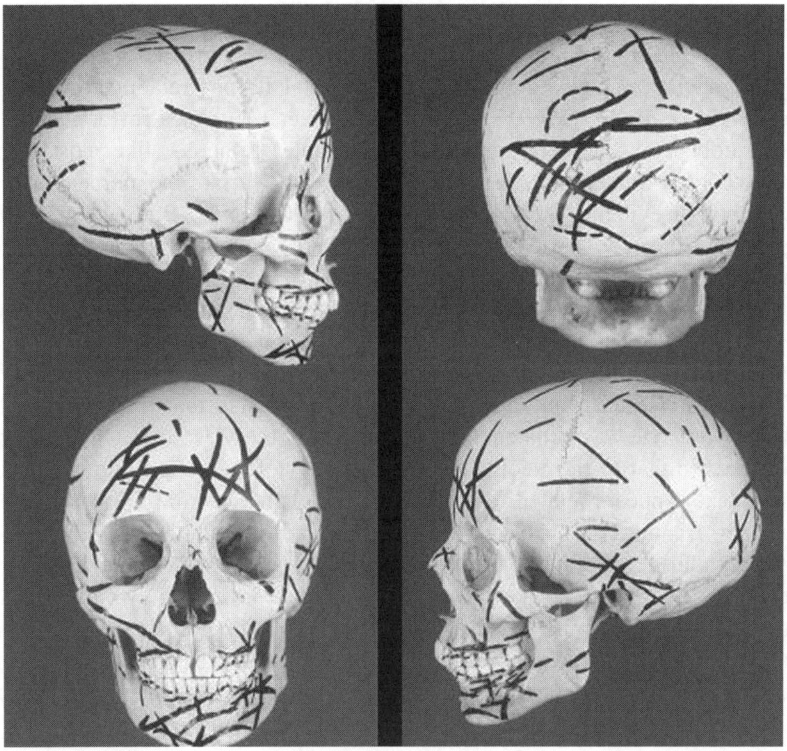

Figure 14.10 Skull showing location of all 122 lesions on the skulls excavated from the mass grave at Sandbjerget, Næstved, Denmark. The distribution of the lesions does not appear to form any particular pattern

of weapon and the armour used at the time. However, the pattern of lesions on the skulls (Figure 14.10) and postcranial bones seemed rather to reflect, to a great extent, patterns of bone preservation. For example the dearth of facial and rib wounds reflected the poor survival of these parts rather than a genuine pattern of injury (Bennike, 2006).

TRAUMA IN THE EARLIEST EUROPEAN POPULATIONS

Analysis of traumatic lesions carried out on bones from 17 Neanderthal individuals by Berger and Trinkaus (1993) revealed a total of 27 healed injuries. These were mainly distributed on the skull and upper body, with fewer on the lower limbs. This pattern of injuries was compared with that found in a variety of active modern human groups. Only rodeo riders, who incur the most injuries among modern humans, fitted the Neanderthal pattern. As the Neanderthals had heavier and more robust bones, they must have received injuries of considerable violence. No doubt Neanderthals ran significant risks while hunting large animals such as bison and elk, because they had to get close to their prey in order to use their heavy spears (Frayer, 1981).

Unlike Neanderthal skeletons, Upper Palaeolithic humans are rarely described as showing evidence of multiple traumas, suggesting that their culture provided a far more effective shield against environmentally induced injury. Skeletal lesions relating to cause of death are particularly rare. Roper (1969) suggested that skeletal evidence for deliberate injury is uncommon, probably because Upper Palaeolithic hunter-gatherers, like most other ethnographically recorded hunter-gatherers, rarely engaged in warfare or interpersonal violence. However, although evidence of violence is rare, it does exist. Both the Old Man from Cro-Magnon and the man buried at Chancelade were probably too disabled to fend for themselves.

In the Mesolithic, traumatic injuries were more frequent, but in most cases non-lethal (Bennike 1997, 2002). The majority of injuries relate to interpersonal violence and seem to have been inflicted in the later phases of the period when settlements apparently became larger. Constandse-Westermann and Newell (1984) suggested that the transition toward a more sedentary lifestyle was initiated in the last phases of the Mesolithic when the population density approached that of the Neolithic. One of the very few western European Mesolithic cases where death appears related to interpersonal violence is from a 7000-year-old grave at Bøgebakken, Denmark (Constandse-Westermann and Newell, 1984). A male adult was buried with an adult female and a 1-year-old child. An arrow was embedded in one of the man's vertebrae. He was most likely shot, but it is unclear why the woman and the child were buried with him (Bennike, 1985). Danish early Neolithic skeletons clearly illustrate that such events were by no means restricted to the Mesolithic (Bennike, 1999; Bennike and Alexandersen, 2002). A male skeleton was found with two arrows lodged in the nasal region and in the sternum, leaving no doubt that the injuries were fatal.

ADVANCES IN THE STUDY OF TRAUMA

It is only a few decades since systematic studies of trauma were virtually absent in the palaeopathological literature. New analyses of skeletal samples, including material from mass graves, have certainly helped to rectify this. The mass graves in Visby, Sweden, were unique in Europe until a decade ago. Since then several studies have added information about traumatic injuries on skeletons from mass graves in England, Portugal, Sweden and Denmark and provided a clearer picture of warfare, and of the occupation/activity patterns of some of the people involved. There has been increasing interest in this area of research, and the excavated material will undoubtedly encourage many more studies, including much-needed studies of various aspects of taphonomy.

Many systematic studies of trauma have been published with a focus on specific populations. For example, Roberts and co-workers studied British palaeopopulations (Roberts, 1991; Roberts, 2000; Roberts and Cox, 2003). Webb (1995) included analysis of trauma in his study of aboriginal Australian skeletal material. Studies on skeletal material from North America cover a large area of research into trauma (e.g. Walker, 1989, 1997; Lovell, 1997; Jurmain, 2001; Smith and Ostendorf, 2003). Martin and Frayer (1997) present anthropological studies of trauma in North America and Europe covering a range of different periods.

A new and much closer collaboration between the disciplines of anthropology and forensic medicine has resulted in a number of experimental studies of trauma (e.g. Maples, 1986; Banks and Brown, 2001). Forensic cases help us to understand the mechanism and patterns of trauma inflicted by various weapons and accidents, and to increase our knowledge of both interpersonal conflict and environmentally induced injury in past populations. Detailed studies have shown how the form, thickness and plasticity of bone are related to the pattern of fractures. Studies of healing and decomposition of bone have provided us with more knowledge of the peri-mortem period.

Interest in trauma specifically related to interpersonal violence has also increased in archaeological research, as witnessed by the growing amount of literature (Keeley, 1996; Mielke, 2002; Parker Pearson and Thorpe, 2005; Otto *et al.*, 2006). From an archaeological point of view, studies of warfare in social anthropology seem to have opened up a whole new field. Parallels with modern warfare, be it in Bosnia or New Guinea, are also being used to help interpret aspects of prehistoric warfare and aggression (Otto *et al.*, 2006).

Interdisciplinary collaboration has also been established between medical specialists in osteoporosis and anthropologists/osteologists who are dealing with questions of bone quality and the pattern of fracture in the past.

In spite of new technologies and interdisciplinary collaboration, there are still many problems to be resolved when studying trauma in the past. For one thing, distinguishing between ante-, peri- and post-mortem injuries, violence or accident, treatment or lack of treatment, sacrifice or punishment may still present major challenges in palaeopathology.

REFERENCES

Adams JC. 1972. *Outline of Fractures*. Churchill Livingstone: Edinburgh.

Arnott R, Finger S, Smith CUM (eds). 2003. *Trepanation. History, Discovery, Theory*. Swets & Zeitlinger: Lisse.

Aufderheide AC, Rodriguez-Martin C. 1998. *The Cambridge Encyclopedia of Human Paleopathology*. Cambridge University Press: Cambridge.

Banks P, Brown A. 2001. *Fractures of the Facial Skeleton*. Wright: Edinburgh.

Bennike P. 1985. *Palaeopathology of Danish Skeletons. A Comparative Study of Demography, Diseases and Injuries*. Akademisk Forlag: København.

Bennike P. 1997. Death in the Mesolithic. Two old men from Korsør Nor. In *The Danish Storebælt since the Ice Age*, Pedersen L, Fischer A, Aaby B (eds). Storebaeltsforbindelsen: København; 99–105.

Bennike P. 1999. The early Neolithic Danish bog finds: a strange group of people! In *Bog Bodies, Sacred Sites and Wetland Archaeology*, Coles B, Coles J, Jørgensen MS (eds). WARP Occasional Paper 12. Short Run Press: Exeter; 27–32.

Bennike P. 2002. Palaeopathology as a tool for the study of survival of past populations. In *Human Biology and History*, Smith M (ed.). Society for the Study of Human Biology Series, No. 42. Taylor and Francis: London; 165–177.

Bennike P. 2003. Ancient trepanations and differential diagnosis: a re-evaluation of skeletal remains from Denmark. In *Trepanation: History, Discovery, Theory*, Arnott R, Finger S, Smith CUM (eds). Swets & Zeitlinger: Lisse; 95–116.

Bennike P. 2006. Rebellion, combat, and massacre: a medieval mass grave at Sandbjerg. In *Warfare and Society. Archaeological and Social Anthropological Perspectives*, Otto T, Thrane H, Vandkilde H (eds). Aarhus University Press: Aarhus; 305–318.

Bennike P, Alexandersen V. 2002. Population plasticity in Stone Age Denmark. In *The Neolithicisation of Denmark. 150 Years of Debate*, Fischer A, Kristiansen K (eds). Collins: Sheffield; 291–301.

Bennike P, Bro-Rasmussen F. 1989. Contracture and anchylosis of the hip joint in a right angle. Five palaeopathological cases. *Int J Anthropol* **4**: 137–151.

Bennike P, Bro-Rasmussen F, Bro-Rasmussen P. 1987. Dislocation and/or congenital malformation of the shoulder joint. Observations on a mediaeval skeleton from Denmark. *Anthropol Anz* **45**: 117–129.

Berger TD, Trinkaus E. 1993. Patterns of trauma among the Neandertals. *J Archaeol Sci* **22**: 841–852.

Berryman HE, Haun SJ. 1996. Applying forensic techniques to interpret cranial fracture patterns in an archaeological specimen. *Int J Osteoarchaeol* **6**: 2–9.

Berryman HE, Symes SA. 1998 Recognizing gunshot and blunt cranial trauma through fracture interpretation. In *Forensic Osteology*, Reichs K (ed.). Charles C. Thomas: Springfield, IL; 333–352.

Boylston A. 2000. Evidence for weapon-related trauma in British archaeological samples. In *Human Osteology in Archaeology and Forensic Science*, Cox M, Mays S (eds). Greenwich Medical Media: London; 357–380.

Boylston A. 2004. Recording of weapon trauma. In *Guidelines to the Standards for Recording Human Remains*, Brickley M, McKinley JI (eds). IFA Paper No. 7. BABAO; Institute of Field Archaeologists: Southampton/Reading; 40–42.

Brothwell D. 2003. The future direction of research. In *Trepanation. History, Discovery, Theory*, Arnott R, Finger S, Smith CUM (eds). Swets & Zeitlinger: Lisse; 365–372.

Buhr AJ, Cooke AM. 1959. Fracture patterns. Lancet **14**: 531–536.

Buikstra JE, Ubelaker D (eds). 1994. *Standards for Data Collection from Human Skeletal Remains*. Archaeological Survey, Research Seminar Series 44. Arkansas Archaeological Survey: Fayetteville.

Cunha E, Silva AM. 1997. War lesions from the famous Portuguese medieval battle of Aljubarrota. *Int J Osteoarchaeol* **7**: 595–599.

Constandse-Westermann TS, Newell RR. 1984. Human biological background of population dynamics in the western European Mesolithic. *Hum Palaeontol Proc B* **87**: 139–223.

Fiorato V, Boylston A, Knüsel C (eds). 2000. *Blood Red Roses. The Archaeology of a Mass Grave from the Battle of Towton AD 1461*. Oxbow Books: Oxford.

Frayer DW. 1981. Body size, weapon use, and natural selection in the European Upper Paleolithic and Mesolithic. *Am Anthropol* **83**: 57–73.

Frayer DW. 1997. Ofnet: evidence for a Mesolithic massacre. In *Troubled Times. Violence and Warfare in the Past. War and Society*, Martin DL, Frayer DW (eds). Gordon and Breach: Amsterdam; 181–216.

Gurdjian ES, Webster JE, Lissner HR. 1950. The mechanism of skull fracture. *Radiology* **54**: 313–339.

Haglund DW, Sorg MH (eds). 1997. *Forensic Taphonomy. The Postmortem Fate of Human Remains*. CRC Press: Boca Raton.

Ingelmark BE. 1939. The skeletons. In *Armour from the Battle of Visby 1361*, Thordeman B (ed.). Almquist & Wiksells: Stockholm; 149–209.

Judd M. 2002 Ancient injury recidivism: an example from the Kerma period of ancient Nubia. *Int J Osteoarchaeol* **12**: 89–106.

Judd M, Roberts CA. 1998 Fracture patterns at the medieval leper hospital in Chichester. *Am J Phys Anthropol* **105**: 43–55.

Judd M, Roberts CA. 1999. Fracture trauma in a medieval British farming village. *Am J Phys Anthropol* **109**: 229–243.

Jurmain R. 2001. Paleoepidemiological patterns of trauma in a prehistoric population from central California. *Am J Phys Anthropol* **115**: 13–23.

Jurmain R, Kilgore L. 1998. Sex-related patterns of trauma in humans of African apes. In *Sex and Gender in Paleopathological Perspective*, Grauer AL, Stuart-Mcadam P (eds). Cambridge: Cambridge University Press; 11–26.

Keeley LH. 1996. *War Before Civilization: The Myth of the Peaceful Savage*. Oxford University Press: Oxford.

Kilgore L, Jurmain R, van Gerven D. 1997 Palaeoepidemiological patterns of trauma in a medieval Nubian skeletal population. *Int J Osteoarchaeol* **7**: 103–114.

Kjellström A. 2005. A sixteenth-century warrior grave from Uppsala, Sweden: the Battle of Good Friday. *Int J Osteoarchaeol* **15**: 23–50.

Knowles AK. 1983. Acute traumatic lesions. In *Disease in Ancient Man*, Hart GG (ed.). Clarke Irwin: Toronto; 61–83.

Knüsel CJ. 2005. The physical evidence of warfare – subtle stigmata? In *Warfare, Violence and Slavery in Prehistory*, Parker Pearson M, Thorpe IJN (eds). BAR International Series 1374. Archaeopress: Oxford; 49–65.

Lovejoy CO, Heiple KG. 1981 The analysis of fractures in skeletal populations with an example from the Libben Site, Ottowa, Ohio. *Am J Phys Anthropol* **55**: 529–541.

Lovell N. 1990 *Patterns of Injury and Illness in Great Apes: A Skeletal Analysis*. Smithsonian Institution: Washington, DC.

Lovell N. 1997. Trauma analysis in paleopathology. *Ybk Phys Anthropol* **40**: 139–170.

Maples WR. 1986. Trauma analysis by the forensic anthropologist. In *Forensic Osteology. Advances in the Identification of Human Remains*, Reichs KJ (ed.). Charles C. Thomas: Springfield, IL; 218–228.

Martin DL, Frayer DW (eds). 1997. Tr*oubled Times. Violence and Warfare in the Past. War and Society 3*. Gordon and Breach: Amsterdam.

Merbs CF. 1989. Trauma. In *Reconstruction of Life from the Skeleton*, Iscan MY, Kennedy KAR (eds). Liss: New York; 161–189.

Mielke JH. 2002. Warfare and population structure. In *Human Biology and History*, Smith M (ed.). Society for the Study of Human Biology Series 42. Taylor and Francis: London; 98–111.

Nilsson SL. 2003. *Embodied rituals and ritualized bodies. Tracing ritual practices in late Mesolithic burials*. Acta Archaeol Lundensia Series 8, No. 46. Almqvist & Wiksell International: Stockholm.

Novak SA. 2000 Battle-related trauma. In *Blood Red Roses. The Archaeology of a Mass Grave from the Battle of Towton AD 1461*, Fiorato V, Boylston A, Knüsel C (eds). Oxbow Books: Oxford; 90–102.

Orschiedt J. 2005. The head burials from Ofnet cave: an example of warlike conflict in the Mesolithic. In *Warfare Violence and Slavery in Prehistory*, Parker Pearson M, Thorpe IJN (eds). BAR International Series 1374. Archaeopress: Oxford; 67–74.

Ortner DJ, Putschar WGJ. 1985. Identification *of Pathological Conditions in Human Skeletal Remains*. Smithsonian Institution: Washington, DC.

Otto T, Thrane H, Vandkilde H (eds). 2006. *Warfare and Society. Archaeological and Social Anthropological Perspectives*. Aarhus University Press: Aarhus.

Parker Pearson M, Thorpe IJN (eds). 2005. *Warfare Violence and Slavery in Prehistory*. BAR International Series 1374. Archaeopress: Oxford.

Roberts C. 1991. Trauma and treatment in the British Isles in the historic period: a design for multidisciplinary research. In *Human Paleopathology. Current Syntheses and Future Options*, Ortner DJ, Aufderheide AC (eds). Smithsonian Institution Press: Washington, DC; 225–240.

Roberts C. 2000. Trauma in biocultural perspective: past, present and future work in Britain. In *Human Osteology in Archaeology and Forensic Science*, Cox M, Mays S (eds). Greenwich Medical Media: London; 337–356.

Roberts C, Cox M. 2003. *Health and Disease in Britain*. Sutton: Stroud.

Rogers J, Waldron T. 1995. *A Field Guide to Joint Disease in Archaeology*. Wiley: Chichester.

Roper MK. 1969. A survey of the evidence for intrahuman killing in the Pleistocene. *Curr Anthropol* **10**: 427–459.

Sauer NJ. 1998 The timing of injuries and manner of death: distinguishing among antemortem, perimortem and post-mortem trauma. In *Forensic Osteology* (2nd edn), Reichs KJ (ed.). Charles C. Thomas: Springfield, IL; 321–332.

Schultz AH. 1967. Notes on diseases and healed fractures of wild apes. In *Diseases in Antiquity*, Brothwell DR, Sandison AT (eds). Charles C. Thomas: Springfield, IL; 47–55.

Schulting RJ, Wysocki M. 2005. 'In this Chambered Tumulus were Found Cleft Skulls…': an assessment of the evidence for cranial trauma in the British Neolithic. *Proc Prehist Soc* **71**: 107–138.

Smith M, Ostendorf. 2003. Beyond palisades: the nature and frequency of late prehistoric deliberate violent trauma in the Chickamauga Reservoir of east Tennessee. *Am J Phys Anthropol* **121**: 303–318.

Steinbock RT. 1976. *Paleopathological Diagnosis and Interpretation*. Charles C. Thomas: Springfield, IL.

Syse B (ed.). 2003. *Långfredagsslaget. En Arkæologisk Historia*. Upplandsmuseet: Uppsala.

Taylor T. 2003. *Buried Soul: How Humans Invented Death*. Fourth Estate: London.

Walker PL. 1989. Cranial injuries as evidence of violence in prehistoric southern California. *Am J Phys Anthropol* **80**: 313–323.

Walker PL. 1997. Wife beating, boxing, and broken noses: skeletal evidence for the cultural patterning of violence. In *Troubled Times. Violence and Warfare in the Past*, Martin DL, Frayer DW (eds). Gordon and Breach: Amsterdam; 145–179.

Webb S. 1995. *Paleopathology of Aboriginal Australians*. Cambridge: Cambridge University Press.

15

Congenital Anomalies

Ethne Barnes

PO Box 90840, Tucson, AZ 85752-0840, USA

INTRODUCTION

Various congenital anomalies in ancient human remains have been identified and reported in the medical and anthropological literature for over a century on a case-by-case basis. In the past 20 years or so, there has been greater interest in determining frequency rates of specific anomalies, generally of a minor nature. This has proved to be difficult without a standardized approach for defining and classifying what to look for among the many anomalies and their variations.

First, we have to define what constitutes congenital anomalies. Any physical condition deviating from what is acceptable as normal that began before birth can be called a congenital anomaly. This includes acquired prenatal infections, such as syphilis or rubella, that affect the developing foetus. Epigenetic, extrinsic or intrinsic factors affecting a genetically susceptible background, or genetic variations/mutations, can lead to developmental defects by interfering with the developing embryo at crucial genetically marked threshold times. For example, a mother with inherent faulty maternal folate metabolism lacking sufficient folic acid in her diet causes harm to the developing embryo. Depleted maternal folic acid levels occurring at the time of programmed closure of the developing neural tube of a genetically susceptible embryo can disrupt closure, resulting in a neural tube defect such as spina bifida or meningocoele. Lack of sufficient maternal iodine can lead to cretinism in the susceptible developing foetus. Other external factors, like chemical poisoning (e.g. with mercury or the drug thalidomide), can interfere with the genetically programmed development of the embryo. Sometimes the congenital disorder does not appear until months or years following birth, depending on the gestational timing of the insult to the developing foetus.

Most often, genetic factors alone cause developmental variations or anomalies that may or may not result in pathology (Larsen, 2001; Sadler, 2006). Our genes represent evolution in action with inherent variability to adapt to change, as found with genetic drift. The developing embryo is a microcosm of evolution.

Disturbances by mutant genes can be localized or tissue specific. Sometimes a number of linked genetic disturbances affect more than one developmental field and are combined into

Advances in Human Palaeopathology Edited by Ron Pinhasi and Simon Mays

a set of defects known as a syndrome. Genetic actions can be modified by mutant genes or, sometimes, extrinsic factors can act as gene modifiers by interfering with molecular signals within susceptible genetic backgrounds.

The defective mutant globin proteins in haemoglobin associated with sickle cell, thalassaemia and other congenital haemoglobin defects reflect population genetic responses to endemic malaria. Genetic defects in developing cartilaginous tissues produce various chondrodysplasias such as achondroplastic dwarfism. Other forms of dwarfism occur with a variety of metabolic defects, including pituitary dwarfism. Developmental defects in connective tissues such as collagen can lead to a variety of disorders, such as osteogenesis imperfecta. Developing skeletal tissue abnormalities produce varying types of skeletal dysplasia (Ortner and Putschar, 1981; Ortner, 2003). Chromosomal aberrations such as trisomy 21 (Down's syndrome), and primary metabolic abnormalities such as the mucopolysaccharidoses, are other causes of congenital abnormalities (Spranger *et al.*, 1974). Most major defects have disastrous effects on the developing foetus and frequently end with spontaneous abortion or cause death soon after birth. Therefore, finding the more serious congenital anomalies in ancient human remains is rare.

Sometimes injury or infection in the growing child or adult can mimic a congenital disorder, especially tissue-specific disorders. On the other hand, congenital disorders can be confused with injury or complications from disease. Careful analysis can identify the true cause. For example, ventral hypoplasia of a vertebral body can mimic compression fracture, but there will be no evidence of compression, only the smaller ventral height of the vertebra. The effects of congenital anaemia caused by thalassaemia and sickle cell can be differentiated from acquired anaemia based on a number of skeletal disturbances plus knowing whether the affected individual comes from a region of the world where these disorders originated.

THE MORPHOGENETIC APPROACH

Since major or severe congenital anomalies are rarely found in ancient human remains, we should focus more on minor or less disturbing developmental defects, particularly those affecting the skeleton. The morphogenetic approach to deciphering anomalies developing from the embryonic precursor of the skeleton and adjacent tissues is useful for defining and classifying various developmental anomalies. The range of possible expressions, both minor and major, for each type of anomaly can also be defined by this approach (Barnes, 1994a). Recent molecular studies in embryology have shown that most of these anomalies are genetically determined (Larsen, 2001; Sadler, 2006).

The developing embryo sets the stage for all that comes later, as the various regulatory genes act in concert with cascading induction and enhancing signals and cross talk between developing tissues as they come together. Within the first 8 to 10 weeks of life, the precursor tissues of the skeleton and all other organs of the body organize into their respective body parts. Whatever is laid down in the anlage determines the final outcome.

Most developmental disturbances arise from altered molecular signals that delay the point at which specific primordial tissues within a designated developmental field or adjacent fields meet during critical threshold events. Delay in the meeting of precursor embryonic vertebral arches leads to delayed and incomplete ossification, resulting in a cleft bony arch as opposed to spina bifida. The latter is caused by disturbance in another developmental field, the neural tube, which interferes with development of the adjacent vertebral arch. Genetically programmed cranial–caudal shifting of borders within the developing embryonic vertebral column produces a number of developmental anomalies. Genetic or epigenetic changes in regulatory signals that alter the timing of threshold events in the developing limbs can lead to a number of anomalies affecting either the whole limb or specific parts of a limb.

The morphogenetic approach to the analyses of skeletal developmental anomalies can produce a wealth of information. Many minor developmental anomalies, such as solitary block vertebra, cleft vertebral arch, sternal aperture, brachydactyly or polydactyly, do not interfere with normal function of the affected part and, thus, are retained in a given genetic population. This also includes minor expressions of rare major defects that may not be present, such as notched palate versus cleft palate. Occasionally, a rare pathological defect occurs unrelated to other anomalies within any population. However, the overall frequency pattern for various anomalies within a population reflects the genetic trends of that population (Anderson, 1968; Barnes, 1994a). This can be valuable where invasive DNA studies are not possible. Ancient migration patterns can be tracked by this method, as well as determining kinship linkages and marriage patterns from clustering of specific anomalies within ancient cemeteries.

Hauser and De Stefano (1989) describe a number of developmental variations of the skull, many of them used as non-metric traits for biological distance studies (Buikstra and Ubelaker, 1994: 85). While these minor developmental variations of the skull are important trait markers, there are other developmental variants of the skull important to the study of congenital anomalies best understood by the morphogenetic approach (Barnes, 1994a).

THE SKULL

Morphogenesis

The primordial embryonic tissues that form the skull evolve from four interacting developmental fields with some contributions from the paraxial mesoderm that meets the base of the skull. The cranial vault, or calvarium, develops from the blastemal mesocranium derived from neural crest cells migrating from the cranial neural plate folds that interface with ectodermal tissue (Figure 15.1a). The final cranial bones ossify directly from membranous tissue. The base of the skull, or chondrocranium, develops from the prechordal cranial base, and the mandible, maxilla, zygomatics and palatine bones come from the first branchial arch (Figure 15.1b). The midface region forms from the frontonasal process (Figure 15.1c).

The prechordal cranial base forming the chondrocranium generates from dense mesenchymal cell concentrations above the primordial cervical spine. Three pairs of cartilaginous plates form and arrange themselves in series beneath the forming brain from the interorbitonasal region to the cervical junction. The caudal-most pair, the parachordal cartilages, forms the base of the skull with contributions from the first cervical pair of somites developing from the paraxial mesoderm of the primordial vertebral column. The other two pairs of cartilaginous plates, along with neural crest cells, contribute to the formation of the mastoids, petrous portions of the temporal bones, most of the sphenoids, the ethmoid and basilar portion of the occipital. The chondrocranium ossifies directly from cartilaginous tissue.

Five branchial or pharyngeal arches form in the cranial front end of the embryo, consisting of internal pouches separated by clefts. Each arch is lined with membrane consisting of ectoderm, mesoderm and endoderm. Neural crest and lateral plate cells migrate into the

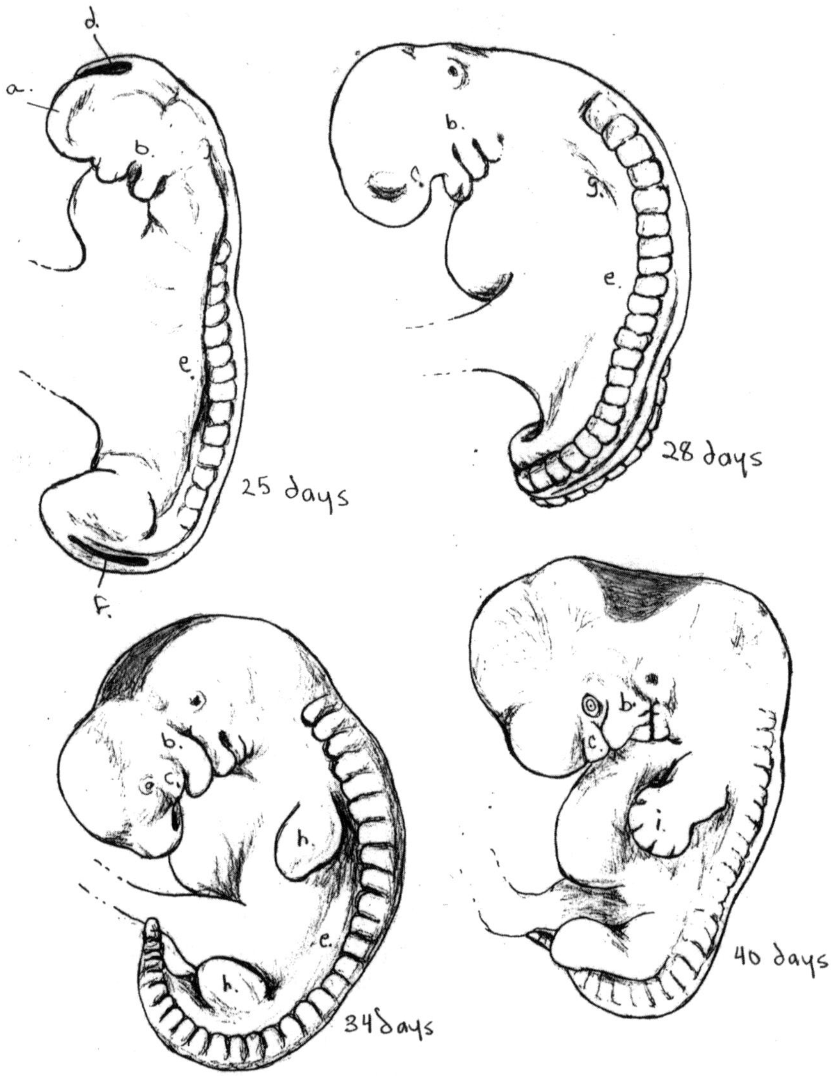

Figure 15.1 Embryonic morphogenesis: (a) cranial neural plate fold; (b) branchial arches; (c) midface region-frontonasal process; (d) anterior neuropore of neural tube; (e) somites; (f) posterior neuropore of neural tube; (g) lateral plate mesoderm; (h) limb buds; (i) digital plate

arches to form the developing skeletal elements and laryngeal cartilages. The first arch (Figure 15.1b) develops into the maxilla and mandible, plus the zygomatics, incus, malleus, and palate. The external auditory meatus derives from the dorsal end of the ectodermal groove of the first branchial arch, while the tympanic plate evolves from the closing membranes separating the groove from its endodermal pouch. As the first branchial arch develops, it forms two swellings that form into two halves of the maxilla and two halves of the mandible, with each mandible half containing a central cartilaginous element. Bones of the maxilla and mandible develop directly from membranous tissue, with membranous bone enveloping the mandibular cartilages. The developing mandible retains within its core a remnant of its precursor cartilage known as Meckel's cartilage. Secondary cartilage later forms the articulating condyles. The second branchial arch produces the stapes, styloid process, stylohyoid ligament and hyoid lesser cornua, while the hyoid body and greater cornua derive from the third branchial arch.

The frontonasal process (Figure 15.1c) swells between the first branchial arch and bulging forebrain as neural crest cells converge on it to form the nose and premaxilla. Placodes for the eyes and nares form on the sides of the frontonasal process and move towards each other as the adjacent maxillary processes expand toward the frontonasal process. The nasal placodes divide into paired lateral and medial processes with a groove between each lateral nasal process and the encroaching maxillary process that eventually becomes the nasolacrimal duct. The medial nasal processes move towards each other and fuse to form the nasal bridge and septum. Then the inferior tips of the nasal processes expand laterally and fuse to form the premaxilla (intermaxillary) processes that later fuse with the tips of the maxillary processes. The medial walls of the maxillary processes thin out to form extensions, the palatine shelves that form the secondary palate.

Anomalies

Cranial Vault

While extra cranial ossicles and metopism of the calvarium are well known, less attention has been given to other blastemal mesocranial developmental anomalies. The so-called Inca bone is actually retention of the foetal mendosa suture (Figure 15.2a) that separates the developing membranous interparietal and cartilaginous supraoccipital portions of the occipital bone during development, programmed to disappear as the two bones fuse into one. Sutural agenesis, the failure of opposing precursor cranial bones to differentiate and form a suture between them, is often referred to incorrectly as premature craniosynostosis. The term craniosynostosis implies premature fusion of sutures anytime after morphogenesis, not the absence of development of the sutures during morphogenesis. The suture never forms, leading to a variety of skull shapes dependent on the location of the missing suture or sutures. Scaphocephaly results from failure of the sagittal suture to form (Figure 15.2b). Total lack of all sutures would result in microcephaly (abnormally small skull), but generally microcephaly relates to underdevelopment of the brain with or without craniosynostosis, and sometimes it occurs along familial lines. The opposite of microcephaly is the greatly enlarged skull associated with hydrocephaly caused by blockage of the normal flow of cerebrospinal fluid within the ventricles of the brain. The blocked fluid accumulates in the lateral ventricles, pressing on the brain and expanding the skull (Sadler, 2006: 309).

Parietal thinning has been related to aging, but it is the result of failure of the diploe to develop properly between the inner and outer tables of the parietals (Figure 15.2c). It becomes more noticeable as the affected individual ages, and it can appear bilaterally or unilaterally.

Although rare, greatly enlarged parietal foramina (Figure 15.2d) and variations of this defect have been noted in family lineages. Programmed movement of the membranous precursors for ossification of the parietals is usually slowed as they near the posterior portions of these bones to allow for Santorini's emissary vessels to pass through, leaving small foramina in the final bones. Not all individuals develop foramina for these tiny vessels,

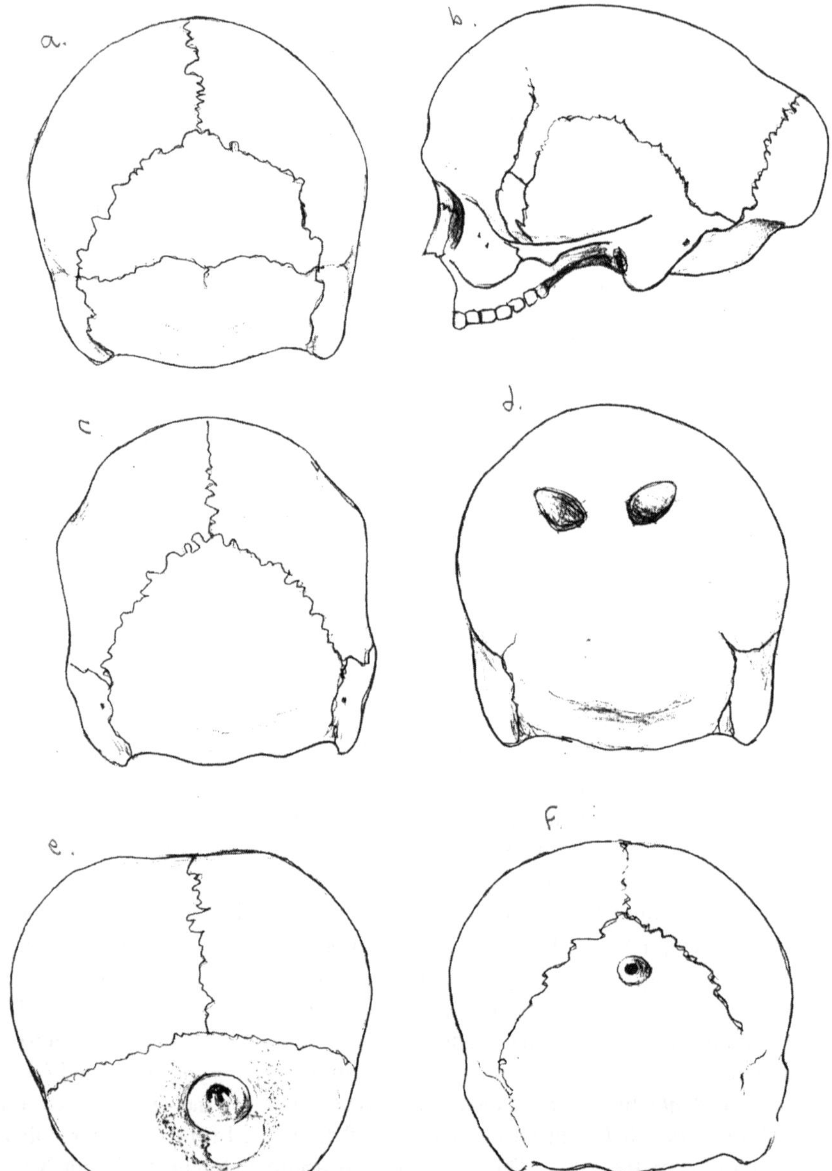

Figure 15.2 Cranial anomalies. (a) Retained mendosa suture, Lake Eufaula, Oklahoma adolescent. (b) Scaphocephaly (agenesis sagittal suture), adult male, Cinco Cerros, Peru. (c) Bilateral parietal thinning, adult male, Corinth, Greece. (d) Bilateral enlarged parietal foramina, young adult, Palo Alto, California; (e) Meningocoele, child, Ancon, Peru. (f) Dermoid cyst, adult male, Corinth, Greece

and sometimes only one develops. Development of enlarged parietal foramina occurs when the movement of membranous precursors for ossification is markedly slowed or they do not reach the final borders, leaving large areas reserved for and around the parietal foramina without the necessary programming to ossify. This anomaly can be expressed as large slits, enlarged foramina connected by sutures, or lozenge-shaped or large ovoid openings covered by membranous tissue.

Occasionally the developing cranial bones entrap overlying ectodermal cells, creating inclusion cysts that tend to occur at the midline or sagittal plane of the calvarium from the nasal region to the base of the skull. They most often consist of keratohylin within a capsule lined with squamosal epithelium as an epidermoid cyst, or the cyst will contain various components of dermoid tissue, forming a dermoid cyst. Timing of the event determines the type of inclusion cyst. Epidermal cysts frequently affect the dura covering the brain, while dermoid cysts tend to develop in the diploe of the cranium and sometimes between the periosteum and the scalp, varying in size from a few millimetres (Figure 15.2f) up to 10 cm, particularly when an ectodermal cyst occurs within the angle of the eye orbit or the anterior fontanelle. Imprints of these rounded cysts are left on the cranium. Sometimes entrapped ectoderm forms a small rounded dermal sinus communicating between the external and internal cranial surfaces connecting to either an internal or external dermoid cyst. Dermoid sinuses most often occur between the foramen magnum and the occipital protuberance. Small dermoid cysts frequently go undetected, whereas epidermoid cysts continue to grow with epithelial desquamation into large cysts.

The most serious anomaly affecting the developing skull occurs when the cranial end of the neural tube (Figure 15.1d), precursor to the developing brain and spinal cord, fails to close properly. If the folds of the neural plate fail to fuse in the formation of the neural tube, then the precursor brain and spinal cord are exposed to amniotic fluid and abnormal development (cranorachischisis) and death, usually during the first trimester. Anencephaly results from failure of the anterior opening (neuropore) of the neural tube to close, leaving the developing embryo without brain development beyond the brain stem. Death occurs either *in utero* or shortly after birth.

Delay in closure of the neuropore can still lead to a neural tube defect known as a postneuralization defect covered by ectodermal tissue. Depending on timing, the defect can just involve a portion of the meningeal covering of the brain protruding through the developing cranium, encased in a skin-covered cyst known as a meningocoele, or result in protrusion of brain tissue within a skin filled sac known as an encephalocoele. If both brain and meninges are involved it is referred to as a meningoencephalocoele. Encephalocoeles generally occur in the occipital region, producing a large bony opening that makes the occipital bone appear bifid. Those born with this condition usually do not survive beyond birth.

Individuals born with the less serious skin-covered meningocoele can survive into adulthood. The defect usually locates in the sagittal plane from the root of the nasion to the base of the occipital. It can occur in the root of the sella turcica or the roof of the orbital angle. The protruding meningeal tissue disrupts development of the affected cranial area. Meningocoeles most often appear in the anterior fontanelle or bregma region; their cystic imprints on the external surface can vary in size, but they always have an irregular opening at the base for the stem that connects the cystic meningeal tissue with the underlying meninges (Figure 15.2e). The bony borders of these saucer-like impressions on the top of the cranium are sharply defined and surrounded by an outer flange of bony build up that responds to pulsations from the meningocoele (Webb and Thorne, 1985).

Cranial Base

The developing chondrocranium skull base is most affected by genetically programmed shifting of the occipitocervical border. When the border shifts cranially (upward) from its normal location, the base of the occipital takes on characteristics of the first cervical vertebra (Figure 15.3a). However, it never separates from the skull base and it lacks a complete posterior arch and rarely forms an anterior arch. Minor expressions can be a precondylar tubercle, bipartite occipital condylar facets, double hypoglossal canals, or horizontal clefts in the basioccipital. Caudal (downward) shifting of the border leads to the atlas being partially or completely incorporated into the base of the occipital (Figure 15.3b). Minor expressions include precondylar processes, hypoplasia of occipital facets, and precondylar facets. Precondylar facets develop in response to the primordial dens or odontoid process failing to move down from its primordial position with the foramen magnum, leaving the tip of the dens and atlas anterior border articulating with the rim of the occipital (Figure 15.3c).

Mandible

The first branchial arch is greatly affected by delays in development, as paired parts sometimes do not meet in a timely fashion or one half of a pair suffers setbacks while the other half does not. Developmental delay in the meeting of the two halves of the mandible can lead to failure of the two to fuse, but complete failure of fusion leaving the two mandibular

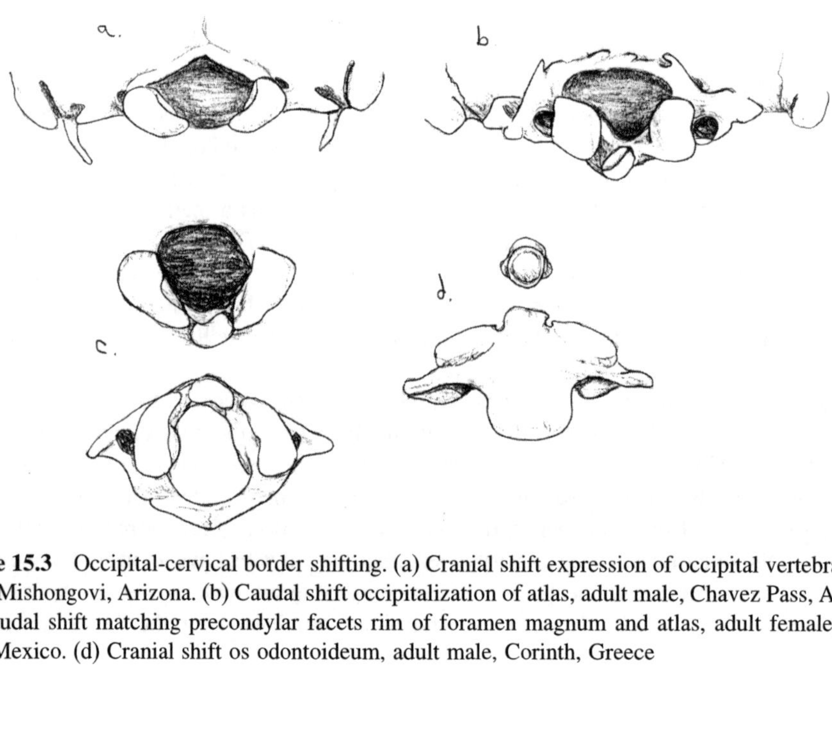

Figure 15.3 Occipital-cervical border shifting. (a) Cranial shift expression of occipital vertebra, adult male, Mishongovi, Arizona. (b) Caudal shift occipitalization of atlas, adult male, Chavez Pass, Arizona. (c) Caudal shift matching precondylar facets rim of foramen magnum and atlas, adult female, Puye, New Mexico. (d) Cranial shift os odontoideum, adult male, Corinth, Greece

Figure 15.4 Mandibular anomalies. (a) Unilateral right mandibular hypoplasia with normal left side, adult male, La Playa, Mexico. (b) Coronoid hyperplasia, adult male, Corinth, Greece. (c) Bifid condyle (bilateral), adult female, Petras, Crete. (d) Bifid condyle (unilateral right) adult female, Heshatauthla, New Mexico. (e) Stafne defect (unilateral right), adult male, La Playa, Mexico

halves as separate bones is very rare. More often there is only a partial cleft or an indentation with a space that separates the central incisors.

One or both precursors for the developing ramus portions of the mandible can be affected by slow development leading to aplasia or hypoplasia of the ramus. Severe unilateral hypoplasia (Figure 15.4a) or the rarer degree of aplasia leading to marked asymmetry of the face is recognized in medicine as hemifacial microsomia. Sometimes the zygomatic of the affected side is also malformed.

On occasion, only the mandibular condyle is affected by developmental disturbance. This part of the mandible develops from secondary cartilaginous tissue to form the head of the growing condyle. This allows the condyles to continue growing into foetal life, vascularized from overlying fibrous tissue extending into the developing heads of the condyles. Once the condyles complete development, the fibrous tissue septa are programmed to recede. Otherwise, the condylar head mediolaterally bifurcates or appears double (Figure 15.4c and d).

Bifid condyles generally appear unilaterally. Sometimes, excessive condylar chondrification can produce a larger than normal condyle that usually is unilateral.

Coronoid hyperplasia of the mandible (Figure 15.4b), usually appearing bilaterally and most often in males, results from morphogenesis programming that causes this part of the mandible to continue growing until adolescence. Greatly elongated coronoids can limit how wide the mouth opens as they impinge on the posterior aspects of the zygomatics.

The anomaly known as Stafne defect results from the sublingual salivary gland developing prematurely within the region of the developing submandibular fossa. The end product is a shallow oval depression (Figure 15.4e) below the mylohyoid line and generally in the region of the third molar, usually appearing unilaterally.

Maxillo-Facial Skeleton

Development of the frontonasal process and paired halves of the maxilla depend on synchronized timing in coming together and forming the upper face. Developmental delay with hypoplasia or aplasia of one or more of the evolving precursors can lead to facial clefting. Severe forms of facial clefting are often associated with malformation of the brain and affected individuals are not viable after birth. Clefting can occur wherever two adjacent parts are supposed to come together and fail to fuse, either partially or completely. This can occur between the lateral nares and eye orbit, between the lateral nares and maxilla, between the medial halves of the nares or between the two halves of the premaxilla.

Cleft lip is the most common viable form of facial clefting. One or both maxillary processes may fail to fuse with the premaxilla either completely or partially. One half of the premaxilla may suffer hypoplasia or aplasia, resulting in a unilateral cleft lip. When the two halves of the premaxilla fail to fuse, a midline cleft results. More severe forms of cleft lip will interfere with the adjacent developing palate, creating an associated cleft palate (Figure 15.5a and b). Sometimes the palate alone is affected by clefting (Figure 15.5c). The most severe form of bilateral cleft lip leaves the premaxillary process rounded, almost ball-like. Minor expressions of cleft lip can appear as small fissures or notches hidden by soft tissue in the alveolar ridge between the canine and lateral incisor, the boundary between the maxilla and premaxilla, or between the central incisors with a midline cleft. The precursors to these teeth can be disrupted by more severe clefting, especially the lateral incisors with lateral clefts. The affected teeth can be rotated, misplaced, peg shaped or absent.

Developmental inclusion cysts involving the overlying ectodermal tissue of the premaxilla and palatine processes occur when ectoderm becomes entrapped by the underlying developing bony precursors as they move towards fusion. The resulting fluid or semisolid cyst lined with epithelium usually locates in the midline of the palate, leaving an oval-shaped impression on the affected area of the bony palate. Most commonly, ectoderm becomes entrapped within the incisive canal, resulting in a greatly enlarged canal opening with one or more cysts involved.

Nasal bone development often is affected by either bilateral or unilateral hypoplasia or aplasia, creating very small or absent nasal bones. Similar developmental deficiency can occur with the lacrimal bones. Hypoplasia of the median nasal prominence results in a flat glabella, short flat nose missing the anterior nasal spine, small premaxilla and frontal sinuses with a normal prognathic-appearing mandible. This has proven to be familial and is known as Binder's syndrome.

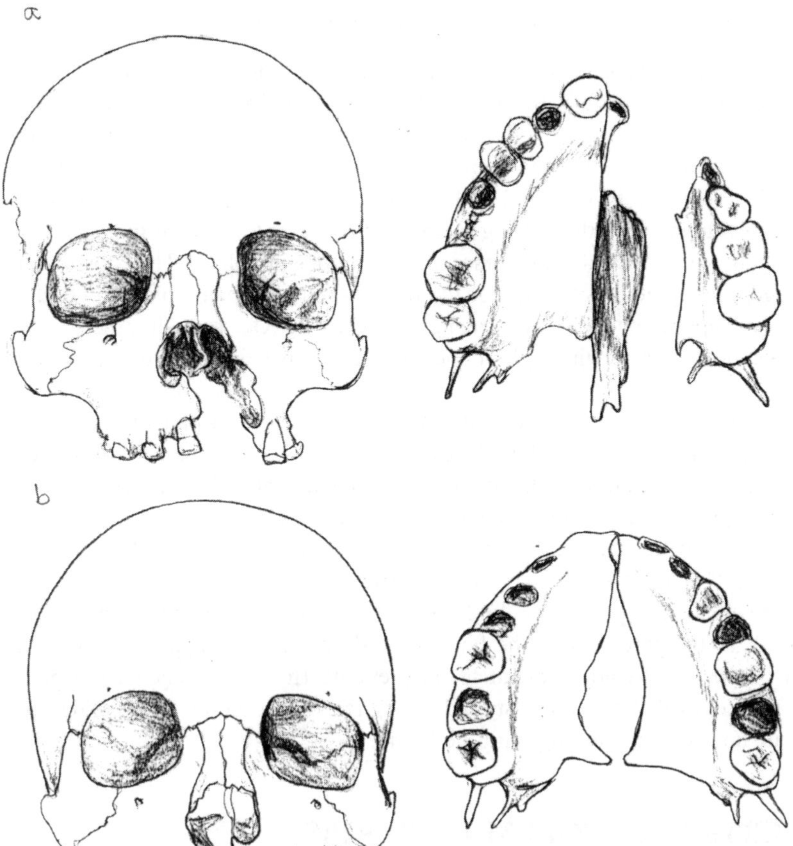

Figure 15.5 Facial clefting. (a) Unilateral left cleft lip and palate, adult female, Navajo Reservoir, New Mexico. (b) Unilateral right incomplete cleft lip and cleft palate, adult female, Pueblo, Colorado. (c) Unilateral left hypoplasia premaxilla and complete cleft palate, adult male, Kentucky

Temporal Bone: Auditory Meatus, Tympanic Plate and Styloid

Complete or partial absence of the external auditory meatus occurs when the dorsal end of the ectodermal groove fails to appear, either unilaterally or bilaterally. With complete absence of the ectodermal groove the tympanic plate does not develop, the styloid either is missing or is rudimentary, and the petrous portion is usually smaller. The inner part of the ear also fails to develop or is partially affected, as is development of the external ear.

Delayed development in the closing membrane between the ectodermal groove and endodermal pouch will affect the precursor of the tympanic plate. The slowdown leads to programmed incomplete ossification of the mature bony tympanic plate in childhood, leaving a tympanic aperture covered with membranous tissue. Occasionally, the extension of the styloid plate precursor for the styloid sheath is excessive, leading to a very large styloid sheath.

Mesenchymal tissue from the second branchial arch forms a pair of cartilaginous bands known as Reichert's cartilages below the first branchial arch and ectodermal groove, growing forward to meet in the midline at the hyoid. While the dorsal ends split off to form the stapes, the rest develops into the stylohyoid chain. This chain, in turn, segments into the bony styloid process, the stylohyoid ligament connecting the styloid process to the hyoid, and forms the lesser cornua. The tip of the bony styloid ossifies separately from its base. Sometimes the precursor for the bony tip of the styloid does not develop, producing a short styloid process (hypoplasia), or the precursor of the bony styloid base fails to appear (aplasia). Failure of a portion of the stylohyoid ligament to differentiate from the bony styloid percursor can lead to an excessively long bony styloid. Very rarely the entire stylohyoid chain fails to differentiate and ossifies, uniting with the hyoid.

THE VERTEBRAL COLUMN AND RIBS

Morphogenesis

Development of the primordial vertebral column depends on the formation of the notochord that provides the structural framework and inductive signals for its development. The notochord literally is the 'back bone' structure of the developing embryo. Without it the embryo becomes a mass of aimless cells. This flexible rod of cells enclosed by a thick membranous sheath forming beneath the neural tube acts as the main guiding force in the developing vertebral column and skull base. Mesenchymal cells line up along the sides of the notochord to form a pair of cylindric condensations known as the paraxial mesoderm. They quickly develop into a series of rounded whorls of cells called somitomeres, beginning at the cranial end and progressing to the caudal end. Somitomeres flanking the notochord from the base of the skull to the embryonic tail follow a molecular cranial–caudal segmentation clock to form discrete paired blocks of segmented mesoderm called somites (Figure 15.1e). Generally, 42 to 44 pairs develop, with the most cranial somites joining with the development of the cranial base. Precursors for the development of the vertebral column include 8 cervical, 12 thoracic, 5 lumbar, 5 sacral and 8–10 coccygeal somites, with most of the later somites quickly disappearing. Responding to inductive signals from the notochord, the ventromedial portions of the paired somites shift into sclerotomes that surround the notochord and neural tube, with the ventral sclerotome portion developing into the precursor vertebral body and the dorsal portion responding to overlying ectodermal tissue and forming the rudimentary vertebral or neural arch around the neural tube. In response to the developing spinal nerves emerging at the same level as corresponding somites, the developing sclerotomes split horizontally and recombine. The cranial half of each sclerotome combines with the caudal half of the preceding one in a cranial–caudal direction. Seven cervicalvertebral segments arise from eight sclerotomes, with the cranial half of the first one fusing with the adjacent occipital sclerotome tissue to form the cranial base. The resegmented sclerotomes are separated by developing fibrous intervertebral fissures. Notochord cells become trapped in the central nucleus pulposus as the notochord regresses and then degenerates following development of the rudimentary vertebral column. Its job is done and it is no longer needed. The costal processes of the developing thoracic vertebral segments receive molecular signals to extend and grow into the ribs that separate from the vertebral segments with the formation of costovertebral joints.

Anomalies

The complexity of the developing vertebral column invites a number of developmental variations, some minor and others more severe. The most common variations occur with shifting of the borders separating the different types of vertebral segment. These shifts are genetically programmed (Larsen, 2001: 103–104). Shifting at the occipitocervical border, discussed under skull anomalies (Section 15.3.2), is part of this phenomenon. Border shifting occurs during development of the sclerotomes as they separate into cranial and caudal halves, with each half joining its neighbouring half. Vertebral segments at the borders that differentiate between the different segments of the vertebral column develop transitional characteristics to accommodate border changes. Variations occur at the borders when they move up a segment (cranial shifting) or down a segment (caudal shifting). Shifting can occur at only one border or at more than one and it may involve just one side of the border or both sides, resulting in unilateral or bilateral changes in the affected vertebral segment. Border shifting can result in a variety of expressions, from minor to major changes.

Cranial shifting can, on rare occasions, affect development of the dens or odontoid of the axis before it descends to its final position. The bony dens may be programmed to develop separately from the axis body, or the apical or tip segment may present as a separate ossicle that usually remains a separate bone (Figure 15.3d) or it may attach to the anterior rim of the foramen magnum. Very rarely, agenesis of the dens, apical segment or both may also occur. Whatever form this defect takes, the unstable atlanto-axial junction often leads to neurological disturbances, such as headache, neck pain, disturbed muscular coordination, intermittent loss of consciousness, transient vertigo and numbness of extremities. Minor deviations may not cause major symptoms until a traumatic event causes hyperflexion of the neck that further destabilizes the junction. Often, the separated tip or apical segment ossicle is lost in archaeological collections, but it can be verified by the presence of a facet for it on the atlas vertebrae. If it had never formed, then there would not be a facet for it on the atlas.

The border between the cervical and thoracic spine is generally marked by the absence of transverse foramina and the development of ribs from the costal processes on the first thoracic vertebral segment. However, the costal processes of the lower cervical segments retain the potential to develop into rudimentary ribs, primarily in the last cervical segment, if programmed to do so with cranial shifting of the cervicothoracic border. This causes the last cervical vertebra to take on thoracic characteristics by developing rudimentary or complete ribs. Expressions of rudimentary cervical ribs range from small tubercles on the tips of the transverse processes, through blunt bony extensions up to 50 mm long, to rib-like extensions articulating with the first rib or attached to the sternum via ligamentous bands. Most often, a complete separate rib develops that may or may not articulate with the first rib (Figure 15.6d). Cervical ribs can be unilateral or bilateral, and they differ in appearance from true first thoracic ribs. With caudal shifting of the border, the first thoracic rib is most affected and is underdeveloped, less than 30 mm long and abnormal in shape yet wider than a cervical rib. This usually occurs unilaterally. The malformed rib frequently articulates or fuses with the second rib near the scalene tubercle or connects to the manubrium by a ligamentous band of tissue.

The border between the thoracic and lumbar vertebral segments is frequently affected by minor expressions of border shifting that affect the twelfth rib and/or the transitional facets between the last thoracic and first lumbar vertebrae. With cranial shifting of the border to between the eleventh and twelfth thoracic vertebrae, the twelfth ribs will be rudimentary or absent. Often, the lower ribs are damaged or missing from archaeological skeletal material, but the twelfth thoracic vertebra will exhibit hypoplasia or aplasia of its rib facets. As the border moves cranially, the transitional facets also shift upwards to between the eleventh and twelfth thoracic vertebrae. Mild caudal shifting of the border to between the first and second lumbar vertebrae shifts the transitional facets down between them instead of between the twelfth thoracic and first lumbar segments. Greater caudal shifting at this border results in a variety of lumbar rib forms, as the costal portions of the first lumbar vertebra are programmed to develop into ribs (Figure 15.6e). Lumbar ribs are more common than cervical ribs. They usually occur bilaterally and often asymmetrically. They vary in length from costal extensions to separate articulating ribs of varying sizes.

Shifting of the lumbosacral border is not unusual. The most commonly recorded expression is sacralization of the last lumbar vertebra that represents complete cranial shifting of the border to between the 4th and 5th lumbar vertebrae. The affected sacrum appears to have six segments instead of five. This can occur bilaterally or unilaterally (Figure 15.7a). Sometimes only the fifth lumbar vertebral transverse processes are affected, as they take on the appearance of the sacral alae and they may or may not articulate with the sacrum. With caudal shifting of the border to between the first and second sacral segments the first sacral segment becomes lumbarized as a separate vertebra (Figure 15.7b), either bilaterally or unilaterally. There appear to be six lumbar vertebrae instead of five and the sacrum has only four segments. Sometimes the separated sacral segment articulates with the alae of the sacrum, other times it does not. Minor expressions appear as an anterior cleft between the first and second sacral segments or the presence of dorsal apophyseal facets between them.

The sacrocaudal border can also be affected by border shifting between the last sacral segment and the first caudal segment. Cranial shifting of the border to between the fourth and fifth sacral segment can result in a complete or partially separated last sacral segment that becomes caudalized. Complete or partial sacralization of the first caudal segment occurs with caudal shifting of the border to between the first and second caudal segments.

Border shifting can be difficult to define when both lumbosacral and sacrocaudal borders are affected, but careful observation can clarify the situation along with a count of the thoracic and lumbar vertebrae. Adding to the confusion can be the presence of an extra vertebral segment developing during morphogenesis, usually at the thoracolumbar or lumbosacral borders. This transitional vertebra can take on the characteristics of either border segment.

The developing precursor of the vertebral body can be affected by delayed withdrawal of the notochord, producing a depressed sagittal cleft or a bifurcation commonly known as a

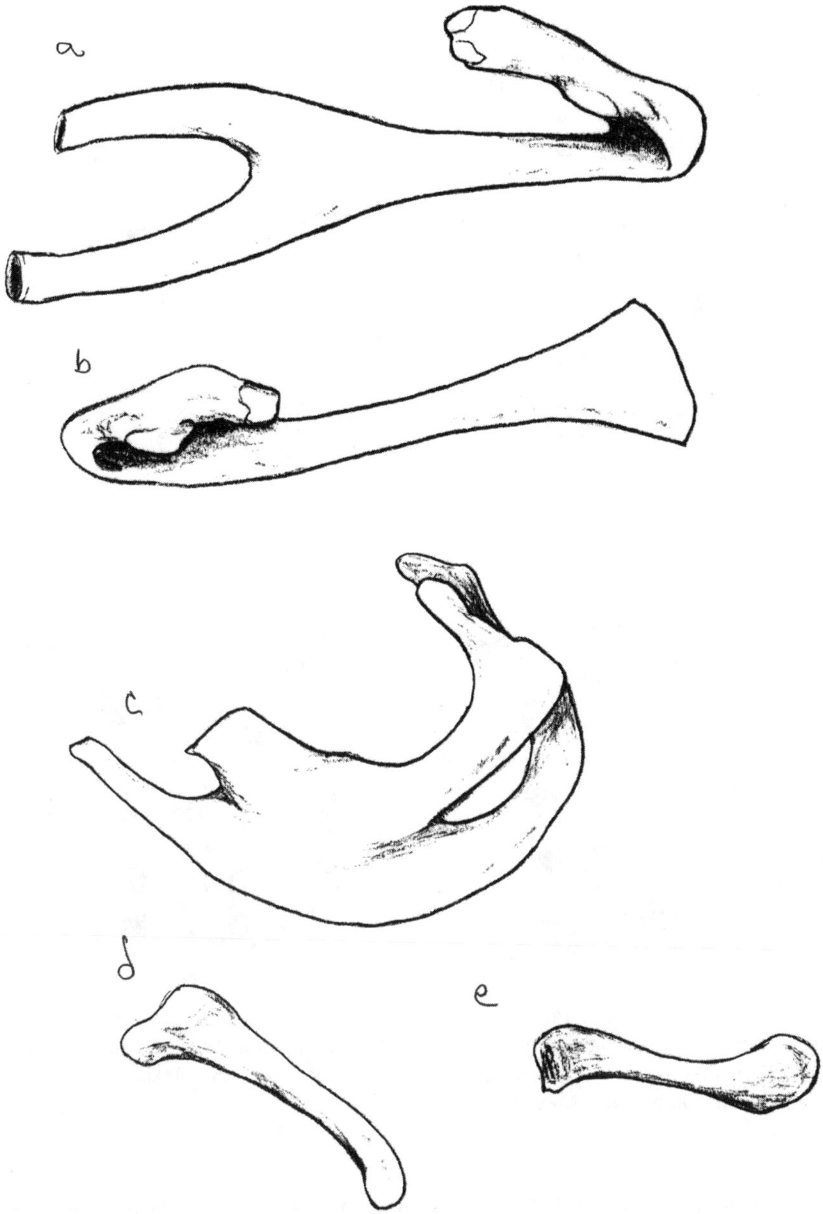

Figure 15.6 Rib anomalies. (a) Bifurcated rib, adult female, Hawikku, New Mexico. (b) Flared rib end, adult female, Hawikku, New Mexico. (c) Merged first and second ribs, adult male, Heshatauthla, New Mexico. (d) Unilateral left cervical rib, adult female, Corinth, Greece. (e) One of asymmetrical pair of lumbar ribs, adult male, Corinth, Greece

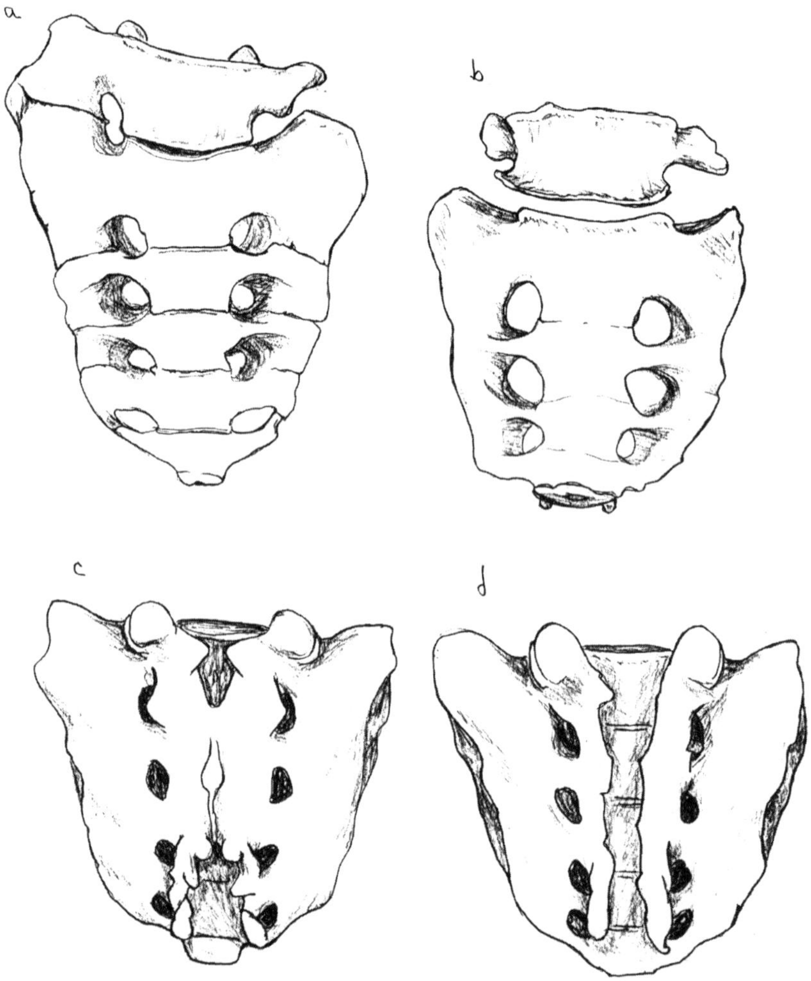

Figure 15.7 Sacral anomalies. (a) Unilateral right sacralization of fifth lumbar with lateral left hypoplasia, adult female, Corinth, Greece. (b) Lumbarized first sacral segment, adult female, Corinth, Greece. (c) Cleft first sacral segment, adult female, Pueblo Bonito, New Mexico. (d) Complete cleft sacrum, adult male, Otowi, New Mexico

butterfly vertebra (Figure 15.8a). The gap is filled with cartilaginous tissue. This anomaly is very rare and usually affects only one thoracic or lumbar vertebra.

Segmentation errors occur in a variety of ways as the primordial vertebral column develops. The matching somite pairs flanking the notochord undergo synchronous but independent development before fusing along the midline. Faulty signalling with delayed development of one somite hemimetamere slows its movement to the midline for fusion, causing it to miss its connection with the matching hemimetamere. So it shifts downward to pair up with the next opposing hemimetamere approaching the midline, leaving its half unpaired to develop into a lateral wedge-shaped hemivertebra (Figure 15.8b). Sometimes extra somites are involved. Sometimes the hemivertebra is incorporated into an adjacent vertebral segment. Solitary

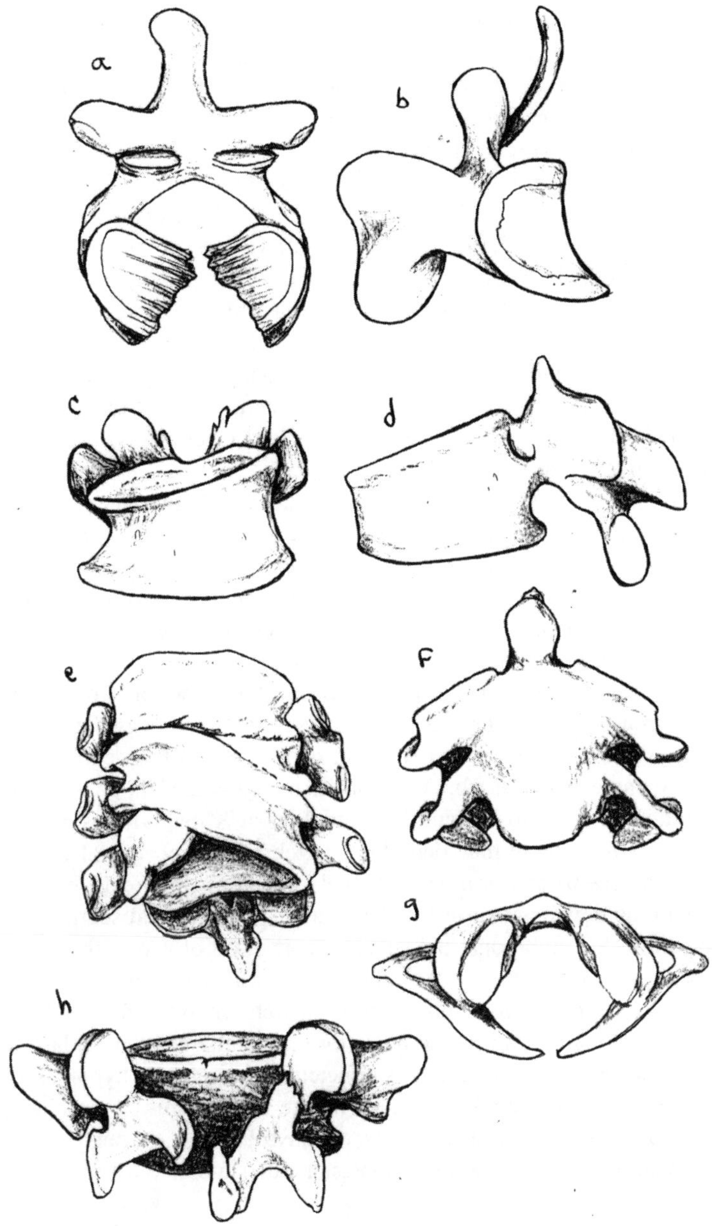

Figure 15.8 Vertebral anomalies. (a) Sagittal cleft 'butterfly' eleventh thoracic vertebra, adult, Trigg site, Virginia. (b) Hemivertebra right side fifth lumbar incompletely sacralized, adult female, Quarai, New Mexico. (c) Lateral hypoplasia (right) tenth thoracic vertebra, adult male, Corinth, Greece. (d) Ventral hypoplasia twelfth thoracic vertebra, adult female, Corinth, Greece. (e) Contralateral balanced hemimetamere shifting fourth and fifth thoracic vertebrae forming block vertebra with third thoracic, adult female, Puye, New Mexico. (f) Block vertebra second and third cervicals, adult female, Elden Pueblo, Arizona. (g) Cleft posterior arch atlas, adult male, Puye, New Mexico. (h) Cleft neural arch from left lamina hypoplasia fifth lumbar vertebra producing asymmetry leading to unilateral right spondylolysis, adult male, Petras, Crete

hemimetamere shifting is rare and is usually due to agenesis of the other half as it simply fails to develop. Most often multiple hemimetamere shifts occur. The shifts can all be on one side (unilateral), leading to scoliosis, or more commonly on both sides (contralateral). Shifts involving two opposing somite pairs separated by normal vertebral segments are balanced and offset spinal deformity (Figure 15.8e).

Hypoplasia of one of the hemimetamere somite pair before becoming a sclerotome leads to lateral hypoplasia or wedging of the vertebral body. This usually leads to a congenital scoliosis. Lateral hypoplasia (Figure 15.8c) can affect one or more vertebral segments, including sacral segments. It can be mild or severe. Severe congenital scoliosis occurs when some segments are affected by hemimetamere aplasia while other segments are affected by hypoplasia, usually affecting the same side (unilateral). When multiple segments are affected in this manner they frequently fuse together. The apophyseal joints fail to develop, while the laminae fuse into a bony mass known as a postlateral bar that expands with time. The affected ribs also tend to fuse together near the vertebral junctions on the affected side, and sometimes the costovertebral joints fail to develop.

Ventral hypoplasia of the vertebral body (Figure 15.8d) occurs when programmed chondrification of the ventral portion of the developing centrum is slow. One vertebral segment is generally all that is involved, often the last thoracic; but more than one can be affected, and most often this occurs in the lower thoracic region and leads to mild kyphosis.

The final patterning for the vertebral column arises from the recombined sclerotomes that develop from the fused somite pairs with transverse fissures forming between them for the intervertebral discs. Failure of this fissure to develop between two precursor vertebral segments results in a block vertebra consisting of two vertebral segments combined as one (Figure 15.8f). This can be partial or complete, depending on the degree of failure of the fissure to form. A single block vertebra condition is not pathological, generally occurs within the cervical region and tends to be familial. Pathological Klippel–Feil syndrome is another matter: it always involves more than one block vertebra, usually several cervical and upper thoracic vertebral segments and/or other vertebral anomalies (Figure 15.9).

Disturbances in the development of the dorsal aspect of vertebral segments often involve tardy development of the embryonic precursor of part or all of a vertebral arch half. Sometimes the anterior or ventral arch of the atlas can be affected with hypoplasia or aplasia, leading to a small or thin anterior arch, bifid bony arch, or tough fibrous band in place of an absent bony arch. Most often it is the posterior or dorsal arch of the atlas that is affected (Figure 15.8g). When a pedicle or lamina is affected unilaterally by hypoplasia the vertebral arch will appear asymmetrical and be unstable (Figure 15.8h). This can lead to spondylolysis in the weight-bearing lumbar area. Sometimes, apophyseal joints or transverse processes will appear smaller than normal because of hypoplasia, either unilaterally or bilaterally, or they may be absent.

When one half of a precursor vertebral arch suffers from hypoplasia and fails to meet the other half in time for programmed bony fusion after birth, the vertebral arch appears cleft or bifurcated with or without the development of the spinous process (Figure 15.8h). Usually no more than one or two vertebral segments are affected, most often the last lumbar or first sacral segment (Figure 15.7c). However, it is not unusual for all of the sacral segments to be affected, producing a completely cleft sacrum (Figure 15.7d). Thoracic vertebral segments are rarely affected; and of the cervical vertebrae, usually only the atlas develops clefting. No pathology results, since the two parts remain joined by tough fibrous tissue.

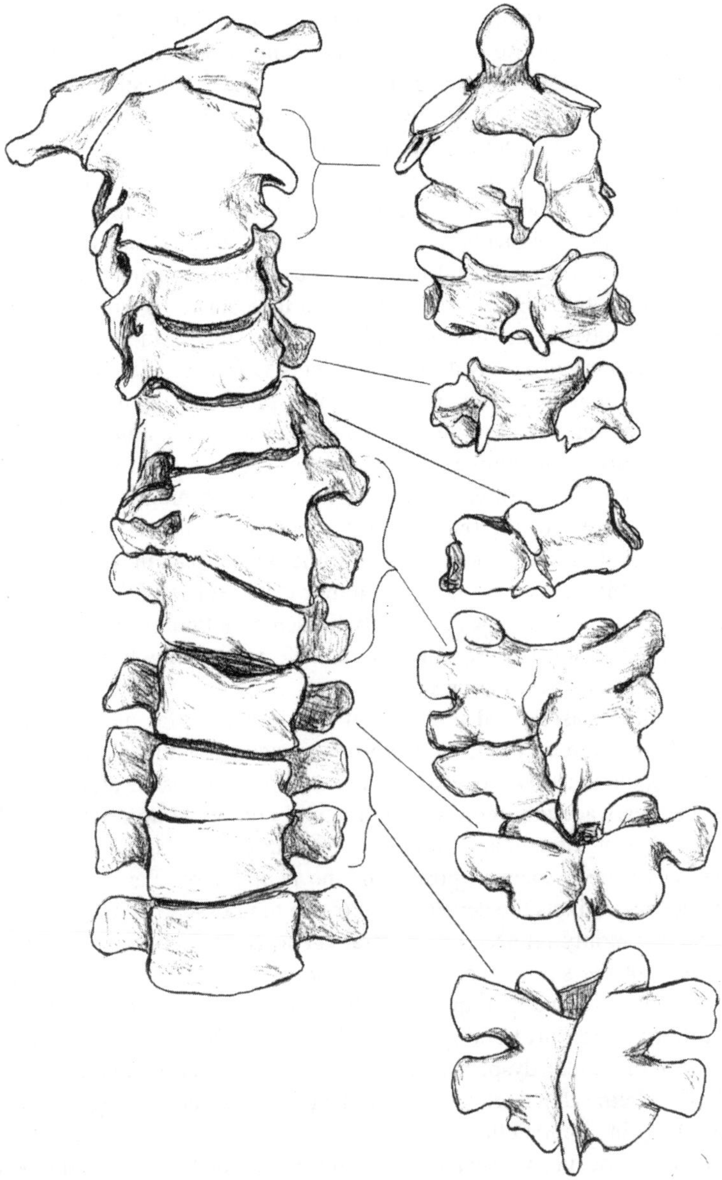

Figure 15.9 Klippel–Feil syndrome. Atlas normal, block vertebra second and third cervicals, asymmetrical neural arch with left hypoplasia fourth cervical, cleft neural arch missing right side fifth cervical that has shifted down to fuse with sixth cervical, block vertebra seventh cervical, first and second thoracics with right lateral hypoplasia of seventh cervical and first thoracic, mild left lateral hypoplasia second thoracic, mild left hypoplasia neural arch third thoracic, block fourth and fifth thoracics with lateral left hypoplasia, sixth thoracic normal

Failure of the caudal end (neuropore) of the neural tube (Figure 15.1f) to close properly results in displacement of a portion of the spinal cord and nerve roots outside the neural canal. The formation of this spinal cord defect impacts the adjacent developing vertebral arches, causing the two halves to become splayed outwards and slowed in development with the pressure from the displaced spinal tissue. The pedicles are thin, the laminae are deformed or absent, and the spinous process fails to develop. The resulting spina bifida is directly related to the neural tube defect, known as meningomyelocoele (myleomeningocoele), and it usually involves several vertebral segments. Hydrocephaly is often associated with this defect, as the normal flow of cerebrospinal fluid is obstructed by the deformity. Most frequently this severe defect occurs in the lumbosacral region, and the chances of survival beyond birth in ancient times would have been slim. Individuals born with postneurulation defects would have had a better chance of survival. Spinal defects occurring after the caudal neuropore closes involve a portion of the meninges covering the spinal cord, again, most often occurring in the lumbosacral region. The resulting spina bifida cystica containing the meningocoele is covered with epidermis tissue, unlike the open cyst of the meningomyelocoele. The opening through the vertebral arches is often smaller in protruding meningocoele; and when several vertebral arches are affected, the bulging meningocoele often produces a fusiform defect in the bony arches that tapers at both ends with the widest part in the middle (Dickel and Doran, 1989). Spina bifida occulta refers to a minor neural tube defect that does not produce a visible skin-covered cyst but which is usually hidden by a lipoma. Neurological symptoms vary with the type and severity of spina bifida defects, ranging from paraplegia to mild symptoms with minor postneurulation defect.

Cleft neural arch not associated with neural tube defect is far more common than that caused by neural tube defects. In fact, spina bifida associated with neural tube defect is quite rare in ancient populations. The many accounts of spina bifida in archaeological skeletal material over the years actually represent cleft neural arch without neural tube defect.

Following closure of the caudal neuropore, the lumbosacral spine continues to develop with canalization of the associated portion of the neural tube. Failure of this part of the neural tube to develop affects the development of the lumbosacral spine, leading to a variety of bony defects commonly referred to as sacral agenesis. Defects can range from complete to partial agenesis of the sacrum and coccyx, and/or lumbar spine. The defects are part of a caudal regression syndrome that appears to be related to defective mesoderm cell growth and migration during the thirrd embryonic week (Larsen, 2001: 66). The faulty mesoderm leads to a variety of caudal dysplasias and associated defects, including the variations of sacral agenesis. Extreme deficiency in caudal development can lead to fusion of the lower limb buds, resulting in sirenomelia.

Most rib anomalies occur in association with developmental defects of thoracic vertebrae, particularly those vertebral defects leading to severe scoliosis. Such ribs are usually fused together at the vertebral end and often not separated from the deformed vertebral segment. Disruption of the sclerotome tissue forming the costal processes can lead to variant rib shapes at the ventral end, including bifurcation (Figure 15.6a) or spurring, flaring (Figure 15.6b), merging (Figure 15.6c), abnormal wideness, and bridging, either partial or complete with articulation between the affected ribs. If the sternum develops abnormally, then the ventral end of the associated ribs will be affected. As mentioned before, cranial – caudal shifting of the vertebral borders can also affect development of ribs.

Very rarely a portion of a precursor rib splits off from its underside as it emerges from the vertebral costal process to develop into a supernumerary or extra rib, known as an intrathoracic rib. The anomalous rib is not a true rib, appearing much smaller than neighbouring ribs, often projecting downwards from a fibrous attachment of its parent rib. It may have a fibrous connection with the diaphragm, which can interfere with full lung expansion. Intrathroacic rib usually appears unilaterally and often appears on the thoracic right side (Resnick, 1989: 1081). Intrathoracic ribs are very rare and can easily be lost or not recognized during excavation of ancient skeletal material.

THE STERNUM

Morphogenesis

Bands of mesenchymal condensations that line up along each side of the ventrolateral body wall of the developing embryo form sternal plates or bands that move forward of the growing primordial ribs. They join at midline to form the sternum as they merge with the precostal process mesenchymal condensation that forms between the ventral ends of the developing clavicles and cranial ends of the sternal plates. The costal process is joined by a pair of suprasternal mesenchymal cell clusters interfacing between it and the ends of the clavicles. Fusion begins at the cranial end of the sternal plates with incorporation of the costal process and its suprasternal cell clusters, and it proceeds caudally. Once the sternal precursors fuse and the embryonic rib cartilages attach, the primordial sternum segments into the manubrium and four mesosternal sternebrae, with remnant mesenchymal tissue extending caudally from each sternal plate for the development of the xiphoid. The anlages for sternal ossification centres are programmed within the embryonic tissue, determined by the timing for merging of the sternal plates. Generally, the manubrium ossifies from three centres, the first two sternebrae from one centre and the last two usually from two centres, while the xiphoid remains cartilaginous until long after birth. The sternebrae are programmed to begin fusion from the caudal end to the cranial end by puberty.

Anomalies

The majority of developmental anomalies of the sternum are not pathological. Excessive delay in fusion of the sternal plates with consequential lack of fusion resulting in a bifid sternum is very rare. Some fusion of the caudal end usually occurs with this defect, and fibrous tissue usually holds the bifid sternum together. Minor expressions of delayed fusion at the cranial end are rare, but can result in a small cleft, fissure or aperture in the manubrium, while more commonly delayed fusion at the caudal end can lead to cleft, fissure or aperture (Figure 15.10a) in the mesosternum. The final shape of the mesosternum is influenced by the genetic timing of fusion of the sternal plates that governs the number of ossification centres. Extra ossification centres within a sternebra lead to a wide sternebra, whereas single centres produce a narrow sternebra. While the first two sternebrae usually contain single centres and remain narrow, the last two are frequently slower to meet and may have up to four or more, becoming wider than the upper two. Single ossification centres in all sternebrae producing a narrow sternum are common. Rarely, all four sternebrae have multiple ossification centres, producing a very wide sternum. Hypoplasia or aplasia of one or both sides of sternebrae precursors, usually the last one, also influence the final shape of the mesosternum (Figure 15.10c). The form of the xiphoid process is quite variable, depending

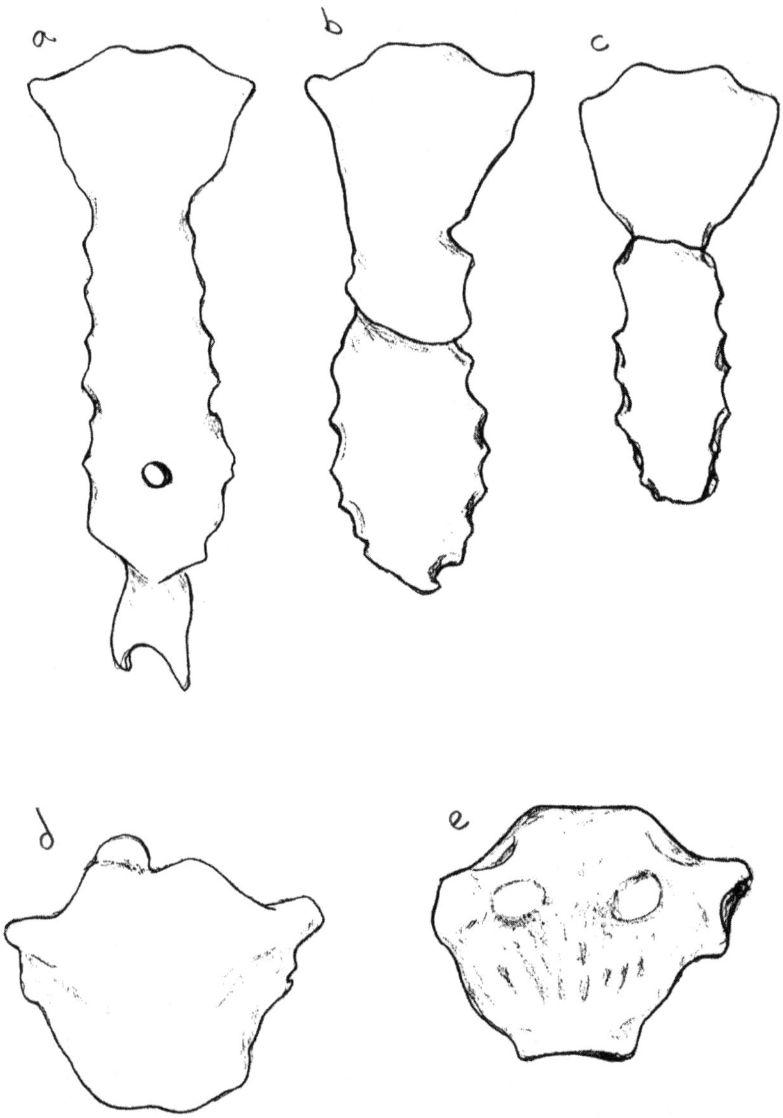

Figure 15.10 Sternal anomalies. (a) Sternal aperture, fused manubrium and xiphoid, adult male, Puye, New Mexico. (b) Misplaced manubrio-mesosternal joint with unilateral right hypoplasia first mesosternal segment, adult, Corinth, Greece. (c) Hypoplasia third mesosternal segment and aplasia last segment, adult female, Aphrodisias, Turkey. (d) Suprasternal ossicle fused to right side suprasternal notch of manubrium, adult male, Corinth, Greece. (e) Bilateral suprasternal ossicles fused to dorsal aspect of manubrium, adult male, Mishongovi, Arizona

on the amount of remnant primordial tissue present at the caudal end of the precursor mesosternum and the timing of fusion of the remnants.

Delayed fusion of the suprasternal structures to the developing mesosternum can lead to development into separate pea-sized or smaller ossicles. Separated suprasternal ossicles are easily lost during recovery, but although these ossicles develop separately they normally fuse to the manubrium during adolescence. This phenomenon occurs bilaterally or unilaterally (Figure 15.10d and e).

Frequently, the fibrous lamina that separates the manubrium from the mesosternum fails to develop, resulting in fusion of the manubrium to the sternum with the process of unification as adolescence approaches (Figure 15.10a). Sometimes the fibrous lamina forms at the wrong level, between the first and second sternebrae instead of the manubrium and first sternebra, leading to a misplaced manubrio-mesosternal joint when unification occurs (Figure 15.10b). The xiphoid process can also fuse with the mesosternum (Figure 15.10a).

THE LIMBS

Morphogenesis

As the primordial axial skeleton takes form along the cranial–caudal axis, thin layers of mesoderm develop laterally to the paraxial mesoderm condensations to form lateral plate mesoderm (Figure 15.1g). This tissue divides into two layers, with one layer contributing mesenchymal cells to the formation of the limbs of the appendicular skeleton. The upper limb buds appear between the levels of the fifth and eighth cervical somites by the 24th day, followed by the lower limb buds forming between the levels of the third to fifth lumbar somites (Figure 15.1h). Each limb bud contains a core of mesenchymal cells from the lateral mesoderm capped by overlying ectodermal cells. The ectodermal cap thickens to form the apical ectodermal ridge at the distal end of the developing limb. The apical ectodermal ridge is responsible for inducing proximal – distal growth of the developing limb, as well as participating in digit formation. Mesenchymal cells within the progressive zone next to the ectodermal ridge respond to induction by expanding in a proximal direction. Cascading genetic signals with overlapping patterns of expression from within the mesenchymal core itself determine differential development of the limb segments in a set pattern as the developing core moves away from the apical ectodermal ridge. Different sets of genes direct development of the upper limb segments and the lower limb segments. The development of each segment within a limb is governed by its own set of genetic expressions. Therefore, a genetic disturbance of one set of governing genetic expressions can alter development of one segment within a limb without disturbing the other segments. Patterning of limb segments along the antero-posterior (cranial–caudal) and dorso-ventral axes is also regulated by various genetic signals in concert with proximal – distal growth. The antero-posterior axis patterning is regulated by genetic molecular actions within a cluster of cells designated as the zone of polarizing activity on the posterior border of the limb, whereas dorso-ventral axis patterning is regulated by genetic molecular action coming from the ventral ectoderm.

As the limb buds grow, the distal portions flatten out to form digital plates separated from the growing limb by circular constrictions (Figure 15.1i). By the sixth week, precursors of the limb bones are modelled in hyaline cartilage arising from the condensed mesenchymal cores. Thickening rods of mesenchyme form digital rays within the digital plates with their tips projecting slightly from the distal border, first in the hand plates and then in the foot plates. The digits of the hand appear in order, with the thumb on the radial or anterior side under genetic molecular direction from the zone of polarizing activity as it moves distally with the posterior border of the apical ectodermal ridge. Similar action positions the sequence of digits in the foot. Programmed cell death between the tips of the digital rays along the apical ectodermal ridge gradually carves them fromthe digital plates as they grow distally to form the precursor fingers and toes. Continuous outgrowth of the rays is influenced by ectodermal ridge tissue, while the mesenchymal core of each ray evolves into the pattern of hyaline cartilage for each of the precursor bones of the hands and feet.

Intervening joints between limb bones arise where condensed mesenchyme growth into bone-forming cartilage is arrested. Instead, mesenchymal cells form a joint disc within these interzones that evolves into the various joints. Little is known of how these interzones are programmed to position and form joints correlating with interacting bone ends.

The developing upper limbs rotate 90° laterally so that the radial thumb sides are placed in a lateral position, whereas the lower limbs rotate 90° medially to place the tibial big toe positions medially. Primary ossification centres for the upper limb bones begin to appear by the end of the embryonic period at 12 weeks. All bones except the clavicles ossify from cartilage. The clavicles ossify directly from membranous precursors.

Anomalies

Development of the right and left limbs is synchronized to mirror each other in form and length along the proximal – distal axes. However, sometimes the timing for proximal – distal growth of one side lags behind the other, leading to asymmetrical limb lengths that can involve one segment or all long-bone segments on one side of a developing limb. Differences up to a few centimetres are not unusual and generally go unnoticed. The clavicle and arm bones appear to be more affected than the leg bones. Generally when the leg bones are asymmetrical, so are the upper limb long bones on the same side. Marked length differences in the lower limbs can affect gait and lead to compensatory measures in the spine.

Aside from the systemic skeletal anomalies such as achondroplasia and other syndromes that affect limb development and defects due to constricting amniotic bands, there are a wide variety of limb anomalies resulting from errors in limb morphogenesis. Interferences in the development of the apical ectodermal cap are known to result in the absence of a part of a limb (micromelia) or the entire limb (amelia). The long bones can fail to develop, with rudimentary hands or feet attached to the trunk by small irregularly shaped bones (phocomelia). The thalidomide incident in the early 1960s showed how certain ingested chemical substances can cause this type of defect. But most limb defect aetiologies are unknown, although a number of them appear to be hereditary. The complex molecular interaction involved in limb development may require more than one faulty interaction or signal to produce a defect. Disturbances in the developing anlages of the limbs produce a wide range of limb anomalies, ranging from minor to severe, that can affect one or more segments of a limb, the whole limb or just one or a few elements of a limb segment.

Upper Limb

The developing clavicle can be affected by hypoplasia or aplasia without involving the rest of the upper limb bones. Absence of both clavicles is often associated with cleidocranial dysplasia, which involves delayed and faulty membranous ossification of the skull. Ossification for both the skull and clavicles arises from membranous tissue, suggesting a disturbance affecting membranous ossification in both developmental fields at the same time. Usually the middle or lateral portion of the clavicle that mirrors the two primary ossification centres is affected. Failure of the lateral portion to develop results in absence of the acromial end (Huber, 1936: 259). Rarely, programmed union of the two ossification areas is delayed, leading to a bifurcated clavicle joined by fibrous tissue instead of bone (Moore, 1985: 628).

The scapula is not usually affected by major developmental disturbances. The most common minor scapular anomaly found in archaeological skeletal collections is the os acromiale that develops separately from the lateral end of the scapular spine (Figure 15.11a). Although the acromion ossifies independently from the scapular spine, it generally is programmed to unite with it by late adolescence. Sometimes this does not happen: a fibrous joint is programmed to develop between the acromion and scapular spine instead of bony union, either bilaterally or unilaterally. Another minor scapular anomaly involves the glenoid neck. Hypoplasia of the developing glenoid neck leads to an irregular glenoid cavity with widening of the lower joint space. This usually occurs bilaterally and the humeral head and neck may be undeveloped (Resnick, 1989: 1076).

The genetic expressions governing the development of the bones in the upper limb segment for the forearm and proximal wrist appear to be vulnerable to disturbances leading to bone hypoplasia or aplasia, particularly the radius and the radial carpal bones. Bones of the thumb may also be affected (Resnick, 1989: 1074). Radioulnar synostosis refers to the failure of the primordial tissue of the proximal forearm segment to separate completely for development of the precursor ulna and radius. The analogous bones continue to develop as programmed, but the proximal ends remain united. Sometimes a separate ossification centre programmed into the ulnar styloid process leads to development of a separate styloid ossicle that can attach to the distal end of the ulna by fibrous tissue or remain a separate bone. Or the styloid process may fail to develop (Figure 15.11b).

Anomalies related to disrupted gene interaction governing the development of the hand and fingers are the most commonly occurring anomalies of the upper limbs, and it is not unusual for more than one anomaly to be present. Delay in the appearance of the distal notches between two or more developing digits interferes with the orderly process of cell death between the affected digits. Depending on the timing of the disruption, union between the digits can be expressed partially or completely. Most often, only soft-tissue webbing results, but occasionally bony union (syndactyly) of two or more digits will occur (Case, 1996: 79–80). Metacarpal growth associated with the affected digits can also be altered (Larsen, 2001: 331).

Shortened digits (brachydactyly) result from hypoplasia or aplasia of one or more developing bones of a digital ray, with most expressions hereditary (Figure 15.11c). Rarely, one or more digits fail to develop for lack of a precursor digital ray (ectrodactyly). Extra digits (polydactyly) appear more often than other digit anomalies and often occur with other digital defects. Extra digits form with disturbances in the patterning of cell death within the notches between the developing digital rays. Polydactyly usually follows heritable genetic lines with a variety of expressions, depending on the timing of the disruptive event. Polydactyly occurs most often on the ulnar side as postaxial polydactyly of the hand and frequently consists of soft tissue only in the form of a rudimentary digit. Expressions affecting the bones of the digit range from bifurcation of the metacarpal or phalanx with duplication of one or more phalangeal bones to complete duplication of the entire bony digit. Complete or partial duplication occurs less often on the radial side in the thumb as preaxial polydactyly, and this defect rarely occurs in the central digits (Case *et al.*, 2006).

Defective development of the central portion of the digital group between the second and fourth digits can result in a variety of cleft hand anomalies commonly known as lobster claw. Generally both hands and feet are affected. This anomaly usually involves aplasia of the third metacarpal and phalangeal bones of the third finger with fusion of the other digits.

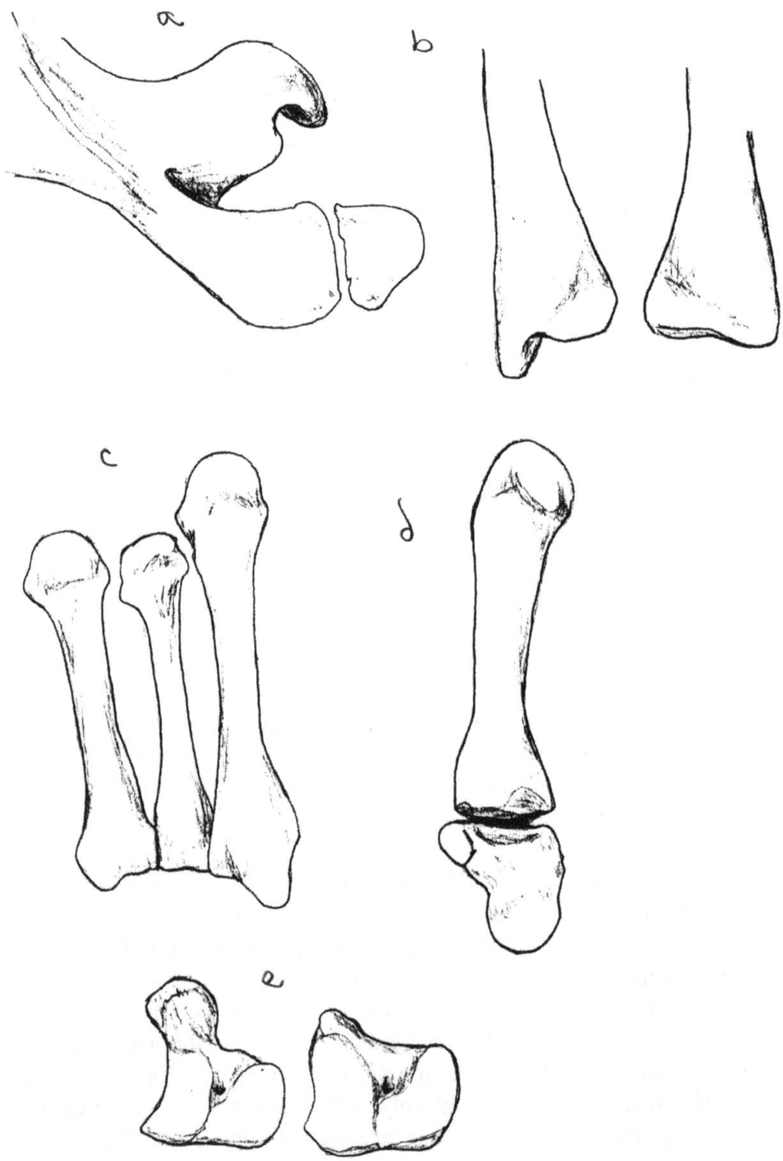

Figure 15.11 Upper limb anomalies. (a) Os acromiale, adult male, Cornith, Greece. (b) Normal right ulna styloid process and left agenesis styloid process, adult female, Corinth, Greece. (c) Brachydactyly from short left fourth metacarpal, adult female, La Playa, Mexico. (d) Os metastyloideum third metacarpal fused to capitate, adult female, Corinth, Greece. (e) Normal hamulus on hamate and hypoplasia of hamulus, adult female, Corinth, Greece

Sometimes the affected bones develop out of position. Lobster claw deformity is usually bilateral and often familial (Smith, 2002).

Sometimes programmed joint space between developing bones fails to take place either completely or partially. Lack of development or incomplete formation of the interzone leaves the adjacent bony anlages united as one bone or partially unified. Sometimes disruption of the developing interzone creates a fibrous union (non-osseous coalition) instead of a bony union connecting the adjacent bones. Affected bones often reflect this fibrous union with porous lesions at the attachment area. Interzone defects can appear in the carpals of the wrist, particularly in the proximal row between the lunate and triquetrum, followed by the capitate and hamate in the distal row (Case, 1996: 87–88).

Symphalangism refers to the bony union of digital bones, most often involving the proximal interphalangeal joint of the fifth digit on the ulnar side of the hand (Smith, 2002). On rare occasions programmed embryonic development of a carpal bone is disrupted to form two completely or partially separated parts. This results in a bipartite or bifurcated bone, and the scaphoid is most often affected (Case, 1996: 88).

Os metastyloideum represents the developmental separation of the styloid process from the third metacarpal leaving it as a separate ossicle. Most often it is attached by fibrous tissue to the metacarpal or the base of the capitate (Figure 15.11d). Sometimes the tubercle of the scaphoid and the hamulus of the hamate are similarly affected. The hamulus can also undergo hypoplasia or aplasia (Figure 15.11e), and occasionally hyperplasia. Occasionally the thumb digit is programmed to develop an extra phalange, producing a triphalangeal thumb.

Lower Limb

The bones of the pelvic girdle, similar to the scapula of the shoulder girdle, are rarely affected by major disturbances in morphogenesis. Pelvic defects that are known to occur are usually related to other developmental disturbances, such as the caudal regression syndromes. The hip joint does appear to be susceptible to developmental instability, either bilaterally or unilaterally, leading to congenital hip dislocation. The defect is not usually apparent at birth and may not be noticed until the infant puts weight on the affected hip. Despite controversy regarding the aetiology of this defect, development of the embryonic hip joint appears to be at fault. There appears to be inadequate formation of the primordial fibrocartilaginous rim forming the acetabular labrum that normally acts as a buttress to the femoral head within the acetabular joint space. Without it, pressure from the iliopsoas tendon causes the capsule to constrict between the femoral head and joint cavity as it stretches to accommodate the forced relocation of the femoral head (Ferguson, 1968: 133). The capsule often adheres to the ilium above the acetabulum or below it on the ischium, where the femoral head presses it against the bone, creating a secondary joint while the acetabulum remains underdeveloped.

Proximal femoral focal deficiency refers to a range of rare expressions of hypoplasia and aplasia of the femoral head, neck and or upper portion of the proximal femur while the distal end develops normally (Resnick, 1989: 1075). Much less severe and more common is asymmetrical development of the femoral necks with one side developing at a normal angle and the other side at much less of an angle (anteversion) (Figure 15.12a). This suggests that timing in development was affected on one side. This anomaly does not appear to affect the spine and hips, but it probably does affect the individual's gait.

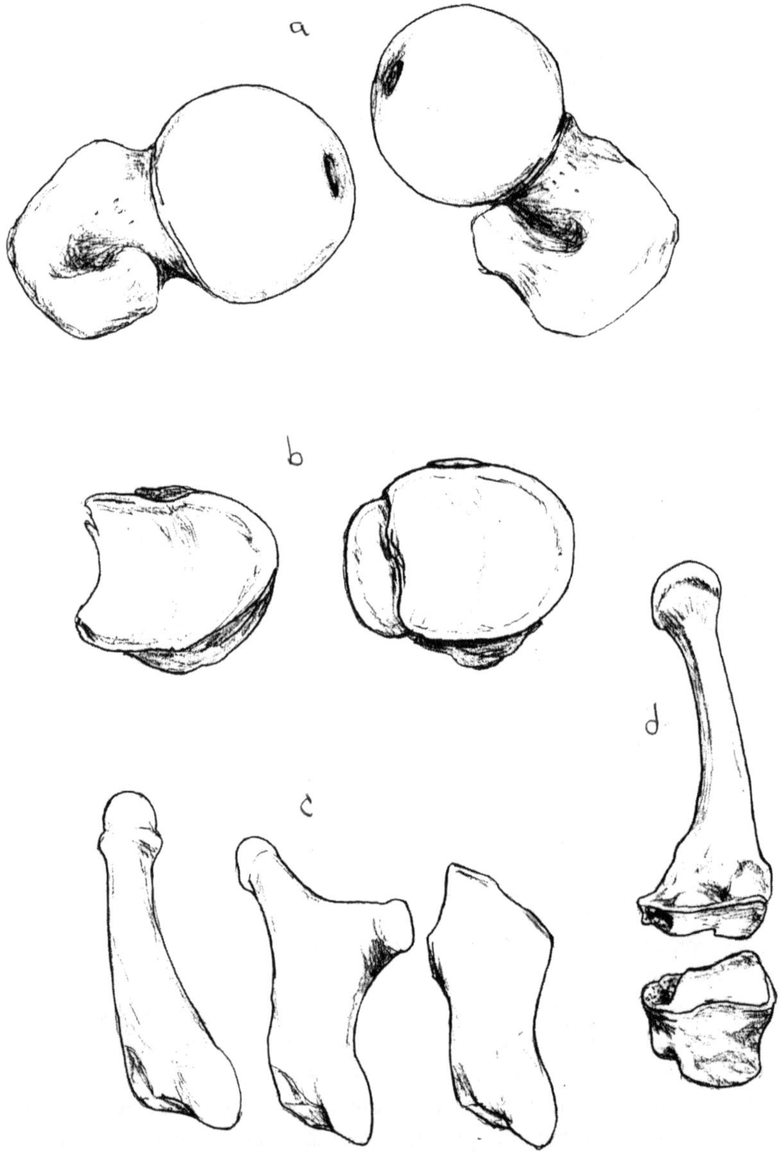

Figure 15.12 Lower limb anomalies. (a) Asymmetrical torsion of femoral necks, adult female, Corinth, Greece. (b) Bipartite left patellae, adult male missing ossicle, Hawikku, New Mexico, and adult male with ossicle, Da Shan Qian, China. (c) Right fifth metatarsals, normal, branching polydactyly, adult, Pueblo Bonito, New Mexico, and double-headed polydactyly, adolescent male, Trincheras, Mexico. (d) Non-osseous coalition (matching lesions) third metacarpal and cuneiform, adult female, Hawikku, New Mexico

Sometimes the supero-lateral angle of the patella is programmed to ossify independently to form a separate ossicle, a condition known as bipartite patella (Figure 15.12b). The ossicle is usually attached by fibrous tissue or this area becomes depressed, forming a patellar fossa or notch. Congenital absence of one or both patellae is known to occur in some families (Huber, 1936: 416). The lower leg appears to be more vulnerable to developmental disturbance than the upper leg, much like the forearm in the upper limb. Hypoplasia or aplasia of the developing lower leg, most often the fibula, results from disturbed development of the embryonic precursor.

Similar to the wrist and hand of the upper limb, the ankle and foot are affected the most by developmental disturbances in the lower limb with many similar anomalies. Cleft hand often is accompanied by cleft foot, with clefting deformity of the mid foot between the second and fourth digits with the third metatarsal and digit often missing. Syndactyly and brachydactyly (Figure 15.13a) also occur in the foot. Hypoplasia of the first metatarsal causes the foot to pronate in order to put the head of the first metatarsal into a weight-bearing position (Ferguson, 1968: 56). Ectrodactyly or missing digits rarely occur in the foot. Polydactyly in the foot, with or without polydactyly in the hand, is not uncommon and follows hereditary lines. Similar to the hand, various expressions of postaxial polydactyly of the fifth toe (Figure 15.12c) occur far more often than preaxial polydactyly of the first toe, whereas central expressions are very rare. Symphalangism is more common in the foot (Case and Heilman, 2005) than in the hand and frequently involves the interphalangeal joint of the fifth toe (Figure 15.13f), sometimes including the interphalangeal joint of the fourth toe. Fifth-toe symphalangism probably reflects a generalized evolutionary trend toward eliminating the joint space between the distal small bones that no longer plays a role in foot function.

The tarsals are subject to interzone failures similar to the carpals of the wrist. The most common interzone defect occurs as a fibrous union (non-osseous coalition) between the third cuneiform and metatarsal (Regan *et al.*, 1999) (Figure 15.12d). Complete or partial osseous union between these two bones is much less common. Similar union or coalition can occur between any of the tarsals and very rarely between the second cuneiform and metatarsal (Resnick, 1989: 1078–1080; Case, 1996: 93–101). Bipartite or bifurcated tarsals occasionally occur. The first cuneiform can be divided into dorsal and plantar bones (Huber, 1936: 427). Rarely a metatarsal base may be bifurcated.

More often the marginal parts of a tarsal are programmed to develop and ossify separately but attach to the parent bone by fibrous tissue or remain as an independent ossicle, or not develop at all. The tubercle projecting from the dorsal border of the talus is sometimes programmed to ossify separately to form an ossicle known as the os trigonum (Figure 15.13c) that may or may not be attached to the bone. Similarly, the border of the anterior articular area for the talus on the calcaneus can form separately (Figure 15.13e) and is usually referred to as calcaneus secundarius (Anderson, 1988). The tuberosity of the navicular occasionally forms as a separate ossicle known as the os tibiale or accessory navicular, although it is not a true accessory bone (Figure 15.13d). The os intermetatarseum is a true anomalous accessory bone that develops within the border between the bases of the first and second metatarsals and the first cuneiform. Although it can appear as an independent ossicle, it more often attaches to one of the three bones involved (Figure 15.13b), appearing as a broad bony spur that can overlap one or both of the other bones (Case *et al.*, 1998).

Club-foot deformity has many expressions that can occur unilaterally or bilaterally. Much less often, the hand can be affected by club deformity known as talipomanus. Talipes equinovarus is the most common expression of club-foot. The heel is turned inward with

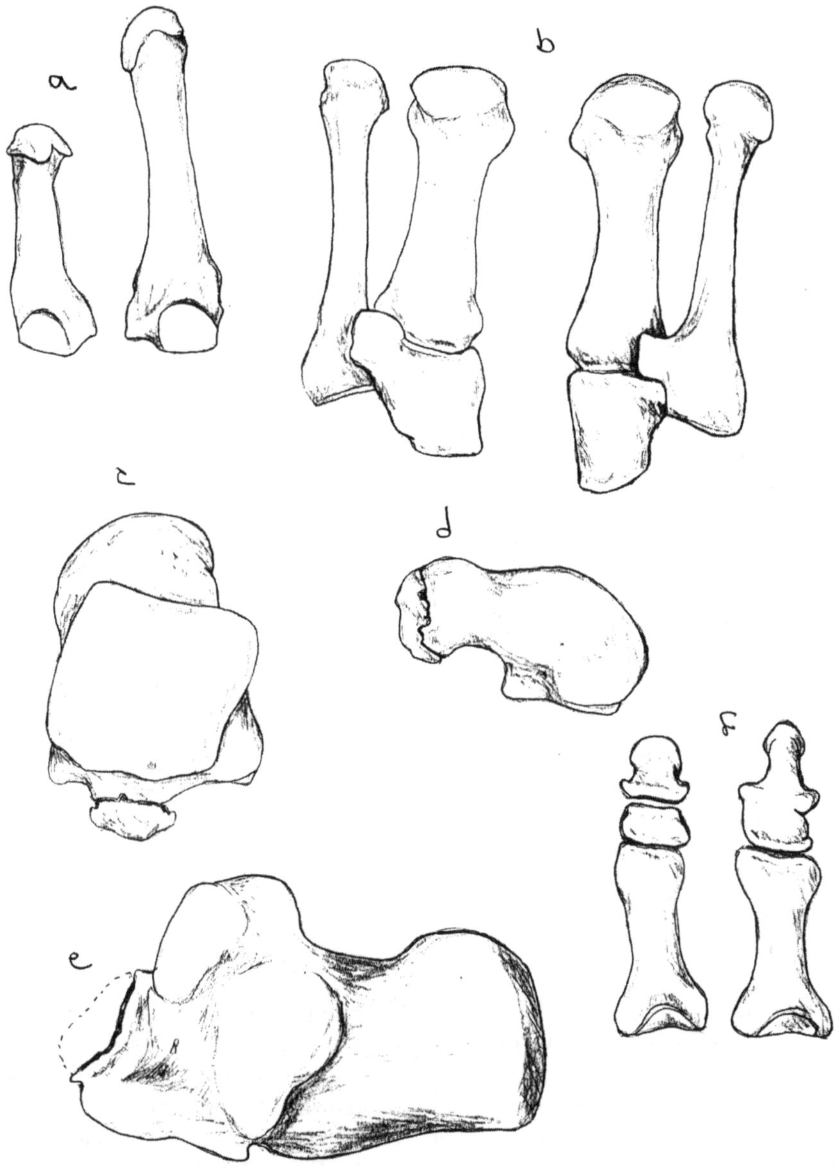

Figure 15.13 Lower limb anomalies. (a) Brachydactyly left fourth metatarsal and normal right fourth metatarsal, Pueblo, New Mexico. (b) Os intermetatarsium attached to left first cuneiform, adult male, Corinth, Greece, and attached to right second metatarsal, adult female, Pueblo Bonito, New Mexico. (c) Os trigonum, adult male, Corinth, Greece. (d) Os navicular (os tibiale), adult male, Corinth, Greece. (e) Calcaneus secundaris (missing ossicle), adult male, Corinth, Greece. (f) Normal fifth toe bones and fifth to symphalangism, adult male, Corinth, Greece

plantar flexion causing the inner border of the foot to supinate as the anterior foot is adducted. Less common expressions of club-foot include talipes calcaneovalgus with the heel turned outward and the anterior foot elevated, talipes calcaneus with the foot dorsiflexed, talipes varus with the heel turned inward and talipes valgus with the heel turned outward. The heel can be elevated and turned outward with talipes equinovalgus or the heel can be turned outward with the outer border of the anterior foot higher than the inner border with talipes planovalgus. Plantar flexion deformity with talipes equines forces walking on the toes. The twisting format of the foot distorts the shapes of the affected bones, which can be identified in dry bones (Brothwell, 1967: 423–428). The aetiology of club-foot is not known, but the evidence suggests disturbance within the mesenchymal tissue of the forming embryonic foot that may involve development of certain muscles or ligaments, thus affecting bones of the foot.

PATTERNS OF CONGENITAL ANOMALIES IN ARCHAEOLOGICAL POPULATIONS

Articles describing isolated examples of congenital anomalies in archaeological skeletal material can be found in the literature, but few studies concerning frequencies for congenital anomalies in archaeological populations have been reported. Every archaeological population has its own pattern of congenital anomalies (Barnes, 1994a: 320). Many local populations within the same region may share similar patterns, such as the prehistoric Puebloan groups living on the Pajarito Plateau of northwest New Mexico (Barnes, 1994a: 231–317). Overall patterns of regional populations that differ from each other suggest different genetic groups. Recent studies of a skeletal collection from the ancient Puebloan site of Hawikku near Zuni, western New Mexico, show some similarities in vertebral anomalies with the skeletal collection from the Puye village of the Pajarito Plateau population, but there are significant differences in lumbosacral border shifting. Shifts at this border occur more frequently in the Puye collection than in the Hawikku collection, with a higher prevalence of cranial shifting for Puye than for Hawikku (London *et al.*, 2005). The differences support the historic and archaeological data that the two areas were populated by different ethnic groups.

Patterns of congenital anomalies may differ through time within the same archaeological site with shifts in population. An example is the three cemeteries located in the same area of Corinth that were used during different time periods of occupation: the 13th century, the 14th–early 15th century and the 16th century AD. The earliest cemetery that was used during the Frankish occupation and the following cemetery used after the Franks abandoned the site show some continuity in the pattern of congenital anomalies. The later cemetery was used under the Ottoman occupation and shows a very different pattern of anomalies, particularly among the men. Some significant differences include the presence of metopism, lumbar ribs, cleft atlas and sacral segments in males from the earlier cemeteries not present in males from the later cemetery. While sacralized fifth lumbars with lumbosacral border shifting occur in males from the Frankish cemetery and not in males from the Ottoman cemetery, lumbarized first sacral segments are found in the Ottoman males but not in the Frankish males.

Frequencies for certain congenital anomalies, especially very rare defects, often stand out in a particular population. Merbs' (2004) study of vertebral anomalies of Canadian Arctic Thule – historic Inuit and the historic Sadlermuit Inuit found an unusually high frequency of the rare congenital anomaly sagittal cleft vertebra among both groups. The smaller Sadlermuit group shows greater intensity of the anomaly that most likely represents genetic drift as this small group became isolated from the others. Snow's (1974) study of pre-contact skeletal remains from Mokapu, Oahu, Hawaii, showed an unusually high frequency of nasal bone aplasia among the ancient Hawaiians. Anderson (1988) studied the frequency of calcaneus secundarius occurring in the medieval cemetery at Trondheim, Norway. Finnegan and Marcisk (1980) analysed the frequency for the manibular Stafne's defect among Plains Archaic and Middle Mississippian populations. Extra long styloids and inclusion cysts of the palate were among some of the congenital anomalies noted by Gregg and Gregg (1987) in proto-Arikara groups of the Northern Plains of the Upper Missouri River basin. My own study of Bronze Age Upper Xia Jia Dian skeletons found in a sacrificial pit at Da Shan Qian, eastern Inner Mongolia, China (Rohn and Barnes, 2004), shows an unusually high frequency for ventral hypoplasia affecting the lower thoracic and upper lumbar spine that argues for close genetic relationships of the sacrificial victims.

Case (1996) studied brachydactyly of the metatarsals among Arikara skeletal remains from South Dakota for genetic links. Cybulski (1988) found that brachydactyly of the metacarpals supported familial linkages in prehistoric burials from Prince Rupert Harbour, British Columbia. Symphalangism of the fifth toe was found to be common in a medieval Slavic population by Teegen and Schultz (1999). My own studies of the 13th-century AD cemetery at Corinth, Greece, show a clustering of third metacarpal os metastyloideum and other clusters of congenital anomalies within this same cemetery that reflect familial linkages.

Depictions of extra fingers and toes are often seen in prehistoric rock art throughout the southwest USA. Some cases of actual polydactyly of the fifth metatarsal have been identified in human remains (Barnes, 1994b; Case *et al.*, 2006). Similar expressions of polydactyly of the fifth toe with a bifurcated fifth metatarsal reflecting genetic linkage were found at Pueblo Bonito in Chaco Canyon, New Mexico, and at Sand Canyon pueblo 50 miles west in the northern San Juan region of southwest Colorado (Barnes, 1994b). A totally different expression of polydactyly with a double-headed fifth metatarsal was noted by Reed (1981: 115) at Las Humanas in Gran Quivira National Monument in the southern Rio Grande area that is similar to one that I recently found among human skeletal remains from an adolescent Trincheras burial in northwestern Mexico. Although only one individual from each skeletal collection from these four sites shows distinct signature expressions of polydactyly, these data reflect population movements between the northern San Juan and Pueblo Bonito, and between Gran Quivira, New Mexico, and Trincheras in northwest Mexico.

THEORETICAL ASPECTS, FUTURE PROBLEMS AND DIRECTIONS

This outline of various congenital anomalies arising during morphogenesis does not cover all the possible developmental variations. However, the morphogenetic approach to analysing congenital anomalies in the skeleton is an open-ended method that allows for deciphering virtually any type of developmental anomaly arising from disturbances occurring during morphogenesis. This includes developmental disturbances in muscles and ligaments or other adjacent soft tissues that can affect associated skeletal development. Palaeopathologists have a unique opportunity to further the understanding of developmental anomalies affecting the skeleton that clinicians do not see, as well as to contribute important data for genetic studies of ancient populations.

Frequency analyses of developmental anomalies in population studies utilizing statistical analysis with tests for significance such as those used for non-metric trait studies can distort findings, since some anomalies appear to be common while others are very rare yet important even if only one or a few individuals are affected. This is especially true when a neighbouring population has a similar rare anomaly. Frequencies are best represented by percentages for comparison studies, emphasizing the type of expression of the anomaly, complete or incomplete, minor or severe, bilateral, asymmetrical or unilateral. Does the anomaly lead to pathology or is it non-pathological? The morphogenetic approach to the study of congenital anomalies in population studies has just begun.

REFERENCES

Anderson JE. 1968. Skeletal 'anomalies' as genetic indicators. In *The Skeletal Biology of Earlier Human Populations*, Brothwell DR (ed.). Pergamon: London; 135–147.

Anderson T. 1988. Calcaneus secundarius: an osteo-archaeological note. *Am J Phys Anthropol* **77**: 529–531.

Barnes E. 1994a. *Developmental Defects of the Axial Skeleton in Paleopathology*. University Press of Colorado: Niwot, CO.

Barnes E. 1994b. Polydactyly in the southwest. *Kiva* **59**: 419–431.

Brothwell D. 1967. Major congenital anomalies of the skeleton: evidence from earlier populations. In *Diseases in Antiquity*, Brothwell DR, Sandison AT (eds). Charles C. Thomas: Springfield, IL; 423–443.

Buikstra JE, Ubelaker D. 1994. *Standards For Data Collection From Human Skeletal Remains*. Arkansas Archaeological Survey Research Series No. 44. Arkansas Archaeological Survey: Fayetteville, AR.

Case DT. 1996. Developmental defects of the hands and feet in paleopathology. Master of Arts thesis, Arizona State University, Tempe, AZ.

Case DT, Heilman J. 2005. Pedal symphalangism in modern American and Japanese skeletons. *Homo* **55**: 251–262.

Case DT, Ossenberg NS, Burnett SE. 1998. Os intermetatarsium: a heritable accessory bone of the foot. *Am J Phys Anthropol* **107**: 199–209.

Case DT, Hill RJ, Merbs CF, Fong M. 2006. Polydactyly in the prehistoric American Southwest. *Int J Osteoarchaeol* **16**: 221–235.

Cybulski JS. 1988. Brachydactyly, a possible inherited anomaly at prehistoric Prince Rupert Harbour. *Am J Phys Anthropol* **76**: 363–376.

Dickel DN, Doran GH. 1989. Severe neural tube defect syndrome from the early Archaic of Florida. *Am J Phys Anthropol* **80**: 325–334.

Ferguson AB. 1968. *Orthopaedic Surgery in Infancy and Childhood*, 3rd edition. Williams & Wilkins: Baltimore, MD.

Finnegan M, Marcisk A. 1980. Anomaly or pathology: the Stafne defect as seen in archaeological material and modern clinical practice. *J Hum Evol* **9**: 19–31.

Gregg JB, Gregg PS. 1987. *Dry Bones: Dakota Territory Reflected*. Sioux Printing Company: Sioux Falls, SD.

Hauser G, De Stefano GF. 1989. *Epigenetic Variants of the Human Skull*. Schweizerbart: Stuttgart.

Huber GC. 1936. *Piersol's Human Anatomy*, Volume I. Lippincott: Philadelphia, PA.

Larsen WJ. 2001. *Human Embryology*, 3rd edition. Churchill Livingstone: Philadelphia, PA.

London MR, Wilczak CA, Dudar JC, Ousley SD. 2005. Congenital and developmental defects of the vertebral column in samples from Hawikku and Puye, New Mexico. *Am J Phys Anthropol* 40(Suppl): 140.

Merbs CF. 2004. Sagittal clefting of the body and other vertebral developmental errors in Canadian Inuit skeletons. *Am J Phys Anthropol* **123**: 236–249.

Moore, KL. 1985. *Clinically Oriented Anatomy*, 2nd edition. Williams & Wilkins: Baltimore, MD.

Ortner DJ. 2003. *Identification of Pathological Conditions in Human Skeletal Remains*, 2nd edition. Academic Press: Amsterdam.

Ortner DJ, Putschar WGJ. 1981. *Identification of Pathological Conditions in Human Skeletal Remains.* Smithsonian Institution Press: Washington, DC.

Reed EK. 1981. Human skeletal material. In *Contributions to Gran Quivira Archaeology*, Hayes AC (ed.). Publications in Archaeology No. 17. National Park Service, US Department of the Interior: Washington, DC; 75–118.

Regan MH, Case DT, Brundige JC. 1999. Articular surface defects in the third metatarsal and third cuneiform: nonosseous tarsal coalition. *Am J Phys Anthropol* **109**: 53–65.

Resnick D. 1989. *Bone and Joint Imaging*. WB Saunders: Philadelphia, PA.

Rohn AH, Barnes E. 2004. Da Shan Qian reconstruction of funerary behavior. *Res China's Frontier Archaeol* **2**: 337–352 (in Chinese).

Sadler TW. 2006. *Langman's Medical Embryology*, 10th edition. Lippincott Williams & Wilkins: Philadelphia, PA.

Smith P. 2002. *The Hand Diagnosis and Indications*, 4th edition. Churchill Livingstone: London.

Snow CE. 1974. *An Initial Study of Skeletal Remains from Mokapu, Oahu.* University Press of Kentucky: Lexington, KY.

Spranger JW, Langer LO, Wiedmann H-R. 1974. *Bone Dysplasias*. W.B. Saunders: Philadelphia, PA.

Teegen WR, Schultz M. 1999. Symphalangism of the 5th toe in the early medieval Slavic population from Starigard/Olendburg (north Germany). *Homo* **50**: 244–248.

Webb SG, Thorne AG. 1985. A congenital meningocele in prehistoric Australia. *Am J Phys Anthropol* **68**: 525–533.

16

Growth in Archaeological Populations

Ron Pinhasi

Department of Archaeology, University College Cork, Cork, Ireland

INTRODUCTION

Growth is defined as a process that involves the increase in size of a body dimension and stands somewhat in distinction to development, which usually implies a qualitative change (Molinari and Gasser, 2004). Although the growth process is usually assumed to be continuous, it in fact involves a series of saltations and stases as growth velocity changes with age (Lampl *et al.*, 1992; Hauspie and Moliniari, 2004). However, the growth process for most body measurements is best modelled as a smooth non-linear curve with a particularly steep slope during the first 2 years of life.

Sub-adults, and particularly infants, are susceptible to the effects of various environmental stressors. Bioarchaeologists who study the skeletal remains of sub-adults investigate the effect of both non-specific (without a known aetiology) and specific stressors as indicators of morbidity, mortality and the individual's health profile (Lewis and Roberts, 1997; Lewis, 2000). At the time of writing, only a few bioarchaeological studies have attempted to examine the interface between these stressors and both endochondral and appositional growth among past populations (e.g. Ribot and Roberts, 1996; Lewis, 2002a).

This chapter begins with a review of the main theoretical and methodological aspects that are involved in the study of growth in past populations. Particular emphasis is placed on a discussion of the limitations of growth studies that rely on archaeological skeletal samples, and the choice of appropriate quantitative methods. The following section discusses growth velocity and intra-skeletal variations and stresses the potential of studies of growth that do not only examine diaphyseal length of long-bones, but also appositional growth and variations in growth of skeletal elements other than long-bones.

The next sections focus the relationship between non-specific and specific stress and growth in past populations, with a particular emphasis on the early post-natal period and early childhood, during which growth is most prone to disruption. A case study

Advances in Human Palaeopathology Edited by Ron Pinhasi and Simon Mays

is presented of non-specific stress and rickets and their effect on the growth pattern of a post-medieval population from London. The chapter concludes with a brief outline of some future research directions.

THEORETICAL AND METHODOLOGICAL ASPECTS

The Nature of Growth

Growth is regulated and affected by a combination of genetic and environmental factors (Tanner, 1986, 1989). Human growth is a target-seeking process in which growth disruption is followed by a 'catch-up' period following amelioration of conditions (Tanner, 1986; Binns, 1998). The 'catch-up' phenomenon occurs following an improvement in health and nutrition after a period of growth faltering due to stress during any of the growth stages (e.g. Johnston and MacVean, 1995).

Human growth and development from birth is characterized by five life-history stages, each characterized by specific biological and behavioural attributes (Bogin and Smith 1996; Bogin 1997, 1998, 2000, 2002):

1. Infancy (0–36 months) – a period of rapid growth and development.
2. Childhood (3–7 years) – slow physical growth, quick mental and behavioural development.
3. Juvenility (girls 7–10 years, boys 7–12 years) – pre-puberty, no longer solely dependent on mother for food and care.
4. Adolescence (ends at 19 years for girls and 21 years for boys) – sexual maturation, rapid growth.
5. Adulthood – cessation of growth.

The growth pattern, in terms of differential growth rate of various skeletal parts and their development, differs for each stage. The growth rate declines sharply during infancy, and is followed by a period of slower velocity decline in childhood. The end of the childhood stage is often marked by a small increase in velocity, which is called the mid-childhood growth spurt (Bogin, 1997). It is then followed by adolescent growth that is associated with sexual maturation and, hence, with hormonal changes. Finally, at maturity, growth is complete, with the fusion of all epiphyses and most cranial sutures.

Skeletal growth patterns, therefore, must be examined in the context of specific growth stages. It is also necessary to examine growth of the various skeletal parts (intra-skeletal growth) as they may vary in both velocities and growth faltering and catch-up due to variable susceptibility to disruptions induced by environmental stressors. Two of the main groups of stressors that are most often discussed in relation to growth are infection and nutrition. A major problem in the study of these stressors is that both are entwined in a vicious circle in which undernutrition leads to infection and vice versa (Briend, 1998; King and Ulijaszek, 1999). Malnutrition lowers the resistance of the host to infection and infectious diseases exaggerate malnutrition, particularly in the case of intestinal infections (Scrimshaw *et al.*, 1968; Scrimshaw, 2003). Neither the presence of an infectious agent, nor the deficiency of an essential nutrient alone will determine whether disease occurs

(Scrimshaw *et al.*, 1968). Exposure to a pathogen, such as *Mycobacterium tuberculosis*, does not by itself determine the prevalence of disease in a population, as exposure will be resisted by most persons and will cause disease in only a few. The attack is more likely to be severe or fatal in the case of malnourished individuals. In addition, the effects of the attack are more severe in specific age groups, such as newly weaned infants or newborns.

Environmental and host factors will render a specific intake of a given nutrient either wholly adequate or grossly deficient (Scrimshaw *et al.*, 1968; Ulijaszek and Strickland, 1993; King and Ulijaszek, 1999). All infectious diseases have adverse metabolic affects and often also influence the amount and type of food consumed. In both past and modern societies, the type and quantity of food consumed, hygiene and general health are affected by gender-based and socioeconomic-based variations. Consequently, growth stunting that is often assumed to be the direct consequence of malnutrition may, in fact, be the outcome of a more complex nutrition–infection system and, therefore, it needs to be addressed from a system-based epidemiological perspective.

Epidemiological studies on the effect of infection on growth rely, for the most part, on indirect evidence, since it is usually impossible to isolate the primary infective or nutritional cause of growth retardation in a given population (Briend, 1998). Most modern developing populations under nutritional and infectious disease stress are provided with food supplements by international aid agencies, so longitudinal studies of the effects of frequent infections on the growth pattern may not isolate the factors involved in the observed growth patterns. However, a study in the Mirzapur area of Bangladesh carefully monitored water and sanitation programmes that successfully reduced the prevalence of diarrhoea in children compared with a control area (Briend, 1998: 334). This gave an opportunity to measure the extent to which growth would improve after the reduction in diarrhoea. There was no measurable improvement in growth of children living in the intervention area. However, a longitudinal study carried out in 1977 in Gambia, in order to assess the effect of different types of infection on weight gain of children, showed that growth was depressed in months of the year with seasonally high prevalence of infections, and indicated that diarrhoea had the most important effect on growth (Briend, 1998: 334–336). The Gambian study indicates that growth was reduced during specific periods in which diarrhoea and other infections prevailed. The failure of the Mirzapur study to demonstrate any effect of infection on growth may be because its effects were obscured by the occurrence of catch-up growth.

Issues in the Investigation of Growth in Skeletal Populations

Research that examines the effects of specific and non-specific stressors on growth patterns of past populations is currently in its infancy. The majority of bioarchaeological growth studies focus on the overall growth profile of a given skeletal population which is often compared with modern growth standards or with other studies of archaeological samples. Most studies focus on a comparative analysis of the growth curves of long-bone diaphyseal dimensions of sub-adults (e.g. Y'Edynak, 1976; Hoppa, 1992), mainly since it is assumed that these best reflect the growth variations. Studies generally interpret interpopulation variations in growth curves in an ad hoc manner, in which deviations from the 'normal' growth pattern are interpreted as indicating some type of exposure to systemic stress or, alternatively, as reflecting cross-population genetically based growth variation. As these studies often do not involve a systematic examination of both non-specific and specific stress indicators, the interpretation of the observed growth curves can provide but limited biocultural information. In other words, without studying stressors, one cannot associate differences in observed growth patterns of populations with either genetic or environmental aspects, as it is impossible to derive an aetiological assessment of the interpopulation variations.

There are at least four major types of study on human archaeological populations that involve aspects of growth:

1. Palaeoanthropological studies that focus on the evolution of human growth, particularly in relation to various life-history changes. The major difficulty with such studies is that their age assessment is exogenous as it is based on non-referent standards taken either from modern humans or from great apes (review in Hoppa and Fitzgerald (1999)).
2. Descriptive and/or comparative studies. Archaeological populations are often compared with a modern population (e.g. Johnston, 1962; Y'Edynak, 1976; Lovejoy *et al.*, 1990; Mays, 1999; see Humphrey (2000: 25–26) for a review). Most of these studies focus on long-bone diaphyseal length in relation to dental age rather than examining other bone dimensions, appositional growth or cross-sectional geometry.
3. Anthropological studies that examine growth patterns in an archaeological sample as an additional source of information about the health status of past populations. In such studies, growth faltering and reduced height-for-age are interpreted in relation to biocultural aspects, such as socioeconomic status, diet, or housing (Cook, 1984; Kennedy, 1984; Rathburn, 1984; see Larsen (1997) and Hoppa and Fitzgerald (1999: 12–15) for reviews).
4. Analytical growth studies that focus on the relationship between growth and one or more specific or non-specific skeletal stress indicators. Such studies either examine growth patterns as a central theme rather than being just proxy for the health status of past populations (e.g. Hummert and van Gerven, 1983; Mays, 1985, 1995; Ribot and Roberts, 1996; Pinhasi *et al.*, 2006), or alternatively infer the influence of nutrition and infection from observed growth patterns (e.g. Jantz and Owsley, 1984; Mensforth, 1985; Saunders *et al.*, 1993).

Bone growth may be divided into endochondral and appositional growth. Endochondral growth involves the replacement of hyaline cartilage with bony tissue. The bones are first formed as hyaline cartilage models. During growth, primary ossification centres appear first in long-bones. Following vascular invasion, the cartilage in the centre of the diaphysis begins to disintegrate as osteoclasts break down the newly formed bone to open up the medullary cavity (Carter *et al.*, 1996; Stini, 1998).

Growth of width of the diaphysis is achieved by direct bone apposition and resorption on the periosteal and endosteal surfaces respectively (Carter *et al.*, 1996). A significant portion of the compact bone in the diaphysis initially develops by appositional bone formation, but the conical regions extending from the centre of the bone toward each of the bone ends are initially formed by progressive endochondral ossification at the growth plate. With increasing age, secondary bone remodelling throughout the skeleton progressively diminishes the distinction between primary bone formed by appositional or endochondral growth (Carter *et al.*, 1996). Unlike endochondral growth that ceases once the secondary ossification centres are fused, appositional growth continues throughout life. It involves an active process of secondary bone remodelling and changes in bone diameter that may be triggered by normal activity, stress, trauma or disease (Stini, 1998). The vast majority of bioarchaeological studies focus on endochondral growth, as they examine age-specific changes in long-bone diaphyseal dimensions. Some studies have examined appositional growth and focus on changes in cortical thickness (van Gerven *et al.*, 1985; Mays 1999; Pinhasi *et al.*, 2005) or density (McEwan *et al.*, 2004).

Utilizing archaeological collections in studies of growth in past populations is fraught with both practical and theoretical difficulties (Humphrey, 2000). These include sampling and preservation biases that affect most archaeological populations (Waldron, 1994; Chapter 2). Sub-adult skeletal samples from a single stratigraphic phase at a given site may be rather poorly representative of the true mortality profile of the past population. This issue is of minor concern to anthropologists that carry out intra-population analysis on variations in the growth velocities of various skeletal features (Humphrey, 1998), but it is a major issue to those who wish to compare growth in several populations. Underrepresentation of certain age cohorts (such as infants and adolescents) will have an effect on the fit between the interpolated growth curve (see below) and the actual raw data. Underrepresentation of infants is usually largely due to differential burial practices, or to archaeological recovery or taphonomic factors (Saunders and Barrans, 1999). A dearth of sub-adults aged between 6 and 18 years is a natural product of the normal mortality profile of a human population. Personal experience, working extensively with European skeletal collections, indicates that one is very unlikely ever to find an archaeological population with adequate representation of adolescents unless the sample size is very large (>500).

Estimation of Age at Death in Juvenile Skeletal Remains

The investigator who wishes to assess the effect of environmental stressors on growth must be cautious as to how they assess the age of the sub-adults in the sample. The vast majority of archaeological samples are of undocumented age (with the exception of a few reference samples (Usher, 2002)). The method of age determination selected by the investigator must be minimally affected by environmental stress, and must not be based on bone length, as each will result in the lack of independence between the variables (individual age, stress indicator, bone dimensions).

The most common methods used for ageing skeletal sub-adults are, in order of precision, dental development, epiphyseal closure and diaphyseal length (Hoppa and Fitzgerald, 1999; Saunders, 1992, 2000). Epiphyseal closure gives a wide age interval and involves the combination of data from various bones. Therefore, it is more effective for forensic studies or for general assessment of minimum/maximum age rather than for the derivation of a more accurate figure. As stated above, bone length is inappropriate as an ageing method for most growth studies. Because, in growth studies, age estimation must be based on a process that is, as far as is possible, independent of bone size, most workers use dental development. The most commonly used standard in growth studies is that of Schour and Massler (1941). This combines crown and root formation stages with dental eruption and displays the combined pattern in chart form in association with a given age in years and a corresponding standard deviation. This standard was modified by Ubelaker (1989) specifically for American Indians. Dental eruption patterns are more affected by environmental aspects, such as diet and health, than is dental calcification, so it is preferable to use the latter to estimate age. However, it may also be useful to assess whether there is a great discrepancy in age estimates derived from dental calcification and dental eruption. If the age estimates based on dental eruption are significantly lower than those obtained by crown and root formation, implying delayed eruption, then this may in itself indicate a high level of non-specific stress.

Although crown and root formation is minimally affected by environmental fluctuations, there is some interpopulation variation (Tompkins, 1996; Liversidge, 2003). Variation is greatest between populations that have been separated for many millennia (e.g. Europeans, Amerindians and Africans), and is less between more closely related groups. It is probably reasonable, therefore, to assume that dental development standards from one population can be applied to other populations provided that they share a close genetic affinity. For the purposes of estimating age in sub-adult skeletal remains it is necessary to derive the estimates and their standard errors from tables, such as those published by Smith (1991), in which age estimates are modified for average age at death rather than age of onset for a given developmental stage.

Successful morphological determination of sex in sub-adult skeletons is restricted to the post-adolescent age. Although various researchers have attempted to detect sexually dimorphic skeletal traits in infants and juveniles, success has been limited; so, at present, most researchers prefer to use unsexed pooled samples (Hoppa and Fitzgerald, 1999: 2–6).

Comparing Growth in Skeletal and Living Populations

Data on archaeological samples is by necessity cross-sectional (Humphrey, 2000), so it is not possible directly to compare stature or other growth aspects in an archaeological sample with longitudinal studies carried out on living children. Cross-sectional studies cannot yield direct information about growth velocity, growth faltering, and catch-up growth. All growth curves that are based on average figures of dimension-for-age will tend to mask or reduce intra-population differences, and this limits the degree to which useful biocultural information can be derived from them.

Most growth studies on living populations examine cross-sectional or longitudinal changes in standing height (for world variations in growth, see Eveleth and Tanner (1990) and Walker *et al.* (2006)) and do not include either bone or limb dimensions. Unless the bioarchaeologist is particularly interested in the comparison of stature of past populations in respect to published historical and modern standards, it is best to compare limb dimensions directly and not to introduce further bias incurred by the application of stature estimation formulae (Humphrey, 2000). Bioarchaeologists who are particularly interested in stature variations in past populations in relation to the historical secular trend should use modern radiographically assessed standards for stature estimation, such as the femur:stature ratios of modern US children published by Feldesman (1992); these have been applied to estimations of stature of medieval English archaeological sub-adults (Mays, 1999).

When researchers wish to compare long-bone growth dimensions-for-age for archaeological samples with data for modern populations they, in most cases, utilize the data charts of Maresh (1955). These are based on a mixed longitudinal and cross-sectional study of 175 Caucasian subjects from Denver, Colorado. Measurements for all limb long-bones are presented in dimensions per age, without epiphyses from 2 months to 12 years and with epiphyses from 10 to 16 years (Humphrey, 2000). Separate data are presented for males and females, both as median dimensions and as percentiles.

The following issues should be considered concerning the use of the Maresh (1955) data as a comparative reference population. First, archaeological populations are never true representatives of the palaeodemographic age and sex structure of the living population to which they once belonged (Waldron, 1994). Moreover, the mortality profile of an archaeological population can never provide a true image of mortality in a past population, since the archaeological sample has been accumulated in most cases during a period measured in decades or even centuries and, therefore, it is not a simple period estimate of mortality. Second, the Maresh (1955) data are sex specific; but, as there is no reliable morphological method for sexing archaeological sub-adults, male and female data must be combined. Third, Maresh (1955) measures limb dimensions using radiography; hence, the data cannot be compared directly with measurements made on dry bones. It is necessary to make corrections for magnification, which will vary according to the type and dimension of the bone and can reach a figure of 5–7 % in the case of the femur (Maresh, 1955). Fourth, it is best to try to use the Maresh (1955) data as a comparative population only for those archaeological groups with which they share a similar ethnic background. Fifth, measures of central tendency, such as the median, tend to blur age-specific variations and ideally, in the case of archaeological samples, one would wish to apply statistical methods that assess raw data rather than median (or mean) dimensions-for-age. The last point will be elaborated upon in the next section.

A key issue in growth studies is the mathematical function that is used for the derivation of a growth curve. The two main types of growth model are structural (or parametric) and non-structural. Non-structural models do not postulate a particular form of the growth curve, usually have a large number of parameters with no biological interpretation, do not tend to an asymptotic value, are unstable in the extremities of the data range, and are easy to fit (Hauspie and Molinari, 2004). Structural models imply a basic functional form of the growth model, usually have fewer parameters, and tend to an upper asymptote and require sophisticated curve-fitting techniques, applying non-linear regression analysis (Hauspie and Molinari, 2004). There are important reasons why the researcher should apply a non-linear (structural) curve function rather than plotting a graph of average successive mean dimensions-per-age. The human growth curve is non-linear, since deceleration, acceleration and stunting occur at various stages, resulting in great variations in growth velocity for different age intervals and even for different individuals from the same population. Thus, the appropriate non-linear function will provide a best fit for the growth curve of any bone dimension. The most commonly used growth curves are the logistic and the Gompertz. The latter should be modified to constrain the lower point of inflection to zero (cf. Humphrey 1998: 60). However, various other structural models have been used in growth studies of living populations (see review in Hauspie and Molinari (2004)).

Frequently in archaeological skeletal series, there are few data points for sub-adults older than about 8 years of age. This has a significant negative effect on the iterative growth curve obtained when adopting an asymptotic function. This is illustrated in Figure 16.1, which displays iliac length versus dental age for three early medieval Austrian populations. All three curves comprise points that represent estimated values of the Gompertz function for each specimen. In the case of the Gars – Thunau sample, most children were under 5 years of age, less than 10 ilia were available for specimens between 5 and 7.5 years of age, and no specimens were older than 7.5 years of age. Consequently, the Gompertz function curves asymptotically in a manner that clearly cannot reflect the true growth pattern of this population. The Zwentendorf curve differs from the curve of Zwölfaxing as no point of inflection occurs at the age of 8 years, so that the curve appears to be nearly linear, which is a more realistic growth pattern.

How can one test whether or not significant variation exists for bone dimension-for-age for a number of archaeological populations? One possibility is to apply analysis of covariance (Ancova), in which regression slopes for each dimension-per-age of each of the populations

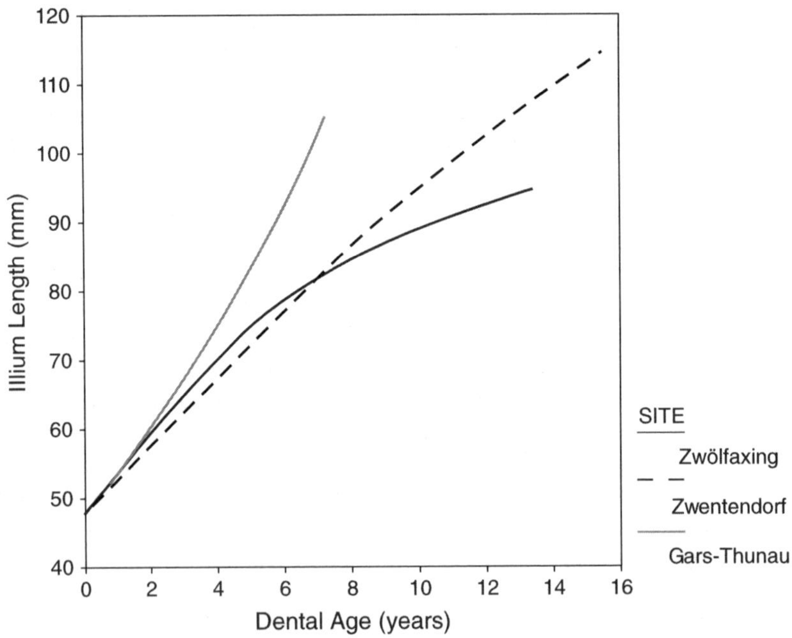

Figure 16.1 Growth curves of ilium length for three early medieval Austrian populations (8–10th centuries AD). Age is assessed on the basis of dental calcification, and Gompertz function parameters were iteratively fitted for each growth curve

are studied, while controlling for the influence of dental age as a covariate. When using Ancova it is necessary to test for the following assumptions: (1) linearity of the relationship between the covariate(s) and the dependent variable; (2) homogeneity of covariate-dependent variable slopes (Stevens, 1996). It is also important to note that the regressions, as they are linear, do not fit the growth curves, but their slopes and intersection nevertheless provide valuable information about the differences in growth between the samples that can be statistically tested (Figure 16.2a and b).

Growth Velocity and Intra-Skeletal Variations

Variations in growth velocity provide important information about the health of infants and children, as changes in the expected rates may indicate environmental stress. Growth retardation during gestation and the first 2 years of life has a powerful influence on adult body size (Li *et al.*, 2003). On average, infants gain 25 cm in total body length during the first year of life, followed by about a 12 cm gain during the second year (Gokhale and Kirschner, 2003). The second period of intensive growth is the adolescent growth spurt; in modern Americans this occurs in boys at a mean age of 11 years and in girls at the mean age of 9 years (Abassi, 1998). The contribution of adolescent growth to the total adult height is on average 17 % for males and 18 % for females, so most growth occurs prior to adolescence.

Although the great majority of bioarchaeological research on interpopulation variation in growth has focused on diaphyseal lengths of the main long-bones, studies of other dimensions, such as metaphyseal widths, and dimensions of the scapula, clavicle, os coxa,

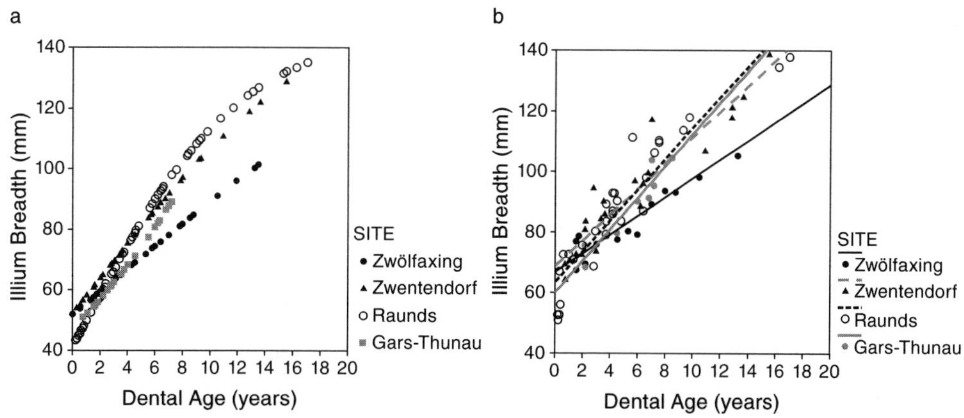

Figure 16.2 (a) Gompertz growth curves for three early medieval Austrian populations (Gars – Thunau, Zwölfaxing, Zwentendorf) and a medieval English population (Raunds); (b) linear functions were applied for the same data, using Ancova, in order to test for significant differences between the slopes of the four populations indicating lower attainment of ilium breadth dimensions per age for Zwölfaxing, for children and adolescents

etc., can provide important information about intra-skeletal variations in growth. A study on longitudinal growth changes in the humerus, radius, femur and tibia in 31 boys and 36 girls aged between 3 and 10 years from Denver, Colorado, showed that both sex- and limb-specific growth velocities vary within and between limbs (Smith and Buschang, 2004). In fact, inter-bone differences in growth velocity parallel inter-bone differences in size – velocity and size are greatest for the femur and least for the radius. These results suggest a childhood proximal–distal growth gradient within each limb.

Intra-skeletal variations in growth were noted when comparing the growth plots of ilium dimensions with those for the diaphyseal length of the long-bones in three early medieval Austrian populations (Pinhasi *et al.*, 2005). Interpopulation variation in the growth patterns of the ilium contrasted with a lack of difference in long-bones. This bone-specific pattern may be due to a combination of genetic mechanisms and sensitivity to environmental factors that differentially affect the growth of specific bones. Moreover, interpopulation variation in growth may be more pronounced for the metaphyseal width dimension of a long-bone than for its diaphyseal length, thus pointing to different growth patterns for the length and breadth dimensions of limb bones (Figure 16.3a and b).

Most archaeological populations fall below the modern Denver standards for all long-bone dimensions-per-age. This trend reflects a general secular trend of increase in limb dimensions since the middle of the 19th century (Tanner *et al.*, 1982; Jantz and Jantz, 1999; Cole, 2003). The increase is usually more pronounced in the case of the lower limbs (Figure 16.4a and b).

The Relationship Between Long-Bone Growth and Non-Specific Stressors

Only a small number of growth studies have focused on the effect of one or several specific or non-specific stressors on growth in past populations. Mays (1985) examined the relationship between Harris line formation and bone appositional and diaphyseal growth in sub-adult samples the Romano-British cemetery at Poundbury, England, but did not detect

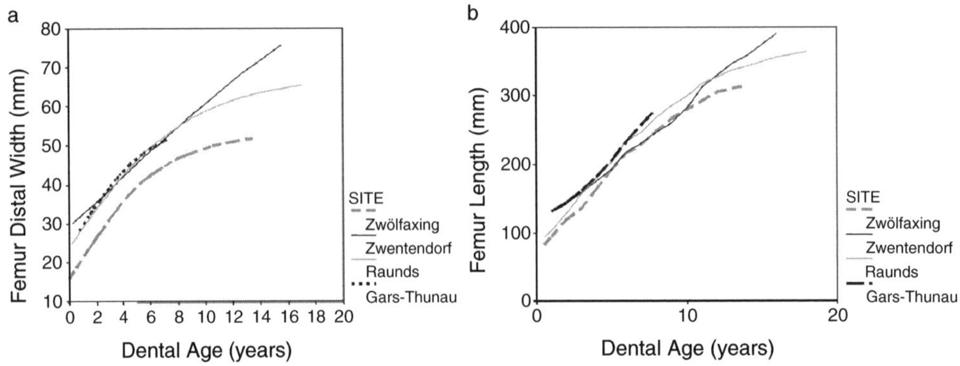

Figure 16.3 Growth differences among four medieval populations (three from Austria, one from England) are more pronounced in femoral distal width (a) than in femoral diaphyseal length (b)

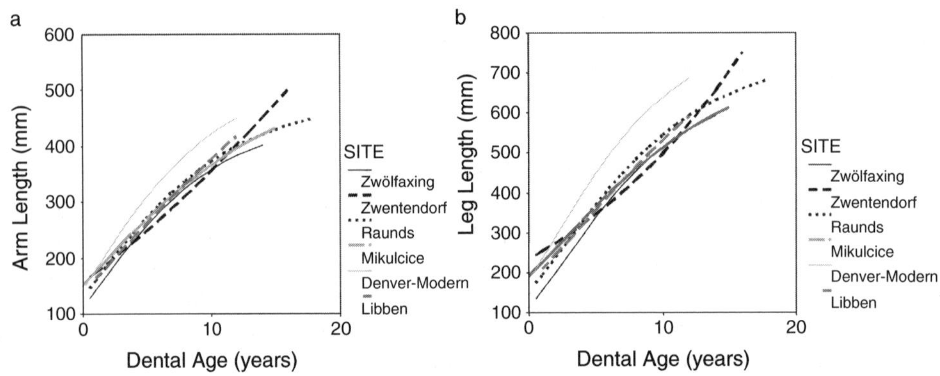

Figure 16.4 An interpopulation comparison of overall growth of (a) arm length (humerus + radius length) and (b) leg length (tibia + femur length) of various archaeological populations, and a modern reference population from Denver, Colorado (Maresh, 1955). Note that the differences between the archaeological and modern samples are more pronounced in the case of leg length

any difference in growth between sub-adults with Harris lines and those without. Mays (1985) points out that it is impossible to detect whether the individuals with Harris lines suffered from an acute or chronic stress episode; but, in any case, it is very likely that catch-up growth would have led to these individuals attaining 'normal' bone length and cortical thickness dimensions.

A second study by Mays (1995) examined the relationship between the occurrence of Harris lines and long-bone length, cortical thickness and enamel hypoplasia in both adults and juveniles from the English medieval site of Wharram Percy. His study showed a significant association among sub-adults between the presence/absence of Harris lines and both cortical bone thickness and enamel hypoplasia. However, Harris lines showed no significant association with any of other variables in the case of the adult sample. These results indicate that Harris lines should be studied in conjunction with additional non-specific and specific stress indicators. Such studies can attempt to detect the actual cause of the stress (nutritional, infectious, etc.) and its nature (acute or chronic) and then lead to a more comprehensive biocultural interpretation of the health status of the population studied.

Ribot and Roberts (1996) examined data on pathologies and growth in early medieval (Raunds) and late Medieval (Chichester) English skeletons. The non-specific stress indicators recorded included enamel hypoplasia, porosity on the skull (orbital and vault), subperiosteal new bone formation on long-bones and the ectocranial surface of the skull, and Harris lines. Growth curves of femoral, tibial and humeral length by age were plotted for the total (stress and non-stressed) individuals from each population. The individual positioning of cases with and without the stress indicators was then examined in relation to the growth curve. The results seemed to show no difference in dimensions-for-age for the stressed and non-stressed individuals, although no statistical test was performed on the data. Nearly all the cases with non-specific stress were of sub-adult infants and children (aged 3 years or under). This suggests that the period of severe stress occurred during the early years of life and that, since the age of death of these individuals was more than likely relatively close to the age period during which they suffered from stress, catch-up growth was less likely to have occurred.

The above studies suggest that the relationship between growth and non-specific stress indicators is not straightforward, but that stressors and linear and appositional growth interact in a complex manner. Hence, it may be necessary to take into consideration the effect of confounding variables when analysing the relationship between two or more elements.

The Effect of Rickets and Non-Specific Stress on Growth: A Case Study

Pinhasi *et al.* (2006) studied growth variation in relation to vitamin D deficiency, non-specific environmental stress (dental enamel defects), industrialization, urbanization and socioeconomic status during infancy (birth to 3 years) and early childhood (3–7 years). They used data on long-bone diaphyseal length dimensions and stress indicators from 234 sub-adults from Anglo-Saxon (Raunds), late medieval (St-Helen-on-the-Walls, York, England) and post-medieval (Broadgate and Christ Church Spitalfields, London) English archaeological skeletal series and analysed the data using both linear and non-linear growth models.

Twenty-one sub-adults from Broadgate and Christ Church Spitalfields showed rickets. The expectation was that they should have lower long-bone dimension-for-age than those without the condition, as rickets is expected to cause growth faltering. However, no statistically significant difference was detected between the long-bone diaphyseal length dimensions of infants with rickets and those without during the age interval of birth to 3 years (Figure 16.5a and b). When examining all sub-adults, pronounced variations in growth between the four populations were noted, mainly during infancy. The diaphyseal length dimensions of long-bones of Broadgate were significantly smaller-for-age than those of Spitalfields and the other samples up to the age of 4 years.

The Broadgate sub-adult population displayed a very high prevalence of non-linear dental enamel hypoplasia (DEH). Non-linear hypoplastic defects affect enamel formation and result in either inadequate enamel deposition or areas on the crown surface without enamel in which the dentine is therefore exposed. Similar pathological defects were noted in early experiments on dogs and rats by Lady Mellanby (1929, 1930, 1934) and this work showed that both vitamin A and vitamin D deficiencies can cause hypoplastic defects. However, in archaeological skeletal samples, it is usually impossible to isolate the main cause of defects.

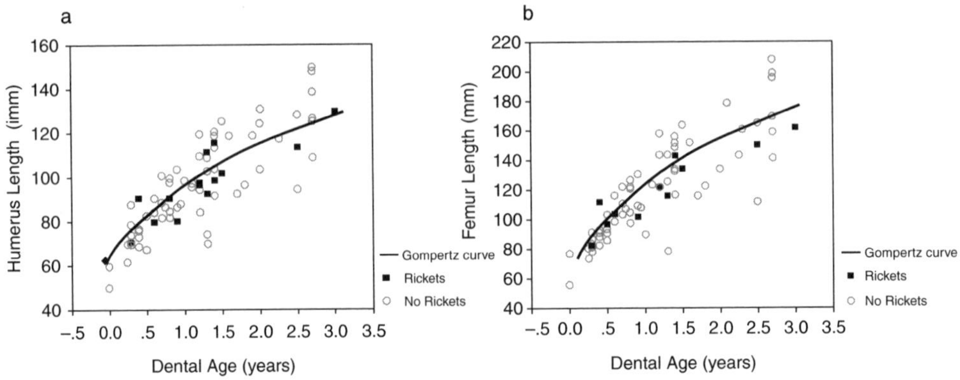

Figure 16.5 Analysis of the effect of rickets in a post-medieval population from London Broadgate on (a) humeral diaphyseal length and (b) femoral diaphyseal length. The graphs compare the dimensions of individuals with rickets and those without. The Gompertz curves illustrate the pooled growth function of the total sample. No significant differences were found between the two groups

As there is no current method that allows one to identify stress markers caused by general malnutrition, it remains impossible to ascertain whether observed hypoplastic defects, even if detected in association with evidence of rickets, are indeed directly caused by vitamin D deficiency.

Out of the 34 Broadgate individuals studied, 32 (94.1%) showed evidence of DEH, 12 (35.3%) showed evidence of hypoplastic and cuspal formation defects starting before birth (Pinhasi *et al.*, 2006), and 17 (50%) had defects starting between birth and 6 months. These results contrast with those of Lewis (2002b), who reports no DEH among infants younger than 6 months from York (St-Helen-on-the-Walls), London (Christ Church Spitalfields), or Raunds. At the age intervals of 0.5–2.5 years, DEH was absent among infants from Raunds, low in frequency among those from York (St-Helen-on-the-Walls: 5%) and moderate among infants from London (Spitalfields: 20%). In a more recent study, King *et al.* (2005) found that the earliest age at which enamel defects first occurred among the populations of Spitalfields and St Bride's, London, was at 1.2 years and the highest frequencies of enamel growth disruptions occurred at 2–4 years.

The Broadgate data showed growth faltering during infancy and early childhood (Pinhasi *et al.*, 2006). It is possible that there is an association between lower growth-for-age values during infancy and early childhood and enamel defects. In other words, a significant growth faltering is expected among the Broadgate sub-adults who had DEH. Notable variations in the degree of growth faltering were observed for different bones and dimensions. Figure 16.6a-c illustrates variations in humerus diaphyseal length, femur diaphyseal length and ilium breadth. The ilium breadth dimension shows particularly pronounced interpopulation variation. Although the comparative Gompertz curves cannot be tested for significance (as they are non-linear), the visual inspection suggests a low dimensions-for-age in the case of Broadgate sub-adults during the postnatal period from birth to about 4 years of age. After this period, the Broadgate sub-adults 'catch up', in the cross-sectional sense, with the rest of the populations and their curves merge. That the most pronounced growth faltering is in

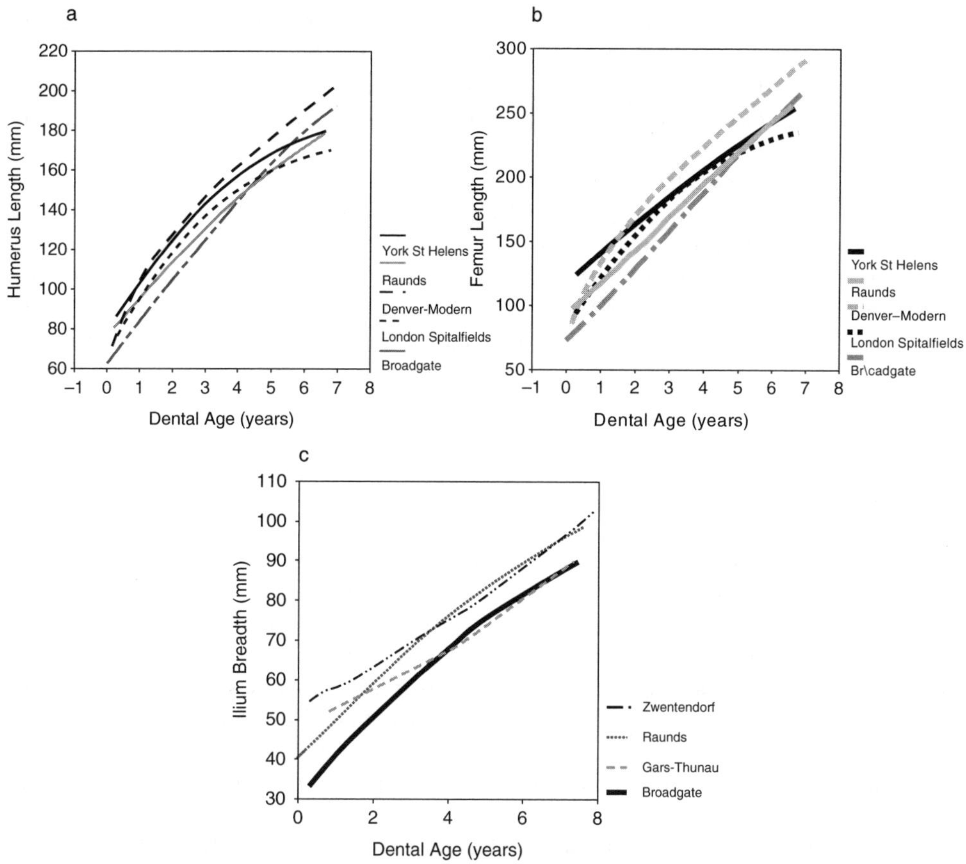

Figure 16.6 Growth curves of (a) humeral diaphyseal length and (b) femoral diaphyseal length of the Broadgate individuals with DEH compared with York (St-Helen-on-the-Walls), London (Spitalfields), Denver (modern), and Raunds; (c) iliac breadth dimensions compared with those of the medieval populations of Zwentendorf, Gars – Thunau and Raunds

iliac breadth supports the concept of different growth processes in the case of the ilium and the diaphyses of long-bones.

The effects of poor nutrition on growth appear to be gravest during the period from about 6 to 36 months of life, as this is the period of maximal growth velocity (Malina and Himes, 1978; Martorell and Habicht, 1986). Growth retardation or stunting during this period is associated primarily with high rates of infection and inadequate nutrition related to poor weaning practices and poor dietary quality (Adair, 1999). The Broadgate mothers were more than likely immunodeficient and malnourished to the extent that their colostrum could not buffer these infants from severe environmental stress even during the period between birth and 6 months during which breastfed infants' stress is customarily buffered (see King and Ulijaszek (1999: 168–170)). The high prevalence of DEH in this population is not due to post-weaning stress. The growth curves of Spitalfields and Broadgate differed greatly, particularly during infancy (Pinhasi *et al.*, 2006). As both these populations are from nearby neighbourhoods of post-medieval London and, thus, presumably share a rather similar genetic structure, it seems that these variations should be attributed to environmental rather than genetic factors. The high prevalence of severe hypoplasia with an early age of onset among the Broadgate infants points to the possibility that the infants suffered from poor overall health.

Studies on modern populations indicate that low socioeconomic status is a major factor that can significantly affect the growth velocity and growth pattern of a given population (Herngreen *et al.*, 1994). By 1650, east London had already acquired a working-class complexion with new neighbourhoods of timber shacks erected outside the city walls (Inwood, 1998). This was high-density housing, lacking sufficient heating and with poor sanitation – environmental conditions that likely affected the overall growth status of the inhabitants.

FUTURE DIRECTIONS

Proper synergy between cross-sectional studies on living populations and those from archaeological collections requires interdisciplinary collaboration between archaeologists and biological anthropologists on the one hand, and nutritionists, paediatricians, human biologists and medical specialists on the other hand. It entails the development of a methodology that should involve the analysis of dimensions obtained from radiographs. Many anthropological/archaeological research laboratories have portable digital X-ray units that can produce a high number of digital images in a relatively short period. The radiographs can be measured with available medical software and then directly compared with those obtained on living individuals.

A better understanding of the nutrition–infection interaction, particularly during infancy, is essential for interpretations and research design of archaeological growth studies (King and Ulijaszek, 1999). This is mainly since the interaction must have played a dominant role among past populations, as it does among those from contemporary populations from developing countries.

More research needs to be carried out on the growth of elements other than major long-bones, such as the scapula, sternum, os-coxae and vertebrae. The growth pattern of these bones under stress may differ from the patterns observed for diaphyseal long-bone growth, and the growth profiles of these bones may show more identifiable traces of growth faltering and stunting.

A focus on growth variations among both contemporaneous and diachronic samples from areas in which the historical and archaeological records indicate indigenous development with minimal exogenous gene flow can lead to a better separation between environmental and genetic influences on growth profiles. Further study of appositional growth in relation to infectious and metabolic disease is needed. This would shed more light on the complex synergistic relationship between growth, nutrition and infection.

Growth studies will greatly benefit from the application of structural growth models and from the adoption of an epidemiological approach. This would involve measurements of age- and sex-specific prevalence of specific and non-specific stress indicators with the application of age-correction methods and assessing bias, interaction and confounding among the variables (Chapter 3).

REFERENCES

Abassi V. 1998. Growth and normal puberty. *Pediatrics* **102**: 507–511.

Adair LS. 1999. Filipino children exhibit catch-up growth from age 2 to 12 years. *J Nutr* **129**: 1140–1148.

Binns CW. 1998. Infant-feeding and growth. In *The Cambridge Encyclopaedia of Human Growth and Development*, Ulijaszek SJ, Johnston FE, Preece MA (eds). Cambridge University Press: Cambridge; 320–323.

Bogin BA. 1997. Evolutionary hypotheses for human childhood. *Ybk Phys Anthropol* **40**: 63–89.

Bogin BA. 1998. The evolution of human nutrition. In *The Anthropology of Medicine*, 3rd edition, Romanucci-Ross L, Moerman D, Tancredi LR (eds). Bergen and Garvey South Hadley; 96–142.

Bogin BA. 2000. Evolution of the human life cycle. In *Human Biology: An Evolutionary and Biocultural Approach*, Stinson S, Bogin B, Huss-Ashmore R, O'Rourke D (eds). Wiley: New York; 377–424.

Bogin BA. 2002. The evolution of human growth. In *Human Growth and Development*, Cameron N (ed.). Academic Press: San Diego, CA; 295–320.

Bogin BA, Smith BH. 1996. Evolution of the human life cycle. *Am J Hum Biol* **8**: 703–716.

Briend A. 1998. Infection. In *The Cambridge Encyclopaedia of Human Growth and Development*, Ulijaszek SJ, Johnston FE, Preece MA (eds). Cambridge University Press: Cambridge; 334–346.

Carter DR, van der Meulen MC, Beaupre GS. 1996. Mechanical factors in bone growth and development. *Bone* **18**: 5S–10S.

Cole TJ. 2003. Secular trend in human physical growth: a biological view. *Econ Hum Biol* **1**: 161–168.

Cook DC. 1984. Subsistence and health in the lower Illinois Valley: osteological evidence. In *Paleopathology at the Origins of Agriculture*, Cohen MN, Armelagos GJ (eds). Academic Press: Orlando, FL; 237–270.

Eveleth PB, Tanner JM. 1990. *Worldwide Variations in Human Growth*. Cambridge University Press: Cambridge.

Feldesman MR. 1992. Femur/stature ratio and estimates of stature in children. *Am J Phys Anthropol* **87**: 447–459.

Gokhale R, Kirschner BS. 2003. Transition of care between paediatric and adult gastroenterology. Assessment of growth and nutrition. *Best Pract Res Clin Gastroenterol* **17**: 153–162.

Hauspie RC, Molinari L. 2004. Parametric models for postanatal growth. In *Methods in Human Growth Research*, Hauspie RC, Cameron N, Molinari L (eds). Cambridge University Press: Cambridge; 27–53.

Herngreen WP, van Buuren S, van Wieringen JC, Reerink JD, Verloove-Vanhorick SP, Ruys JH. 1994. Growth in length and weight from birth to 2 years of a representative sample of Netherlands children (born in 1988–89) related to socioeconomic status and other background characteristics. *Ann Hum Biol* **21**: 449–463.

Hoppa RD. 1992. Evaluating human skeletal growth: an Anglo-Saxon example. *Int J Osteoarchaeol* **2**: 275–288.

Hoppa RD, Fitzgerald CM. 1999. *Human Growth in the Past: Studies from Bones and Teeth*. Cambridge University Press: Cambridge.

Hummert JR, van Gerven JR. 1983. Skeletal growth in a medieval population from Sudanese Nubia. *Am J Phys Anthropol* **60**: 471–478.

Humphrey LT. 1998. Growth patterns in the modern human skeleton. *Am J Phys Anthropol* **105**: 57–72.

Humphrey LT. 2000. Growth studies of past populations: an overview and an example. In *Human Osteology in Archaeology and Forensic Science*, Cox M, Mays S (eds). Greenwich Medical Media: London; 23–38.

Inwood S. 1998. *A History of London*. Macmillan: London.

Jantz LM, Jantz RL. 1999. Secular change in long bone length and proportion in the United States, 1800–1970. *Am J Phys Anthropol* **110**: 57–67.

Jantz RL, Owsley D. 1984. Long bone growth variations among Arikara skeletal populations. *Am J Phys Anthropol* **63**: 13–20.

Johnston FE. 1962. Growth of the long bones of infants and young children at Indian Knoll. *Am J Phys Anthropol* **20**: 249–254.

Johnston FE, MacVean RB. 1995. Growth faltering and catch-up growth in relation to environmental change in children of a disadvantaged community from Guatemala City. *Am J Hum Biol* **7**: 731–740.

Kennedy KAR 1984. Growth, nutrition, and pathology in changing paleodemographic settings in South Asia. In *Paleopathology at the Origins of Agriculture*, Cohen MN, Armelagos GJ (eds). Academic Press: Orlando; 169–192.

King S, Ulijaszek SJ. 1999. Invisible insults during growth and development: contemporary theories and past populations. In *Human Growth in the Past*, Hoppa RD, Fitzgerald CM (eds). Cambridge University Press: Cambridge; 161–181.

King T, Humphrey LT, Hillson S. 2005. Linear enamel hypoplasias as indicators of systemic physiological stress: evidence from two known age-at-death and sex populations from postmedieval London. *Am J Phys Anthropol* **128**: 547–559.

Lampl M, Veldhuis JD, Johnson ML. 1992. Saltation and stasis: a model of human growth. *Science* **258**: 801–803.

Larsen CS. 1997. *Bioarchaeology*. Cambridge University Press: Cambridge.

Lewis M. 2000. Non-adult palaeopathology: current status and future potential. In *Human Osteology in Archaeology and Forensic Science*, Cox M, Mays S (eds). Greenwich Medical Media: London; 39–57.

Lewis M, Roberts CA. 1997. Growing pains: the interpretation of stress indicators. *Int J Osteoarchaeol* **7**: 581–586.

Lewis ME. 2002a. The impact of industrialisation: comparative study of child health in four sites from medieval and post-medieval England (AD850–1859). *Am J Phys Anthropol* **119**: 211–223.

Lewis ME. 2002b. *Urbanisation and Child Health in Medieval and Post-Medieval England*. British Archaeological Reports, British Series 339. Archaeopress: Oxford.

Li H, Stein AD, Barnhart HX, Ramakrishnan U, Martorell R. 2003. Associations between prenatal and postnatal growth and adult body size and composition. *Am J Clin Nutr* **77**: 1498–1505.

Liversidge H. 2003. Variations in modern human dental development. In *Patterns of Growth and Development in the Genus* Homo, Thompson JL, Krovitz GE, Nelson AJ (eds). Cambridge University Press: Cambridge; 73–113.

Lovejoy CO, Russell KF, Harrison ML. 1990. Long bone growth velocity in the Libben population. *Am J Hum Biol* **2**: 533–541.

Malina RM, Himes HE. 1978. Patterns of childhood mortality and growth status in a rural Zapotec community. *Ann Hum Biol* **5**: 517–531.

Maresh MM. 1955. Linear growth of long bones of extremities from infancy through adolescence. *Am J Dis Child* **89**: 725–742.

Martorell R, Habicht J-P. 1986. Growth in early childhood in developing countries. In *Human Growth: A Comprehensive Treatise, Volume 3: Methodology, Ecologic, Genetic, and Nutritional Influences on Growth*, 2nd edition, Falkner F, Tanner JM (eds). Plenum: New York; 241–262.

Mays SA. 1985. The relationship between Harris line formation and bone growth and development. *J Archaeol Sci* **12**: 207–220.

Mays SA. 1995. The relationship between Harris lines and other aspects of skeletal development in adults and juveniles. *J Archaeol Sci* **22**: 511–520.

Mays SA. 1999. Linear and appositional long bone growth in earlier human populations: a case study from medieval England. In *Human Growth in the Past*, Hoppa RD, Fitzgerald CM (eds). Cambridge University Press: Cambridge; 290–313.

McEwan JM, Mays S, Blake GM. 2004. Measurements of bone mineral density of the radius in a medieval population. *Calcif Tissue Int* **74**: 157–61.

Mensforth R. 1985. Relative tibia long bone growth in the Libben and Bt-5 prehistoric skeletal populations. *Am J Phys Anthropol* **68**: 247–262.

Mellanby M. 1929. *Diet and Teeth: An Experimental Study. Part I: Dental Structure in Dogs*. Medical Research Council Special Reports Series 140. HMSO: London.

Mellanby M. 1930. *Diet and Teeth: An Experimental Study. Part II: A. Diet and Dental Disease, B. Diet and Dental Structure in Mammals other than Dogs*. Medical Research Council Special Reports Series 153. HMSO: London.

Mellanby M. 1934. *Diet and Teeth: an Experimental Study. Part III: The Effect of Diet on the Dental Structure and Disease in Man*. Medical Research Council Special Reports Series 191. HMSO: London.

Molinari L, Gasser T. 2004. The human growth curve: distance, velocity and acceleration. In *Methods in Human Growth Research*, Hauspie RC, Cameron N, Molinari L (eds). Cambridge University Press: Cambridge; 27–53.

Pinhasi R, Teschler-Nicola M, Knaus A, Shaw P. 2005. A cross-population analysis of the growth of long bones and the os coxae of three early medieval Austrian populations. *Am J Hum Biol* **17**: 470–488.

Pinhasi R, Shaw P, White B, Ogden A. 2006. Morbidity, rickets and long-bone growth in post-medieval Britain – a cross-population analysis. *Ann Hum Biol* **33**: 372–389.

Rathburn TA. 1984. Skeletal pathology from the Paleolithic through the Metal Ages in Iran and Iraq. In *Paleopathology at the Origins of Agriculture*, Cohen MN, Armelagos GJ (eds). Academic Press: Orlando; 137–168.

Ribot I, Roberts CA. 1996. A study of non-specific stress indicators and skeletal growth in two mediaeval sub-adult populations. *J Archaeol Sci* **23**: 67–79.

Saunders SR. 1992. Sub-adult skeletons and growth related studies. In *Skeletal Biology of Past Peoples: Research Methods*, Saunders SR, Katzenberg MA (eds). Wiley-Liss: New York; 1–20.

Saunders SR. 2000. Sub-adult skeletons and growth-related studies. In *Biological Anthropology of the Human Skeleton*, Katzenberg MA, Saunders SR (eds). Wiley-Liss: New York; 135–161.

Saunders SR, Barrans L. 1999. What can be done about the infant category in skeletal samples? In *Human Growth in the Past*, Hoppa RD, Fitzgerald CM (eds). Cambridge University Press: Cambridge; 182–209.

Saunders SR, Hoppa, RD, Southern R. 1993. Diaphyseal growth in a nineteenth century skeletal sample of sub-adults from St. Thomas' Church, Belleville, Ontario. *Int J Osteoarchaeol* **3**: 265–281.

Schour I, Massler M. 1941. The development of the human dentition. *J Am Dent Assoc* **28**: 1153–1160.

Scrimshaw NS. 2003. Historical concepts of interactions, synergism and antagonism between nutrition and infection. *J Nutr* **133**: 316S–321S.

Scrimshaw NS, Taylor CE,Gordon JE. 1968. *Interaction of Nutrition and Infection*. World Health Organization Monograph Series No. 57. WHO: Geneva.

Smith BH. 1991. Standards of human tooth formation and dental age assessment. In *Advances in Dental Anthropology*, Kelley MA, Larsen CS (eds). Wiley-Liss: New York; 143–168.

Smith SL, Buschang PH. 2004. Variation in longitudinal diaphyseal long bone growth in children three to ten years of age. *Am J Hum Biol* **16**: 648–657.

Stevens J. 1996. *Applied Multivariate Statistics for the Social Sciences*, 3rd edition. Lawrence Erlbaum: Mahwah, NJ.

Stini WA. 1998. Skeletal development. In *The Cambridge Encyclopaedia of Human Growth and Development*, Ulijaszek SJ, Johnston FE, Preece MA (eds). Cambridge University Press: Cambridge; 196–197.

Tanner JM. 1986. Growth as a target-seeking function: catch-up and catch-down growth in man. In *Human Growth: A Comprehensive Treatise, Volume 1. Developmental Biology: Prenatal Growth*, 2nd edition. Falkner F, Tanner JM (eds). Plenum: New York; 167–179

Tanner JM. 1989. *Foetus into Man*, 2nd edition. Castlemead: Ware.

Tanner JM, Hayashi T, Preece MA, Cameron N. 1982. Increase in length of leg relative to trunk in Japanese children and adults from 1957 to 1977: comparison with British and Japanese Americans. *Ann Hum Biol* **9**: 411–423.

Tompkins RL. 1996. Human population variability relative to dental development. *Am J Phys Anthropol* **99**: 79–102.

Ubelaker DH. 1989. *Human Skeletal Remains: Excavation, Analysis, Interpretation*. Taraxacum: Washington, DC.

Ulijaszek SJ, Strickland SS. 1993. *Nutritional Anthropology: Prospects and Perspectives*. Smith Gordon: London.

Usher BM. 2002. Reference samples: the first step in linking biology and age in the human skeleton. In *Palaeodemography: Age Distributions from Skeletal Samples*, Hoppa RD, Vaupel JW (eds). Cambridge University Press: Cambridge; 29–47.

Van Gerven DP, Hummert JR, Burr DB. 1985. Cortical bone maintenance and geometry of the tibia in prehistoric children from Nubia's Batn el Hajar. *Am J Phys Anthropol* **66**: 275–280.

Waldron T. 1994. *Counting the Dead: the Epidemiology of Skeletal Populations*. Wiley: Chichester.

Walker R, Gurven M, Hill K, Migliano A, Chagnon N. 2006. Growth rates and life histories twenty-two small-scale societies. *Am J Hum Biol* **18**: 295–311.

Y'Edynak G. 1976. Long bone growth in western Eskimo and Aleut skeletons. *Am J Phys Anthropol* **45**: 569–574.

Index

accidental trauma 314–15
actinomycosis 211–12
ADS *see* Archaeology Data Service
AFM *see* atomic force microscopy (AFM)
age- and sex- specific prevalence 51–2
age-at-death 38–40, 367–8
age determination 129–31
amputations 311
analysis of covariance (ANCOVA) 369–71
analytical growth studies 366
ancestral sequence inference 167
ancient DNA
  authentication 155–9
  bacterial versus human 157–9
  degradation 154
  molecular palaeopathology 147–68
  study strategies 155–7
ANCOVA *see* analysis of covariance (ANCOVA)
angiosarcoma 267
ankylosis 83
anthracosis 257
aplasia 337–8, 349–50, 352–5, 357, 360
atomic force microscopy (AFM) 121, 141
auditory meatus anomalies 340
auricular surface ageing technique 110
autolysis 15–16
autopsy samples 123, 126
avulsion fractures 315, 317

BABAO *see* British Association for Biological Anthropology and Osteoarchaeology (BABAO)
backscatter scanning electron microscopy (BSEM) 17–19, 121, 128, 133–42
  diagenetic processes 17–20
Bartonella quintana 160
bejel 205
benign bone tumours 261–2, 271–4
bilharzia 149, 161
Binder's syndrome 338
bioapatite 7
bipartite patella 357
Black Death *see* bubonic plague
blade injury 313–14
blastic lesions 83
blunt injury 312–16
BMD *see* bone mineral density (BMD)
bog bodies
  computed tomography 101–17
  diagenetic processes 13, 17–20, 25
  facial reconstructions 116
  fractures 111–12, 114
  lesions 111–12
  neoplasms 256
  preservation 108
  X-rays 102–3
bone inventories 62–3
bone mineral density (BMD) 90–5, 228, 235–6, 243
bone structure 5–9
bone survival/recovery rates 31, 32–3
Borremose Woman 104, 112, 114, 116–17
bowing 239
brachydactyly 353–4, 357, 358, 360
British Association for Biological Anthropology and Osteoarchaeology (BABAO) 61, 178, 186
brucellosis
  differential diagnosis 191, 193, 211–12
  molecular palaeopathology 161–2

brushite 13
BSEM *see* backscatter scanning electron microscopy (BSEM)
bubonic plague 147, 148–9, 151, 182
burial practices 35–6
Burkitt's lymphoma 254
butterfly vertebrae 344–5

carcinoma
  differential diagnosis 201
  metastatic 81, 83
  nasopharyngeal 254–5, 257
caries sicca 205–6
case-control studies 45, 53
catastrophic events 36
catch-up growth 365, 368, 372–3
caudal shifts 336, 340–2
CEH *see* cuspal enamel hypoplasia (CEH)
Chagas' disease 47, 160
chemical diagenesis 4, 11–15
chondroblastoma 259, 262, 271–2
chondroma 259, 260, 271
chondrosarcoma 265–7, 275
chordoma 265–7, 275–6
chromosomal aberrations 330
cillini 35
Clony Cavan Man 108, 111, 113
CLSM *see* confocal laser scanning microscopy (CLSM)
club-foot 357
Codman's triangles 266–7
collagen
  diagenetic processes 4–5, 11–25
  histological studies 131
  metabolic bone disease 215
  trauma 309–25
comminuted fractures 310
compensatory eruption 290–3
compound fractures 310
computed tomography (CT)
  bog bodies 101–2, 107–10
  diagenesis 113–15
  foreign objects 106
  fractures 106, 111–12
  mummies 102–3, 114–16
  pathology/pseudopathology 112–14
  post-capture image processing 109–10, 115
  post-mortem degradation 112–14
  three-dimensional visualization 101–17
confidence intervals 52
confocal laser scanning microscopy (CLSM) 121, 140
congenital anomalies 329–61
  border shifts 341–2
  caudal shifts 336, 341–2
  cranial base 336, 340–1
  cranial shifts 341, 342, 348
  cranial vault 333–6
  embryonic morphogenesis 331–2
  limbs 351–9
  mandible 336–8
  maxillo-facial skeleton 338–9
  morphogenetics 330–1, 340–1, 349, 351–2
  populations 359–60, 361
  ribs 340–9
  sacrum 342, 344
  skull 331–40
  sternum 349–51
  syphilis 205–9
  temporal bone 340
  vertebral column 340–9
conodonts 5
continuous eruption 290–1
contrecoup fractures 312
coronoid hyperplasia 337–8
cortical bone
  age-related loss 228, 232–5
  thickness 87–90
cranial base anomalies 336, 340–1
cranial vault anomalies 333–6
cribra orbitalia 201
cross-sectional studies 368, 376
crystallinity index 13
CT *see* computed tomography (CT)
cuspal enamel hypoplasia (CEH) 285–8, 298
cyanobacteria 21–2
cysts 262, 296–8

dactylitis 199
data collection
  academic/applied palaeopathology 66–7
  changing criteria 67, 71
  demography 63
  differential diagnosis 63–4
  future directions 70–2
  historical development 58–61
  inventories 62–3
  methodologies 61–4
  multiple conditions 63
  multiple tests 64, 70
  publishing strategies 72

standardization 60, 70
static/evolutionary models 69
terminology 58, 60, 69
databases 177–87
British collections 178–9
data collection 178
development 178–9
historical development 58
regional 180–3
Royal Mint site case study 182
types 179–80, 184–7
Wellcome Osteological Research Database 181–2, 187
DCL *see* defective cement lines (DCL)
decalcification 123
decapitation 320–1
defective cement lines (DCL) 219
degenerative disc disease 209
degenerative joint disease (DJD) 67
DEH *see* dental enamel hypoplasia (DEH)
demography
age adjustment 38–9, 40
data collection 61
macroscopic analysis/data collection 61
population analysis 31–41
standardized mortality ratios 38–40
densitometry 91–4
dental attrition 290–1
dental caries 286–7, 298
dental enamel hypoplasia (DEH) 284–8, 298, 373–5
dentine 5, 90–11, 20, 283
dentition, *see* teeth
dermoid cysts 335
Derry Cashel 108
developmental inclusion cysts 338
diagenetic processes 4
bone 3–5, 11–25
chemical 4, 11–15
computed tomography 114–15
crystallinity index 13
definition 4
fungal 4, 16–17
histological studies 121–43
hydrology 11
microbial 3–25
mummified remains 14
pathways 22–5
tanning 14–15, 22
taphonomy 4
teeth 3, 9–25
tunnelling 16–20
water 5, 11–15, 21–3
diaphysial bending deformity 218
diaphysial long-bone growth 371–3, 376
digital X-ray radiogrammetry (DXR) 90, 95
dinosaur bones 122
disease clusters 60
dislocations 318–19
DJD *see* degenerative joint disease (DJD)
dual X-ray absorptiometry (DXA)
osteoporosis 230–6
skeletal remains 77–95
DXA *see* dual X-ray absorptiometry (DXA)
DXR *see* digital X-ray radiogrammetry (DXR)

echinococcosis 192, 194210
EDLAXS *see* energy-dispersive low-angle X-ray scattering (EDLAXS)
ELISA *see* enzyme-linked immunosorbent assay (ELISA)
embryonic morphogenesis 332
enamel 5, 9–11
dental enamel hypoplasia 284–8, 298, 373–5
erosion 299–300
enchondromatosis 261, 263
endosteal sclerosis 83
energy-dispersive low-angle X-ray scattering (EDLAXS) 94
Entamoeba histolytica 152
enzyme-linked immunosorbent assay (ELISA) 152
epidemiology 45–55
age- and sex- specific prevalence 51–2
case-control studies 53
case studies 49–53
confidence intervals 52–3
definition 45–6
differentially weighted criteria 50–1
epidemiological transitions 47
error assessment 49
future directions 53–5
growth profiles 365–73, 376
methodology 46–8
microbiology 45, 53–4
missing data 49
population analysis 31–41
population size 54
preservation status 49
prevalence rates 47, 50–3
research directions 46–8
risk 48

epidemiology (*Continued*)
  time-scale 49
  transitions 47
Escherichia coli 154
Ewing's sarcoma 267–70

facial reconstructions 116
fibroma 260–2, 272
fibrosarcoma 265–7, 275
fibrous dysplasia 263, 265
foreign objects
  computed tomography 103, 114–15
  radiography 86, 87
forensic medicine 324
fractures 310–20
  accidental 314–15
  computed tomography 101–17
  interpersonal violence 312–14, 324
  metabolic bone disease 215–43
  pathological 315
  peri-mortem 311, 312, 320–3
  quantitation 317
  radiography 85–7
  spondylolysis 318
  stress 317
  trepanation 319–20
Frer foot 110
fungal infections 194210–12

gel electrophoresis 152, 162, 164
general adaptation syndrome 59
geographic information systems (GISs) 181, 183
giant cell tumours 259, 261, 272–3
gingivitis 289
GISs *see* geographic information systems (GISs)
Global History of Health Project 184
Gompertz curves 369–71, 374
gout
  case study 38–40
  epidemiology 48
  radiography 77–95
granuloma
  differential diagnosis 191–214
  periapical 293–8
Grauballe man 107–8, 109–13, 114, 309–10
gross dental enamel hypoplasia (DEH) 284–8, 298
growth deficits 222
growth profiles 365–82
  age at death estimations 367–8
  curve-fitting methods 369
  future directions 376
  Gompertz curves 369–71, 374
  growth velocity 371, 376
  intra-skeletal variations 370–1
  living populations 368–70
  long-bone growth 371–3, 376
  methodology 364–76
  models 369
  nature of growth 364–5
  non-specific stressors 365, 371–3, 376
  rickets 364, 373–6
  subadult populations 365, 367–73, 374
growth velocity 370-1, 376
gunshot wounds 314, 316
gypsum 21

haemangioma 261, 273
haematoma 310
haemorrhage
  metabolic bone disease 223
  neoplasms 258
HAP *see* hydroxyapatite (HAP)
Harris lines 82
health indices
  epidemiology 45-6
  population analysis 36
  status of past populations 38–40
hepatitis B virus 168
high-performance liquid chromatography (HPLC) 152, 153
histological studies
  age determination 129–31
  ancient bone 121–43
  applications 128–40
  ethical considerations 132
  grinding/polishing 124, 127–8
  histomorphometry 129, 139
  historical development 122–3
  Maltese cross patterns 132–3
  methodology 123–8
  mosaic bone patterns 135–6
  palaeopathological diagnostic investigations 131–40
  quantitation 129–30, 139
  rapid manual method 124
  sample cleaning 126
  sample dehydration 126
  sample embedding 126–7

sampling strategies 128
sectioning techniques 123
histomorphometry 129, 139
Hodgkin's disease 255
host–pathogen interactions 166
HPLC *see* high-performance liquid chromatography (HPLC)
Hunterian Museum 178, 185
hydrology 11
hydroxyapatite (HAP)
bone 5–9
diagenetic processes 11–15, 15–54,
histological studies 131–40
teeth 9
hypernephroma 256
hyperparathyroidism 218
hypoplasia 336–40, 344–6, 352–3, 355–7, 360
see also dental enamel hypoplasia (DEH)

immunology 152
impacted fractures 315
Inca bone 333
incomplete fractures 310
infants *see* subadult populations
Infectious disease 191–214
bone formation/destruction 192–4, 196–7, 198–9, 204–5
brucellosis 192, 194, 309–10
echinococcosis 192, 194, 210
leprosy 192, 194, 200–4
mycosis 192, 194, 210–12
osteomyelitis 192–4, 195–6, 203–4, 213
periostitis/periostosis 192, 194, 196–9, 208
rhinomaxillary abnormalities 201, 205
skeletal lesions 191–214
smallpox 192, 194, 195, 212
treponematosis 192–4, 197, 199, 205–9
tuberculosis 192, 194, 199–200
influenza virus 168
interpersonal violence 312–14, 324-5
inventories 62–3

Klippel–Feil syndrome 346, 347
kyphosis 199–200, 218, 346

Langerhans cell histiocytosis (LCH) 166
leiomyoma 256, 270
leishmaniasis 160
leprosy
differential diagnosis 192, 194, 200–4
epidemiology 46, 48
molecular palaeopathology 147–68
population analysis 35–6
leukaemia 270
light microscopy 121
limb anomalies 352–59
linear enamel hypoplasias 287
lipids 152–3
lobster claw deformity 355
long-bone growth 371–3, 376
long-term infections 149
longitudinal studies 368
Looser's zones 218
lower limb anomalies 355–9
lytic lesions
differential diagnosis 193, 206
metabolic bone disease 215
population analysis 31

macroscopic analysis
academic/applied palaeopathology 64–5
application of techniques 64–5
assessment/recording 61–2
case studies 60
changing criteria 67, 71
collaboration 72
demography 63
differential diagnosis 63–4
future directions 70–2
historical development 58–61
inventories 62–3
methodologies 61–4
publishing strategies 72
standardization 60–1, 70
static/evolutionary models 69
terminology 60, 62, 69
malaria 152, 160–1
malignancies *see* neoplasms
malnutrition 364–5
Maltese cross patterns 132–3
mandibular anomalies 336–8
marrow
hyperplasia 241
tumours 267–70
mass graves 321–3
maxillo-facial skeleton anomalies 338–9
measles virus 236–7
meningocoeles 335, 336, 348

metabolic bone disease 215–43
  diagnosis 216–19, 223–5, 228–30, 237–9
  documentary sources 219–20, 225–6
  osteomalacia 215, 218–21, 242
  Paget's disease 83, 134–6, 215, 236–41
  populations 219–22, 226–8, 230–7, 239–43
  scurvy 65, 215, 223–7, 241–2
  social factors 219, 225–6
  see also osteoporosis; rickets
metastatic carcinoma 81, 83, 267, 269–70, 276–7
micro-computed tomography 228
microbial pathogens
  ancestral sequence inference 167
  diagenesis 4, 13, 15–25
  evolution 166
  genomics 166–7
  histological studies 131–40
  host–pathogen interactions 166
  long-term infections 149, 165
  opportunistic infections 149–51
  survival mechanisms 148–51
  vector-borne 159–62
microbiology 154–65
  ancient DNA 154–9
  bacterial versus human DNA 157–9
  epidemiology 45–7, 53–4
  host–pathogen interactions 166
  methodology 152–5
  microbial genomics 166–7
  specific infections 159–65
microevolution 254–5
microscopical focal destruction 16–17
microtomography 121, 141
MIH *see* molar incisal hypomineralization (MIH)
molar incisal hypomineralization (MIH) 298
molecular palaeopathology 147–68
  ancestral sequence inference 167
  ancient DNA 154–9, 163, 168
  bacterial versus human DNA 157–9
  direct observation 152
  host–pathogen interactions 166
  immunological methods 152
  methodology 155–7
  microbial genomics 166–7
  microbiology 154–68
  non-DNA methods 152–3
  physico-chemical methods 152–3
  specific infections 159–65
  survival mechanisms of pathogens 148–51
morbidity status 31, 36–7
morphogenetics 330–1
  limbs 351–9
  ribs 340
  skull 331–40
  sternum 349
  vertebral column 340–9
mosaic bone 135–6, 237, 239
Mount Llullaillaco mummies 115
  multiple myeloma 257–8, 268, 274–6
mummies
  computed tomography 101–17
  diagenetic processes 13
  molecular palaeopathology 152, 161
  neoplasms 255–7, 270–1
  X-rays 102–3
Mycobacterium bovis 165
Mycobacterium leprae 48, 158–9, 163–5, 200, 204
Mycobacterium tuberculosis 54, 152–3, 155–8, 165, 167
mycosis 192, 194, 210–12
myeloma 257–8, 267, 268, 274–6

nasopharyngeal carcinoma 254, 257
Native American Graves Protection and Repatriation Act 178
near-field scanning optical microscopy (NSOM) 121, 140–1
neoplasms 258–84
  archaeological evidence 270–
  benign bone tumours 259–65, 271–4
  causes 253–4
  classification 258–9, 267–8
  malignant bone tumours 265–71, 271–5
  microevolution 254–5
  mummies 255–7, 270–1
  occupational tumours 257–8
  skeletal 258–9
  soft-tissue 255, 258, 270–1
  tumour-like processes 253–78
neural tube defects 329, 348
neuroma 274
non-DNA biomolecular methods 152–4
non-osseous bone coalition 355–7
non-ossifying fibroma 259, 264
non-specific stressors 365, 371–3, 376
NSOM *see* near-field scanning optical microscopy (NSOM)
nutrition
  growth 366
  osteoporosis 237–8

occupational tumours 257–8
Old Man Croghan 108–9
opportunistic infections 149–51
os metastyloideum 354–5
ossified haematomas 223
osteoarthritis 53, 67, 86–7
osteoblastoma 259, 261,
osteocalcin 6, 8
osteological paradox 34, 36–7
osteoma 260, 262, 271–2
osteomalacia 215, 218–21, 243
osteomyelitis
  differential diagnosis 191–3, 196, 201, 211–12
  radiography 83
osteon counts 129–34, 138
osteopetrosis 81
osteoporosis 215, 227–36, 242–3
  bone quality 228
  bone quantity 228, 242
  cortical bone loss 230, 232–5
  fractures 230, 235, 236
  histological studies 140
  lifestyle factors 243
  populations 230–6
  radiography 81, 87, 89–93, 95
  trabecular bone loss 230, 232
osteosarcoma 261, 265, 267, 274
osteosclerosis
  differential diagnosis 191–214
  metabolic bone disease 236
  radiography 83

Paget's disease of bone (PDB) 215–18, 236–43
  histological studies 131–40
  radiography 77
Palaeopathology Association 78
parietal
  foramina 333–5
  thinning 333–4
pathogen DNA 47–8, 54–5
  see also microbial pathogens
pathological fractures 315
PCR *see* polymerase chain reaction (PCR)
PDB *see* Paget's disease of bone (PDB)
peri-mortem fractures 311, 312, 320–3
periapical
  granuloma 294–298
  voids in jaw bones 283, 293–298
periodontal disease 286, 288–3, 299–300
periodontitis 289
periostitis/periostosis 192, 194, 196–9, 208
physico-chemical methods 152–3
pitted enamel hypoplasias 287
plain-film radiography *see* radiography
plane-form enamel hypoplasias 287
Plasmodium falciparum 152, 160, 167
Plasmodium vivax 167
polydactyly 353–4, 357, 360
polymerase chain reaction (PCR) 154–6, 163–4
population analysis 31–44
  age estimation methods 31, 33–6
  bone survival/recovery 31, 32–4
  burial practices 35–6
  catastrophic events 36
  demography 31, 37–40
  epidemiology 31, 33, 37, 53
  gout case study 38–40
  morbidity status 31, 36–7
  osteological paradox 31, 36–7
  pseudopathology 34–5
  representativeness 32–7
  subadult underrepresentation 33
  taphonomy 33–4
porosity measures 8, 15
porotic hyperostosis 65, 68–9
post-capture image processing 108, 109–10, 115–16
post-depositional perforations 35
post-mortem
  degradation 112–14
  fractures 34–6, 312, 314
Pott's deformity 199
prevalence rates 40, 49–53
primary bone malignancies 274–5
projectiles 312, 314
prostatic hyperplasia 106
protein detection 152–3
proteolytic enzymes 19
proximal femoral focal deficiency 355
pseudofractures 218
pseudopathology
  computed tomography 114–15
  population analysis 31–6
putrefaction 16, 20
pyrite framboids 21

Qilakitsoq mummies 115
quantitative computed tomography (qCT) 77,94

radiocarbon dating 3–5
radiogrammetry 87–92

radiography 77–100
  blastic lesions 83–5
  bone densitometry 91–4
  cortical thickness 87–91
  dual X-ray absorptiometry 77, 89–90, 92–3
  foreign objects 87,
  fracture calluses 86–7
  historical development 77–8
  lytic lesions 83–4
  methodology 79–87
  osteoporosis 82, 87, 89–92, 95
  post-depositional changes 81, 83, 93
  quantitation 77, 94
  radiogrammetry 87–92
  regions of interest 90, 92
recording strategies 62
recovery/survival rates 31, 32–3
remodelled bone
  histological studies 121–43
  macroscopic study 197, 200–4
renal osteodystrophy 135–6
rheumatoid arthritis 83, 106–7
rhinomaxillary abnormalities 201–2, 205
rib anomalies 343, 348
rickets 209, 215, 218–24, 241–2
  cuspal enamel hypoplasia 285
  growth profiles 365, 376
  radiography 86
Rickettsia prowazekii 160

sacral anomalies 344, 346
sand bodies 12
scanning electron microscopy (SEM) 93
  ancient bone 121, 132–3, 140–1
  diagenetic processes 17–20
  teeth 283–300
  see also backscatter scanning electron microscopy (BSEM)
scaphocephaly 333–4
Schistosoma haematobium 152, 161–2
Schistosoma mansoni 161–2
schistosomiasis 149, 161–2
Schmorl's nodes 315
sclerosis
  metabolic bone disease 215–43
  radiography 83
scurvy 65, 215, 223–7, 241–2
SEM *see* scanning electron microscopy (SEM)
single nucleotide polymorphisms (SNPs) 165
skin ulceration 199
skull anomalies 331–41
smallpox 192, 194, 195, 212–13
SMR *see* standardized mortality ratios (SMR)
SNPs *see* single nucleotide polymorphisms (SNPs)
soft-tissue
  lesions 297
  neoplasms 253–5, 270–1
  trauma 309–10
spatial clusters 35–6
sphagnan 14–15
spiculation 261–2, 267
spina bifida 330, 348
spondylolysis 318, 345, 346
SQSTM1 gene 243
SRB *see* sulphate-reducing bacteria (SRB)
Stafne defect 337–8, 360
standardized mortality ratios (SMR) 38–40
Staphylococcus spp. 195
stereolithography 116–17
sternal anomalies 350–1
Streptococcus spp. 195
stress fractures 317
stress indicators 60
subadult populations
  growth profiles 365, 368–9, 373
  metabolic bone disease 215–43
  population analysis 31
  underrepresentation 33, 367
sulphate-reducing bacteria (SRB) 20–1
symphalangism 355, 357
syphilis
  dental enamel hypoplasia 284
  differential diagnosis 201
  molecular palaeopathology 147

talipes equinovarus 357
tanning 14–15, 22
taphonomy
  definition 4
  population analysis 31–2
  trauma 309, 324
teeth 283–309
  abscesses 293–297
  age determination 129–30
  alveolar responses 283
  attrition eruption 290–1
  compensatory/continuous eruption 290–1, 292–3
  computed tomography 103
  cuspal malformations 284–88, 298

dental enamel hypoplasia 284–88, 298, 372–3
diagenetic processes 4, 11–25
disease classification 293–8
enamel erosion 299–300
epidemiology 45
growth profiles 371–3, 373–6
metabolic bone disease 215
molar incisal hypomineralization 298
periapical voids in jaw bones 283, 293–298
periodontal disease 288–93, 299–300
structure 9–10
temporal styloid anomalies 341
terminology 61, 69
tetanus 149
thalidomide 329, 352
Tollund Man 112–13, 114
trabecular bone loss 227, 229
trauma 309–25
accidental 314–17
amputations 311
decapitation 320
dislocations 318–19
early European populations 323
interpersonal violence 312–15, 324
mass graves 321–3
pathological 315
quantitation 317
spondylolysis 318
trepanation 319–20
see also fractures
trench fever 160
trepanation 319–20
treponematosis
differential diagnosis 191–214
radiography 81
Trypanosoma cruzi 47, 160–1
tuberculosis
differential diagnosis 191, 193, 198–200
molecular palaeopathology 152–3, 155–7, 162–5
tumour-like processes 253–78
tumours *see* neoplasms
tunnelling 16–20
Tuthmosis IV 103
tympanic plate anomalies 340
typhoid fever 148, 160
Tyrolean Iceman 106, 115

ultrasonication 126
undulant fever 161–2
upper limb anomalies 352–5

variable-number tandem repeat (VNTR) 163–4
vertebral border shifts 342
vertebral column anomalies 340–9
viral infections 236
visual observation *see* macroscopic analysis
vitamin C deficiency *see* scurvy
vitamin D deficiency *see* osteomalacia; rickets
VNTR *see* variable-number tandem repeat (VNTR)

water
bone 5, 11
diagenetic processes 4, 11–16, 22–3
Wellcome Osteological Research Database (WORD) 181–2, 183, 187
Western Hemisphere Project 65
WORD *see* Wellcome Osteological Research Database (WORD)

X-ray microtomography 121
see also computed tomography (CT); radiography

yaws 205
Yersinia pestis 159–60, 167, 182
Yersinia pseudotuberculosis 167

zoonoses 148, 168, 257

This index was prepared by Neil Manley